2011年6月8日，庄巧生先生与王连铮研究员、董玉琛院士、程顺和院士、刘旭院士、万建民院士、辛志勇研究员在天津武清参加中麦175现场会

2016年6月初，庄巧生先生与育种团队成员在试验地观摩新品种

何中虎◎主编

中麦 175 选育与主要特性解析

中国农业出版社
农村设物出版社
北　京

序

何中虎于1993年从国外学成归来，到中国农业科学院作物科学研究所工作，“九五”期间开始主持小麦品质育种课题组的工作。在庄巧生先生的指导下，经过团队同事们的不懈努力，在小麦品质研究与育种可用分子标记发掘应用、中国-CIMMYT合作育种两个方面都取得了较大进展，先后获得2008年国家科技进步一等奖和2015年国家科技进步二等奖。令人高兴的是，近10多年来育种工作也取得了显著进展，育成并大面积推广了中麦175、中麦895和中麦578等新品种（后两个品种与中国农业科学院棉花研究所合作育成），扭转了中国农业科学院作物科学研究所从20世纪80年代中期以来小麦育种下滑的局面，并为全国小麦育种和生产持续发展做出了较大贡献。

在将育种重点逐步转向黄淮麦区的同时，团队还加强了北部冬麦区的新品种选育与推广工作。在北京育种站，团队选用北京农业大学杨作民教授用法国材料Fr81-4（VPM1衍生后代，含2NS易位系）创制的矮秆抗病亲本BPM27与京411杂交，2001年育成的中麦175在矮秆、抗病性和广泛适应性方面表现十分突出，是我国首个同时通过北部冬麦区水地和黄淮旱肥地国家审定的水旱兼用型品种。中麦175审定后的第二年即被选为北部冬麦区、北京市、山西省与河北省冀中北地区区域试验的对照品种，2010年至2020年连续10年成为北部冬麦区第一大品种，也成为黄淮旱肥地第一大品种，还是青海、甘肃春麦区发展冬麦的第一大品种，累计推广5 000多万亩，带动北部冬麦区和黄淮旱肥地实现第7次品种更新换代，为小麦生产做出了显著贡献。“高产节水多抗广适冬小麦新品种中麦175选育与应用”获得2017年度农业部中华农业科技奖科研成果一等奖。

在过去10多年系统研究的基础上，课题组团队从已发表的论文中挑选25篇编成《中麦175选育与主要特性解析》一书，以便与同行进行交流，听取意见，为今后进一步提高育种水平和研究能力做些准备。本书包括矮秆基因与产量潜力、节水节肥与抗病性、加工品质与营养特性共3部分，其中创新性较明显的有3个方面：（1）系统解析矮秆基因*Rht24*的起源、演化与功能，矮秆、株型紧凑、穗数多是中麦175高产稳产的主要原因，为高产节水型乃至旱地新品种选育提供了新思路；（2）中麦175是国内用VPM1育成的少数几个推广品种之一，抗麦瘟病，含抗条锈病新基因*Yrzm175*和抗叶锈病基因*Lr37*，抗病性好；（3）面筋延展性好，面粉颜色亮白，为面条和馒头兼用型品质奠定基础，含锌量较一般品种高30%，营养价值高。上述评价是否成立有待读者评说。

总之，团队自1996年成立至今，能在品质研究、兼抗型育种方法建立和新品种选育方面都有较大收获，应该说是难能可贵的。团队不仅传承了庄先生的优良学风，而且在新的历史条件下，善于开拓国内外合作网络，及时利用各种新技术，坚持理论联系实际，实现了品种选育与论文的双丰收；同时还形成了一支人员和专业搭配合理、分工明确、团结协作、高效运行的研究队伍，这是研究工作能取得进展的根本原因。期望他们继续努力，为提高我国小麦育种水平做出更大的贡献。

刘　旭

2023年5月

前　言

在国际玉米小麦改良中心（CIMMYT）完成小麦育种博士后研究和堪萨斯州立大学的访问学习后，我于1993年5月来到中国农业科学院作物育种栽培研究所（后合并更名为作物科学研究所）工作，主要承担了三项研究任务，即小麦新品种选育、品质研究与分子标记发掘应用、中国-CIMMYT合作育种。在团队成员的共同努力下，后两项很快取得较大进展，有幸获得2008年国家科技进步一等奖和2015年国家科技进步二等奖，但新品种选育直到2010年前后才取得显著进展。

20世纪90年代，由于北部冬麦区小麦面积大幅度减少等原因，作物科学研究所的小麦育种研究进展缓慢，对生产贡献较小。在庄巧生先生和时任所长辛志勇研究员的指导下，考虑到品质改良的紧迫性，将原来的冬麦育种组、硬粒小麦组和品质实验室合并组建了小麦品质育种组，任命我为负责人。2000年后，在时任所长万建民院士的支持下，课题组的人员配置、实验室和试验地条件快速完善，逐步建立了分工明确、团结协作的研究团队，形成了品质研究、育种可用分子标记发掘应用支撑新品种选育的新格局。在北京、石家庄和安阳建立了育种站，分别面向北部冬麦区、黄淮北片冬麦区和黄淮南片冬麦区开展育种工作，其中安阳育种站由作物科学研究所和棉花研究所合作共建。

根据生产发展需要，团队将育种工作重点逐步转向黄淮麦区，同时加强了北部冬麦区的育种工作。为了实现高产矮秆与节水广适的有效结合，北京育种站采取了一些新做法。新品系初级产量比较试验和产量比较试验同时在水浇地（三水）和旱地（冬灌一水，防冻害）进行，产量比较试验则在北京、天津、河北北部和山西多点进行；在分离世代和品种产量比较阶段则加强了品质测试工作，以培育优质高产兼顾的新品种。期望育成的新品种在北部冬麦区东片（京、津、冀等）大面积推广的同时，还能在西片（陇东等）、新疆及黄淮旱肥地扩大阵地，为小麦生产做出应有贡献。我们育成并推广了以中麦175为代表的高产优质新品种，连续10年被选为北部冬麦区水浇地第一大品种，还成为黄淮旱肥地第一大品种，也是陇东和青海冬麦的主栽品种，累计推广5 000多万亩，基本扭转了中国农业科学院作物科学研究所从20世纪80年代中期以来小麦育种下滑的局面。

在加强品种推广的同时，我们还采用育种、栽培、生理、谷物化学与基因组学相结合的方法，对中麦175的株高、灌浆速率、水肥利用效率、抗病性、加工品质与营养特性等做了较为系统深入的研究，主要创新性可归纳为三点：(1) 中麦175的株高由*Rht8*、*Rht24*等矮秆基因控制，后者为我们报道的新基因，株型紧凑、穗数多是高产稳产的主要原因。(2) 良好的根系为肥水高效利用、耐热性和广适性提供保障；中麦175含2NS易位系，是国内用法国VPM 1育成的少数几个推广品种之一，抗麦瘟病，含抗条锈病新基因*Yrzm175*、抗叶锈病基因*Lr37*等，在陇东和四川等地一直保持良好的抗病性。（3）中麦175面筋强度中等偏弱，但延展性好，面粉颜色亮白，为面条和馒头兼用型品质奠定基础；锌高效吸收运转使籽粒锌含量高，具备较高营养价值。

为了便于虚心向各部门同行请教和学习，全面提高我们的育种和研究水平，同时也向主管部门和资助机构汇报工作进展，我们从撰写的论文中挑选了学术性与应用价值兼顾的25篇编成《中麦175选育与主要特性解析》。在这25篇论文中，有6篇是主要合作单位中国农业大学、首都师范大学、甘肃农业科学院、天津科技大学等与我们协作完成的。本书包括矮秆基因与产量潜力、节水节肥与抗病性、加工品质与营养特性共3部分。这些论文已在国内外SCI期刊、《中国农业科学》《作物学报》等发表。为了保证相对统一的格式，文章基本保持原貌，只对文字和个别差错之处做了修改。

我们的小麦育种工作一直得到中国农业科学院和作物科学研究所历任领导的大力支持，先后获得国家科技攻关计划项目、863项目、948重大国际合作项目（2003—2015）、国家重点研发计划项目、中国农业科学院创新工程项目等资助，在此表示衷心感谢。北京育种站的陈新民研究员、王德森副研究员、李思敏副研究员等为中麦175的培育和推广做出了突出贡献，夏先春研究员、张艳研究员、肖永贵副研究员、曹双和副研究员、郝元峰研究员等为中麦175主要特性解析付出了不懈努力，一并致谢。特别感谢李振声院士、董玉琛院士（已故）、程顺和院士、万建民院士、赵振东院士等的鼓励和指导，感谢李鸣博士在文件转换、文字核对和内容审读等方面做了不可或缺的工作，还要感谢中国农业出版社杨天桥编审为本书出版所做的努力。

感谢刘旭院士的长期指导与帮助，并在百忙中为本书作序，这是对我们的鼓励和鞭策，将激励我们继续努力，为小麦育种做出更大贡献。

在本文集编辑期间，庄巧生先生不幸去世。庄先生是我国小麦遗传育种学科的主要奠基人之一，他建立并发展了中国农业科学院冬小麦育种团队，包括原来的冬麦组和现在的品质育种团队，为我国小麦生产和育种技术发展做出了巨大贡献。很荣幸与庄先生一起工作近30年，衷心感谢他给予的指导、鼓励和帮助，他经常提醒我们研究要为生产服务，品种论文双丰收是团队的目标。庄先生亲自参加了中麦175选育和推广工作，即便在他90高龄以后，每年还到昌平和中圃场试验基地参加育种工作。庄先生的逝世使我们失去了一位好老师、好朋友，努力工作是我们对庄先生最好的缅怀和纪念。他创新求实、服务生产的学风将永远激励我们继续前进，为小麦种业科技发展和国家粮食安全做出更大的贡献。

由于时间较短，加上作者水平有限，疏漏、错误之处在所难免，敬请指正。

何中虎
2023年5月

目　录

中麦 175 高产高效广适特性解析与育种方法思考

何中虎[1,2]，陈新民[1]，王德森[1]，张　艳[1]，肖永贵[1]，李法计[1]
张　勇[1]，李思敏[1]，夏先春[1]，张运宏[1]，庄巧生[1]

([1]中国农业科学院作物科学研究所，北京 100081；
[2]CIMMYT 中国办事处，北京 100081)

摘要：解析中麦 175 的高产潜力、水肥高效特性、优良品质及广泛适应性机理将为培育突破性新品种提供理论指导和实用信息。(1) 中麦 175 穗数容易达到 750 个/m^2，收获指数高达 0.49，穗数多和收获指数高为实现高产潜力奠定基础；矮秆及茎秆弹性好与株型紧凑、叶片小且直立为穗多、不倒伏提供保障。(2) 千粒重与株高对水分反应敏感性低，灌浆中后期叶片衰老速度慢，灌浆速度快，这是其水分利用效率高和在旱肥地表现突出的主要原因。中麦 175 的产量水分敏感指数为 0.86，而京冬 8 号为 1.13。(3) 氮肥吸收和利用效率皆高，穗粒数和千粒重对肥料敏感性低，为肥料高效利用奠定基础；在 6 种不同施氮水平下，中麦 175 的产量均高于京冬 17，氮肥施用量对穗粒数和千粒重影响小。良好的根系为肥水高效利用提供保障。(4) 籽粒软质，面筋强度中等偏弱、延展性好，面粉颜色亮白，为面条和馒头兼用型品质奠定基础；锌高效吸收运转使籽粒锌含量高，具备较高营养价值。(5) 含有光周期不敏感基因 *Ppd-D1b*、肥水高效、抗寒、耐高温和抗病性等为中麦 175 的广适性和跨区域种植提供保障。本研究为高产高效广适性新品种培育提供了重要经验和理论支撑。

关键词：小麦品种；中麦 175；高产潜力；水肥高效；广泛适应性

Characterization of Wheat Cultivar Zhongmai 175 with High Yielding Potential, High Water and Fertilizer Use Efficiency, and Broad Adaptability

HE Zhong-hu[1,2], CHEN Xin-min[1], WANG De-sen[1], ZHANG Yan[1], XIAO Yong-gui[1], LI Fa-ji[1], ZHANG Yong[1], LI Si-min[1], XIA Xian-chun[1], ZHANG Yun-hong[1], ZHUANG Qiao-sheng[1]

([1]*Institute of Crop Science, Chinese Academy of Agricultural Sciences, Beijing* 100081;
[2]*CIMMYT China Office, Beijing* 100081)

Abstract: Documentation of leading cultivars will provide crucially important information for cultivar development. The objective of this study is to characterize high yield potential, water and fertilizer use efficiency, excellent quality and broad adaptation in wheat cultivar Zhongmai 175. High yield potential was largely due to

原文发表在《中国农业科学》，2015，48（17）：3394-3403.

the increased spike number and high harvest index, while short (around 80 cm) and erect plant type with small leaves contributed to its outstanding lodging resistance. Spike number could easily reach 750/m^2, 20%-25% increase in comparison with check cultivar Jingdong 8, while harvest index was 0.49. Insensitivity of plant height and thousand kernel weight to water stress, due to slow leaf senescence and fast grain filling rate, made it high water use efficiency and better performance under rainfed condition. Water sensitive index for yield in Zhongmai 175 was 0.86, while that of Jingdong 8 was 1.13. Nutrient intake and use efficiency contributed to insensitivity of kernel number per spike and thousand kernel weight, and thus better performance under different fertilizer applications. Zhongmai 175 outyielded Jingdong 17 at six different nitrogen levels. It was characterized with soft kernel, weak dough strength and excellent extensibility, and bright flour color, thus conferred excellent dual qualities for Chinese noodles and steamed bread. High Zn content was due to its efficiency in Zn intake and transportation, thus excellent nutritional quality. Presence of *Ppd-D1b*, water and fertilizer use efficiency, tolerance to cold in winter and high temperature during grain filling stage, and resistance to yellow rust, provided basis for its broad adaptation in three wheat ecological zones. This study will provide very important information and experience for developing new cultivars with high yield potential and broad adaptability.

Keywords: wheat cultivar; Zhongmai 175; high yield potential; water and fertilizer use efficiency; broad adaptation

0 引言

北部冬麦区曾是我国小麦育种与研究的中心，中国农业科学院与中国农业大学等单位不仅在育种方法研究方面走在国内前列，培育的新品种也曾在生产中发挥过重大作用。但自1990年以来，本麦区的育种工作遇到前所未有的挑战，主要表现在4个方面。一是北部冬麦区东片包括北京、天津、冀东与冀中北和晋中地区（基本为水浇地）受产业结构调整及减少农业用水等政策的影响，小麦面积由过去的200多万hm^{2}下降到目前的不足100万hm^2，其中北京市的冬麦种植面积由过去的18.7万hm^{2}降到目前的3.3万hm^2；同时，节水、节肥、易管理已成为生产的普遍要求，育种力量明显下降。北京市农林科学院于1990年育成的京冬8号曾为生产发展作出了重要贡献，但其产量偏低、秆高易倒伏、品质较差、感白粉病等，难以满足市场需求。1995—2005年虽然各单位审定了不少品种，但大面积推广很少。总体来说，小麦育种没有实质性突破，与主产区的差距有所拉大。二是由于气候变化的影响日益显著，冷冬年份出现频率减少，山东和石家庄等地育成的新品种矮秆高产特性突出，颇受农民青睐，尽管存在冬季冻害等风险，仍然形成了南部品种北移的明显趋势，如济麦20和济麦22在天津大面积推广。三是以旱地为主的北部冬麦区西片及黄淮旱肥地育种力量有所下降，造成现有品种抗病性差，植株偏高，虽然在旱灾年份表现较好，但丰水年份倒伏十分严重，往往是丰产不丰收，这些地区迫切需要茎秆偏矮、抗旱、抗寒和灌浆期抗高温的高产广适性品种。虽然历史上本区东片育成的品种也曾在西片大面积推广，如丰抗8号和中优9507先后成为陇东地区的主栽品种，但总体来说东片品种的抗条锈病能力偏差，水分利用效率偏低，不易在西片推广。四是20世纪90年代初，针对气候变化的可能影响，农业部曾启动了冬麦北移和西延计划，但受气候和品种的双重影响，冬麦替代春麦的设想在当时未能实现[1]；但近10多年来的气候变暖进一步加速，春麦改冬麦技术在不少地区已大面积推广，生产上迫切需要抗寒、高产、矮秆抗倒、抗病、广适性新品种。

针对这一情况，我们将育种工作的重点逐步转向主产区如河北中南部和河南省，同时不放弃甚至加强北部冬麦区的育种工作，期望育成的新品种能在北部冬麦区东片大面积推广的同时，还能在西片和新疆及黄淮旱肥地扩大种植面积，为小麦生产作出应有贡献。为此，育成并推广了以中麦175为代表的高产高效优质广适性新品种，基本扭转了中国农业科学院作物科学研究所从20世纪80年代中期以来小麦育种下滑的局面。通过株型等的改良，2001年育成的中麦175实现了高产潜力、水肥高效、抗病抗逆、优良面条品质、广适性与早熟性的良好结合，分别通过北部冬麦区水地（国审麦2008016）、黄淮旱肥地（国审麦 2011018)、北京市(京审麦 2007001)、山西省(晋审麦 2007007)、河北省(冀审麦 2009017)、青海省(青审麦 2011001)和甘肃省(甘审麦 2012010)的品种审定，且已成为上述地区的主栽品种及北部冬麦区、北京、山西与河北冀中北地区区试的对照品种。本文采用育种、生理、栽培、谷物化学与分子标记相结合的方法，较全面地解析了中麦 175 的优异特性，目的是为培育突破性新品种提供理论支撑。

1 产量潜力高

中麦 175 的高产潜力主要表现在两个方面。一是在各级区域试验中均比对照增产，平均增产 4.8%，增产点率为 67%～100%，在多数试验中名列前三位，在所有生产试验中均名列第一，平均较对照增产 8.9%(表 1)。株型紧凑和小区面积小是区试增幅小于生产试验增幅的重要原因。二是大面积生产示范表现更突出，河北、天津和北京等地将中麦 175 作为高产创建的主要品种，大面积产量容易达到 8 250kg·hm^{-2}，部分测产和实收结果列于表 2。如河北省保定市徐水农场二分场示范 6.7hm^2，2005—2009 年连续 4 年度超过 8 250kg·hm^{-2}，这在北部冬麦区是很不容易的，或者说以前的品种不具备这一产量潜力。2007—2008 年度北京市房山区琉璃河镇实收产量高达 9 169.5kg·hm^{-2}，创本麦区高产纪录。另外，在青海省小面积产量达到 10 665kg·hm^{-2}，在石家庄全生育期仅浇 1 水的条件下产量达 10 609kg·hm^{-2}，在新疆阿克苏产量高达 10 440kg·hm^{-2}。

表 1 中麦 175 在两省市和国家区域试验中的产量表现

Table 1 Yield performance of Zhongmai 175 in various regional trials

试验年份与类型 Season and trial type	产量 Yield (kg·hm^{-2})	±CK (%)	增产点率 YIS (%)	位次 Rank
2004—2006 年北京市区域试验 2004—2006 Beijing YT	6 693.0	1.5	78	1, 3
2005—2006 年北京市生产试验 2005—2006 Beijing PT	5 953.5	9.4	67	1
2005—2007 年河北省区域试验 2005—2007 Hebei YT	6 798.0	4.3	86	3, 3
2007—2008 年河北省生产试验 2007—2008 Hebei PT	7 209.0	7.8	100	1
2006—2008 年北部冬麦区区域试验 2006—2008 NCP YT	7 375.5	9.1	91	2, 1
2007—2008 年北部冬麦区生产试验 2006—2008 NCP PT	7 324.5	6.7	83	1
2008—2010 年黄淮旱肥地区域试验 2008—2010 YRVRR YT	5 634.0	4.3	71	6, 3
2010—2011 年黄淮旱肥地生产试验 2010—2011 YRVRR PT	5 521.5	8.6	88	1
2005—2012 年北部冬麦区新品种展示 2005—2012 NCP NVDT	8 283.0	11.9	100	1, 1, 1, 2, 3, 2, 2

YIS: Yield increase site; YT: Yield trial; PT: Pilot trial; NCP: Northern China Plain; YRVRR: Yellow River Valley Rainfed Region; NVDT: New Variety Demonstration Trial

资料来源：省市和全国区域试验资料汇总(2004—2012) Data source: Provincial and national regional trials from 2004—2012

从山西省、河北省、北部冬麦区区域试验和品种展示结果来看，中麦 175 的平均成穗率为 49.2%，比对照京冬 8 号高 4.6%；每平方米 687 穗，较对照多 62.9 个；穗粒数 31.2 个，较对照多 1.7 粒；千粒重 39.9g，较对照低 4.0g；成穗率高、穗数显著增多和穗粒数较多是其增产的主要原因。在大面积高产示范中，中麦 175 的穗数容易达到 750 个/m^2，较京冬 8 号等多 20%～25%；由于其株型紧凑、叶片较小且直立，同时株高较京冬 8 矮 15cm 左右，秆矮弹性好，虽然穗多，但大面积很少出现倒伏，为实现高产潜力提供了保障。为了进一步明确中麦 175 的高产机制，于 2012—2014 年对北部冬麦区的 5 个主栽品种进行了系统比较(表 3)，尽管试验点因后期分别遇到大风倒伏和高温胁迫，两年皆为灾年，但总体趋势仍可供参考。中麦 175 的产量居第一位，显著高于京冬 8 号和农大 211，也高于公认的高产品种轮选 987；中麦 175 和轮选 987 穗数显著多于京冬 8 号和京冬 17；中麦 175 的收获指数为 42.1%，显著高于除轮选 987 外的其他 3 个品种。中麦 175 的株高与现有品种接近，但比对照京冬 8 号降低约 10～15cm，茎秆弹性好，抗倒伏能力显著提高。进一步分析表明，中麦 175 成熟期干物质分配到籽粒中的比例高达 39.9%，显著高于其他品种，如轮选 987 和农大 211 仅为 34.7%和 33.1%。由于中麦 175 的灌浆期较其他品种仅长 1～2d，灌浆速度快是收获指数高的主要原因。在充分灌溉的条件下，由于中麦 175 株型紧凑、叶片较小且直立，在抽穗前和灌浆中期的植被指数、抽穗前的叶面积指数显著低于京冬8号，但灌浆中期的叶绿素含量显著高于京冬8号，而灌浆中期的叶片衰老速度则显著低于京冬8号，这可能是中麦175收获指数和产量皆高的生理基础[2]。分子检测表明，中麦175不含矮秆基因 *Rht1*、*Rht2* 和 *Rht8*，根据其株高判断，应含有其他矮秆基因，这有待将来验证；另外，它含有3个可以增加粒重的等位基因 *TaCwi-A1a*、*TaSAP1-A1-2606C* 和 *TaGS-D1a*（国家小麦改良中心资料）。因此，株高显著降低、株型直立、穗数多和收获指数高是其高产的主要原因。

综上所述，株型紧凑、叶片较小且直立，成穗率高、穗数较一般品种多10%～25%，株高80cm左右，抗倒伏能力强，后期叶片衰老速度慢，灌浆速度快，收获指数高，这是中麦175高产的主要原因。

表2 中麦175在大面积生产中的产量表现

Table 2 Performance of Zhongmai 175 in pilot fields under irrigated environments

地点 Location	年份 Season	面积 Area (hm^2)	产量 Yield ($kg \cdot hm^{-2}$)	备注 Note
河北省高碑店市方官村 Gaobeidian, Hebei Province	2013—2014	73.5	8 363	实收，春季仅浇一水 Harvested yield, one irrigation
河北省香河县吴庄村 Xianghe, Hebei Province	2008—2009	6.7	8 613	测产 Predicted yield based on yield components
河北省固安县牛驼镇 Guan, Hebei	2011—2012	133.3	8 382	测产 Predicted yield based on yield components
河北省徐水农场二分场 Xushui, Hebei	2008—2009	6.9	8 273	实收 Harvested yield
河北省保定市5个县 Baoding, Hebei	2011—2012	2 000.0	8 046	测产，比前3年增产20.3% Predicted yield, increase 20.3%
北京市房山区扬户屯村 Fangshan, Beijing	2008—2009	13.3	9 169	实收，创北部冬麦区高产纪录 Harvested yield, new record
天津市武清区南蔡村 Wuqing, Tianjin	2010—2011	6.7	8 702	测产，当年最高产量 Predicted yield
天津市武清区 Wuqing, Tianjin	2012—2013	8.0	7 949	实收，当年最高产量 Harvested yield
山西省晋城地区北石店 Jincheng, Shanxi	2009—2010	86.7	8 046	实收 Harvested yield
青海省平安县小峡镇 Pingan, Qinghai	2011—2012	0.3	10 665	实收，创本省冬麦高产纪录 Harvested yield, new record

资料来源：本单位收集的相关数据 Data source: Collected from our collaborators

表3 北部冬麦区5个主栽品种产量相关性状比较

Table 3 Trait comparison of five major cultivars in North Winter Wheat Region

品种 Cultivar	产量 Yield ($kg \cdot hm^{-2}$)	穗数 Spikes/m^2	穗粒数 Grains/spike	千粒重 TKW (g)	株高 PH (cm)	生物量 Biomass ($kg \cdot hm^{-2}$)	收获指数 HI
中麦175 Zhongmai 175	4 913a	627a	29.3a	36.4b	77.7b	11 040a	42.1a
轮选987 Lunxuan 987	4 759ab	623a	29.1a	36.6b	75.3bcd	11 072a	38.3bc
京冬17 Jingdong 17	4 569ab	532b	30.1a	36.2b	76.8bc	9 951a	39.8abc
京冬8号 Jingdong 8	4 480b	534b	27.3a	42.2a	89.2a	10 557a	37.4c
农大211 Nongda 211	4 445b	597ab	28.0a	36.6b	74.7cd	11 209a	38.2bc

TKW：千粒重；PH：株高；HI：收获指数。下同。同一列不同字母表示差异达5%显著水平

TKW: Thousand kernel weight; PH: Plant height; HI: Harvest index. The same as below. Different letters in the same column indicate significant difference at 5% probability level

2 水分利用效率高，为水旱兼用型品种

中麦175的水分利用效率高主要表现在3个方面。一是在黄淮旱肥地区试中增产显著（表1），分别通过北部冬麦区水地和黄淮旱肥地两次国家审定，在建立国家品种审定制度的30多年中，它是唯一一个同时通过水地和旱地国家审定的品种。二是在石家庄仅浇1水的条件下，产量高达10 609.5kg·hm^{-2}。石家庄市农业科学院于2012—2013年和2013—2014年在赵县高肥力条件下，对近10年国家审定的100个冬性及半冬性品种进行了0水（全生育期不浇水）、1水（拔节期）和2水（拔节、抽穗期）试验（3次重复，小区面积6m^2），在全生育期仅浇1水的条件下，中麦175较对照石4185增幅高达9.7%，居100个参试品种首位，石麦19、西农558和石麦22的增幅也在7%以上，分居2～4位。在全生育期浇2水的条件下，中麦175增幅为5.3%，居100个参试品种的第12位（郭进考，2014，个人交流）。这说明中麦175不仅产量潜力达到了黄淮麦区的高产水平，而且实现了高产潜力与节水性能的良好结合，其产量水平和节水能力皆居参试品种前列。三是在甘肃和陕西等地的旱地大面积示范中突破8 000kg·hm^{-2}，2014年还创造了陕西省（9 079.5kg·hm^{-2}）和甘肃省（9 262.5kg·hm^{-2}）旱地高产纪录（表4）。甘肃省泾川县大面积示范表明，中麦175在大旱之年和丰水年份皆表现突出，灾年减产少，丰年抗倒伏、增产幅度大，是解决地膜小麦倒伏的理想品种，非常适合陇东地区种植[3]。当地老百姓说，种植中麦175，灾年有粮吃，丰年创高产。在一般干旱的2010—2011年，中麦175与当地主栽品种西峰27产量接近，增产1.4%。在丰水年份如2011—2012年，在地膜覆土栽培条件下，比当地种植多年的长6359增产17.2%；在露地栽培条件下，比当地主栽品种西峰27增产29.3%，增产的主要原因是株高较低，抗倒性显著优于当地抗旱品种。在严重干旱的2012—2013年（60年不遇），露地条件下的产量与对照品种西峰27接近，减产1.7%；但在地膜覆土栽培条件下，比对照长6 359增产16.1%。据计算，2012—2013年中麦175在地膜栽培条件下的水分利用效率为1.1kg·mm^{-1}，而当地主栽品种泾川1号在相同条件下为0.92kg·mm^{-1}，西峰27在露地栽培下仅为0.54kg·mm^{-1}；2013—2014年中麦175最高水分利用效率达1.50kg·mm^{-1}，对照品种仅为0.85kg·mm^{-1}。这充分说明中麦175为水分高效及水旱兼用型品种。为了进一步明确中麦175的水分高效利用机理，在不浇水（W0）、仅浇拔节水（W1）和同时浇拔节水和开花水（W2）3种条件下，比较了中麦175与北部冬麦区代表性国审品种京冬17的产量等性状（表5）。在3种处理中，中麦175的产量和水分利用率皆高于京冬17（图1，表5）；在不浇水和仅浇拔节水时，中麦175的千粒重和收获指数差异不显著，穗粒数在3种条件下差异不显著，而京冬17的所有性状差异都达显著水平，说明中麦175对水分反应相对不敏感，这与郭进考的结果一致（郭进考，2014，个人交流）。大面积示范表明，中麦175为水分高效型品种，如河北省高碑店市在2013—2014年度种植73.5hm^2，春季浇1水和春季浇2水的产量（8 363kg·hm^{-2}和8 520kg·hm^{-2}）相差无几，但前者的水分利用效率为1.21kg·mm^{-1}，而后者仅为1.05kg·mm^{-1}。

在充分灌溉和仅浇冬水条件下的系统比较表明，中麦175产量的水分敏感指数为0.86，而京冬8号为1.13，主要原因是千粒重与株高对水分反应敏感性低，虽然穗数对水分的敏感性较高，但其数值仍高于京冬8号，这是其在有限灌溉条件下高产的主要原因[2]。中麦175携带较多的京411优异根系遗传区段，主根长和根干物质重等的改良较为显著，3DL和5BL携带控制根长的主效位点[4]。优良的生理和根系特性为水分高效利用奠定了基础。

综上所述，中麦175穗数多，株高、千粒重、穗粒数和收获指数对水分胁迫敏感性低，灌浆中后期叶绿素含量高，叶片衰老速度慢，这是其水分高效、在节水条件下表现突出的原因，中麦175为水旱（肥地）兼用型高产新品种。

表4 中麦175在旱肥地的高产表现

Table 4 Performance of Zhongmai 175 in pilot fields under rainfed environments

地点 Location	年份 Season	面积 Area (hm^2)	产量 Yield ($kg \cdot hm^{-2}$)	备注 Note
陕西省永寿县永寿村 Yongshou, Shaanxi	2011—2012	3.0	8 745.0	实收，创陕西省旱地高产纪录 Harvested yield, new record
陕西省省永寿县 Yongshou, Shaanxi	2013—2014	0.2	9 079.5	实收，创陕西省旱地高产纪录 Harvested yield, new record
陕西麟游县九成宫镇 Linyou, Shaanxi	2013—2014	6.7	8 053.5	实收，比对照晋麦47增32.7% Harvested yield, increase 32.7%
甘肃省泾川县 Jingchuan, Gansu	2012—2013	6.7	6 090.0	实收，比对照泾麦1号增20.8% Harvested yield, increase 20.8%
甘肃省泾川县太平乡 Jingchuan, Gansu	2013—2014	200	7 951.5	测产，比对照晋麦79增42.9% Predicted yield, increase 42.9%
甘肃崇信县黄寨乡 Chongxin, Gansu	2013—2014	0.2	9 262.5	实收，比对照晋麦79增产51.4%，省旱地高产纪录 Harvested yield, increase 51.4%, new record

资料来源：本单位收集的相关数据 Data source: Collected from our collaborators

表5 不同灌溉量对中麦175和京冬17产量相关性状的影响

Table 5 Effect of different irrigations on yield and related traits of Zhongmai 175 and Jingdong 17

灌溉量 Irrigation	穗数 Spikes/m^2	穗粒数 Grains/spike	千粒重 TKW (g)	收获指数 HI	生物量 Biomass ($kg \cdot hm^{-2}$)	产量 Yield ($kg \cdot hm^{-2}$)	水分利用效率 WUE ($kg \cdot mm^{-1}$)
0, W0	520b/488b	30.2a/32.4b	45.7b/48.4b	42.9b/39.5c	15827c/14949c	7510c/6620c	21.9b/18.9b
1, W1	637a/619a	30.4a/33.2ab	45.2b/47.7c	44.5b/43.4b	17942b/17352b	8830b/8160b	23.8a/21.3a
2, W2	653a/635a	30.5a/33.9a	46.9a/49.7a	49.4a/45.1a	19078a/18341a	9350a/9050a	22.4b/22.2a

"/"前为中麦175，后为京冬17。WUE：水分利用效率。同一列不同字母表示差异达5%显著水平

Zhongmai 175 is before "/" and Jingdong 17 is below. WUE: Water use efficiency. Different letters in the same column indicate significant difference at 5% probability level

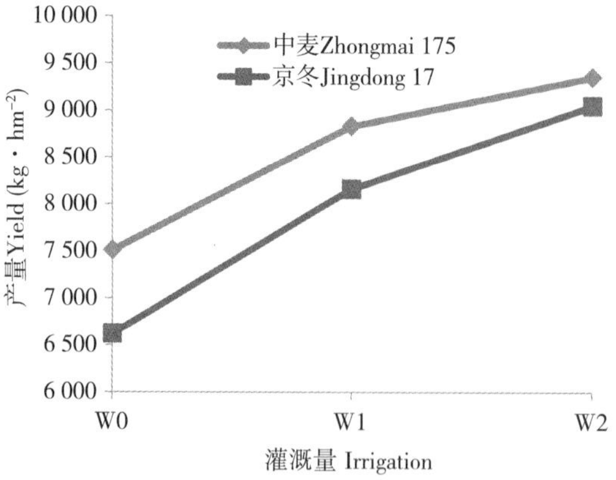

图1 不同灌溉条件下中麦175和京冬17产量比较

Fig. 1 Yield comparison between Zhongmai 175 and Jingdong 17 at different irrigations

3 肥料利用效率高

据中国科学院遗传发育研究所对北部和黄淮麦区64份主要品种氮和磷利用效率的研究，中麦175在不施氮、不施磷及氮和磷皆不施的3种处理中，产量分别居参试品种的第1、第9和第1位，说明中麦175为肥料高效型品种（童依平，2012，个人交流）。

为了进一步确认这一特性并明确其机理，设置6种氮肥处理，对中麦175与京冬17进行系统比较。在不同施氮水平下，中麦175的产量均高于京冬17（图2）。从表6可以看出，氮肥施用量对中麦175的穗数影响最大，但穗粒数和千粒重在不同处理间差异不显著，而氮肥施用量对京冬17的产量三要素均有显著影响。在不同施氮水平下，中麦175氮肥利用率即吸收效率都高于京冬17；在60kg和120kg 2个低氮条件下，中麦175氮肥利用效率和农学效率显著高于京冬17，其中前者氮肥利用效率分别为25.2和18.3kg·kg^{-1}，后者分别为6.6和7.3kg·kg^{-1}。另有研究表明，京411及其14个衍生品种在正常施肥和常年不施肥条件下进行比较，中麦175的产量皆为最高，主要原因是穗数多，产量三因素对肥料的敏感性相对较低[5]。上述结果皆说明中麦175的确是一个肥料高效型品种，其氮肥吸收和利用效率都较高。研究还表明，中麦175在苗期具有较强的耐低磷能力，6A、3B和4BS染色体对苗期氮和磷利用效率均有影响，1B和2AL染色体位点表现较强的耐低磷能力，4B染色体 *BS00022177 _ 5 1* 附近的标记密集区与氮利用效率、耐低氮能力和磷利用效率等有关[6]。

综上所述，中麦175氮肥吸收即利用率高，在低氮条件下的氮肥利用效率和农学效率都高；在不同施肥水平下的产量都高于对照京冬17，说明它实现了高产潜力与肥料高效利用率的良好结合，是肥料高效型高产品种。

表6 不同氮肥水平对中麦175和京冬17产量及氮肥利用效率相关性状的影响

Table 6 Effect of different N fertilizer levels on yield and nitrogen use efficiency related traits of Zhongmai 175 and Jingdong 17

施氮肥量 N rate（kg·hm^{-2}）	穗数 Spikes/m^2	穗粒数 KPS	千粒重 TKW（g）	产量 Yield（kg·hm^{-2}）	氮肥利用率 NRE（%）	氮肥利用效率 NUtE（kg·kg^{-1}）	氮肥农学效率 NAE（kg·kg^{-1}）
0	634c/606b	30.5a/32.4c	47.3a/49.9ab	8 970c/8 410d	0/0	0/0	0/0
60	692a/613ab	30.8a/33.9ab	47.2a/50.3ab	9 380b/8 510cd	0.27a/0.24a	25.2a/6.6c	6.9a/1.6b
120	703a/609ab	31.5a/33.2abc	47.4a/49.7b	9 410b/8 570cd	0.20bc/0.18b	18.3b/7.3c	3.7b/1.3bc
120+60	653bc/635a	30.5a/33.9ab	46.9a/49.7b	9 350b/9 050b	0.21b/0.18b	9.9c/19.3b	2.1bc/3.5a
120+120	704a/624ab	31.5a/34.4a	47.1a/51.0a	9 720a/9 390a	0.18cd/0.17b	17.5/b23.4a	3.1bc/4.1a
120+180	678ab/599b	30.5a/32.9bc	47.3a/50.2ab	9 330b/8 650c	0.16d/0.15b	7.4c/5.2c	1.2c/0.8c

同一品种的不同处理间的不同字母表示在0.05水平差异显著。KPS：穗粒数；NRE：氮肥利用率；NUtE：氮肥利用效率；NAE：氮肥农学效率 Values followed by different letters are significantly different at the 0.05 probability level. KPS: Kernels per spike; NRE: Apparent recovery efficiency of applied N; NUtE: N utilization efficiency; NAE: N agronomy efficiency

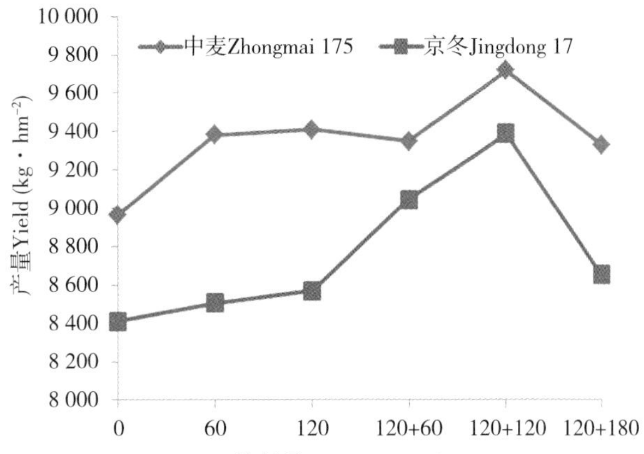

图2 中麦175和京冬17在不同施肥条件下的产量

Fig. 2 Grain yield of Zhongmai 175 and Jingdong 17 in different N treatments

4 加工品质和营养品质优良

4.1 面条和馒头兼用型品质

14份大田样品测试表明，中麦175籽粒蛋白质含量（11.6%，14%湿基）中等，籽粒硬度较低（21.1SKCS单位），出粉率较高（平均73.7%），吸水率、形成时间、稳定时间、延展性、最大抗延阻力分别为52.7%、1.7min、2.6min、183.4mm、77.0BU，属于软质中弱筋、延展性好的类型[7]。分子标记检测表明，中麦175不含1BL·1RS易位系，高分子量麦谷蛋白亚基分别为null、7+9和2+12，低分子量麦谷蛋白亚基分别为Glu-A3c和Glu-B3h，这是面筋强度弱、延展性好的主要原因。面粉L^*值较高（92.26）、a^*值（−1.27）和b^*值（7.13）较低，面片L^*值较高（82.74）、a^*值（0.30）和b^*值（21.23）较低，说明面粉和面片白度均较高。中麦175黄色素含量较高，含有较低基因*Psy-A1a*和*Psy-B1b*，表现良好的面粉和面片色泽。中麦175馒头品质优异，特点是体积大、形状好、质地优良，14份样品平均总分为85.3分，而对照品种京411馒头总分为75.0分。面条品质优良，特点是口感、颜色和黏弹性均好，平均总分70.9分，高于对照样品商业雪花粉（70.0分）[7]。从表7可以看出，在同一环境下，中麦175的馒头和面条品质均显著优于北部冬麦区的代表性品种京冬8号和轮选987。

中麦175的糯蛋白基因皆为野生型，即*Wx-A1a*、*Wx-B1a*和*Wx-D1a*，但其大淀粉粒重量占61.5%，小淀粉粒重量占38.5%，适宜的淀粉粒度分布可能是其馒头和面条品质优良的重要原因。用中麦175进行的大小淀粉粒重组试验表明，小淀粉粒含量为30%～35%的面粉制作的馒头内部结构最好、总评分最高，小淀粉粒含量为30%～40%的面粉制作的面条软硬度最适宜[8]。中麦175的抗性淀粉（63.0%）显著高于北京0045（57.3%）[9]和其他品种，说明它具有一定的保健功能。

4.2 锌含量高，营养健康价值高

缺铁性贫血在很大程度上是由于缺锌引起的，锌缺乏是终极意义上的"隐性饥饿"，培育和推广富锌作物品种具有重要意义[10]。将锌含量较高的24个品种在7个地点种植两年，河农326（58.2±7.3）、冀麦26（51.9±4.1）、农大3197（47.8±3.5）、京冬8号（46.1±2.5）和中麦175（45.8±2.7）的锌含量居前5位[11]。除中麦175外，其他4个品种目前的种植面积皆很小。

在不施锌肥的条件下，中麦175和良星99开花期的根长和根表面积及籽粒锌收获指数差异不显著，但在施锌肥的条件下，中麦175开花期的根长和根表面积及籽粒锌收获指数均显著高于良星99，说明土壤施锌肥可显著提高锌向其籽粒转移，而良星99则变化不明显，因而中麦175籽粒锌浓度（51.7mg·kg^{-1}）高于良星99（46.8mg·kg^{-1}）（邹春琴，个人交流）。这表明中麦175对土壤中锌的吸收转运效率较高。

总之，中麦175的特点是软质，面筋强度中等偏弱、延展性好，大小淀粉粒分布比例适宜，面粉颜色白，为面条和馒头兼用型优质品种，对土壤中锌的吸收转运效率高，营养健康价值高。

5 适应性广泛

中麦175的审定和推广区域包括北京、天津、河北、山西、山东、河南西部、陕西、甘肃及青海，跨9个省（市）4个麦区，已成为北部冬麦区水地及甘肃和青海春麦改冬麦地区的第一大品种，为黄淮旱肥地及陇东旱地的主栽品种，累计推广面积近120万hm^2，其中，2014年秋播约33万hm^2，这足以说明它具有非常广泛的适应性。分子检测表明，中麦175含有效应较大的光周期不敏感基因*Ppd-D1b*，这可能是其广适性的分子基础。中麦175为强冬性品种，春化基因组成为*vrn-A1*、*vrn-B1*、*vrn-D1*和*vrn-B3*，抗寒性居目前本麦区推广品种之首。2009—2010年北部冬麦区遇到了罕见的低温危害，大幅度减产，但中麦175仍获好收成，天津武清区6.7hm^{2}示范方实收产量高达7 890kg·hm^{-2}，为天津市当年最高纪录。随着全球气候变暖，在春麦区改种冬麦不仅能大幅度提高产量，而且熟期显著提前，更重要的是能避开春小麦用水高峰，便于统筹安排作物生产。春麦改种冬麦的基本要求是品种越冬性过关[12]。由于抗寒性和产量等表现突出，中麦175已成为青海省黄河谷地和湟水河流域的川水地区及甘肃张掖、武威等地的第一大冬麦品种。

中麦175含*Eps*，早熟性突出，比黄淮旱肥地品种早熟3～5d；籽粒偏小，灌浆速度特别快，能避开后期高温胁迫。采用花后12d覆盖塑料薄膜进行高温胁迫处理，在高温处理与非处理间白天（6：00—18：00）温差为 2.2℃（中午极端温差最高可达 10℃以上，一般可达 5℃以上）的情况下，中麦 175 灌浆速度受高温影响很小（图 3），其热敏感指数显著低于石优 17 和衡观 35，与耐高温品种石麦 15 和京冬 8 号接近。条锈病、叶锈病和白粉病轻，同时株高较当地品种矮 20～30cm 且在不同水分条件下变幅小，茎秆强度好，在灾年和丰水年皆表现突出，丰水年抗倒伏、创高产，这是它在黄淮旱肥地及陇东大面积推广的重要原因。还表现很强的耐晚播能力，2013—2014 年度河南洛宁县播期较正常年份推迟 1 个月，株高仍达 75cm，较其他品种增产 10%以上。

中麦 175 苗期对当前流行小种条中 32、条中 29 表现高抗，成株期对水 4、条中 29、条中 32 表现免疫至中抗，对水 7、HY8、条中 33 表现中抗。对条中 29 的抗性由 1 对位于 2AS 染色体的显性基因控制，目前正在寻找与其紧密连锁的 SNP 标记。

表 7 中麦 175、京冬 8 号和轮选 987 的面条和馒头加工品质比较

Table 7 Comparison of noodle and steamed bread qualities of Zhongmai 175, Jingdong 8, and Lunxuan 987

类型 Food type	特性 Character	中麦 175 Zhongmai 175	京冬 8 号 Jingdong 8	轮选 987 Lunxuan 987	雪花粉 Xuehua
面条 Noodle	色泽（15）Color	9.8	7.5	9.0	10.5
	表面状况（10）Appearance	7.5	7.0	6.0	7.0
	软硬度（20）Firmness	13.0	13.0	14.0	14.0
	黏弹性（30）Viscoelasticity	21.0	16.5	18.0	21.0
	光滑性（15）Smoothness	11.3	9.0	8.3	10.5
	食味（10）Flavor	8.0	6.5	8.5	7.0
	总分（100）Total score	70.6	59.5	63.8	70.0
馒头 Steamed Bread	比容（20）Specific volume	19.0	19.0	16.0	15.0
	外形（10）Shape	9.0	7.0	8.0	8.0
	表面光滑（10）Smoothness	8.0	9.0	8.0	8.0
	表面色泽（10）Skin color	10.0	8.0	4.0	10.0
	压缩张弛性（35）Stress relaxation	31.0	33.0	33.0	35.0
	结构（15）Structure	13.5	10.5	10.5	12.0
	总分（100）Total score	90.5	86.5	79.5	88.0

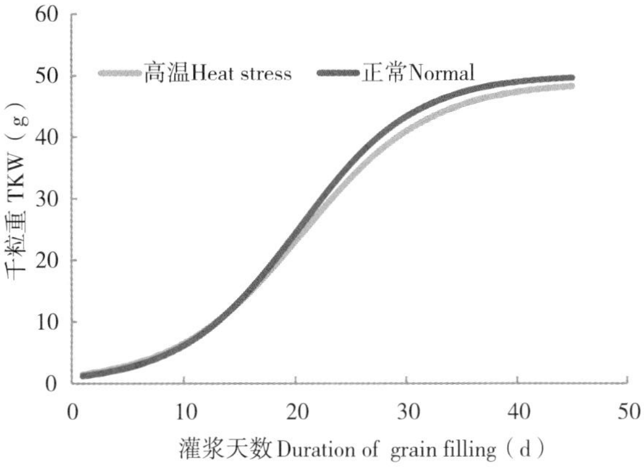

图 3 中麦 175 在田间正常温度和高温胁迫下灌浆速率的 Logistic 拟合曲线

Fig. 3 Logistic curve of grain filling rate of Zhongmai 175 under normal and heat stress conditions

6 育种经验分析与未来设想

6.1 育种经验分析

自 20 世纪 80 年代末育成京 411 和京冬 8 号以后，北部冬麦区的育种工作没有太大进展，主要表现在新品种植株偏高易倒伏，产量没有明显突破，同时对肥水要求过高。本区灌浆期光温充足，有利于增加粒重；但穗分化时间短，提高穗粒数主要依靠减少不孕小穗数和提高结实性来实现；同时对抗寒性和后期抗高温的要求很高，白粉病和条锈病则是主要病害[13-14]。据此，确定的具体策略是，以中高水肥地区为主，通过降低株高和株型改良来显著提高抗倒性；在保持较多穗数的基础上，通过增加粒数或粒重来提高产量潜力；加强育成品系的节水节肥性能筛选，保证新品种既能大面积创高产，又能够适应目前的粗放管理。要实现这一目标，需抓住 2 个关键环节，一是组合配置，二是后代选择与鉴定。

京 411 的抗寒性居本区推广品种的前列，产量潜力也高于京冬 8 号，穗数多（600～680 穗 · m^{-2}），株高 90～95cm，株型较紧凑，穗层整齐，适应性广[14]。虽然其面筋强度较弱，但手工制作的馒头和面条品质优良。缺点是植株偏高，在高产地块有倒伏现象，后期易早衰，造成粒重不稳，同时感白粉病和条锈病，机器收获时籽粒破碎率较高。自 20 世纪 80 年代后期的 10 多年，京 411 一直是高产骨干亲本[13]。针对京 411 的不足，选用了中国农业大学培育的新品系 BPM 27（组合为 20102//津 441/Fr81-4）为母本，其突出特点是秆硬、矮秆抗倒伏（约 70cm）、株型紧凑、穗较大且码密、早熟、兼抗白粉和条锈病，于 1995 年配置单交组合，双亲的优点多且互补性很好。采用系谱法经 6 年选择育成，于 2000—2001 年度参加产量鉴定，编号为 CA 0175 即中麦 175，它聚合了双亲的主要优点，基本实现了预期目标[14]。意外收获是水肥利用效率和适应性远远超过了双亲，原因还有待进一步研究。当然，中麦 175 也继承了京 411 的一些缺点，如机器收获时籽粒破碎率较高和白粒易穗发芽。

在早期世代，注重株型和穗育性的选择，即要求株型紧凑、叶片小且直立，以便容纳更多穗数；穗部性状突出码多、结实性和多粒的选择，为高产稳产提供保障。重点加强了育成品系的多点鉴定与大面积示范，不断扩大审定和推广地区。2002—2003 年度的产量比较试验在 4 个地点进行，中麦 175 全部增产，比对照京 411 平均增产 9.9%，故将其作为重点品系参加各级区域试验。由于在北京、山西和北部冬麦区的区域试验中皆表现突出，均提前一年参加生产试验，在尚未审定之前，在河北徐水农场进行大面积示范，发现其产量表现和适应性明显超过同期的其他品种。另外，在山西晋城的大面积示范中发现其水分利用效率高，有可能在黄淮旱肥地利用，于是当年秋播试着参加黄淮旱肥地区域试验并最终通过审定，使推广区域显著扩大。中麦 175 在青海省的发展则是借鉴京 411 的推广经验，在甘肃省河西的发展主要得益于当地春麦改冬麦对品种抗寒性的要求。基于中麦 175 的成功经验，目前的产量比较试验在 6 个地点进行，覆盖北京、天津、河北和山西 4 省区，还在昌平增设 1 个仅浇越冬水的旱地环境，最近又在新疆增加了 2 个试验点，这样就形成了较完善的品种比较网络，这是近年来育种工作取得较快进展的重要因素。中麦 175 的育种和推广实践表明，通过株型改良培育多穗型高产抗倒伏广适性品种完全可行，培育水旱兼用的半矮秆品种也是可行的，其关键指标是株型直立、株高适中和穗数多，这与玉米的直立耐密性杂交种有其相似之处[15]。

6.2 高产广适性品种培育的初步设想

中国小麦生产面临严峻挑战，在气候变化和纹枯病及赤霉病日益加剧的情况下[16]，既要保证较高产量和可接受的品质，还要降低水肥等投入，以提高产业竞争力并保护环境，这在缺水十分突出的北部冬麦区和黄淮北片显得更为重要。因此，除部分地区继续加强高产更高产甚至超高产品种培育之外，大部分地区应把培育高产、水肥高效、抗病、抗逆、广适性品种作为主要任务[17]，当然优质面条和馒头及专用面包品质等还要继续重视。就育种技术而言，尽管小麦测序已取得较大进展，用于品质和抗病性的分子育种技术已经成熟并在国内外开始应用[18]，全基因组选择在其他作物中也已取得阶段性进展[19]，但考虑到产量、水肥利用效率及抗热性等的复杂性及表型鉴定的巨大工作量，在近期内发掘出效应较大、育种家可用的基因特异性标记的难度相当大，因此高产高效广适性品种的培育仍需主要依靠常规育种技术。如前所述，常规育种成功的关键是组合配置和后代的准确选择与鉴定，考虑到实际育种中水肥管理的调节难度不小，鉴于国内外的广适性品种包括适于旱地的大面积品种多是从水地品种中筛选出来的现实，加强亲本和育成品系品种的系统鉴定至关重要。初步设想是，首先根据大面积生产表现和系统可靠的鉴定，明确高产、水肥高效、抗热及抗病等亲本，在此基础上合理配置组合。需要说明的是，石家庄市农业科学院对主要品种节水性能的系统鉴定值得学习和参考，产量、品质和抗病性等都需做类似的工作。除适当扩大分离群体的规模外，在早期分离世代如 F_2～F_4 仍可沿用过去的成熟做法。其次，要尽量扩大产量鉴定和品种比较试验的规模和点次，可在分离世代的后期如 F_5～F_6 尤其是产量鉴定及品种比较阶段加强多点和多环境鉴定，至少一个乃至几个点可为 1～2 水的有限灌溉和中低肥/不施肥条件，目的是发现既高产又高效广适的"超级品种"。这是 CIMMYT 及国际小麦育种的成功经验，国内应借鉴学习。第三，在高世代材料鉴定中，应及时大胆采用一些可用的分子标记，以便为表型鉴定补充有益信息。经过长期不懈努力，全国定能培育出一批高产高效广适性新品种，为小麦产业发展作出新的贡献。

参考文献
References

[1] 万富世．冬小麦品种越冬性改良理论与实践．北京：中国农业科技出版社，2006.
WanF S. *Theory and Practice of Cold Tolerance Improvement in Winter Wheat*. Beijing: China Agri-technology Press, 2006. (in Chinese)

[2] 李兴茂，倪胜利．不同水分条件下广适性小麦品种中麦 175 的农艺和生理特性解析．中国农业科学，2015，已接受．
Li X M, Ni S L. Agronomic and physiological characterization of wide adaptable wheat cultivar Zhongmai 175 under two different irrigation conditions. *Scientia Agricultura Sinica*, 2015, in press. (in Chinese)

[3] 史聚宝，李兴茂，孟治岳．小麦全膜覆土穴播技术存在的问题与对策．作物杂志，2014，1：129-131.
Shi J B, Li X M, Meng Z Y. Constrains and strategy for plastic mulch technology in wheat. *Crops*, 2014, 1: 129-131. (in Chinese)

[4] 肖永贵，路亚明，闻伟锷，陈新民，夏先春，王德森，李思敏，童依平，何中虎．小麦骨干亲本京 411 及衍生品种苗期根部性状的遗传．中国农业科学，2014，47（15）：2916-2926.
Xiao YG, Lu Y M, Wen W E, Chen X M, Xia X C, Wang D S, Li S M, Tong Y P, He Z H. Genetic contribution of seedling root traits among elite wheat parent Jing 411 to its derivatives. *Scientia Agricultura Sinica*, 2014, 47 (15): 2916-2926. (in Chinese)

[5] 肖永贵，李思敏，李法计，张宏燕，陈新民，王德森，夏先春，何中虎．两种施肥环境下冬小麦京 411 衍生系的产量和生理性状分析．作物学报，2015，41（9）：1333-1342.
Xiao YG, Li S M, Li F J, Zhang H Y, Chen X M, Wang D S, Xia X C, He Z H. Genetic analysis of yield and physiological traits among genotypes derived from elite parent Jing 411 under two under fertilizer environments. *Acta Agronomica Sinica*, 2015, 41 (9): 1333-1342. (in Chinese)

[6] 李法计，肖永贵，金松灿，夏先春，陈新民，汪洪，何中虎．京 411 衍生系苗期氮和磷利用效率相关性状的遗传解析．麦类作物学报，2015，35（6）：737-746.
Li F J, Xiao Y G, Jin S C, Xia X C, Chen X M, Wang H, He Z H, Genetic analysis of nitrogen and phosphorus use efficiency related traits at seeding stage of Jing 411 and its derivatives. *Journal of Triticeae Crops*, 2015, 35 (6): 737-746. (in Chinese)

[7] 张艳，陈新民，绍凤成，王德森，何中虎．中麦 175 馒头和面条品质稳定性分析．麦类作物学报，2012，32（3）：440-447.
Zhang Y, Chen X M, Shao F C, Wang D S, He Z H. Analysis on quality stability of steamed bread and noodle by wheat cultivar Zhongmai 175. *Journal of Triticeae Crops*, 2012, 32 (3): 440-447. (in Chinese)

[8] Guo Q, He Z H, Xia X C, Qu Y Y, Zhang Y. Effects of wheat starch granule size distribution on qualities of Chinese steamed bread and raw white noodle. *Cereal Chemistry*, 2014, 91 (6): 623-630.

[9] Yu J L, Wang S J, Wang J R, Li C L, Xin Q W, Huang W, Zhang Y, He Z H, Wang S. Effect of laboratory milling on properties of starches isolated from different flour millstreams of hard and soft wheat. *Food Chemistry*, 2015, 172: 504-514.

[10] 郝元峰，张勇，何中虎．作物锌生物强化研究进展．生命科学，2015，27（8）：1047-1054.
Hao Y F, Zhang Y, He Z H. Progress in Zinc biofortification of crops. *Life Science*, 2015, 27 (8): 1047-1054. (in Chinese)

[11] Zhang Y, Song Q C, Yan J, Tang J W, Zhao R R, Zhang Y Q, He Z H, Zou C Q, Ortiz-Monasterio

I. Mineral element concentrations in grains of Chinese wheat cultivars. *Euphytica*, 2010, 174: 303-313.

[12] 何中虎，夏先春，陈新民，庄巧生. 中国小麦育种进展与展望. 作物学报，2011，37（2）：202-215.
He Z H, Xia X C, Chen X M, Zhuang Q S. Progress of wheat breeding in China and the future perspective. *Acta Agronomica Sinica*, 2011, 37 (2): 202-215. (in Chinese)

[13] 周阳，何中虎，陈新民，王德森，张勇，张改生. 30余年来北部冬麦区小麦品种产量改良遗传进展. 作物学报，2007，33（9）：1530-1535.
Zhou Y, He Z H, Chen X M, Wang D S, Zhang Y, Zhang G S. Genetic gain of wheat breeding for yield in northern winter wheat zone over 30 years. *Acta Agronomica Sinica*, 2007, 33 (9): 1530-1535. (in Chinese)

[14] 陈新民，何中虎，王德森，庄巧生，张运宏，张艳，张勇，夏先春. 利用京411为骨干亲本培育高产小麦新品种. 作物杂志，2009，4：1-5.
Chen X M, He Z H, Wang D S, Zhuang Q S, Zhang Y H, Zhang Y, Zhang Y, Xia X C. Experience in developing high yielding varieties from core parent Jing 411. *Crops*, 2009, 4: 1-5. (in Chinese)

[15] LiY, Ma X L, Wang T Y, Li Y X, Liu C, Liu Z Z, Sun B C, Shi Y S, Song Y C, Carlone M, Bubeck D, Bhardwaj H, Whitaker D, Wilson W, Jones E, Wright K, Sun S K, Niebur W, Smith S. Increasing maize productivity in China by planting hybrids with germplasm that responds favorably to higher planting densities. *Crop Science*, 2011, 51: 2391-2400.

[16] He Z H, Joshi A K, Zhang W J. *Climate Vulnerability and Wheat Production*. Climate Vulnerability: Understanding and Addressing Threats to Essential Resources, Elsevier Inc., Academic Press, Waltham, USA, 2013: 57-67.

[17] He Z H, Xia X C, Peng S B, LumpkinT. Meeting demands for increased cereal production in China. *Journal of Cereal Sciences*, 2014, 59: 235-244.

[18] Liu Y N, He Z H, Appels R, Xia X C. Functional markers in wheat: Current status and future prospects. *Theoretical and Applied Genetics*, 2012, 125: 1-10.

[19] Huang X H, Wei X H, Sang T, Zhao Q, Feng Q, Zhao Y, Li C Y, Zhu C R, Lu T T, Zhang Z W, Li M, Fan D L, Guo Y L, Wang A H, Wang L, Deng L W, Li W J, Lu Y Q, Weng Q J, Liu K Y, Huang T, Zhou T Y, Jing Y F, Li W, Lin Z, Buckler E S, Qian Q, Zhang Q F, Li J Y, Han B. Genome-wide association studies of 14 agronomic traits in rice landraces. *Nature Genetics*, 2010, 42: 961-969.

矮秆基因与产量潜力

QTL mapping for plant height and yield components in common wheat under water-limited and full irrigation environments

Xingmao Li [A,B], Xianchun Xia[A], Yonggui Xiao [A], Zhonghu He [A,C], Desen Wang [A], Richard Trethowan [D], Huajun Wang [E], and Xinmin Chen [A,F]

[A] *Institute of Crop Science/National Wheat Improvement Center/Chinese Academy of Agricultural Sciences (CAAS), Beijing 100081, China.*

[B] *Key Laboratory of High Efficiency Water Utilisation in Dry Farming Region, Gansu Academy of Agricultural Sciences, Lanzhou 730070, China.*

[C] *CIMMYT China Office, Beijing 100081, China.*

[D] *Plant Breeding Institute, University of Sydney, Private Bag 4011, Narellan, NSW 2567, Australia.*

[E] *Gansu Provincial Key Laboratory of Aridland Crop Science, Lanzhou 730070, China.*

[F] *Corresponding author. Email: chenxinmin@caas.cn*

Abstract: Plant height (PH) and yield components are important traits for yield improvement in wheat breeding. In this study, 207 $F_{2:4}$ recombinant inbred lines (RILs) derived from the cross Jingdong 8/Aikang 58 were investigated under limited and full irrigation environments at Beijing and Gaoyi, Hebei province, during the 2011-12 and 2012-13 cropping seasons. The RILs were genotyped with 149 polymorphic simple sequence repeat (SSR) markers, and quantitative trait loci (QTLs) for PH and yield components were analysed by inclusive composite interval mapping. All traits in the experiment showed significant genetic variation and interaction with environments. The range of broad-sense heritabilities of PH, 1000-kernel weight (TKW), number of kernels per spike (KNS), number of spikes per m^2 (NS), and grain yield (GY) were 0.97-0.97, 0.87-0.89, 0.59-0.61, 0.58-0.68, and 0.23-0.48. The numbers of QTLs detected for PH, TKW, KNS, NS, and GY were 3, 10, 8, 7 and 9, respectively, across all eight environments. PH QTLs on chromosomes 4D and 6A, explaining 61.3-80.2% of the phenotypic variation, were stably expressed in all environments. *QPH.caas-4D* is assumed to be the *Rht-D1b* locus, whereas *QPH.caas-6A* is likely to be a newly discovered gene. The allele from Aikang 58 at *QPH.caas-4D* reduced PH by 11.5-18.2% and TKW by 2.6-3.8%; however, KNS increased (1.2-3.7%) as did NS (2.8-4.1%). The *QPH.caas-6A* allele from Aikang 58 reduced PH by 8.0-11.5% and TKW by 6.9-8.5%, whereas KNS increased by 1.2-3.6% and NS by 0.9-4.5%. Genotypes carrying both *QPH.caas-4D* and *QPH.caas-6A* alleles from Aikang 58 showed reduced PH by 28.6-30.6%, simultaneously reducing TKW (13.8-15.2%) and increasing KNS (3.4-4.9%) and NS (6.5-10%). *QTKW.caas-4B* and *QTKW.caas-5B.1* were stably detected and significantly associated with either KNS or NS. Major KNS QTLs *QKNS.caas-4B* and *QKNS.caas-5B.1* and the GY QTL

Published in Crop & Pasture Science, 2015, 66, 660-770.

QGY. caas-3B. 2 were detected only in water-limited environments. The major TKW *QTKW. caas-6D* had no significant effect on either KNS or NS and it could have potential for improving yield.

Additional keywords: drought tolerance, plant height, quantitative trait locus, *Triticum aestivum*, yield

Introduction

Grain yield is a major objective for wheat (*Triticum aestivum* L.) production, particularly for developing countries such as China with a high population and limited areas for cultivation. Yield improvement has been achieved through the integration of new cultivars and improved crop management technology. Genetic improvement of yield has been achieved by increasing harvest index and reducing plant height (PH) (Zheng *et al*. 2011; Xiao *et al*. 2012). Dwarfing genes, named from *Rht1* to *Rht22*, are on 10 different chromosomes of durum and bread wheat (Ellis *et al*. 2005; Haque *et al*. 2011; Peng *et al*. 2011; McIntosh *et al*. 2012). However, only *Rht8c*, *Rht9*, *Rht-B1b*, *Rht-B1d* and *Rht-D1b* have been successfully used in wheat breeding. The dwarfing genes *Rht8c* and *Rht9* are used commercially in Europe, Russia, China and Japan (Borojevic and Borojevic 2005), and *Rht-B1b* and *Rht-D1b*, the basis of the Green Revolution, were distributed worldwide through the International Maize and Wheat Improvement Center (CIMMYT). The dwarfing genes most widely distributed in Chinese cultivars are *Rht-D1b*, *Rht-B1b* and *Rht8* (Zhang *et al*. 2006). However, more recently, a gene associated with reduced PH, *TaSTE*, on chromosome 3A, was also found to be widely distributed in Chinese cultivars (Zhang *et al*. 2013*b*). *Rht-B1e* has recently been deployed in European wheat-breeding programs (Li *et al*. 2012*a*), and *Rht11* is used commercially in Ukraine and Russia (Divashuk *et al*. 2012). Molecular markers for some of the most important dwarfing genes such as *Rht-D1b*, *Rht-B1b* and *Rht8* have also been developed (Worland *et al*. 1998; Ellis *et al*. 2002) and can be used to identify genetic resources for wheat improvement (Zhang *et al*. 2006). In addition, a recently developed polymerase chain reaction (PCR) marker for *Rht-B1e* may facilitate the use of this dwarfing allele in wheat breeding (Li *et al*. 2012*a*).

Dwarfing genes can be grouped into two categories: those insensitive and those sensitive to exogenous gibberellic acid (GA). The GA-insensitive *Rht-B1b* and *Rht-D1b* alleles increase fertility and grain number per spike, reduce PH by ~15%, and increase yield (Flintham *et al*. 1997; Chapman *et al*. 2007; Xiao *et al*. 2012), particularly under well-managed, irrigated environments (Mathews *et al*. 2006). However, these genes can also reduce seedling leaf size, thus reducing seedling vigour (Rebetzke *et al*. 2007). Moreover, because of higher sensitivity to stress before heading, these GA-insensitive genes tend to reduce yield under high-temperature or drought conditions (Richards 1992; Chapman *et al*. 2007). However, several GA-sensitive major dwarfing genes such as *Rht8*, *Rht9*, *Rht12* and *Rht13* have potential to reduce PH without affecting seedling growth (Rebetzke *et al*. 2012), thus reducing the impact of stresses. *Rht8* reduces PH by ~10% without significant negative effects on yield (Worland *et al*. 1998) and only minimal effects on other agronomic traits (Worland *et al*. 2001). *Rht8* is linked with the photoperiod insensitivity gene *Ppd-D1a* and can therefore increase carbon-partitioning to the grain and, also, grain number.

Compared with *Rht-D1b* genotypes, *Rht-D1c* originating from a tandem segmental duplication of *Rht-D1b*, when deployed in combination with *Rht-D1b*, *Rht-B1b* or *Rht-B1c*, can significantly reduce PH (Richards 1992; Flintham *et al*. 1997; Li *et al*. 2012*b*). *Rht-D1b* + *Rht-B1b* cultivars have significantly higher kernel number per spike (KNS) than *Rht-D1b*+*Rht-B1c* genotypes (Flintham *et al*. 1997), and *Rht-D1b* +*Rht8c* types have significantly higher harvest index, KNS and grain weight per spike than *Rht-D1b* materials (Xiao *et al*. 2012). Therefore, selection for a specific combination of dwarfing genes is important for further improving yield potential.

Previous reports indicated that quantitative trait loci (QTLs) for yield components are located on almost all chromosomes (Börner *et al*. 2002; Huang *et al*. 2004; Quarrie *et al*. 2005; Kumar *et al*. 2007; Cuthbert *et al*. 2008; McIntyre *et al*. 2010; Deng *et al*. 2011; Golabadi *et al*. 2011; Bennett *et al*. 2012). QTL meta-analysis was introduced to identify QTLs for yield components located in the same chromosomal regions but in different genetic backgrounds. Zhang *et al*. (2010) reported that QTLs for yield components were distributed on chromosomes 1A, 1B, 2A, 2D, 3B, 4A, 4B, 4D and 5A, and important gene groups such as *Rht* and *Vrn* were associated with yield components (Cuthbert *et al*. 2008; Quarrie *et al*. 2005). The contribution of the 1B/1R translocation to increased yield was also confirmed by QTL analysis (Pinto *et al*. 2010). Major QTLs for 1000-kernel weight (TKW) and number of kernels per spike (KNS) were positively associated with *Rht-B1a* and *Rht-D1a* loci (Zhang *et al*. 2013*a*), whereas Cuthbert *et al*. (2008) found that *Rht-B1b* was associated with increased TKW. In addition, stable QTLs for yield components on chromosomes 7AL and 7BL were identified in 11 environments (Quarrie *et al*. 2005). *TaSAP1-A1*, *TaCwi-A1*, *TaCKX6-D1* and *TaGW2*, all associated with TKW, were cloned and found to be located on chromosomes 7A, 2A, 3D and 6A, respectively (Su *et al*. 2011; Ma *et al*. 2012; Zhang *et al*. 2012; Chang *et al*. 2013). By contrast, *TaCKX2.1* and *TaCKX2.2* were associated with KNS and located on 3DS (Zhang *et al*. 2011). Quarrie *et al*. (2005) identified yield QTLs in 24 environments; QTLs on chromosomes 1AL, 4BS/L, 4DL and 5DL were largely associated with drought stress. However, few studies on drought stress in defined and contrasting environments have been reported (Pinto *et al*. 2010; Bennett *et al*. 2012).

Grain yield improvement under full irrigation in Chinahas been achieved by increasing TKW and/or KNS (Zheng *et al*. 2011; Xiao *et al*. 2012). However, water shortage is a major factor limiting wheat production in northern China, where 80% of Chinese wheat is produced. During the last 10 years, the number of irrigations for wheat has declined from four or five to two per season. Thus, it is vital that new cultivars perform well under differing levels of irrigation. The identification and deployment of QTLs for major yield components and PH could prove very useful in developing these more responsive cultivars. In this study, a genetic population originating from a cross between the short-statured cultivar Aikang 58 and a tall cultivar Jingdong 8 was used to identify QTLs for PH and yield components and their expression in water-limited and fully irrigated environments with the ultimate aim of identifying trait-associated DNA markers suitable for wheat breeding.

Materials and methods

Plant materials

Wheat genotypes (210), including 207 $F_{2:4}$ lines derived from a cross between Jingdong 8 and Aikang 58, the two parents, and a check variety Zhongmai 175, were evaluated in field trials. Aikang 58, carrying dwarfing genes *Rht-D1b* and *Rht8* and the 1B/1R translocation (Wang *et al*. 2012), was released in the Yellow and Huai River Valleys Facultative Wheat Region in 2005. Jingdong 8, carrying *Rht8* and the 1B/1R translocation, was released in the Northern Winter Wheat Region in 1995. The parents carried the same vernalisation genes *vrn-A1*, *vrn-B1*, *vrn-D1* and *vrn-B3*, and photoperiod insensitivity gene *Ppd-D1a* (Zhang *et al*. 2008; Zhongwei Wang, unpubl. data from our laboratory).

Field trials and phenotypic evaluation

Field trials were conducted in the 2011-12 and 2012-13 cropping seasons at the CAAS Beijing Experimental Station, and at Gaoyi in Hebei province. In both locations, pre-sowing irrigation was provided to ensure good germination. The field trials at each location included two water treatments: limited irrigation, where one irrigation was applied at pre-overwintering to ensure winter survival; and full irrigation applied at pre-overwintering, green recovery in early spring, post stem elongation, and at grain-filling. The water supply was ~70 mm for each irrigation. Hereafter, the eight field trials are denoted as BD1, BI1, BD2, BI2 (representing Beijing 2011-12 limited and full irrigation, Beijing 2012-13 limited and full irrigation) and GD1, GI1, GD2, GI2 (representing Gaoyi 2011-12 limited and full irrigation, and Gaoyi 2012-13 limited and full irrigation) (for details see Table 1). All experiments were completed using a randomised complete block design with two replications. Each plot consisted of four 2-m rows with 30 cm between rows. Weed, disease and pest control was conducted as required to minimise other yield limitations.

The PH was measured at the group level from the ground to the tip of the spike, excluding awns at late grain-filling. Spike number in two 0.5-m centre rows was scored and was transformed to number of spikes per m^2 (NS). KNS was calculated from the mean of 30 randomly selected spikes in each plot. After harvest, TKW was measured by weighing two samples of 500 kernels from each plot. Grain yield (GY) was determined as the weight of grain harvested per unit area ($g \cdot m^{-2}$). The heading date (HD) was determined as the number of days from sowing to heading.

Statistical analyses

Analysis of variance (ANOVA) of phenotypic data and correlation coefficients among environments were performed with the Statistical Analysis System (SAS Institute 2000). PROC GLM was used in ANOVA for all traits. The PROC MIXED procedure was used in ANOVA to estimate the effect of markers on traits. Correlation analysis between parameters was performed using the PROC CORR procedure. Broad-sense heritabilities were estimated in all environments as: $h^2=\sigma_g^2/(\sigma_g^2+\sigma_{ge}^2/e+\sigma_\varepsilon^2/re)$, where the genetic variance $\sigma_g^2=(MS_f-MS_e)/re$, genotype×environment interaction variance $\sigma_{ge}^2=(MS_{fe}-MS_e)/r$, error variance $\sigma_e^2=MS_e$, MS_f is genotype mean square, MS_{fe} is genotype × environment interaction mean square, MS_e is error mean square, and r and e are the numbers of replicates and environments, respectively.

Genetic maps and QTL analysis

Genomic DNA was extracted from young leaves collected in bulks of 50 plants from each line in the field, by using the modified CTAB method (Saghai-Maroof *et al.* 1984). In total, 1240 simple sequence repeat (SSR) markers were used to screen for polymorphism between Aikang 58 and Jingdong 8. Polymorphic markers were subsequently used to genotype the entire $F_{2:4}$ population. Relevant information regarding the SSR markers (BARC, GWM, WMC, and CFE codes) was taken from the GrainGenes Website (http://wheat.pw.usda.gov). PCR amplification of SSR markers was performed in a MyCycler Thermal Cycler (BIO-RAD, Hercules, CA, USA) in total volumes of 15 μL containing 50 ng of template DNA, 1×PCR buffer, 4 pmol of each primer, 200 mM of each dNTP and 1 U of *Taq* DNA polymerase (Takara, Beijing). The PCR program was an initial denaturation at 95℃ for 5 min, followed by 38 cycles of denaturing at 95℃ for 1 min, annealing at 55-61℃ for 1 min, extension at 72℃ for 1 min, and a final extension at 72℃ for 10 min. PCR products were separated in 6.0% denaturing polyacrylamide gel electrophoresis (PAGE) and visualised by silver staining. PCR of the *Rht-D1* gene-specific marker followed Ellis *et al.* (2002).

Linkage maps were constructed using QTL IciMapping ver. 3.2 software (Quantitative Genetics Group ICS-CAAS, www.isbreeding.net) with the criterion > 3.0 logarithm of odds (LOD) (Li *et al.* 2007) for significance. Based on the consensus map, 149 SSR markers were located on 20 wheat chromosomes. QTL identification and QTL × environment (E) interaction was performed using inclusive composite interval mapping (ICIM). Walking speed chosen for all QTL was 1.0 cM, with $P=0.001$ in stepwise regression. A LOD threshold of 3.0 was chosen for declaration of putative QTLs. The phenotypic variance explained (PVE) was estimated through stepwise regression (Li *et al.* 2007), and QTL epistatic effects were calculated following Li *et al.* (2008).

Results

Phenotypic variation

On average, the lowest GY (3 186 kg • ha^{-1}) and the shortest PH (64 cm) were obtained in the BD2 environment because of the low spring rainfall in Beijing 2013; the highest GY, 6 100 kg • ha^{-1}, was achieved under BI1 because of good distribution of rainfall and full irrigation (Table 1). The temperature in spring 2013 was lower than in 2012, so that the heading date in 2013 was 6-11 days later than in 2012. This led to a shorter grain-filling duration and lower GY in 2013. The reduction of average GY and PH was 1 438. 3 kg • ha^{-1} and 11. 3 cm from water stress. The HD of Aikang 58 was 0. 6-1. 4 days later than that of Jingdong 8 (Li *et al.* 2015).

Table 1 Water supply and average grain yield, plant height and heading date of the trials in the 2011-12 and 2012-13 crop seasons

BD, GD: Beijing and Gaoyi sites under limited (drought) irrigation (LI);
BI, GI: Beijing and Gaoyi sites under full irrigation (FI)

Location	Code	Rainfall (mm)		Irrigation (mm)	Heading date (days)	Plant height (cm)	Grain yield (kg • ha^{-1})
		Spring	Sowing to harvest				
			2011—12				
Beijing	BD1, LI	64	150	70	207	76. 7	4613. 4
	BI1, FI	64	150	280	209	83. 6	6100. 1
Gaoyi	GD1, LI	44	106	70	194	75. 3	5086. 2
	GI1, FI	44	106	210	197	81. 9	5497. 4
			2012-13				
Beijing	BD2, LI	26	174	70	213	64. 0	3186. 1
	BI2, FI	26	174	280	214	83. 9	5621. 6
Gaoyi	GD2, LI	37	127	70	205	71. 7	4328. 6
	GI2, FI	37	127	210	207	83. 5	5748. 5

Analyses of variance were conducted for PH, KNS, NS, TKW and GY across all eight environments. Significant differences were found among the $F_{2:4}$ lines for all traits (Table 2). There was transgressive segregation for all traits, indicating quantitative inheritance. Plant height in the $F_{2:4}$ families showed a bimodaldistribution ranging from 46. 5 to 102. 0 cm in both full and limited irrigation conditions, with two peaks at ~65 and 75 cm in limited irrigation, and two peaks at ~75 and 90 cm in full irrigation (Fig. 1). The average plant height of Aikang 58 was 33 cm shorter than that of Jingdong 8 across all environments.

Table 2 Analysis of variance for plant height (PH), 1000-kernel weight (TKW), number of kernels per spike (KNS), number of spikes per m^2 (NS) and grain yield (GY) in $F_{2:4}$ lines from Jingdong 8/Aikang 58 across eight environments

$^{**}P<0.01$

Source of variance	PH		TKW		KNS		NS		GY	
	d. f.	MS	d. f.	MS	d. f.	MS	d. f.	MS	d. f.	MS
Treatment	1	106 403. 3**	1	2 010. 9**	1	184. 2**	1	4 704 933. 9**	1	12 227 944. 1**
Site	3	5 838. 9**	3	21 375. 1**	3	26 239. 4**	3	5 014 903. 2**	3	3 673 873. 7**
Line	209	1 046. 9**	209	109. 7**	209	34. 0**	209	12 631. 3**	209	6 225. 1**
Rep	1	150. 9**	1	14. 6	1	19. 2	1	4 295. 4	1	48 672. 4**
Line×site	626	16. 2**	626	8. 4**	626	8. 8**	626	2 885. 4	626	3 071. 1**

(continued)

Source of variance	PH		TKW		KNS		NS		GY	
	d. f.	MS	d. f.	MS	d. f.	MS	d. f.	MS	d. f.	MS
Line×treatment×site	838	44.8**	838	6.7**	838	10.7**	838	5 063.0**	838	8 237.1**
Error	1 674	11.2	1 670	4.6	1 674	7.5	1 672	3 075.6	1 671	1 927.0

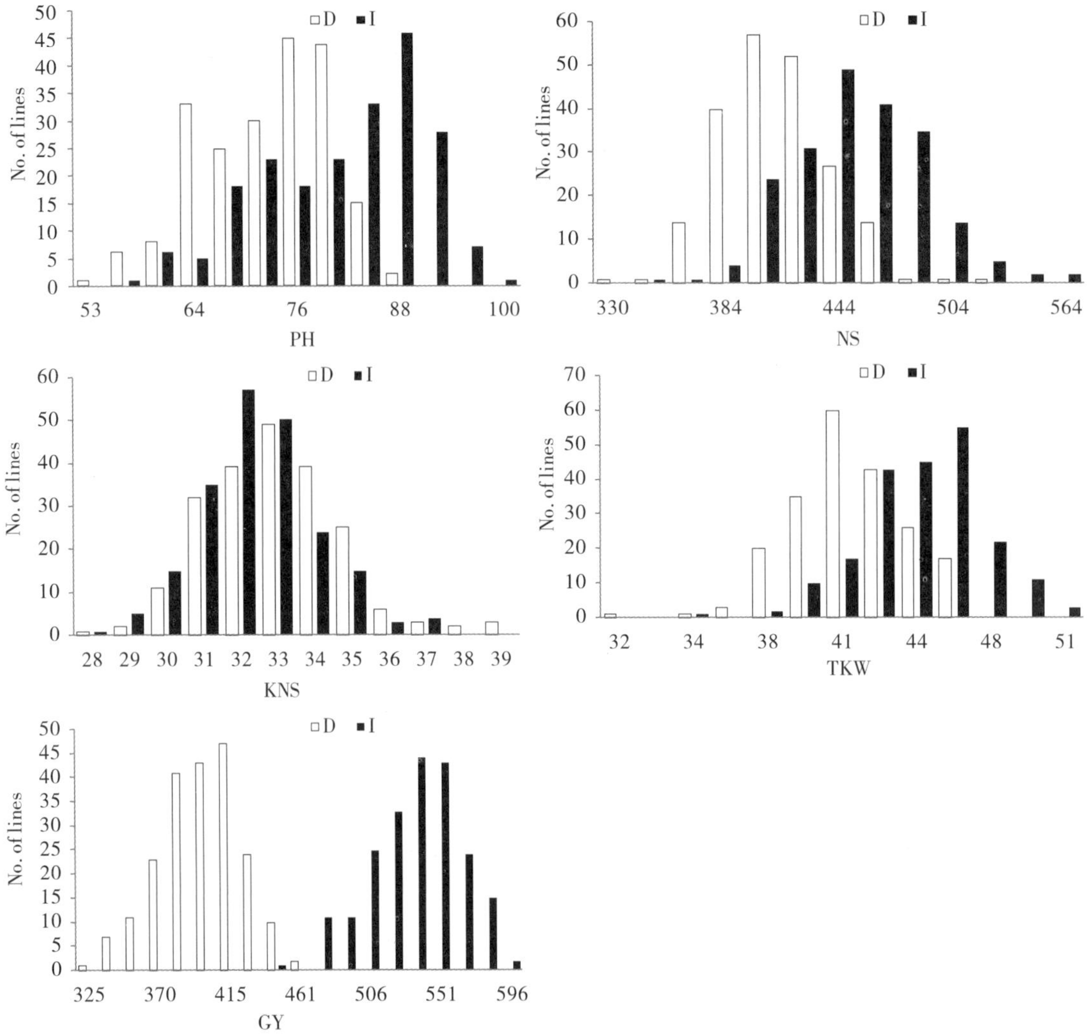

Fig. 1 Distributions of plant height (PH), 1000-kernel weight (TKW), number of kernels per spike (KNS), number of spikes per m^2 (NS) and grain yield (GY) in $F_{2:4}$ progeny from Jingdong 8/Aikang 58. White bars, Average of four limited irrigation environments; black bars, average of four fully irrigated environments.

The values of NS, TKW, PH, and GY under limited irrigation were smaller than those under full irrigation, whereas the values of KNS were similar under both limited and full irrigation (Table 3). The short parent, Aikang 58, showed larger NS, but no significant difference from the tall parent Jingdong 8 in all environments; however, Jingdong 8 had higher TKW and PH in all environments (Table 3). The broad-sense heritabilities of PH and TKW under limited irrigation were 0.97 and 0.87, compared with 0.97 and 0.89, respectively, under full irrigation. By contrast, heritabilities of KNS, NS and GY were lower, at 0.59, 0.58 and 0.48 under limited irrigation, with corresponding values of 0.61, 0.68, and 0.23 under full irrigation.

QTL analyses

Both *Rht8* and the 1B/1R translocation, which are present in both parents, have been reported to affect one or more of the traits investigated; hence effects that may be associated with these genes were not detected in this study.

Plant height

Three QTLs for PH were identified in the Jingdong 8/Aikang 58 population; these were located on chromosomes 4D, 6A and 6B (Table 4). The alleles reducing PH at these three loci came from the short parent, Aikang 58.

The QTLs on 4D and 6A, designated *QPH. caas-4D* and *QPH. caas-6A*, were identified in all environments. *QPH. caas-6B* was detected in two environments (Table 4). In total, the three QTLs explained 67.1-82.3% of the phenotypic variation for PH, and the QTLs at 4D and 6A explained 61.3-80.2%.

Both *QPH. caas-4D* and *QPH. caas-6A* showed significant interactions with environment (Supplementary material table 1, available on journal's website). Although the interaction between *QPH. caas-4D* and *QPH. caas-6A* was not significant, there were significant epistatic effects between *QPH. caas-4D* and other loci such as *QPH. caas-6B* (Supplementary fig. 1).

1 000-Kernel weight

The ICIM identified 10 QTLs for TKW. Three were distributed on chromosome 1A, two on 5B and one on chromosomes 3B, 4B, 4D, 6A and 6D (Table 4). Alleles that increased TKW at six loci on chromosomes 1A (3), 5B (2) and 6D were contributed by Aikang 58, whereas alleles of a further four QTLs for TKW were contributed by Jingdong 8, on chromosomes 3B, 4B, 4D and 6A. The PVE by Jingdong 8 QTLs was 16.5-33.1% compared with 7.5-20.4% for Aikang 58 under limited irrigation. Thus, Jingdong 8 made a larger contribution to TKW than Aikang 58 as predicted from the parental TGW.

Several TKW QTLs appeared to be specific toa smaller number of environments, for example, *QTKW. caas-1A. 2*, *QTKW. caas-1A. 3*, *QTKW. caas-5B. 2* and *QTKW. caas-4B*. However, *QTKW. caas-4D*, *QTKW. caas-5B. 1* and *QTKW. caas-6A* were detected in more than five environments (Table 4). The Aikang 58 alleles of *QTKW. caas-4D* and *QTKW. caas-6A* contributed more TKW and showed pleiotropic effects on PH.

QTKW. caas-1A. 3, *QTKW. caas-4B*, *QTKW. caas-4D*, *QTKW. caas-5B. 1* and *QTKW. caas-6A* showed significant interaction with environment (Supplementary table 1). Significant epistatic effects were detected for *QTKW. caas-4B* and *QTKW. caas-1A. 3* under limited irrigation (Supplementary fig. 1).

Number of kernels per spike

Eight QTLs for KNS were identified on each of chromosomes 1B, 2A, 4B, 4D, 6A and 7D, and two on chromosome 5B (Table 4). Alleles that increased KNS on chromosomes 1B, 4B, 4D, 6A and 7D were contributed by Aikang 58, with additional QTLs for increased KNS from Jingdong 8 on chromosomes 5B (2) and 2A.

No QTL for KNS was found in environments BI2 and GI2. Significant epistatic effects were detected for *QKNS. caas-1B*, *QKNS. caas-5B. 1*, *QKNS. caas-4D* and *QKNS. caas-6A* under limited irrigation and for *QKNS. caas-1B* under full irrigation(Supplementary fig. 1). Five QTLs (*QKNS. caas-1B*, *QKNS. caas-2A*, *QKNS. caas-5B. 1*, *QKNS. caas-6A* and *QKNS. caas-7D*) showed significant interaction with environment (Supplementary table 1).

Number of spikes

Seven QTLs for NS were identified on each of chromosomes 1A, 2D, 4D, 5B, and 6A, and two on chromosome 2A (Table 4). Alleles that increased NS on chromosomes 2A, 2D and 5B were contributed by Jingdong 8, and positive effect alleles from Aikang 58 were on chromosomes 1A, 2A, 4D and 6A. Almost all NS QTLs expressed larger dominance than additive effects. No QTL was detected in environment GD1 or BI2. *QNS. caas-2D*, *QNS. caas-4D* and *QNS. caas-6A* showed significant interaction with environment (Supplementary table 1). An interaction between *QNS. caas-5B* and *QNS. caas-2D* was observed only under limited irrigation (Supplementary fig. 1).

Grain yield

Nine QTLs for GY were identified on each of chromosomes 1A, 2A, 3A, 4A, 4D, 5B, and 7A, and two on chromosome 3B (Table 4). Alleles that increased GY at four loci on chromosomes 1A, 3B (2) and 7A were contributed by the tall parent, Jingdong 8, whereas alleles of two yield-positive QTLs on chromosomes 3A and 5B were contributed by Aikang 58. Both parents contributed alleles that increased GY at *QGY. caas-2A* and *QGY. caas-4D1* loci in different environments.

Two stable QTLs, *QGY. caas-3B. 2* and *QGY. caas-7A*, showed significant interaction with environment (Supplementary table 1). A larger epistatic effect was observed for QTL in full irrigation than in limited irrigation (Supplementary fig. 1).

Comparison of the effects of the **QPH. caas-4D** *and* **QPH. caas-6A** *alleles*

The phenotypic values of PH, TKW, KNS, NS and GY for individual QTLs and QTL combinations as determined by molecular markers are summarised in Supplementary table 2. *QPH. caas-4D*, fianked by *Rht-D1* and *Xbarc105*, and *QPH. caas-6A*, fianked by *Xbarc103* and *Xwmc256*, were associated with PH, TKW, KNS and NS. The *QPH. caas-4D* allele from Aikang 58 reduced PH (11.5-18.2%) and TKW (2.6-3.8%) and increased KNS (1.2-3.7%), NS (2.8-4.1%) and GY (0-1.2%). Similarly, the *QPH. caas-6A* allele from Aikang 58 reduced PH (8.0-11.5%) and TKW (6.9-8.5%) and increased KNS (1.2-3.6%) and NS (0.9-4.5%).

Genotypes carrying both*QPH. caas-4D* and *QPH. caas-6A* alleles from Aikang 58 reduced PH by 28.6% and 30.6% in limited irrigation and full irrigation, respectively, simultaneously reducing TKW (13.8-15.2%) and GY (2.1-2.6%), and increasing KNS (3.4-4.9%) and NS (6.5-10%). *QPH. caas-4D* co-located with GY; however, no association was observed for *QPH. caas-6A*.

Table 3 Summary of mean, maximum, minimum, standard deviation and heritability (h^2) values for plant height (PH), 1000-kernel weight (TKW), number of kernels per spike (KNS), number of spikes per m^2 (NS) and grain yield (GY) measured in the Jingdong 8/Aikang 58 population

LI, limited irrigation; FI, full irrigation

Trait	Environment	Jingdong 8	Aikang 58	Zhongmai 175	Average	Min-max	h^2
PH (cm)	LI	85.3	53.8	72.3	71.9±9.0	46.5-95	0.97
	FI	94.3	59.8	79.4	83.2±9.2	60-102	0.97
TKW (g)	LI	43.7	40.3	40.5	42.1±5.3	29.0-63.3	0.87
	FI	45.5	42.0	41.4	43.7±5.5	26.3-55.7	0.89
KNS	LI	30.5	32.5	29.7	33.0±5.0	22.8-47.9	0.59
	FI	30.5	32.4	29.2	31.6±5.0	21.9-44.2	0.61
NS	LI	423.3	438.2	447.9	408.7±69.6	246.7-623.3	0.58
	FI	438.3	485.1	511.4	483.9±97	261.7-725.0	0.68
GY ($g \cdot m^{-2}$)	LI	438.4	380.4	502.1	390.5±68.2	211.3-605.4	0.48
	FI	556.1	506.7	620.3	526.9±72.3	345.4-762.4	0.23

Table 4 QTLs for plant height (PH) and yield-related traits

TKW, 1000-kernel weight; KNS, number of kernels per spike; NS, number of spikes per m^2; GY, grain yield. BD1, BI1, BD2, BI2: Beijing 2011-12 limited and full irrigation, Beijing 2012-13 limited and full irrigation; GD1, GI1, GD2, GI2: Gaoyi 2011-12 limited and full irrigation, Gaoyi 2012-13 limited and full irrigation. LOD, Logarithm of odds, threshold of 3.0 was used for declaration of QTL; PVE, phenotypic variation explained (%) by QTL; Add, additive effect (positive, increased effect from Jingdong 8; negative, increased effect from Aikang 58); Dom., dominance effect of QTL allele

Trait	Envir.	QTL	Marker	Distance	LOD	PVE	Add.	Dom.
PH	BD1	*QPH. caas-4D*	*Xbarc105*	3.4	33.9	67.1	8.70	2.79
		QPH. caas-6A	*Xbarc103*	3.0	9.9	12.9	3.76	2.69
	BD2	*QPH. caas-4D*	*Xbarc105*	3.4	22.4	51.1	6.14	2.20
		QPH. caas-6A	*Xbarc103*	3.0	7.7	10.5	2.82	1.79
		QPH. caas-6B	*Xbarc247*	7.0	3.3	5.5	1.54	2.38
	BI1	*QPH. caas-4D*	*Xbarc105*	2.4	38.5	67.1	9.81	4.06
		QPH. caas-6A	*Xbarc103*	3.0	10.1	13.1	3.99	3.81
	BI2	*QPH. caas-4D*	*Xbarc105*	2.4	35.7	64.5	10.58	3.66
		QPH. caas-6A	*Xbarc103*	3.0	10.9	14.1	4.41	4.44
		QPH. caas-6B	*Xbarc247*	2	3.3	3.7	2.36	2.11
	GD1	*QPH. caas-4D*	*Xbarc105*	3.4	31.7	60.5	7.55	2.96
		QPH. caas-6A	*Xbarc103*	2.0	9.5	12.4	3.39	2.27
	GD2	*QPH. caas-4D*	*Xbarc105*	3.4	23.0	49.6	6.89	1.37
		QPH. caas-6A	*Xbarc103*	2.0	9.3	11.7	3.52	0.43
	GI1	*QPH. caas-4D*	*Xbarc105*	3.4	33.1	60.6	8.28	3.11
		QPH. caas-6A	*Xbarc103*	2.0	12.4	15.5	4.18	2.82
	GI2	*QPH. caas-4D*	*Xbarc105*	3.4	17.4	42.4	6.96	3.46
		QPH. caas-6A	*Xwmc256*	1.0	13.0	19.2	4.92	3.83
TKW	BD1	*QTKW. caas-4D*	*Rht-D1*	1.4	4.59	18.8	0.78	9.21
		QTKW. caas-6A	*Xbarc103*	0	4.39	7.4	1.42	0.76
		QTKW. caas-6D	*Xbarc365*	3.9	6.02	12.6	−2.02	−0.06
	BD2	*QTKW. caas-4B*	*Xgwm251*	3.7	3.36	5.8	0.93	−0.12
		QTKW. caas-4D	*Xwmc89*	4.5	5.87	9.5	1.11	0.51
		QTKW. caas-5B. 1	*Xbarc74*	0.4	4.89	7.5	−1.13	0.39
		QTKW. caas-6A	*Xwmc256*	1.0	8.96	14.3	1.41	1.08
	BI1	*QTKW. caas-1A. 2*	*Xbarc269*	5.8	3.45	21.7	−0.17	−4.32
		QTKW. caas-1A. 3	*Xwmc84*	4.4	4.03	23.3	−0.38	−5.31
		QTKW. caas-3B	*Xwmc43*	8.6	4.77	9.8	1.32	−0.19
		QTKW. caas-4D	*Xbarc105*	9.5	3.74	10.1	0.70	1.53
		QTKW. caas-5B. 1	*Xgwm335*	5	3.94	8.4	−1.14	0.35
		QTKW. caas-5B. 2	*Xbarc110*	0.9	3.27	17.0	−0.35	−5.81
		QTKW. caas-6D	*Xbarc365*	0.9	8.75	14.1	−1.53	0.36
	BI2	*QTKW. caas-4D*	*Xbarc105*	13.5	14.69	39.9	2.76	1.26
		QTKW. caas-6A	*Xbarc103*	3.0	8.78	13.7	1.70	0.97
	GD1	*QTKW. caas-1A. 1*	*Xgwm135*	5.0	3.55	6.9	−0.84	0.38
		QTKW. caas-6A	*Xbarc103*	1.0	8.87	16.5	1.44	0.38

(continued)

Trait	Envir.	QTL	Marker	Distance	LOD	PVE	Add.	Dom.
		QTKW. caas-6D	*Xbarc365*	5.9	6.54	13.5	−1.30	0.36
	GD2	*QTKW. caas-4B*	*Xgwm251*	7.0	4.35	7.5	1.02	−0.52
		QTKW. caas-4D	*Xwmc89*	1.5	3.74	5.1	0.73	0.62
		QTKW. caas-5B. 1	*Xgwm335*	1.0	7.02	11.0	−1.39	0.09
		QTKW. caas-6A	*Xbarc103*	2.0	13.35	20.5	1.75	0.59
	GI1	*QTKW. caas-5B. 1*	*Xgwm335*	5.0	5.89	13.0	−1.41	0.27
		QTKW. caas-6D	*Xbarc365*	3.9	6.50	11.9	−1.37	0.43
	GI2	*QTKW. caas-5B. 1*	*Xgwm335*	1.0	5.58	8.2	−1.17	0.35
		QTKW. caas-6A	*Xbarc103*	2.0	14.48	24.4	1.91	0.42
		QTKW. caas-6D	*Xbarc54*	4.0	3.19	5.1	−0.85	0.23
KNS	BD1	*QKNS. caas-5B. 1*	*Xgwm544*	0.4	3.5	6.9	0.64	0.08
		QKNS. caas-7D	*Xbarc126*	5.0	3.8	9.5	−0.71	0.08
	BD2	*QKNS. caas-4D*	*Xbarc105*	2.6	5.7	12.2	−1.28	−1.02
		QKNS. caas-6A	*Xwmc256*	0.1	3.8	7.2	−1.17	−0.61
	BI1	*QKNS. caas-4D*	*Xbarc105*	4.4	6.0	12.9	−0.93	0.24
		QKNS. caas-6A	*Xwmc256*	0.1	3.3	6.1	−0.68	−0.34
	GD1	*QKNS. caas-1B*	*Xbarc61*	1.2	4.0	7.6	−0.72	−0.37
		QKNS. caas-2A	*Xgwm95*	0.4	3.7	6.7	0.75	−0.30
		QKNS. caas-5B. 2	*Xbarc110*	7	2.9	13.9	0.47	−1.94
	GD2	*QKNS. caas-4B*	*Xgwm251*	7.7	3.6	8.2	−0.98	0.32
		QKNS. caas-5B. 1	*Xgwm371*	1	4.0	12.9	0.48	−3.19
	GI1	*QKNS. caas-6A*	*Xwmc256*	2.9	5.2	10.7	−1.10	0.02
		QKNS. caas-7D	*Xbarc260*	11	4.4	12.3	−1.15	−0.07
NS	BD1	*QNS. caas-2A. 2*	*Xgwm296*	2	3.0	17.8	11.49	88.08
	BI1	*QNS. caas-4D*	*Xbarc105*	0.6	4.6	9.2	−8.66	−24.72
		QNS. caas-6A	*Xbarc103*	0.1	5.0	9.9	−19.11	−9.68
	BD2	*QNS. caas-1A*	*Xbarc263*	3	3.0	24.1	−4.26	99.50
		QNS. caas-4D	*Rht-D1*	4.4	3.0	32.5	−0.23	104.21
		QNS. caas-5B	*Xgwm371*	2	3.0	23.6	1.46	73.76
	GD1	*QNS. caas-4D*	*Xwmc89*	2.5	4.7	12.6	−18.85	−3.04
	GI1	*QNS. caas-2A. 1*	*Xgwm339*	0.9	3.0	6.4	−15.81	11.70
		QNS. caas-2D	*Xbarc318*	1.1	3.0	6.3	9.38	−27.88
		QNS. caas-4D	*Xwmc89*	0.5	3.0	6.6	−11.90	−15.11
	GI2	*QNS. caas-2D*	*Xgwm539*	5	4.1	11.4	17.45	−8.68
		QNS. caas-6A	*Xbarc103*	0.1	3.0	5.2	−4.44	−16.45
GY	BD1	*QGY. caas-3B. 1*	*Xgwm547*	1	10.7	30.2	5.85	−175.67
		QGY. caas-3B. 2	*Xwmc43*	9.6	3.0	7.7	9.60	−12.36

(continued)

Trait	Envir.	QTL	Marker	Distance	LOD	PVE	Add.	Dom.
		QGY. caas-5B	*Xbarc110*	0. 1	5. 1	14. 4	−0. 40	−181. 03
		QGY. caas-7A	*Xbarc121*	0. 4	3. 0	5. 8	9. 79	9. 61
	BD2	*QGY. caas-3B. 2*	*Xbarc251*	6. 6	3. 6	10. 7	11. 25	−18. 91
	BI1	*QGY. caas-4D*	*Xwmc89*	1. 5	5. 6	14. 2	−17. 52	−16. 61
		QGY. caas-7A	*Xbarc281*	0. 4	3. 0	5. 5	9. 65	15. 93
	BI2	*QGY. caas-3A*	*Xbarc67*	1	4. 5	8. 7	−19. 04	2. 29
		QGY. caas-4D	*Xbarc105*	2. 5	7. 5	15. 2	24. 11	−3. 25
	GD1	*QGY. caas-4A*	*Xwmc468*	0. 3	3. 0	6. 5	−8. 82	−3. 07
	GI1	*QGY. caas-1A*	*Xbarc269*	0. 8	3. 2	12. 1	2. 03	−55. 15
		QGY. caas-2A	*Xbarc201*	0. 1	4. 1	7. 6	10. 34	−0. 80
		QGY. caas-4D	*rht2*	6. 6	3. 5	13. 5	−7. 18	−21. 40
	GI2	*QGY. caas-2A*	*Xgwm95*	0. 6	3. 2	6. 9	−5. 24	24. 52

Discussion

QTL for PH

QPH. caas-6A was proximal to the SSR marker *Xwmc256* and between markers *Xbarc113* and *Xbarc103* on chromosome 6AL. Although several QTLs for PH were previously reported on chromosome 6A (Börner *et al.* 2002; Huang *et al.* 2004; Liu *et al.* 2005; Marza *et al.* 2006; Spielmeyer *et al.* 2007; Griffiths *et al.* 2012), these are either on chromosome 6AS or close to the centromere. A minor QTL for PH on chromosome 6AL was linked with *Xgwm427* (Liu *et al.* 2002; Cui *et al.* 2011), which is 53 cM from *Xwmc256*, based on the wheat consensus map (Somers *et al.* 2004). Similarly, it is unlikely that *Rht14*, *Rht16* and *Rht18*, previously reported on chromosome 6A, influence these results because they originated from durum wheat (Haque *et al.* 2011). Therefore, *QPH. caas-6A* in the Jingdong 8/Aikang 58 population is likely to be a newly discovered gene on chromosome 6AL, based on the linked markers *Xwmc256* and *Xbarc113*. This QTL could have great potential in breeding programs targeting high yield environments because Aikang 58 conferred excellent resistance to lodging even at high seeding density. Thus, further work is needed to identify a more closely linked marker.

QPH. caas-4D was located in approximately the same region as the dwarfing gene *Rht-D1b*, and corresponds to the QTL detected by Huang *et al.* (2006). Aikang 58 carried *Rht-D1b* (Wang *et al.* 2012); therefore, it is assumed that *QPH. caas-4D* is *Rht-D1b*. *QPH. caas-4D* had a larger effect on PH than *QPH. caas-6A* in the materials studied. For example, the average height reduction of *Rht-D1b* was 11 cm, whereas the *QPH. caas-6A* allele in the interval *Xbarc103-Xwmc256* reduced PH by 7 cm. *QPH. caas-4D* and *QPH. caas-6A* showed an additive effect without epistatic interaction on PH when combined (Supplementary fig. 1). Genotypes carrying alleles from Aikang 58 at the *Xbarc105-4D* and *Xbarc103-6A* loci reduced PH by 20 cm. However, the genotype carrying these two genes were 8-11 cm taller than Aikang 58. Therefore, Aikang 58 may carry other dwarfing genes besides *QPH. caas-4D* and *QPH. caas-6A*.

No major gene or QTL for PH on chromosome 6B has previously been reported. In the present study, we detected a minor QTL, *QPH. caas-6B*, which is different from the QTL reported by Griffiths *et al.* (2012) based on *Xwmc105*. Cui *et al.* (2011) also detected a QTL for PH in this interval. However,

QPH. caas-6B was detected in only two environments in the present study. This inconsistency may reflect an interaction between this QTL and major QTLs found on chromosome 4D, but further study is needed to confirm this presumption.

Table 5 Summary of pleiotropic QTLs detected in the Jingdong 8/Aikang 58 $F_{2:4}$ population

Chromosome location	QTL	PH	TKW	KNS	NS	GY
1A	*QGY. caas-1A*		−1 (21.7)			1 (12.1)
2A	*QGY. caas-2A*			1 (6.7)		−1 (6.9) /1 (7.6)
3B	*QGY. caas-3B. 2*		1 (9.8)			2 (7.7−10.7)
4B	*QTKW. caas-4B*		+2 (5.8−7.5)	−1 (8.2)		
4D	*QGY. caas-4D*	+8 (28.5−47.5)	+5 (5.1−39.9)	−2 (12.2−12.9)	−4 (6.6−32.5)	2 (13.5−12.2) / −1 (15.2)
5B	*QTKW. caas-5B. 1*		−5 (7.5−13.0)	+2 (6.9−12.9)	+1 (23.6)	
5B	*QTKW. caas-5B. 2*		−1 (17.0)	+1 (13.9)		−1 (14.4)
6A	*QPH. caas-6A*	+8 (10.5−19.2)	+6 (7.4−24.4)	−3 (6.1−10.7)	−2 (5.2−9.9)	

Values are number of environments with QTL detected, and in parentheses is phenotypic variation explained by the QTL. PH, Plant height; TKW, 1000-kernel weight; KNS, number of kernels per spike; NS, number of spikes per m^2; GY, grain yield. −, Alleles with decreasing effect come from Aikang58; +, alleles with increasing effect come from Jingdong 8

Pleiotropic effect of QTL

The presence of QTL clusters for several traits has been reported in previous studies (Börner *et al*. 2002; Quarrie *et al*. 2005). Our results indicated QTL clusters on chromosomes 1A, 2A, 3B, 4B, 4D, 5B and 6A (Table 4). For example, QTLs located between *Xgwm544* and *Xgwm371* on chromosome 5B were associated with KNS, TKW and NS under limited irrigation, and with heading time (Li *et al*. 2015), frost resistance, fiowering time, kernel weight per spike, and GY in previous reports (Tóth *et al*. 2003; Quarrie *et al*. 2005; Cui *et al*. 2013; Galaeva *et al*. 2013). This region is proximal to *Xgwm335* on the long arm of chromosome 5B, which carries a QTL for TKW reported by Groos *et al*. (2003) and Wang *et al*. (2009) based on the same marker *Xgwm371*. However, Golabadi *et al*. (2011) detected marker *Xwmc28* on chromosome 5BL for NS, which was different from *QNS. caas-5B* based on the wheat consensus map. *QKNS. caas-5B. 1* for KNS and *QNS. caas-5B* for NS were not reported previously. *Xgwm335*, located 0.4-5cM from *QTKW. caas-5B. 1*, can be a potential target for marker-assisted selection in wheat-breeding programs.

Because of pleiotropic effects of genes, breeders need tobe aware of the positive and negative effects of different QTLs in relation to the particular trait being selected. Both *QPH. caas-4D* and *QPH. caas-6A* were associated with KNS, TKW and NS; however, the pleiotropic effect of *QPH. caas-4D* was different from that of *QPH. caas-6A*. *QPH. caas-4D* was associated with GY and HD whereas *QPH. caas-6A* was not (Li *et al*. 2015). *QPH. caas-6A* had a larger effect on TKW than on PH, but no effect on GY. QTLs for photosystem (PS) II efficiency, chlorophyll content, leaf temperature, NS, TKW and stripe-rust resistance were previously reported near *QPH. caas-6A* (Huang *et al*. 2004; Kumar *et al*. 2012; Rosewarne *et al*. 2013). Although *TaGW2*, which is associated with grain width, TKW, days to heading, and maturity is located in this region (Su *et al*. 2011; Zhang *et al*. 2013c), the other observed effects were not previously identified. Clearly, this is an important region for wheat development, and it should be further characterised.

Alleles *QPH. caas-6A* and *QPH. caas-4D* from Aikang 58 decreased PH and TKW and increased KNS and NS, but *QPH. caas-6A* had no effect on GY, and *QPH. caas-4D* increased GY in 2011-12 and reduced it in 2012-13. *QHD. caas-4D* co-segregated with *QPH. caas-4D*, and was detected only in 2011-12 (Li *et al.* 2015). Therefore, differences between years may be associated with HD, which was 6-8 days earlier in 2011-12 than in 2012-13, and GY, which was 108.7 g m^{-2} higher in 2011-12 than in 2012-13, based on the control, Zhongmai 175. However, the genotype containing these two dwarfing alleles from Aikang 58 showed the lowest GY (Supplementary table 2). This may be attributed to an epistatic effect of two dwarfing genes (Supplementary fig. 1).

This co-location of QTLs for yield and yield components is commonly detected on chromosomes 1A, 2A, 3B, 4D and 5B (Table 5). The presence of Jingdong 8 alleles at *QGY. caas-3B. 2* was associated with an increase in TKW, suggesting that the increased GY at *QTKW. caas-3B. 2* results from increased grain weight. Although *QGY. caas-3B. 2* had a lower PVE, it was observed in at least two environments, and QTLs for GY were also reported by Cuthbert *et al.* (2008) and Bennett *et al.* (2012) in a similar region based on *Xwmc43*. Bennett *et al.* (2012) reported that this QTL had a large effect on GY and grain weight. *Rht5* was detected in a similar region based on the linked markers *Xbarc102-Xgwm264-Xwmc43* (Somers *et al.* 2004; Ellis *et al.* 2005). Another stable QTL for GY, *QGY. caas-7A*, is consistent with QTLs on chromosome 7AS previously reported by Marza *et al.* (2006) and Bennett *et al.* (2012), based on the common markers *Xbarc108* and *Xbarc281*. This QTL was associated with fiag leaf glaucousness, TKW, ear emergence time and kernel number per m^2 (Bennett *et al.* 2012); however, it was not co-located with other yield-component QTLs in the current study.

QTKW. caas-6D accounted for >10% of PVE in four environments and was independent of KNS and NS in the current study. A QTL for GY was identified by Huang *et al.* (2004) in a similar region near the centromere. Wang *et al.* (2009) identified QTLs for TKW, grain-filling rate and grain thickness in a similar region based on the common marker *Xbarc54*. Mir *et al.* (2012) also found *Xbarc54* to be associated with grain weight. Therefore, *QTKW. caas-6D* could be valuable in improving GY.

QTLs expressed in limited irrigation and full irrigation

Generally, yield components had lower heritability and phenotypic values under limited irrigation than under full irrigation (Table 3). GY and PH showed significant interaction with environment (Table 2), and the QTLs for these traits also had significant interaction with environment, although the PVE (A/E) value was small (Supplementary table 1). In addition, QTL identification was affected by the irrigation treatment. *QTKW. caas-1A. 2*, *QTKW. caas-1A. 3*, *QTKW. caas-3B* and *QTKW. caas-5B. 2* were detected only under full irrigation. Some stable QTLs (*QKNS. caas-4D*, *QTKW. caas-6A*, *QTKW. caas-5B. 1* and *QKNS. caas-7D*) had a larger effect under full irrigation than under limited irrigation. By contrast, some QTLs showed larger effects under limited irrigation than full irrigation, such as *QTKW. caas-6D* and *QNS. caas-4D* at in Gaoyi 2011-12. If confirmed in further testing, these QTLs could be useful in improving wheat yield potential in China and perhaps elsewhere. Some QTLs were found only under limited irrigation, for example, *QTKW. caas-1A. 1*, *QNS. caas-1A*, *QKNS. caas-1B*, *QKNS. caas-2A*, *QNS. caas-2A. 2*, *QGY. caas-3B. 1*, *QGY. caas-3B. 2*, *QGY. caas-4A*, *QTKW. caas-4B*, *QKNS. caas-5B* and *QGY. caas-5B. 2*. Of particular interest is *QTKW. caas-4B*, proximally located to *Xgwm251* at a genetic distance of 3.7-7.0 cM, because this QTL was consistently detected under limited irrigation and also exhibited a minor effect on KNS. In previous research, this region was reported to infiuence shoot biomass and root length (Kadam *et al.* 2012), TKW (Wang *et al.* 2009), tiller number and KNS (Deng *et al.* 2011; Naruoka *et al.* 2011) and yield (Quarrie *et al.* 2005; Golabadi *et al.* 2011). Another QTL, *QGY. caas-5B. 2*, near vernalisation gene *vrn-B1*, accounted for >10% of PVE for KNS and GY under limited irrigation. These yield-component QTLs detected in at least two environments may be useful for improving yield under limited irrigation.

Acknowledgements

The authors are grateful to Professor Robert McIntosh, University of Sydney, for reviewing this manuscript. This work was supported by the CGIAR Generation Challenge Program (GCP, G7010.02.01), National Natural Science Foundation of China (31161140346), and National Key Technology R&D Program of China (2011BAD35B03).

References

Bennett D, Reynolds M, Mullan D, Izanloo A, Kuchel H, Langridge P, Schnurbusch T (2012) Detection of two major grain yield QTL in bread wheat (*Triticum aestivum* L.) under heat, drought and high yield potential environments. *Theoretical and Applied Genetics* 125, 1473-1485. doi: 10.1007/s00122-012-1927-2

Börner A, Schumann E, Furste A, Coster H, Leithold B, Röder MS, Weber WE (2002) Mapping of quantitative trait loci determining agronomic important characters in hexapioid wheat (*Triticum aestivum* L). *Theoretical and Applied Genetics* 105, 921-936. doi: 10.1007/s00122-002-0994-1

Borojevic K, Borojevic K (2005) The transfer and history of "reduced height genes" (*Rht*) in wheat from Japan to Europe. *The Journal of Heredity* 96, 455-459. doi: 10.1093/jhered/esi060

Chang J, Zhang J, Mao X, Li A, Jia J, Jing R (2013) Polymorphism of *TaSAP1-A1* and its association with agronomic traits in wheat. *Planta* 237, 1495-1508. doi: 10.1007/s00425-013-1860-x

Chapman SC, Mathews KL, Trethowan RM, Singh RP (2007) Relationships between height and yield in near-isogenic spring wheats that contrast for major reduced height genes. *Euphytica* 157, 391-397. doi: 10.1007/s10681-006-9304-3

Cui F, Li J, Ding A, Zhao C, Wang L, Wang X, Li S, Bao Y, Li X, Feng D, Kong L, Wang H (2011) Conditional QTL mapping for plant height with respect to the length of the spike and internode in two mapping populations of wheat. *Theoretical and Applied Genetics* 122, 1517-1536. doi: 10.1007/s00122-011-1551-6

Cui F, Zhao CH, Li J, Ding AM, Li XF, Bao YG, Li JM, Ji J, Wang HG (2013) Kernel weight per spike: What contributes to it at the individual QTL level? *Molecular Breeding* 31, 265-278. doi: 10.1007/s11032-012-9786-8

Cuthbert JL, Somers DJ, Brûlé-Babel AL, Brown PD, Crow GH (2008) Molecular mapping of quantitative trait loci for yield and yield components in spring wheat (*Triticum aestivum* L.). *Theoretical and Applied Genetics* 117, 595-608. doi: 10.1007/s00122-008-0804-5

Deng SM, Wu XR, Wu YY, Zhou RH, Wang HG, Jia JZ, Liu SB (2011) Characterization and precise mapping of a QTL increasing spike number with pleiotropic effects in wheat. *Theoretical and Applied Genetics* 122, 281-289. doi: 10.1007/s00122-010-1443-1

Divashuk MG, Vasilyev AV, Bespalova LA, Karlov GI (2012) Identity of the Rht-11 and Rht-B1e reduced plant height genes. *Russian Journal of Genetics* 48, 761-763. doi: 10.1134/S1022795412050055

Ellis MH, Spielmeyer W, Gale KR, Rebetzke GJ, Richards RA (2002) "Perfect" markers for the *Rht-B1b* and *Rht-D1b* dwarfing genes in wheat. *Theoretical and Applied Genetics* 105, 1038-1042. doi: 10.1007/ s00122-002-1048-4

Ellis MH, Rebetzke GJ, Azanza F, Richards RA, Spielmeyer W (2005) Molecular mapping of gibberellin-responsive dwarfing genes in bread wheat. *Theoretical and Applied Genetics* 111, 423-430. doi: 10.1007/ s00122-005-2008-6

Flintham JE, Börner A, Worland AJ, Gale MD (1997) Optimizing wheat grain yield: effects of Rht (gibberellin-insensitive) dwarfing genes. *The Journal of Agricultural Science* 128, 11-25. doi: 10.1017/S0021859696003942

Galaeva MV, Fayt VI, Chebotar SV, Galaev AV, Sivolap YM (2013) Association of microsatellite loci alleles of the group-5 of chromosomes and the frost resistance of winter wheat. *Cytologyand Genetics* 47, 261-267. doi: 10.3103/S0095452713050046

Golabadi M, Arzani A, Mirmohammadi Maibody SAM, Tabatabaei BES, Mohammadi SA (2011) Identification of microsatellite markers linked with yield components under drought stress at terminal growth stages in durum wheat. *Euphytica* 177, 207-221. doi: 10.1007/s10681-010-0242-8

Griffiths S, Simmonds J, Leverington M, Wang Y, Fish L, Sayers L, Alibert L, Orford S, Wingen L, Snape J

(2012) Meta-QTL analysis of the genetic control of crop height in elite European winter wheat germplasm. *Molecular Breeding* 29, 159-171. doi: 10.1007/s11032-010-9534-x

Groos C, Robert N, Bervas E, Charmet G (2003) Genetic analysis of grain protein-content, grain yield and thousand-kernel weight in bread wheat. *Theoretical and Applied Genetics* 106, 1032-1040.

Haque MA, Martinek P, Watanabe N, Kuboyama T (2011) Genetic mapping of gibberellic acid-insensitive genes for semi-dwarfism in durum wheat. *Cereal Research Communications* 39, 171-178. doi: 10.1556/CRC. 39. 2011. 2. 1

Huang X, Kempf H, Ganal M, Röder M (2004) Advanced backcross QTL analysis in progenies derived from a cross between a German elite winter wheat variety and a synthetic wheat (*Triticum aestivum* L.). *Theoretical and Applied Genetics* 109, 933-943. doi: 10.1007/s00122-004-1708-7

Huang X, Cloutier S, Lycar L, Radovanovic N, Humphreys D, Noll J, Somers D, Brown P (2006) Molecular detection of QTLs for agronomic and quality traits in a doubled haploid population derived from two Canadian wheats (*Triticum aestivum* L.). *Theoretical and Applied Genetics* 113, 753-766. doi: 10.1007/s00122-006-0346-7

Kadam S, Singh K, Shukla S, Goel S, Vikram P, Pawar V, Gaikwad K, Khanna-Chopra R, Singh N (2012) Genomic associations for droughttolerance on the short arm of wheat chromosome 4B. *Functional & Integrative Genomics* 12, 447-464. doi: 10.1007/s10142-012-0276-1

Kumar N, Kulwal PL, Balyan HS, Gupta PK (2007) QTL mapping for yield and yield contributing traits in two mapping populations of bread wheat. *Molecular Breeding* 19, 163-177. doi: 10.1007/s11032-006-9056-8

Kumar S, Sehgal SK, Gill BS, Kumar U, Prasad PVV, Joshi AK, Gill BS (2012) Genomic characterization of drought tolerance-related traits in spring wheat. *Euphytica* 186, 265-276. doi: 10.1007/s10681-012-0675-3

Li HH, Ye GY, Wang JK (2007) A modified algorithm for the improvement of composite interval mapping. *Genetics* 175, 361-374. doi: 10.1534/genetics. 106. 066811

Li HH, Ribaut JM, Li ZL, Wang JK (2008) Inclusive composite interval mapping (ICIM) for digenic epistasis of quantitative traits in biparental populations. *Theoretical and Applied Genetics* 116, 243-260. doi: 10.1007/s00122-007-0663-5

Li A, Yang W, Guo X, Liu D, Sun J, Zhang A (2012*a*) Isolation of a gibberellin-insensitive dwarfing gene, *Rht-B1e*, and development of an allele-specific PCR marker. *Molecular Breeding* 30, 1443-1451. doi: 10.1007/s11032-012-9730-y

Li Y, Xiao J, Wu J, Duan J, Liu Y, Ye X, Zhang X, Guo X, Gu Y, Zhang L, Jia J, Kong X (2012*b*) A tandem segmental duplication (TSD) in green revolution gene *Rht-D1b* region underlies plant height variation. *New Phytologist* 196, 282-291. doi: 10.1111/j.1469-8137. 2012. 04243. x

Li XM, He ZH, Xiao YG, Xia XC, Trethowan R, Wang HJ, Chen XM (2015) QTL mapping for leaf senescence related traits in common wheat under limited and full irrigation environments. *Euphytica* 203, 569-582. doi: 10.1007/s10681-014-1272-4

Liu DC, Gao MQ, Guan RX, Li RZ, Cao SH, Guo XL, Zhang AM (2002) Mapping quantitative trait loci for plant height in wheat (*Triticum aestivum* L.) using a $F_{2:3}$ population. *Acta Genetica Sinica* 29, 706-711.

Liu ZH, Anderson JA, Hu J, Friesen TL, Rasmussen JB, Faris JD (2005) A wheat intervarietal genetic linkage map based on microsatellite and target region amplified polymorphism markers and its utility for detecting quantitative trait loci. *Theoretical and Applied Genetics* 111, 782-794. doi: 10.1007/s00122-005-2064-y

Ma DY, Yan J, He ZH, Wu L, Xia XC (2012) Characterization of a cell wall invertase gene TaCwi-A1 on common wheat chromosome 2A and development of functional markers. *Molecular Breeding* 29, 43-52. doi: 10.1007/s11032-010-9524-z

Marza F, Bai GH, Carver BF, Zhou WC (2006) Quantitative trait loci for yield and related traits in the wheat population Ning7840 Clark. *Theoretical and Applied Genetics* 112, 688-698. doi: 10.1007/s00122-005-0172-3

Mathews KL, Chapman SC, Trethowan R, Singh RP, Crossa J, Pfeiffer W, van Ginkel M, DeLacy I (2006) Global adaptation of spring bread and durum wheat lines near-isogenic for major reduced height genes. *Crop Science* 46, 603-613. doi: 10.2135/cropsci2005. 05-0056

McIntosh RA, Dubcovsky J, Rogers WJ, Morris CF, Appels R, Xia XC (2012) Catalogue of gene symbols for wheat: 2012 supplement. National BioResource Project, Komugi Wheat Genetics Resources Database. Available at: www. shigen. nig. ac. jp/wheat/komugi/genes/macgene/supplement2012. pdf.

McIntyre CL, Mathews KL, Rattey A, Chapman SC, Drenth J, Ghaderi M, Reynolds M, Shorter R (2010) Molecular detection of genomic regions associated with grain yield and yield-related components in an elite bread wheat cross evaluated under irrigated and rainfed conditions. *Theoretical and Applied Genetics* 120, 527-541. doi: 10.1007/s00122-009-1173-4

Mir RR, Kumar N, Jaiswal V, Girdharwal N, Prasad M, Balyan HS, Gupta PK (2012) Genetic dissection of grain weight in bread wheat through quantitative trait locus interval and association mapping. *Molecular Breeding* 29, 963-972. doi: 10.1007/s11032-011-9693-4

Naruoka Y, Talbert LE, Lanning SP, Blake NK, Martin JM, Sherman JD (2011) Identification of quantitative trait loci for productive tiller number and its relationship to agronomic traits in spring wheat. *Theoretical and Applied Genetics* 123, 1043-1053. doi: 10.1007/s00122-011-1646-0

Peng ZS, Li X, Yang ZJ, Liao ML (2011) A new reduced height gene found in the tetraploid semi-dwarf wheat landraceAiganfanmai. *Genetics and Molecular Research* 10, 2349-2357. doi: 10.4238/2011. October. 5. 5

Pinto RS, Reynolds MP, Mathews KL, McIntyre CL, Olivares-Villegas JJ, Chapman SC (2010) Heat and drought adaptive QTL in a wheat population designed to minimize confounding agronomic effects. *Theoretical and Applied Genetics* 121, 1001-1021. doi: 10.1007/s00 122-010-1351-4

Quarrie SA, Steed A, Calestani C, Semikhodskii A, Lebreton C, Chinoy C, Steele N, Pljevljakusi'c D, Waterman E, Weyen J, Schondelmaier J, Habash DZ, Farmer P, Saker L, Clarkson DT, Abugalieva A, Yessimbekova M, Turuspekov Y, Abugalieva S, Tuberosa R, Sanguineti MC, Hollington PA, Aragués R, Royo A, Dodig D (2005) A high-density genetic map of hexaploid wheat (*Triticum aestivum* L.) from the cross Chinese Spring SQ1 and its use to compare QTLs for grain yield across a range of environments. *Theoretical and Applied Genetics* 110, 865-880. doi: 10.1007/s00122-004-1902-7

Rebetzke GJ, Ellis MH, Bonnett DG, Richards RA (2007) Molecular mapping of genes for coleoptile growth in bread wheat (*Triticum aestivum* L.). *Theoretical and Applied Genetics* 114, 1173-1183. doi: 10.1007/s00122-007-0509-1

Rebetzke GJ, Ellis MH, Bonnett DG, Mickelson B, Condon AG, Richards RA (2012) Height reduction and agronomic performance for selected gibberellin-responsive dwarfing genes in bread wheat (*Triticum aestivum* L.). *Field Crops Research* 126, 87-96. doi: 10. 1016/j. fcr. 2011. 09. 022

Richards RA (1992) The effect of dwarfing genes in spring wheat in dry environments. I. Agronomic characteristics. *Australian Journal of Agricultural Research* 43, 517-527. doi: 10.1071/AR9920517

Rosewarne GM, Herrera-Foessel SA, Singh RP, Huerta-Espino J, Lan CX, He ZH (2013) Quantitative trait loci of stripe rust resistance in wheat. *Theoretical and Applied Genetics* 126, 2427-2449. doi: 10.1007/s00122-013-2159-9

Saghai-Maroof MA, Soliman KM, Jorgensen RA, Allard RW (1984) Ribosomal DNA spacer length polymorphisms in barley: Mendelian inheritance, chromosomal location, and population dynamics. *Proceedings of the National Academy of Sciences of the United States of America* 81, 8014-8018. doi: 10.1073/pnas. 81. 24. 8014

SAS Institute (2000)'SAS user's guide: Statistics.' (SAS Institute, Inc.: Cary, NC, USA)

Somers DJ, Isaac P, Edwards K (2004) A high-density microsatellite consensus map for bread wheat (*Triticum aestivum* L.). *Theoreticaland Applied Genetics* 109, 1105-1114. doi: 10.1007/s00122-004-1740-7 Spielmeyer W, Hyles J, Joaquim P, Azanza F, Bonnett D, Ellis ME, Moore C, Richards RA (2007) A QTL on chromosome 6A in bread wheat (*Triticum aestivum* L.) is associated with longer coleoptiles, greater seedling vigour and final plant height. *Theoretical and Applied Genetics* 115, 59-66. doi: 10.1007/s00122-007-0540-2

Su Z, Hao C, Wang L, Dong Y, Zhang X (2011) Identification and development of a functional marker of TaGW2 associated with grain weight in bread wheat (*Triticum aestivum* L.). *Theoretical and Applied Genetics* 122, 211-223. doi: 10.1007/s00122-010-1437-z

Tóth B, Galiba G, Feher E, Sutka Y, Snape W (2003) Mapping genes affecting fiowering time and frost resistance on chromosome 5B of wheat. *Theoretical and Applied Genetics* 107, 509-514. doi: 10.1007/s00122-003-1275-3

Wang RX, Hai L, Zhang XY, You GX, Yan CS, Xiao SH (2009) QTL mapping for grain filling rate and yield-related traits in RILs of the Chinese winter wheat populationHeshangmai Yu8679. *Theoretical and Applied Genetics* 118, 313-325. doi: 10.1007/s00122-008-0901-5

Wang G, Hu T, Li X, Dong N, Feng S, Li G, Zhang L,

Ru Z (2012) Detection of dwarfing genes in wheat variety Aikang 58 and its parents. *Journal of Henan Agricultural Science* 41, 21-25.

Worland AJ, Korzun V, Röder MS, Ganal MW, Law CN (1998) Genetic analysis of the dwarfing gene *Rht8* in wheat. Part II. The distribution and adaptive significance of allelic variants at the *Rht8* locus of wheat as revealed by microsatellite screening. *Theoretical and Applied Genetics* 96, 1110-1120. doi: 10.1007/s001220050846

Worland AJ, Sayers EJ, Korzun V (2001) Allelic variation at the dwarfing gene Rht8 locus and its significance in international breeding programmes. *Euphytica* 119, 157-161. doi: 10.1023/A: 1017582122775

Xiao YG, Qian ZG, Wu K, Liu JJ, Xia XC, Ji WQ, He ZH (2012) Genetic gains in grain yield and physiological traits of winter wheat in Shandong Province, China, from 1969 to 2006. *Crop Science* 52, 44-56. doi: 10.2135/ cropsci2011.05.0246

Zhang XK, Yang SJ, Zhou Y, He ZH, Xia XC (2006) Distribution of the *Rht-B1b*, *Rht-D1b* and *Rht8* reduced height genes in autumn sown Chinese wheat detected by molecular markers. *Euphytica* 152, 109-116. doi: 10.1007/s10681-006-9184-6

Zhang XK, Xia XC, Xiao YG, Zhang Y, He ZH (2008) Allelic variation at the vernalization genes *vrn-A1*, *vrn-B1*, *vrn-D1* and *vrn-B3* in Chinese common wheat cultivars and their association with growth habit. *Crop Science* 48, 458-470. doi: 10.2135/cropsci2007.06.0355

Zhang LY, Liu DC, Guo XL, Yang WL, Sun JZ, Wang DW, Zhang AM (2010) Genomic distribution of quantitative trait loci for yield and yield-related traits in common wheat. *Journal of Integrative Plant Biology* 52, 996-1007. doi: 10.1111/j.1744-7909.2010.00967.x

Zhang J, Liu W, Yang X, Gao A, Li X, Wu X, Li L (2011) Isolation and characterization of two putative cytokinin oxidase genes related to grain number per spike phenotype in wheat. *Molecular Biology Reports* 38, 2337-2347. doi: 10.1007/s11033-010-0367-9

Zhang L, Zhao YL, Gao LF, Zhao GY, Zhou RH, Zhang BS, Jia JZ (2012) *TaCKX6-D1*, the ortholog of rice *OsCKX2*, is associated with grain weight in hexaploid wheat. *New Phytologist* 195, 574-584. doi: 10.1111/j.1469-8137.2012.04194.x

Zhang J, Dell B, Biddulph B, Drake-Brockman F, Walker E, Khan N, Wong D, Hayden M, Appels R (2013a) Wild-type alleles of *Rht-B1* and *Rht-D1* as independent determinants of thousand-grain weight and kernel number per spike in wheat. *Molecular Breeding* 32, 771-783. doi: 10.1007/ s11032-013-9905-1

Zhang W, Zhang L, Qiao L, Wu J, Zhao G, Jing R, Lv W, Jia J (2013b) Cloning and haplotype analysis of *TaSTE*, which is associated with plant height in bread wheat (*Triticum aestivum* L.). *Molecular Breeding* 31, 47-56. doi: 10.1007/s11032-012-9767-y

Zhang X, Chen J, Shi C, Chen J, Zheng F, Tian J (2013c) Function of TaGW2-6A and its effect on grain weight in wheat (*Triticum aestivum* L.). *Euphytica* 192, 347-357. doi: 10.1007/s10681-012-0858-y

Zheng TC, Zhang XK, Yin GH, Wang LN, Han YL, Chen L, Huang F, Tang JW, Xia XC, He ZH (2011) Genetic gains in grain yield, net photosynthesis and stomatal conductance achieved in Henan Province of China between 1981 and 2008. *Field Crops Research* 122, 225-233. doi: 10.1016/j.fcr.2011.03.015

Preliminary Exploration of the Source, Spread, and Distribution of *Rht24* Reducing Height in Bread Wheat

Xiuling Tian, Zhanwang Zhu, Li Xie, Dengan Xu, Jihu Li, Chao Fu, Xinmin Chen, Desen Wang, Xianchun Xia, Zhonghu He*, and Shuanghe Cao*

X. Tian, Z. Zhu, L. Xie, D. Xu, J. Li, C. Fu, X. Chen, D. Wang, X. Xia, Z. He, and S. Cao, Institute of Crop Sciences, National Wheat Improvement Center, Chinese Academy of Agricultural Sciences (CAAS), Beijing 100081, *China*;

Z. Zhu, Institute of Food Crops, Hubei Academy of Agricultural Sciences, Wuhan 430064, *China*;

Z. He, International Maize and Wheat Improvement Center (CIMMYT) China Office, c/o CAAS, Beijing 100081, *China*. *Received* 1 *Dec*. 2017. *Accepted* 7 *Oct*. 2018.

* *Corresponding authors (shcao8@163.com; zhhecaas@163.com).*

Abstract: *Rht24* is a major dwarfing allele that not only reduces plant height but also increases grain weight. The objective of this study was to trace the source, spread, and distribution of *Rht24* and determine its frequency in conjunction with other important dwarfing alleles for *Rht1* (*Rht-B1b*), *Rht2* (*Rht-D1b*), and *Rht8*. Allele-specific cleaved amplified polymorphic sequence (CAPS) markers were developed for accurate and effective genotyping of *Rht1* and *Rht2*, and closely linked flanking markers were used for genotyping of *Rht8* and *Rht24*. Marker analysis showed that *Rht24* occurs at a higher frequency (84.2%) than other important dwarfing alleles in elite wheat (*Triticum aestivum* L.) varieties and usually couples with *Rht2* or *Rht8*. Geno-typing of old varieties and landraces showed that *Rht24* was widely used in wheat breeding before the Green Revolution (GR). 'Akakomugi' and 'Norin 10', the donors of *Rht8* and GR genes *Rht1* and *Rht2*, respectively, harbored *Rht24*, so *Rht24* likely followed the transfer routes of GR genes and *Rht8* to spread worldwide. Pedigree analysis showed that many Chinese elite lines and backbone parents were derived from Akakomugi, contributing to the high frequency of *Rht24* in Chinese varieties. Additionally, Norin 10-derived semidwarf varieties from the CIMMYT were widely used in Chinese wheat breeding. Finally, *Rht24* was also detected in 30 old Chinese varieties and landraces. Overall, these findings show that the presence of *Rht24* in modern Chinese elite varieties originated from Akakomugi, Norin 10, or Chinese landraces. Thus, *Rht24* was introgressed early, spread widely, and became genetically fixed in a wide range of modern Chinese wheat germplasm.

Abbreviations: CAPS, cleaved amplified polymorphic sequence; GR, Green Revolution; PCR, polymerase chain reaction; RE, restriction endonuclease; SNP, single nucleotide polymorphism; SSR, simple sequence repeat.

Published in Crop Science, 2019, 59: 19-24.

Wheat (*Triticum aestivum* L.) is one of the most important staple crops. With the increasing world population, decreasing cultivated land, and intensification of global warming, world food security is increasingly challenged (Shaw, 2007). Therefore, the foremost objective is to increase wheat grain yield through plant breeding (Peng et al., 2011; Chen et al., 2015). Plant height is an important agronomic trait related to yield potential (Ellis et al., 2005). Over recent decades, the breeding and popularization of dwarf and semidwarf wheat varieties has greatly improved wheat production, and semidwarf cereals were the most important symbol of the worldwide Green Revolution (GR) (Li et al., 2013). Thus, control of plant height has been a key target in wheat breeding.

Twenty-four dwarfing loci (*Rht1*-*Rht24*) have been cataloged in wheat (McIntosh et al., 2017). Many studies have shown that dwarfing alleles *Rht1* (or *Rht-B1b*) in chromosome 4BS, *Rht2* (or *Rht-D1b*) in 4DS, *Rht8* in 2DL, and *Rht24* in 6AL are the most widely distributed among Chinese wheat varieties (Yang et al., 2006; Zhang et al., 2006; Tang et al., 2012; Tian et al., 2017). The GR genes *Rht1* and *Rht2* were transferred from the Japanese variety 'Norin 10' into US varieties after World War II, then introduced to CIMMYT in Mexico from the United States (Reitz and Salmon, 1968; Lumpkin, 2015). From the CIMMYT, GR genes were distributed worldwide in wheat breeding programs (Gale et al., 1985). *Rht8* is an important dwarfing allele, located 0.6 cM from the simple sequence repeat (SSR) marker *Xgwm261*, and reduces plant height by 3 to 8 cm (Korzun et al., 1998; Worland et al., 2001). Many studies indicate that *Rht8* was used more frequently than the GR genes (Botwright et al., 2005; Zhang et al., 2006; Tian et al., 2017). *Rht8* was transferred from the Japanese landrace 'Akakomugi' to Italian varieties 'Mentana', 'Villa-Glori', and 'Ardito' at the beginning of 20th century, which were widely used in Italy and South America over the following decade (Worland, 1999; Lorenzetti, 2000; Borojevic and Boro-jevic, 2005). After World War II, *Rht8* was transferred to southern and central European varieties (Borojevic and Potocanac, 1966) and subsequently into > 200 wheat varieties worldwide (Borojevic and Borojevic, 2005). *Rht24* was recently identified as an important dwarfing allele, as it not only reduces plant height by an average of 6.0 to 7.9 cm but also increases thousand-grain weight by 2.0 to 3.4 g (Tian et al., 2017). *Rht24* is flanked by cleaved amplified polymorphic sequence (CAPS) markers *TaAP2* and *TaFAR*, with 1.85 cM of genetic distance (Tian et al., 2017). To date, the origin and spread of *Rht24* remains unclear. Therefore, the objective of this study was to investigate the source, route of spread, and distribution of *Rht24* in comparison with *Rht1*, *Rht2*, and *Rht8*.

MATERIALS AND METHODS

Plant Materials

Three hundred and twenty-nine wheat varieties from China and other countries were used for this study. Of these, 221 Chinese and 56 introduced varieties were used to investigate the frequency of the *Rht* dwarfing alleles (Supplemental Table S1). Among the Chinese varieties, 34 were from the North China winter wheat region (Zone I), 123 from the Yellow and Huai Valley wheat region (Zone II), and 64 from the Middle and Lower Yangtze Valley autumn-sown facultative wheat region (Zone III). Three varieties, 'Shi 4185', 'Aikang 58', and 'Jingdong 8', harboring different alleles of the GR genes, were used to develop allele-specific CAPS markers of *Rht1* and *Rht2*. A second group of 52 varieties was used to trace the source and spread of *Rht24* and included 34 Chinese landrace varieties, three varieties from Japan, one from Korea, three from Italy, and 11 from the CIMMYT (Supplemental Fig. S2, Supple-mental Table S2).

DNA Extractionand Molecular Marker Analysis

Genomic DNA was extracted from young leaves using a modified cetyl trimethylammonium bromide (CTAB) method (Saghai-Maroof et al., 1984) and was used to develop allele-specific CAPS markers for

Rht1 and *Rht2* (Supplemental Table S3). The dwarfing allele of *Rht8* was detected by the closely linked SSR marker *Xgwm261* (Korzun et al., 1998). *Rht24* was genotyped by flanking CAPS markers *Ta-AP2* and *TaFAR* (Tian et al., 2017). The 15-mL polymerase chain reaction (PCR) contained 7.5 mL of 2'*Taq* PCR Mix (Tianwei Biotechnology Company), 5 pmol of each primer, and 100 ng of template DNA. Amplifi-cation was performed at 94℃ for 5 min, followed by 35 cycles at 94℃ for 20 s, 55 to 68℃ (depending on specific primers) for 30 s, and 72℃ for 1 min, with a final extension at 72℃ for 5 min. Restriction endonucleases (REs) *Pvu* II and *Bpm* I (New England Biolabs Company, http://www.neb-china.com/) were used to digest the PCR target fragments from *Rht-B1* and *Rht-D1* loci, respectively, for the CAPS markers (Supplemental Fig. S1). The PCR or RE digested products were separated in 2% agarose or 6% polyacrylamide gels.

Experimental Procedure of Developing CAPS Markers for Green Revolution Genes

To detect GR genes more accurately, we developed allele-specific markers using CAPS technology, a tool to identify single nucleotide polymorphisms (SNPs) effectively and accurately (Shahinnia and Tabatabaei, 2009). Cleaved amplified polymorphic sequence technology combines PCR with RE digestion, with PCR used to obtain subgenome-specific target fragments and RE digestion to differentiate the target alleles with poly-morphic SNP. First, locus-specific primers were designed for *Rht-A1*, *Rht-B1*, and *Rht-D1* loci in Chinese Spring based on the sequence alignment (Supplemental Fig. S1). Target PCR fragments for *Rht-B1* and *Rht-D1* were isolated and sequenced from three representative varieties, Jingdong 8, Shi 4185, and Aikang 58. Sequence analyses showed that Shi 4185 and Aikang 58 contained *Rht1* and *Rht2*, respectively, whereas Jingdong 8 harbored neither of the dwarfing alleles (Supplemental Fig. S1). Based on the single SNP differences, *Pvu* II and *Bpm* I (New England Biolabs Company, http://www.neb-china.com/) were used to digest the relevant PCR fragment from *Rht-B1* and *Rht-D1* loci, respectively.

RESULTS

Development of CAPS Markers for Green Revolution Genes

Molecular markers for GR genes have previously been developed (Ellis et al., 2002; Pearce et al., 2011; Rasheed et al., 2016). The dwarfing alleles *Rht1* and *Rht2* from *Rht-B1* and *Rht-D1* loci, respectively, differ at a single SNP compared with their wild-type alleles, and previous PCR-based markers had difficulty differentiating these SNPs, especially in hexaploid wheat with its three homoeologous subgenomes (A, B, and D). For marker *Rht-B1*-CAPS, the PCR target fragment (2309 bp) of Jingdong 8 with the wild-type allele of *Rht-B1* was digested into 1884-and 425-bp subfragments, whereas the counterpart of Shi 4185 with *Rht1* could not be cut (Supplemental Fig. S1a). For marker *Rht-D1*-CAPS, the PCR target fragment (510 bp) of Jingdong 8 with the wild-type allele of *Rht-D1* was digested into 391-and 119-bp subfragments, whereas that of Aikang 58 with *Rht2* was intact after digestion (Supple-mental Fig. S1b). These allele-specific CAPS markers proved to be efficient and reliable in distinguishing the target alleles (Supplemental Tables S1 and S3, Fig. 1).

Frequencies of Dwarfing Genes *Rht1*, *Rht2*, *Rht8*, and *Rht24*

Among the 221 Chinese and 56 introduced wheat varieties, 95 (34.3%), 130 (47.0%), 163 (58.8%), and 233 (84.2%) contained *Rht1*, *Rht2*, *Rht8*, and *Rht24*, respectively (Table 1). Of them, *Rht24* was present at the highest frequency (91.5%) in Chinese wheat varieties, compared with *Rht1* (30.8%), *Rht2* (52.5%), and *Rht8* (62.5%). Likewise, the frequency of *Rht24* (55.3%) in introduced wheat varieties was also higher than that of *Rht1* (48.3%), *Rht2* (24.9%), and *Rht8* (44.6%) (Table 1). The distribution of the semi-dwarfing alleles differed significantly between ecological zones. *Rht1* occurred at a higher frequency in Zone I (47.1%) compared with

Zones II (27.6%) and III (28.1%); *Rht2* exhibited the highest frequency in Zone II (64.3%) compared with Zones I (14.7%) and III (50.0%); *Rht8* was more frequent in Zones II (65.9%) and III (70.3%) than in Zone I (35.3%) (Table 1). Notably, *Rht24* was present at a high frequency in all three regions, Zone I (82.4%), Zone II (91.9%), and Zone III (95.3%) (Table 1).

Based on conjointanalysis, 232 lines contained more than one dwarfing allele, and 45 harbored only a single allele. The *Rht8* + *Rht24* combination had the highest frequency (50.2%), followed by *Rht24* + *Rht2* (42.2%), *Rht24* + *Rht1* (31.0%), *Rht8* + *Rht2* (27.8%), *Rht8* + *Rht1* (20.2%), and *Rht1* + *Rht2* (2.2%) (Table 1). *Rht1* was rarely found in combination with *Rht2* (Table 1). Among combinations of three dwarfing genes, *Rht2* + *Rht8* and *Rht24* occurred at the highest frequency (25.3%) (Table 1).

The frequency of GR genes (including either *Rht1*, *Rht2*, or both) was 79.1% (Supplemental Table S5). *Rht24* was present in 84.2% of varieties in this study (Supple-mental Table S4). Thus, the frequencies of GR genes and *Rht24* were comparable. The frequency of GR and *Rht24* in combination was 67.6%, whereas the frequency of *Rht24* with *Rht8* was 50.2% (Table 1). Thus, *Rht24* was nearly always present in combination with one of the GR genes or *Rht8*.

Sources of *Rht24*

To investigate the source of *Rht24*, 52 representative wheat varieties produced at different times in various countries were selected and genotyped with the markers closely flanking *Rht24*. Among them, 40 contained *Rht24*, including 20 Chinese landraces, five CIMMYT wheat varieties ('Nadadores', 'Potam S70', 'Nuri F70', 'Mexipak 65', and 'Tanori F71'), and 15 old elite varieties or backbone parents ('Daruma', 'Norin 10', 'Akako-mugi', 'Funo', 'Abbondanza', 'Villa-Glori', 'Nanda 2419', 'Xiannong 39', 'Xinong 6028', 'Jingyang 60', 'Zhengyin 1', 'Fengchan 3', 'Aifeng 3', 'Bima 1', and 'Bima 4') (Supple-mental Table S2). In addition to the Chinese landraces, varieties Jingyang 60, Villa Glori, Xinong 6028, Bima 1, Bima 4, Nanda 2419, Norin10, Akakomugi, 'Huixian-hong', and 'Xiaobaimai' were developed before the 1960s (Supplemental Table S2). These results clearly indicated that there were historically independent sources of *Rht24*.

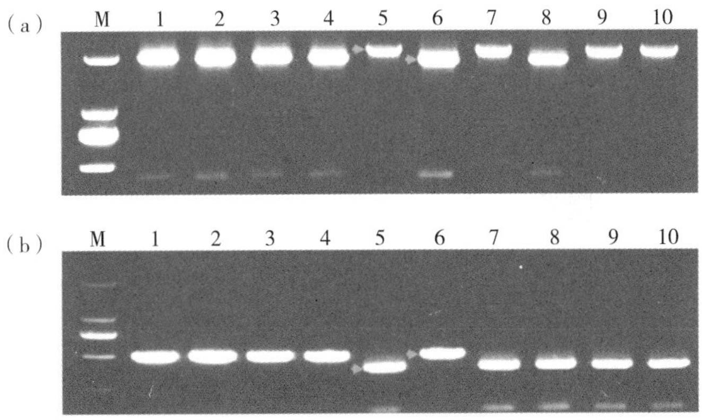

Fig. 1 Amplification patterns of cleaved amplified polymorphic sequence (CAPS) markers for (a) *Rht1* and (b) *Rht2* in representative Chinese wheat varieties. Lane M shows the marker (20-bp DNA ladder, Takara Bio Company). Lanes 1 to 10 show Beijing 0045, Jimai 20, Zhoumai 12, and Aikang 58 (genotype *Rht-B1a*, *Rht-D1b*); Qinnong 142 (*Rht-B1b*, *Rht-D1a*); Liangxing 99 (*Rht-B1a*, *Rht-D1b*); Shan 354 (*Rht-B1b*, *Rht-D1a*), Nongda 139 (*Rht-B1a*, *Rht-D1a*); and Shi 4185 and Gaocheng 8901 (genotype *Rht-B1b*, *Rht-D1a*). Red arrows show the target bands.

DISCUSSION

Rht24 Originated Early and Spread Worldwide through Multiple Routes

In this study, 52 representative old varieties or landraces were chosen to trace the source of *Rht24*. For example, Daruma and Akakomugi, donors of GR genes and *Rht8*, respectively, were developed at the beginning of 20th century. Both Daruma and Akakomugi also harbored *Rht24*, indicating that *Rht24*, like the GR genes and *Rht8*, had an early origin in modern varieties. Eight other varieties (Norin 10, Villa-Glori, Zhengyin 1, Bima 1, Bima 4, Nanda 2419, Xinong 6028, and Jingyang 60), developed prior to the 1960s, contained *Rht24* (Supplemental Table S2). *Rht24* was also detected in 20 of 25 Chinese landraces (Supplemental Table S2). These findings show that *Rht24* was widely distributed in wheat before the GR.

Norin 10 was imported to North America by Salmon in 1948 and used to produce semidwarf winter wheat varieties at Washington State University before being introduced into the CIMMYT from the United States. These varieties were used by Borlaug pre-CIMMYT in Mexico to transfer semidwarf stature to spring wheat varieties that were then distributed worldwide (Borlaug, 1968). Another Japanese dwarf variety, Akakomugi, was imported into Italy and South America and later introduced to southern and central Europe (Borojevic and Borojevic, 2005). Both Norin 10 and Akakomugi, the donors of GR genes and *Rht8*, respectively, harbored *Rht24*, indicating that the latter could have spread with both the GR genes and *Rht8*. As such, the GR and the resulting semidwarf varieties accelerated the spread of *Rht24* worldwide.

Table 1 Frequencies of *Rht1*, *Rht2*, *Rht8*, and *Rht24* as well as their combinations in Chinese and introduced wheat varieties

Dwarfing allele‡	Percentage (no.) of Chinese varieties†				Percentage (no.) of introduced varieties	Total
	Zone I	Zone II	Zone III	Total		
	% (no.)					
None	2.9 (1)	0.8 (1)	0.0 (0)	0.9 (2)	3.6 (2)	1.4 (4)
Rht1	47.1 (16)	27.6 (34)	28.1 (18)	30.8 (68)	48.3 (27)	34.3 (95)
Rht2	14.7 (5)	64.3 (79)	50.0 (32)	52.5 (116)	24.9 (14)	47.0 (140)
Rht8	35.3 (12)	65.9 (81)	70.3 (45)	62.5 (138)	44.6 (25)	58.8 (163)
Rht24	82.4 (28)	91.9 (113)	95.3 (61)	91.5 (202)	55.3 (31)	84.2 (233)
GR	55.8 (19)	91.1 (112)	75.0 (48)	81.0 (179)	71.4 (40)	79.1 (219)
Rht24+*Rht1*	44.1 (15)	25.2 (31)	28.1 (18)	29.0 (64)	21.4 (12)	31.0 (86)
Rht24+*Rht2*	14.7 (5)	59.3 (73)	48.4 (31)	49.3 (109)	14.3 (8)	42.2 (117)
Rht24+ *Rht8*	23.5 (8)	59.4 (73)	65.6 (40)	55.7 (121)	28.5 (16)	50.2 (137)
Rht1+ *Rht8*	8.8 (3)	21.9 (27)	17.2 (11)	18.6 (41)	26.8 (15)	20.2 (56)
Rht2+ *Rht8*	2.9 (1)	39.0 (48)	37.5 (24)	33.0 (73)	7.1 (4)	27.8 (77)
GR+*Rht24*	52.9 (18)	83.8 (103)	73.4 (47)	75.6 (168)	33.9 (19)	67.6 (187)
GR+*Rht8*	11.7 (4)	61.0 (75)	51.6 (33)	50.7 (112)	32.1 (18)	46.9 (130)
Rht1+*Rht2*	5.9 (2)	0.8 (1)	3.1 (2)	2.3 (5)	1.8 (1)	2.2 (6)
Rht24+*Rht1*+*Rht2*	5.9 (2)	0.8 (1)	3.1 (2)	2.3 (5)	1.8 (1)	2.2 (6)
Rht24+*Rht1* + *Rht8*	8.8 (3)	19.5 (24)	17.2 (11)	17.2 (38)	16.1 (9)	17.0 (47)
Rht24+*Rht2* + *Rht8*	2.9 (1)	35.0 (43)	35.9 (23)	30.3 (67)	5.3 (3)	25.3 (70)

(continued)

Dwarfing allele‡	Percentage (no.) of Chinese varieties†				Percentage (no.) of introduced varieties	Total
	Zone I	Zone II	Zone III	Total		
Rht1+*Rht2* + *Rht8*	0.0 (0)	0.0 (0)	3.1 (2)	0.9 (2)	1.8 (1)	1.1 (3)
Rht24+*Rht1*+*Rht2* + *Rht8*	0.0 (0)	0.0 (0)	3.1 (2)	0.9 (2)	1.8 (1)	1.1 (3)
GR +*Rht8* + *Rht24*	11.7 (4)	54.5 (67)	50.0 (32)	46.6 (103)	19.6 (11)	41.2 (114)

† Zone I, Northern China winter wheat region; Zone II, Yellow andHuai River Valley wheat region; Zone III, Middle and Low Yangtze Valleys region.

‡ Each item showed the total frequency of each *Rht* allele or combination (i. e. the percentage of the varieties containing each target *Rht* allele or combination in Chinese varieties from each region or introduced varieties).

Three Putative Sources Contributed to the High Frequency of *Rht24* in China

Genotypicresults based on 221 Chinese and 56 foreign varieties showed that Chinese wheat varieties had a much higher frequency (91.5%) of *Rht24* than introduced varieties (55.3%) (Table 1). Further investigations to understand the basis of the high frequency of *Rht24* in Chinese varieties showed that Xiannong 39, an important backbone parent in China, originated from Daruma, but its intermediate parent, 'Suwon 86', did not have *Rht24* (Fig. 2). However, the Xinong 6028 parent of Xiannong 39 carried *Rht24* (Fig. 2). Italian varieties Funo, Abbondanza, and 'St1472/506' (later designated as Zhengyin 1), derived from Akakomugi, were also widely used as backbone parents in China and carried *Rht24*. Of the 11 CIMMYT varieties analyzed in this study, five contained *Rht24*, including Nadadores, Potam S70, Nuri F70, Mexipak 65, and Tanori F71. These CIMMYT varieties were either commercially grown in China or used as parents to develop new varieties (Liu and Zheng, 1997). Finally, *Rht24* was also detected in~80% of Chinese land-races, including well-known backbone parents, Xiannong 39, Xinong 6028, Bima 1, Bima 4, Fengchan 3, Aifeng 3, and Wangshuibai (Supplemental Table S2). Thus, *Rht24* in modern Chinese varieties was likely introduced along three different routes: first, a Japan-America-CIMMYT route from Norin 10; second, from a Japan-Italy route via Akakomugi and derived varieties; and third, from Chinese landraces used as parents of modern varieties.

***Rht24* Is the Most Frequent of the Four Dwarfing Genes Used in Wheat Breeding**

The GR genes and *Rht8* have been used worldwide since the GR (the late 1960s) and are subject to positive selec-tion in wheat breeding. Previous studies have shown the frequency of *Rht8* to be higher than both *Rht1* and *Rht2* in Chinese wheat varieties (Zhou et al., 2003; Zhang et al., 2006; Tang et al., 2012). A conjoint analysis of *Rht1*, *Rht2*, *Rht8*, and *Rht24* using 277 wheat Chinese and foreign varieties was consistent with this finding (Supplemental Table S1). Although the frequencies of *Rht24* were different in various Chinese ecological zones, as well as in other countries, *Rht24* was present at a higher frequency. Würschum et al. (2017) also showed that *Rht24* had a high distribution frequency (67%) compared with GR genes and *Rht8* in>1000 wheat varieties originating mainly from Europe.

This study also showed that the combination of *Rht24* and GR genes had the highest frequency (67.6%) compared with other combinations of dwarfing alleles, including *Rht8* + *Rht24* (50.2%) and *Rht8* + GR (46.9%) (Supplemental Table S5). *Rht24* and *Rht8* are present in combination with *Rht1* or *Rht2* at frequencies of 15.9 and 24.2%, respectively. *Rht1*, *Rht2*, *Rht8*, and *Rht24* are present alone at frequencies of 3.6, 2.2, 2.9, and 7.6%, respectively. Dwarfing alleles *Rht8* and *Rht24* are GA3 sensitive and have little effect on coleoptile length, which is important for seedling emergence in dryland production systems (Rebetzke and Richards, 2000; Würschum et al., 2017). As arid regions widen in key wheat production areas due to a changing global climate, such as in the midlatitude region, GA3-sensi tive dwarfing alleles *Rht8* and *Rht24* will become preferred in wheat breeding programs.

In conclusion, *Rht24* originated early and was used widely in wheat breeding even before the GR. Akakomugi and Norin 10, the donors of *Rht8* and GR genes, respectively (Yamada, 1990), also harbor *Rht24*, which accelerated its global spread in conjunction with the GR. Taken together, these findings account for the high frequency of *Rht24* in contemporary wheat varieties.

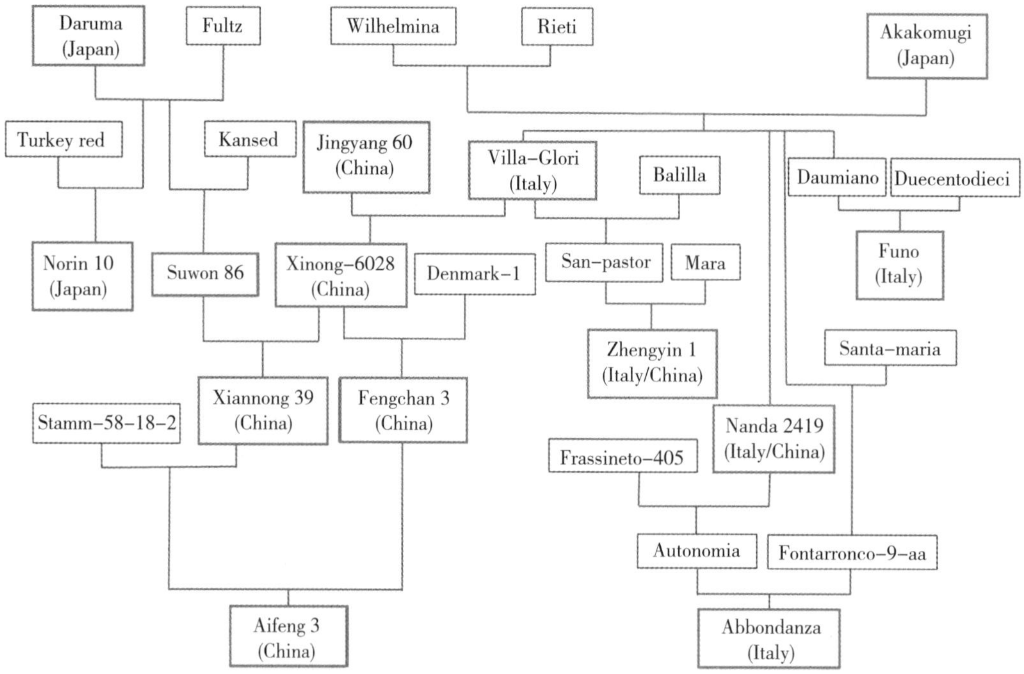

Fig. 2 Pedigree analyses of key varieties. Orange boxes highlight the genotyped varieties; the varieties in red script have *Rht24* based on the molecular marker test. Note: the pedigree is based on the pedigree described by Q. S. Zhuang (2003).

Acknowledgments

The authors are grateful to Professor R. A. McIntosh, Plant Breeding Institute, University of Sydney, for critical review of this manuscript.

References

Borlaug, N. E. 1968. Wheat breeding and its impact on world food supply. In: K. W. Findlay and K. W. Shepherd, editors, Proceedings of the 3rd International Wheat Genetics Symposium, Canberra, Australia. 5-9 Aug. 1968. CIMMYT, Mexico City. p. 1-36.

Borojevic, K., and K. Borojevic. 2005. The transfer and history of "reduced height genes" (*Rht*) in wheat from Japan to Europe. J. Hered. 96: 455-459. doi: 10.1093/jhered/esi060

Borojevic, S., and J. Potocanac. 1966. The development of the Yugoslav programme for creating high-yielding wheat vari-eties. Contemporary Agric. 11-12: 318-335.

Botwright, T. L., G. J. Rebetzke, A. G. Condon, and R. A. Rich-ards. 2005. Influence of the gibberellin-sensitive *Rht8* dwarf-ing gene on leaf epidermal cell dimensions and early vigour in wheat (*Triticum aestivum* L.). Ann. Bot. (Lond.) 95: 631-639. doi: 10.1093/aob/mci069

Chen, S. L., R. H. Gao, H. Y. Wang, M. X. Wen, J. Xiao, N. F. Bian, et al. 2015. Characterization of a novel reduced height gene (*Rht23*) regulating panicle morphology and plant architecture in bread wheat. Euphytica 203: 583-594. doi: 10.1007/s10681-014-1275-1

Ellis, M., W. Spielmeyer, K. Gale, G. Rebetzke, and R. Richards. 2002. "Perfect" markers for the *Rht-B1b* and *Rht-D1b* dwarfing genes in wheat. Theor. Appl. Genet. 105:

1038-1042. doi: 10.1007/s00122-002-1048-4

Ellis, M. H., G. J. Rebetzke, F. Azanza, R. A. Richards, and W. Spielmeyer. 2005. Molecular mapping of gibberellin-respon-sive dwarfing genes in bread wheat. Theor. Appl. Genet. 111: 423-430. doi: 10.1007/s00122-005-2008-6

Gale, M. D., S. Youssefian, and G. E. Russell. 1985. Dwarfing genes in wheat. In: G. E. Russell, editor, Pro-gress in plant breeding. Butterworth, London. p. 1-35. doi: 10.1016/B978-0-407-00780-2.50005-9

Korzun, V., M. S. Röder, M. W. Ganal, A. J. Worland, and C. N. Law. 1998. Genetic analysis of the dwarfing gene (*Rht8*) in wheat. Part I. Molecular mapping of *Rht8* on the short arm of chromosome 2D of bread wheat (*Triticum aestivum* L.). Theor. Appl. Genet. 96: 1104-1109. doi: 10.1007/ s001220050845

Li, A. X., W. L. Yang, X. Y. Lou, D. C. Liu, J. Z. Sun, X. L. Guo, et al. 2013. Novel natural allelic variations at the *Rht1* loci in wheat. J. Integr. Plant Biol. 55: 1026-1037. doi: 10.1111/ jipb.12103

Liu, S. C., and D. S. Zheng. 1997. Utilization of CIMMYT wheat germplasm in China. In: Z. H. He and S. Rajaram, editors, China/CIMMYT collaboration on wheat breeding and germplasm exchange: Results of 10 years of shuttle breeding (1984-1994). CIMMYT, Beijing. p. 11-18. http: //reposi-tory. cimmyt. org/xmlui/handle/10883/1222 (accessed 4 July 1995).

Lorenzetti, R. 2000. Lascienza del grano. Archivi di stato-Saggi 58-Ed-Minist. Beni Attivita Culturali, Rome.

Lumpkin, T. A. 2015. How a gene from Japan revolutionized the world of wheat: CIMMYT's quest for combining genes to mitigate threats to global food security. In: Y. Ogihara, et al., editors, Advances in wheat genetics: From genome to field. Springer, Tokyo. p. 13-20. doi: 10.1007/978-4-431-55675-6_2

McIntosh, R. A., J. Dubcovsky, W. J. Rogers, C. Morris, and X. C. Xia. 2017. Catalogue of gene symbols for wheat: 2017 supplement. SHIGEN. https://shigen. nig. ac. jp/wheat/komugi/genes/macgene/supplement2017. pdf (accessed 2 Feb. 2017).

Pearce, S., R. Saville, S. P. Vaughan, P. M. Chandler, E. P. Wil-helm, C. A. Sparks, et al. 2011. Molecular char-acterization of *Rht1* dwarfing genes in hexaploid wheat. Plant Physiol. 157:1820-1831. doi: 10.1104/pp.111.183657

Peng, Z. S., X. Li, Z. J. Yang, and M. L. Liao. 2011. A new reduced height gene found in the tetraploid semi-dwarf wheat landrace Aiganfanmai. Genet. Mol. Res. 10: 2349-2357. doi: 10.4238/2011. October. 5. 5

Rasheed, A., W. E. Wen, F. M. Gao, S. N. Zhai, H. Jin, J. D. Liu, et al. 2016. Development and validation of KASP assays for genes underpinning key economic traits in bread wheat. Theor. Appl. Genet. 129: 1843-1860. doi: 10.1007/s00122-016-2743-x

Rebetzke, G. J., and R. A. Richards. 2000. Gibberellic acid-sensitive dwarfing genes reduce plant height to increase kernel number and grain yield of wheat. Crop Pasture Sci. 51: 235- 246. doi: 10.1071/AR99043

Reitz, L. P., and S. C. Salmon. 1968. Origin and use of Norin 10 wheat. Crop Sci. 8: 686-689. doi: 10.2135/cropsci1968.00111 83X000800060014x

Saghai-Maroof, M. A., K. M. Soliman, R. A. Jorgensen, and R. W. Allard. 1984. Ribosomal DNA spacer-length polymorphisms in barley: Mendelian inheritance, chro-mosomal location, and population dynamics. Proc. Natl. Acad. Sci. USA 81: 8014- 8018. doi: 10.1073/pnas. 81. 24. 8014

Shahinnia, F., and B. E. Tabatabaei. 2009. Conversion of barley SNPs into PCR-based markers using dCAPS method. Genet. Mol. Biol. 32: 564-567. doi: 10.1590/S1415-47572009005000047

Shaw, D. J. 2007. World food security: A history since 1945. Pal-grave Macmillan, Basingstoke, UK.

Tang, N., B. Li, H. Min, and Y. G. Hu. 2012. Distribution of dwarfing genes *Rht-B1b*, *Rht-D1b* and *Rht8* in Chinese bread wheat cultivars detected by molecular markers. J. China Agric. Univ. 17: 21-26.

Tian, X. L., W. E. Wen, L. Xie, L. Fu, D. A. Xu, C. Fu, et al. 2017. Molecular mapping of reduced plant height gene *Rht24* in bread wheat. Front. Plant Sci. 8: 1397. doi: 10.3389/fpls.2017.01379

Würschum, T., S. M. Langer, C. Longin, H. Friedrich, M. R. Tucker, and W. L. Leiser. 2017. A modern green revolution gene for reduced height in wheat. Plant J. 92: 892-903. doi: 10.1111/tpj.13726

Worland, A. J. 1999. The importance of Italian wheats to world-wide varietal improvement. J. Genet. Breed. 53: 165-173.

Worland, A. J., E. J. Sayers, and V. Korzun. 2001. Allelic variation at the dwarfing gene *Rht8* locus and its signifi-cance in international breeding programmes. Euphytica 119: 157-161. doi: 10.1023/A: 1017582122775

Yamada, T. 1990. Classification of GA response, Rht genes and culm length in Japanese varieties and landraces of wheat. Euphytica 50: 221-239. doi: 10.1007/BF00023648

Yang, S. J., X. K. Zhang, Z. H. He, X. C. Xia, and

Y. Zhou. 2006. Distribution of dwarfing genes *Rht-B1b* and *Rht-D1b* in Chinese bread wheats detected by STS marker. Sci. Agric. Sin. 39：1680-1688.

Zhang，X. K.，S. J. Yang，Y. Zhou，Z. H. He，and X. C. Xia. 2006. Distribution of the *Rht-B1b*，*Rht-D1b* and *Rht8* reduced height genes in autumnsown Chinese wheats detected by molecular markers. Euphytica 152：109-116. doi：10. 1007/s10681-006-9184-6

Zhou，Y.，Z. H. He，G. S. Zhang，X. C. Xia，X. M. Chen，L. P. Zhang，et al. 2003. *Rht8* dwarf gene distribution in Chinese wheats identified by microsatellite marker. Acta Agron. Sin. 29：810-814.

Zhuang，Q. S. 2003. Chinese wheat improvement and pedigree analysis. China Agric. Press，Beijing.

Molecular Mapping of Reduced Plant Height Gene *Rht24* in Bread Wheat

Xiuling Tian[1,2], Weie Wen[1,2], Li Xie[2], Luping Fu[2], Dengan Xu[2], Chao Fu[2], Desen Wang[2], Xinmin Chen[2], ianchun Xia[1,2], Quanjia Chen[1], Zhonghu He[2,3] and Shuanghe Cao[2]

[1] *College of Agronomy, Xinjiang Agricultural University, Urumqi, China*, [2] *Institute of Crop Science, National Wheat Improvement Center, Chinese Academy of Agricultural Sciences, Beijing, China*, [3] *International Maize and Wheat Improvement Center (CIMMYT), Beijing, China*

Abstract: Height is an important trait related to plant architecture and yield potential in bread wheat (*Triticum aestivum* L.). We previously identified a major quantitative trait locus *QPH. caas-6A* flanked by simple sequence repeat markers *Xbarc103* and *Xwmc256* that reduced height by 8.0-10.4%. Here *QPH. caas-6A*, designated as *Rht24*, was confirmed using recombinant inbred lines (RILs) derived from a Jingdong 8/Aikang 58 cross. The target sequences of *Xbarc103* and *Xwmc256* were used as queries to BLAST against International Wheat Genome Sequence Consortium database and hit a super scaffold of approximately 208 Mb. Based on gene annotation of the scaffold, three gene-specific markers were developed to genotype the RILs, and *Rht24* was narrowed to a 1.85 cM interval between *TaAP2* and *TaFAR*. In addition, three single nucleotide polymorphism (SNP) markers linked to *Rht24* were identified from SNP chip-based screening in combination with bulked segregant analysis. The allelic efficacy of *Rht24* was validated in 242 elite wheat varieties using *TaAP2* and *TaFAR* markers. These showed a significant association between genotypes and plant height. *Rht24* reduced plant height by an average of 6.0-7.9 cm across environments and were significantly associated with an increased TGW of 2.0-3.4 g. The findings indicate that *Rht24* is a common dwarfing gene in wheat breeding, and *TaAP2* and *TaFAR* can be used for marker-assisted selection.

Keywords: CAPS marker, genome mining, *Rht24*, SNP chip, *Triticum aestivum*

INTRODUCTION

Bread wheat (*Triticum aestivum* L.) is one of the most important food crops providing a significant proportion of the calories and protein consumed by humans (Mir et al., 2012; Chen et al., 2016). Grain yield in wheat largely depends on plant architecture, particularly plant height, which is significantly associated with biomass production and harvest index that ultimately determine yield potential (Law et al., 1978). Most importantly, appropriately reduced plant height reduces lodging and increases grain yield (Evans, 1998). The introduction of semi-dwarf varieties leading to the "Green Revolution" greatly enhanced crop yields globally (Peng et al., 1999).

To date, 23 dwarfing genes (*Rht1-Rht23*) have been

Published in Frontiers in Plant Science, 2017, 8: 1379.

cataloged in wheat (Supplementary Table S1) (McIntosh et al., 2015). Among them, only four homeologous genes *Rht1* (*Rht-B1b*), *Rht2* (*Rht-D1b*), *Rht3* (*Rht-B1c*), and *Rht10* (*Rht-D1c*) from the chromosome 4BS, 4DS, 4BS, and 4DS, respectively, had been cloned (Peng et al., 1999; Pearce et al., 2011; Wu et al., 2011), 13 were on chromosomes 2AS (2), 2BL (1), 2DL (1), 3BS (1), 5AL (1), 5DL (1), 6AS (3), 7AS (1), and 7BS (2), but the locations of six other dwarfing genes have not been determined yet (Peng et al., 1999, 2011; Ellis et al., 2005; Wu et al., 2011; McIntosh et al., 2013; Chen et al., 2015). So far, more than 50 quantitative trait loci (QTL) for plant height have also been identified on all wheat chromosomes in previous reports (Börner et al., 2002; Peng et al., 2003; Quarrie et al., 2005; Liu et al., 2011; Griffiths et al., 2012; Würschum et al., 2015). For example, Wei et al. (2010) identified quite a few stable plant height QTL located on chromosomes 2A, 2B, 2D, 3B, 4B, 5A, 5D, 7B, and 7D. Griffiths et al. (2012) found that all chromosomes except for 3D, 4A, and 5D, had plant height genes using meta-QTL analysis. However, only a few genes for reduced stature have been used in wheat breeding because most showed negative effects on grain yield (Law et al., 1978; Worland et al., 1998; Chen et al., 2015). *Rht-B1b* (4BS), *Rht-D1b* (4DS) and *Rht8* (2DL) are extensively used in wheat breeding globally and molecular markers have been developed for their marker-assisted selection (MAS) (Korzun et al., 1998; Peng et al., 1999; Rebetzke and Richards, 2000; Ellis et al., 2002; Asplund et al., 2012). *Rht-B1b* (*Rht1*) and *Rht-D1b* (*Rht2*) were two major genes in the Green Revolution, and at present, approximately 70% of wheat varieties throughout the world contain at least one of them (Evans, 1998). *Rht8* reduced plant height around 10% and it has not significant negative effect on grain yield (Worland et al., 1998).

Simple sequence repeat (SSR) markers were favored for research and breeding in the last two decades due to high polymorphism, good repeatability, co-dominance and a simple polymerase chain reaction (PCR) -based system for analysis (Röder et al., 1998). With rapid advancements in sequencing technology, the quality of wheat genome assembly has been significantly improved and large numbers of high-quality scaffolds to facilitate gene isolation are available (Rogers, 2014). Simultaneously, many single nucleotide polymorphisms (SNPs) were identified for developing SNP chips, providing high-throughput genotyping platforms that have been extensively used for genome-wide association studies (GWAS) and QTL mapping at high resolution (Colasuonno et al., 2014; Wang et al., 2014; Sukumaran et al., 2015; Gao et al., 2016; Wu et al., 2016). Many QTL for agronomic traits, grain and industrial quality and disease resistance have been identified in wheat (Buerstmayr et al., 2009; Wang et al., 2014; Naruoka et al., 2015; Jin et al., 2016; Liu et al., 2016; Zhai et al., 2016). Thus, the availability of high quality wheat genome data and high-throughput SNP genotyping platforms greatly advanced wheat genetics and breeding.

We previously identified a plant height major QTL *QPH.caas-6A* between *Xwmc256* and *Xbarc103*, which explained 8.0-10.4% of the phenotypic variance across eight environments using an $F_{2:4}$ population derived from the Jingdong 8/Aikang 58 cross (Li et al., 2015). In addition, *QPH.caas-6A* probably was consistent with *QTL_height_6A_1*, which also linked with *Xwmc256* on chromosome 6AL and explained 6.3-29.1% of the phenotypic variance (PVE) (Griffiths et al., 2012). Furthermore, *QPH.caas-6A* was identified to reduce TGW 6.5-8.2%, increase kernel number per spike and number of spike by 2.4-3.5% and NS by 2.0-4.6%, respectively, which is similar to "green revolution" gene *Rht2* (Li et al., 2015). Thus *QPH.caas-6A* was a potentially useful dwarfing locus in wheat breeding. Here we designate *QPH.caas-6A* as *Rht24*. The previously closest markers flanking *Rht24* were *Xbarc103* and *Xwmc256*, which were 8.2 cM apart. The aim of the present study was to identify markers more closely linked to *Rht24* using the wheat genomic database and a 660K SNP chip and thereby establish a more efficient MAS system for wheat breeding.

MATERIALS AND METHODS

Plant Materials

Two hundred and fifty-six recombinant inbred lines (RILs, $F_{2:6}$) from a cross between Aikang 58 (AK58) Jingdong 8 (JD8) were used for genetic analysis. AK58, a leading variety occupying more than one million ha in the Yellow and Huai Valley, has a short plant height and excellent lodging resistance. Jingdong 8, with relatively tall plant height and excellent resistance to heat during the grain filling stage, was an elite variety in the Northern China Plain Region. Two sets of Chinese and introduced elite varieties (Supplementary Tables S4, S5), comprising 154 (Set I) and 88 (Set II) varieties, were used to validate the efficacy of MAS for *Rht24* and to investigate allelic distributions.

Field Trials and Phenotype Evaluation

The RILs and parents were planted at Gaoyi in Hebei province and Anyang in Henan during the 2014-2015 cropping season, and at Gaoyi during 2015-2016 cropping season. The experimental design was randomized complete blocks and three replications. Each plot was a single 2 m row with 25 cm between rows.

Set I varieties were planted at Anyang in Henan province and Suixi in Anhui, respectively, during the 2012-2013 and 2013-2014 cropping seasons. Set II varieties were grown at Shijiazhuang in Hebei and Beijing, respectively, during the 2012-2013 and 2013-2014 cropping seasons. These were grown in randomized complete blocks with three replications. Each plot consisted of four 2 m rows spaced 30 cm apart and approximately 50 plants in each row. Field management was according to local practice. During the whole wheat growth period in our field trial, there is no extreme weather causing serious damages, such as cold spell in later spring, dry and hot wind in filling stage, and fertilization (bottom fertilizer including 150 kg/H^2 Urea plus 400 kg/H^2 Diammonium Phosphate; 300 kg/H^2 Urea just before elongation stage) is enough for wheat growth and development.

Plant height was measured from the ground to spike (awns excluded) at grain-filling. For each plot, five representative primary tillers on different plants in the middle of each row were selected to measure the plant height, and the averaged value was used for subsequent analysis. TGW was determined by weighing triplicate 200 grain samples.

Genotypic Analysis Using Simple Sequence Repeat (SSR) Markers

Genomic DNA was isolated from young leaves using the modified CTAB method (Saghai-Maroof et al., 1984). Twenty-two SSR markers on chromosome 6AL were chosen for genetic analyses① (Supplementary Table S7).

Polymerase chain reaction was performed in a 15 μl reaction system containing 7.5 μl of 2 × Taq PCR Mix (Tianwei Biotechnology Co., Ltd., Beijing②), 5 pmol of each primer and 100 ng of genomic DNA. Amplification was performed at 94℃ for 5 min, followed by 35 cycles at 94℃ for 20 s, 50-60℃ (depending on specific primers) for 30 s and 72℃ for 1 min, with a final extension at 72℃ for 5 min. The PCR products were separated in 6% polyacrylamide or 2% agarose gels.

Gene-Specific Marker Development by Genome Mining Approach

To confirm the target sequences of the SSR mentioned above, PCR products were purified using theTIANgel MIDI Purification Kit (Tiangen, Biotechnology Co., Ltd., Beijing③). The purified heat shock method (TransGen Biotech Co., Ltd., Beijing④). At least three positive clones from each transformation were randomly selected and sequenced at Shanghai Sangon Biotech Co., Ltd⑤. The SSR target sequences were

---

① http://wheat.pw.usda.gov
② http://www.tw-biotech.com
③ http://tiangen.casmart.com.cn
④ http://www.transgen.com.cn
⑤ http://www.sangon.com/

used as BLAST queries① to identify desirable scaffolds. The genes of interest in the target scaffold were isolated and sequenced following the above procedure. Sequence sites polymorphic between the parents were identified by alignments with DNAMAN software② for designing gene-specific markers. Cleaved amplified polymorphic sequences (CAPS) or derived CAPS (dCAPS) marker for each target gene was developed following Thiel et al. (2004).

SNP Chip-Based Screening

Wheat 660K SNP chip was developed by the Institute of Crop Science, Chinese Academy of Agricultural Sciences, synthesized by Affymetrix and commercially available at CapitalBio Corporation③. Recently, the chip has been efficiently used in our lab (Jin et al., 2016).

To identify more molecular markers linked to *Rht24* the 660K SNP chip was used to test two parents and two contrasting bulks of the RIL population, comprising equal amounts of and then transformed into Trans1-T1 competent cells by the④ PCR products were ligated with cloning vector pEASY-T5 Zero genomic DNA from 10 tall and 10 short RILs, respectively. Genotyping was performed at CapitalBio Technology Company⑤. SNPs polymorphic between parents and between two bulks were selected to develop CAPS markers.

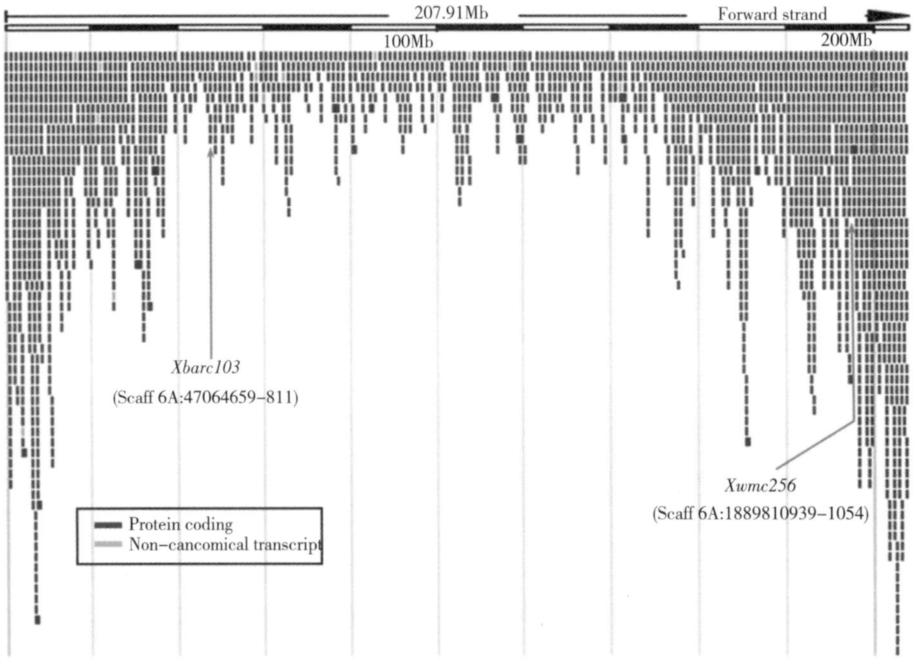

Fig. 1 A super scaffold spanning *Rht24*. Locations of SSR loci *Xbarc103* and *Xwmc256* are shown by arrows.

Genetic Linkage Map Construction and Statistical Analysis

Quantitative trait locusIciMapping 4.0⑥ was used for linkage map construction and QTL analysis using a LOD score threshold of 3.0 (Lincoln et al., 1992). Plant height was reformed with an algorithm SERiation and criterion SARF (sum of adjacent recombination frequencies). Variance analysis and *t*-test (Duncan method) were

① http://plants.ensembl.org/Triticum_aestivum/Info/Index
② http://www.lynnon.com
③ http://wheat.pw.usda.gov/ggpages/topics/Wheat660_SNP_array_developed_by_CAAS.pdf
④ http://wheat.pw.usda.gov
⑤ http://www.capitalbiotech.com
⑥ http://www.isbreeding.net/
⑦ http://www.sas.com

conducted with SAS 9.4[7]. The broad-sense heritability (h_B^2) of the corresponding traits was calculated using the following formula:

$h_B^2 = \sigma_g^2 / (\sigma_g^2 + \sigma_{ge}^2/e + \sigma_\varepsilon^2/re)$, where σ_g^2, σ_{ge}^2, and σ^2_ε were estimates of lines, line × environment interactions and residual error variances, respectively, and e and r represented the numbers of environments and replicates, respectively (Supplementary Table S6; Nyquist and Baker, 1991). T-test was used to perform the association analysis between *Rht24* and plant height as well as TGW.

RESULTS

Confirmation of *Rht24*

In our previous study, *Rht24* was identified with LOD score of approximately 12, explaining 8.0-10.4% of the phenotypic variance; it was mapped within an 8.2 cM interval flanked by *Xbarc103* and *Xwmc256* (Li et al., 2015). To further verify *Rht24*, 22 SSR markers were selected from the targeted region of chromosome 6AL (Supplementary Table S7). Among them, only *Xbarc103* and *Xwmc256* were polymorphic between the parental lines and bulk lines built by 10 tall and 10 short lines (Supplementary Figure S1). *Rht24* was located between the two markers, which were 19.46 cM apart. Although the genetic distance was greater than reported previously, the QTL for reduced plant height was confirmed.

Narrow-Down of *Rht24* by the Genome Mining Approach

To confirm the chromosomal location of *Rht24*, we cloned and sequenced the differential fragments of *Xbarc103* and *Xwmc256* in the tall and short lines, respectively (Supplementary Figure S1). Sequencing results showed that they were consistent with the reported sequences[1]. Using the sequences as queries to BLAST against International Wheat Genome Sequence Consortium database[2], a super scaffold of approximately 208 Mb (approximately 1200 genes) spanning *Rht24* was identified (Fig. 1). Twenty genes involved in plant morphogenesis and hormone metabolism were selected from the scaffold and sequenced (Supplementary Table S8). Polymorphisms between AK58 and JD8 were detected at loci *TaGA3*, *TaFAR*, and *TaAP2* (Supplementary Figures S2A-C and Table S2). Based on these polymorphisms, three gene-specific CAPS or dCAPS markers were developed (Fig. 2A-C and Supplementary Figures S2A-C). Upon genotyping the RIL, the location of *Rht24* was narrowed to a 1.85 cM interval flanked by *TaFAR* and *TaAP2* (Fig. 3).

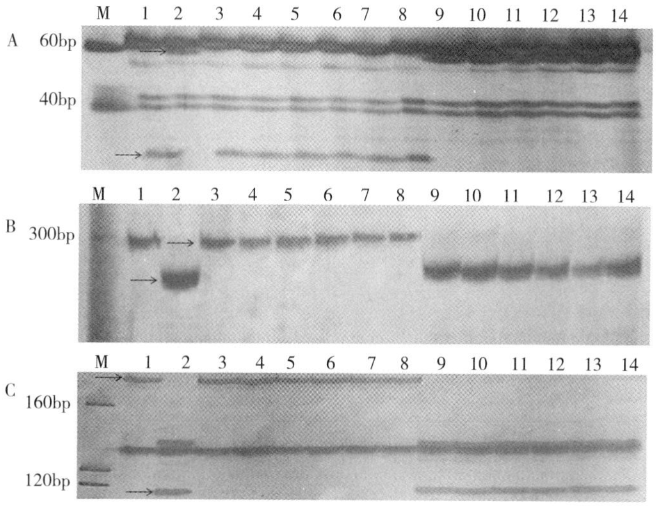

Fig. 2 PCR patterns of polymorphic markers *TaGA3* (A), *TaFAR* (B), and *TaAP2* (C). Polymorphic fragments associated with height differences between short and tall RILs are shown by *arrows*. M, Marker (20 bp DNA ladder, Takara Bio Company). *Lanes 1* and *2* are the parents, AK58 and JD8, respectively; *lanes 3-8* are short RILs; *lanes 9-14* are tall RILs.

① http://wheat.pw.usda.gov

② http://plants.ensembl.org/Triticum_aestivum/Info/Index

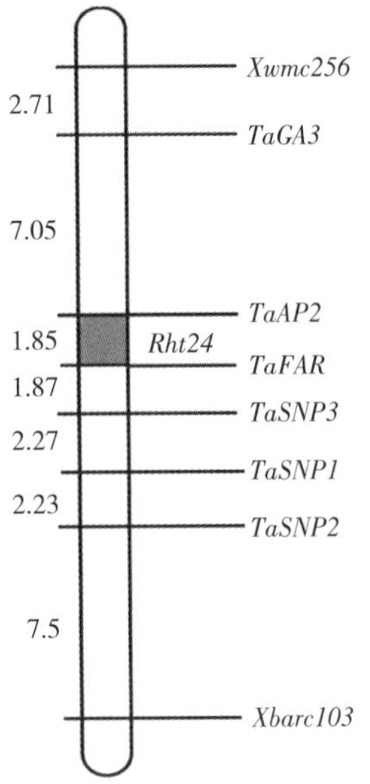

Fig. 3 Linkage map of *Rht24*. The names and corresponding locations of all markers are indicated on the *right* side of the map. Genetic distance (cM) is shown on the *left* side of the map. The region of *Rht24* is highlighted in *blue*.

Enriching Molecular Markers in the *Rht24* Region by a Combination of SNP Genotyping Assays and Genomic Identification

To enrich molecular markers in the *Rht24* region, the 660K SNP chip was used in genotyping assays. In total, more than 400 polymorphic SNPs on chromosome 6A from wheat 660K SNP chips were identified between two contrasting bulks (data not shown). Among them, the target sequences of polymorphic SNPs anchored in the super scaffold above were used to develop site-specific CAPS markers (Supplementary Table S9). Among them, three polymorphic loci, *TaSNP1*, *TaSNP2*, and *TaSNP3*, were mapped in the region of *Rht24* (Supplementary Figures S3A-C, S4A-C, Table S2 and Figure 3). However, they did not further narrow down the *Rht24* locus.

Validation of Molecular Markers Closely Linked to *Rht24*

Among 242 wheat varieties genotyped by markers *TaFAR* and *TaAP2* (Supplementary Figures S5A, B), *Rht24* was present in 185 (76.4%) varieties and absent in 41 (17%) (Supplementary Table S3). The average plant height for the varieties containing *Rht24* was 78.2 and 80.2 cm in Sets I and II, respectively, whereas for those without *Rht24* the comparative values were 86.1 and 86.2 cm, respectively, indicating that varieties with *Rht24* were generally shorter than those without (PI 0.029, PII 0.003, Table 1). *Rht24* was also significantly associated with thousand grain weight (TGW) in Set I (P=0.014, Table 2). Although *Rht24* had no significant association with TGW in Set II, the average TGW of entries with *Rht24* was 2 g higher than those without (P 0.196, Table 2). On average *Rht24* reduced plant height by 6.0-7.9 cm and increased TGW by 2.0-3.4 g in all varieties used in this study (Tables 1, 2).

DISCUSSION

Rht24 Is an Important Locus Affecting Plant Height and Grain Weight in Wheat

Previous studies showed that loci controlling plant height were present in all 21 wheat chromosomes (Mccartney et al., 2005; Gale and Law, 1973; Griffiths et al., 2012). The QTL *QTL _ height _ 6A _ 1* linked with *Xwmc256* on chromosome 6AL explained 6.3-29.1% of the phenotypic variance (PVE) (Griffiths et al., 2012). *Rht24*, previously designated as *QPH.caas-6A*, was flanked by SSR markers *Xwmc256* and *Xbarc103* (Li et al., 2015), indicating that it could be the same as *QTL_height_6A_1*. Here we developed two markers, *TaFAR* and *TaAP2* that closely flanked *Rht24*. The two markers were used to test the genetic effect of *Rht24* and observed similar result with the previously reported (Table 3; Griffiths et al., 2012; Li et al., 2015). *Rht24* genotyped by *TaAP2* and *TaFAR* as markers showed a significant association with reduced height among elite varieties. It was present in about 76% of elite varieties indicating that it was positively selected in wheat breeding programs (Supplementary Table S3). In our previous study, *Rht24* reduced TGW by 6.9-8.5%. However, *Rht24* had a positive effect on thousand kernel weight in the present study (Table 2). We speculate that two reasons probably account for this case. The AK58/JD8 RIL and natural populations have better hereditary stability than the previous $F_{2:4}$ mapping population and are used to define *Rht24* with more accurate effect analyses. Additionally, the interval of *Rht24* was narrowed down and thus more interference effects of other genes were excluded using the closest flanking markers, *TaFAR* and *TaAP2*. In fact, it still cannot be ruled out that other genes disturb the effect of *Rht24* on TGW based on our existing mapping information. Thus it is necessary to further narrow down the interval of *Rht24*. Thus, *Rht24* is an important QTL for plant height and TGW in wheat and MAS mediated by selection for appropriate *TaAP2* and *TaFAR* alleles should be effective.

Distribution of *Rht-B1b*, *Rht-D1b*, and *Rht8* in Chinese germplasm was detected by molecular markers, with frequencies of 24.5, 45.5, and 46.8%, respectively (Zhang et al., 2006). In addition, *Rht-B1b* and *Rht-D1b* were tested by KASP assays in current Chinese leading varieties with frequencies of 36.9 and 38.6%, respectively (Rasheed et al., 2016). Our study showed that *Rht24* was present in about 76% of elite Chinese varieties, a much higher frequency than either *Rht-B1b* or *Rht-D1b* (Supplementary Table S3). The results indicated that *Rht24* had the highest frequency distribution compared with the other three loci, and it was frequently present in combination with *Rht-D1b* or *Rht8*. It was reported that *Rht8* not only to reduce plant height but also significantly increase TGW (Zhang et al., 2016), however, *Rht-B1b* and *Rht-D1b* showed negative effects on TGW (Li et al., 1998). In this study, *Rht24* increased TGW by 2.0-3.4 g in all varieties used (Table 2).

In all, *Rht24* was important and extensively used in Chinese wheat breeding programs. Further study is needed to understand the origin of *Rht24* in Chinese wheat and investigate *Rht24* effect on the other traits, such as tiller number, flowering date, spike length and so on.

Combination of SNP Chip-Based Screening and Genome Mining Was an Effective Approach for Fine Mapping of *Rht24*

As the development of genome sequencing technology, high-quality wheat genome sequencing data has been continuously updated and released①. Based on BLAST against wheat genome databases, the markers from preliminary mapping can be used as entry points to mine new markers linked to the loci of interest. In the present study we used SSR markers flanking *Rht24* as queries and identified a super scaffold. According to gene annotations, we developed three gene-specific markers and narrowed the region of *Rht24* from a 19.46 cM interval (8.2 cM in our previous study) to 1.8 cM. A 660K SNP chip was recently developed and released②. The 660K chip has several advantages compared to the 90K SNP chip, such as higher marker density and higher resolution as well as better distribution on chromosomes. We used this chip to screen candidate markers linked to *Rht24* and a significant number of polymorphic SNPs were obtained (data not shown). It was an obviously effective approach to take advantage of constantly updated genome information and high-throughput SNP platforms.

Table 1 Association analysis between *Rht24* genotypes and plant height in two sets of germplasm

Germplasm	Year	Environment	Genotype[a]	Number of accessions	Mean plant height (cm)[b]	*SD* (cm)	Range (cm)
Set I	2012-2013	Anyang	*A*	17	88.5a	14.7	75.0-138.3
			B	128	81.0a	7.0	63.3-106.7
		Suixi	*A*	17	84.8a	12.4	71.0-125.0
			B	128	77.2b	6.6	55.0-95.7

① http://plants.ensembl.org/Triticum_aestivum

② http://wheat.pw.usda.gov/ggpages/topics/Wheat660_SNP_array_developed_by_CAAS.pdf

(continued)

Germplasm	Year	Environment	Genotype[a]	Number of accessions	Mean plant height (cm)[b]	*SD* (cm)	Range (cm)
	2013-2014	Anyang	*A*	17	83. 6a	15. 2	66. 0-135. 7
			B	128	74. 4b	7. 0	58. 2-92. 1
		Suixi	*A*	17	87. 8a	12. 4	72. 4-125. 8
			B	128	80. 2b	6. 6	59. 8-100. 9
		Average	*A*	17	86. 1a	13. 5	72. 2-131. 2
			B	128	78. 2b	6. 4	59. 5-98. 8
Set II	2012-2013	Beijing	*A*	24	82. 3a	10. 3	62. 9-109. 8
			B	57	76. 0b	8. 0	49. 3-93. 9
		Shijiazhuang	*A*	24	85. 3a	9. 0	68. 2-108. 4
			B	57	80. 5b	7. 6	55. 6-97. 3
	2013-2014	Beijing	*A*	24	82. 3a	10. 3	62. 9-109. 8
			B	57	76. 0b	8. 0	49. 3-93. 9
		Shijiazhuang	*A*	24	85. 3a	9. 0	68. 2-108. 4
			B	57	80. 5b	7. 6	55. 6-97. 3
		Average	*A*	24	86. 2a	9. 5	67. 8-110. 7
			B	57	80. 2b	7. 5	57. 7-95. 7

[a] *A*, *JD8 parental genotypes and B*, *AK58 parental genotypes based on flanking markers TaAP2 and TaFAR. Varieties with recombinant genotypes are not included.*

[b] *Different lowercase letters following the mean plant height indicate significant differences between two genotypes at* $P < 0.05$ (*Dunnett's t-test*); *SD*, *standard deviation.*

Table 2 Association analysis between *Rht24* genotypes and TGW in two sets of germplasm

Germplasm	Year	Environment	Genotype[a]	Number of accessions	Mean TGW (g)[b]	*SD* (g)	Range (g)
Set I	2012-2013	Anyang	*A*	17	38. 7a	6. 0	30. 1-53. 3
			B	128	41. 4b	5. 0	25. 3-56. 1
		Suixi	*A*	17	39. 2a	6. 1	29. 3-49. 8
			B	128	42. 7b	5. 5	25. 7-56. 3
	2013-2014	Anyang	*A*	17	47. 0a	6. 9	35. 3-59. 7
			B	128	50. 3b	5. 2	28. 3-63. 4
		Suixi	*A*	17	42. 4a	7. 4	30. 2-57. 7
			B	128	46. 3b	5. 7	22. 5-62. 2
		Average	*A*	17	41. 8a	6. 3	31. 7-54. 0
			B	128	45. 2b	5. 0	26. 1-59. 2
Set II	2012-2013	Beijing	*A*	24	30. 7a	6. 6	21. 5-43. 3
			B	57	32. 2a	5. 5	21. 4-42. 9
		Shijiazhuang	*A*	24	32. 3a	7. 1	20. 9-48. 1
			B	57	34. 1a	6. 0	21. 1-46. 3
	2013-2014	Beijing	*A*	24	29. 6a	6. 0	20. 5-44. 4
			B	57	30. 9a	4. 9	19. 7-40. 1
		Shijiazhuang	*A*	24	39. 1a	9. 4	25. 5-66. 1
			B	57	42. 3a	7. 7	26. 3-56. 3
		Average	*A*	24	32. 9a	7. 1	22. 1-50. 5
			B	57	34. 9a	5. 7	22. 5-46. 1

[a] *A*, *FAR-a/AP2-a* (*parental JD8 genotype*); *B*, *FAR-b/AP2-b* (*parental AK58 genotype*).[b] *Different lowercase letters following the mean plant height indicate significant differences between two genotypes at* $P < 0.05$ (*Dunnett's t-test*); *SD*, *standard deviation.*

Table 3 Phenotypic effect of *Rht24* on plant height in the AK58/JD8 RIL population.

Environment	Interval	LOD[a]	Add[b]	PVE (%)[c]
Anyang (2014-2015)	*TaAP2-TaFAR*	4. 6	−3. 1	9. 9
Gaoyi (2014-2015)	*TaAP2-TaFAR*	11. 7	−3. 9	23. 7
Gaoyi (2015-2016)	*TaAP2-TaFAR*	5. 6	−3. 1	10. 7
Average	*TaAP2-TaFAR*	8. 0—3. 4	16. 8	

[a] *Logarithm of odds score; QTLs were detected at a LOD threshold of* 3. 0. [b]*Additive effects.* [c]*Percentage of phenotypic variance explained by Rht24.*

ACKNOWLEDGMENT

The authors are grateful to Prof. R. A. McIntosh, Plant Breeding Institute, University of Sydney, for critical review of this manuscript.

REFERENCES

Asplund, L., Leino, M. W., and Hagenblad, J. (2012). Allelic variation at the *Rht8* locus in a 19th century wheat collection. *Sci. World J.* 2012, 146-151. doi: 10.1100/2012/385610

Bazhenov, M. S., Divashuk, M. G., Amagai, Y., Watanabe, N., and Karlov, G. I. (2015). Isolation of the dwarfing *Rht-B1p* (*Rht17*) gene from wheat and the development of an allele-specific PCR marker. *Mol. Breed.* 35: 213. doi: 10.1007/ s11032-015-0407-1

Börner, A., Schumann, E., Fürste, A., Cöster, H., Leithold, B., Röder, S., et al. (2002). Mapping of quantitative trait loci determining agronomic important characters in hexaploid wheat (*Triticum aestivum* L.). *Theor. Appl. Genet.* 105, 921-936. doi: 10.1007/s00122-002-0994-1.

Buerstmayr, H., Ban, T., and Anderson, J. A. (2009). QTL mapping and marker-assisted selection for Fusarium head blight resistance in wheat: a review. *Plant Breed.* 128, 1-26. doi: 10.1111/j.1439-0523.2008.01550.x

Chaudhry, A. (1973). *A Genetic and Cytogenetic Study of Height in Wheat.* Doctoral dissertation, University of Cambridge, Cambridge.

Chen, C., He, Z. H., Lu, J. L., Li, J., Ren, Y., Ma, C. X., et al. (2016). Molecular mapping of stripe rust resistance gene *YrJ22* in Chinese wheat cultivar Jimai 22. *Mol. Breed.* 36: 118. doi: 10.1007/s11032-016-0540-5

Chen, S. L., Gao, R. H., Wang, H. Y., Wen, M. X., Xiao, J., Bian, N. F., et al. (2015). Characterization of a novel reduced height gene (*Rht23*) regulating panicle morphology and plant architecture in bread wheat. *Euphytica* 203, 583-594. doi: 10.1007/s10681-014-1275-1

Colasuonno, P., Gadaleta, A., Giancaspro, A., Nigro, D., Giove, S., Incerti, O., et al. (2014). Development of a high-density SNP-based linkage map and detection of yellow pigment content QTLs in durum wheat. *Mol. Breed.* 34, 1563-1578. doi: 10.1007/s11032-014-0183-3

Daoura, B. G., Liang, C., and Hu, Y. G. (2013). Agronomic traits affected by dwarfing gene *Rht5* in common wheat (*Triticum aestivum* L.). *Aust. J. Crop Sci.* 7, 1270-1276.

Ellis, M., Spielmeyer, W., Gale, K., Rebetzke, G., and Richards, R. (2002). "Perfect" markers for the *Rht-B1b* and *Rht-D1b* dwarfing genes in wheat. *Theor. Appl. Genet.* 105, 1038-1042. doi: 10.1007/s00122-002-1048-4

Ellis, M. H., Rebetzke, G. J., Azanza, F., Richards, R. A., and Spielmeyer, W. (2005). Molecular mapping of gibberellin-responsive dwarfing genes in bread wheat. *Theor. Appl. Genet.* 111, 423-430. doi: 10.1007/s00122-005-2008-6

Evans, L. T. (1998). Crop evolution, adaptation and yield. *Photosynthetica* 34, 56-60. doi: 10.1023/A: 1006889901899

Gale, M. D., and Law, C. N. (1973). Semi-dwarf wheats induced by monosomy and associated changes in gibberellin levels. *Nature* 241, 211-212. doi: 10.1038/ 241211a0

Gao, L., Turner, M. K., Chao, S., Kolmer, J., and Anderson, J. A. (2016). Genome wide association study of seedling and adult plant leaf rust resistance in elite spring wheat breeding lines. *PLoS ONE* 11: e0148671. doi: 10.1371/journal.pone.0148671

Griffiths, S., Simmonds, J., Leverington, M., Wang, Y., Fish, L. J., Sayers, L., et al. (2012). Meta-QTL analysis of the genetic control of crop height in elite European winter wheat germplasm. *Mol. Breed.* 29, 159-171. doi: 10.1007/ s00122-009-1046-x

Haque, M. A., Martinek, P., Watanabe, N., and Kuboyama, T. (2011). Genetic mapping of gibberellic acid-sensitive genes for semi-dwarfism in durum wheat. *Cereal Res. Commun.* 39, 171-178. doi: 10.1556/

CRC. 39. 2011. 2. 1

Jin, H., Wen, W. E., Liu, J. D., Zhai, S. N., Zhang, Y., Yan, J., et al. (2016). Genome-wide QTL mapping for wheat processing quality parameters in a Gaocheng 8901/Zhoumai 16 recombinant inbred line population. *Front. Plant Sci.* 7: 1032. doi: 10. 1007/s00122-006-0346-7

Konzak, C. F. (1976). *A Review of Semid-Warfing Gene Sources, and a Description of Some New Mutants Useful for Breeding Short-Stature Wheats*. Vienna: Induced Mutations in Cross-Breeding.

Korzun, V., Röder, M. S., Ganal, M. W., Worland, A. J., and Law, C. N. (1998). Genetic analysis of the dwarfing gene (*Rht8*) in wheat. Part I. Molecular mapping of *Rht8* on the short arm of chromosome 2D of bread wheat (*Triticum aestivum* L.). *Theor. Appl. Genet*. 96, 1104-1109. doi: 10. 1007/s0012200 50845

Law, C. N., Snape, J. W., and Worland, A. J. (1978). The genetic relationship between height and yield in wheat. *Heredity* 40, 15-20. doi: 10. 1038/hdy. 1978. 13

Li, X. M., Xia, X. C., Xiao, Y. G., He, Z. H., Wang, D. S., Trethowan, R., et al. (2015). QTL mapping for plant height and yield components in common wheat under water-limited and full irrigation environments. *Crop Pasture Sci*. 66, 660-670. doi: 10. 1071/CP14236

Li, X. P., Jiang, C. Z., and Liu, H. L. (1998). Effects of different dwarfing genes on agronomic characteristics of winter wheat. *Acta Agron. Sin*. 24, 475-478.

Lincoln, S. E., Daly, M. J., and Lander, E. S. (1992). *Constructing Genetic Maps with MapMaker/EXP* 3.0, 3rd Edn. Cambridge, MA: Whitehead Institute.

Liu, G., Xu, S. B., Ni, Z. F., Xie, C. J., Qin, D. D., Li, J., et al. (2011). Molecular dissection of plant height QTLs using recombinant inbred lines from hybrids between common wheat (*Triticum aestivum* L.) and spelt wheat (*Triticum spelta L.*). *Chin. Sci. Bull*. 56, 1897-1903. doi: 10. 1007/s11434-011-4506-z

Liu, J. D., He, Z. H., Wu, L., Bai, B., Wen, W. E., Xie, C. J., et al. (2016). Genome-wide linkage mapping of QTL for black point reaction in bread wheat (*Triticum aestivum* L.). *Theor. Appl. Genet*. 129, 2179-2190. doi: 10. 1007/s00122-016-2766-3

Mccartney, C. A., Somers, D. J., Humphreys, D. G., Lukow, O., Ames, N., Noll, J. S., et al. (2005). Mapping quantitative trait loci controlling agronomic traits in the spring wheat cross RL4452× 'AC Domain'. *Genome* 48, 870-883. doi: 10. 1139/g05-055

McIntosh, R. A., Dubcovsky, J., Rogers, W. J., Morris, C., Appels, R., and Xia, X. C. (2015). *Catalogue of Gene Symbols for Wheat*: 2015-2016. *Supplement*. Available at: http://shigen. nig. ac. jp/wheat/komugi/genes/macgene/supplement2015. pdf

McIntosh, R. A., Hart, G. E., and Gale, M. D. (2013). *Catalogue of Gene Symbols for Wheat*: 2013—2014 *Supplement*. Available at: http://shigen. nig. ac. jp/wheat/komugi/genes/macgene/supplement2013-2014. pdf

Mir, R. R., Kumar, N., Jaiswal, V., Girdharwal, N., Prasad, M., Balyan, H. S., et al. (2012). Genetic dissection of grain weight in bread wheat through quantitative trait locus interval and association mapping. *Mol. Breed*. 29, 963-972. doi: 10. 1007/s11032-011-9693-4

Naruoka, Y., Garland-Campbell, K. A., and Carter, A. H. (2015). Genome-wide association mapping for stripe rust (Puccinia striiformis f. sp. *tritici*) in US Pacific Northwest winter wheat (*Triticum aestivum* L.). *Theor. Appl. Genet*. 128, 1083-1101. doi: 10. 1007/s00122-015-2492-2

Nyquist, W. E., and Baker, R. J. (1991). Estimation of heritability and prediction of selection response in plant populations. *Crit. Rev. Plant Sci*. 10, 235-322. doi: 10. 1080/07352689109382313

Pearce, S., Saville, R., Vaughan, S. P., Chandler, P. M., Wilhelm, E. P., Alkaff, N., et al. (2011). Molecular characterization of *Rht-1* dwarfing genes in hexaploid wheat. *Plant Physiol*. 157, 1820-1831. doi: 10. 1104/pp. 111. 183657

Peng, J., Richards, D. E., Hartley, N. M., Murphy, G. P., Devos, K. M., Flintham, J. E., et al. (1999). 'Green revolution' genes encode mutant gibberellin response modulators. *Nature* 400, 256-261. doi: 10. 1038/22307

Peng, J. H., Ronin, Y., and Fahima, T. (2003). Domestication quantitative trait loci in Triticum dicoccoides, the progenitor of wheat. *Proc. Natl. Acad. Sci. U. S. A.* 100, 2489-2494. doi: 10. 1073/pnas. 252763199

Peng, Z. S., Li, X., Yang, Z. J., and Liao, M. L. (2011). A new reduced height gene found in the tetraploid semi-dwarf wheat landrace Aiganfanmai. *Genet. Mol. Res*. 10, 2349-2357. doi: 10. 4238/2011

Quarrie, S. A., Steed, A., and Calestani, C. (2005). High-density genetic map of hexaploid wheat (*Triticum aestivum* L.) from the cross Chinese Spring x SQ1 and its use to compare QTLs for grain yield across a range of environments. *Theor. Appl. Genet*. 110, 865-880. doi: 10. 1007/s00122-004-1902-7

Rasheed, A., Wen, W. E., Gao, F. M., Zhai, S. N., Jin, H., Liu, J. D., et al. (2016). Development and validation of KASP assays for genes underpinning key economic traits in bread wheat. *Theor. Appl. Genet.* 129, 1843-1860. doi: 10.1007/s00122-016-2743-x

Rebetzke, G. J., and Richards, R. A. (2000). Gibberellic acid-sensitive dwarfing genes reduce plant height to increase kernel number and grain yield of wheat. *Crop Pasture Sci.* 51, 235-246. doi: 10.1071/AR99043

Röder, M. S., Korzun, V., Wendehake, K., Plaschke, J., Tixier, M. H., Leroy, P., et al. (1998). A microsatellite map of wheat. *Genetics* 149, 2007-2023. Rogers, J. (2014). "The IWGSC survey sequencing initiative," in *Proceedings of the International Plant and Animal Genome Conference XXII*, San Diego, CA.

Saghai-Maroof, M. A., Soliman, K. M., Jorgensen, R. A., and Allard, R. W. (1984). Ribosomal DNA spacer-length polymorphisms in barley: mendelian inheritance, chromosomal location, and population dynamics. *Proc. Natl. Acad. Sci. U. S. A.* 24, 8014-8018.

Sukumaran, S., Dreisigacker, S., Lopes, M., Chavez, P., and Reynolds, M. P. (2015). Genome-wide association study for grain yield and related traits in an elite spring wheat population grown in temperate irrigated environments. *Theor. Appl. Genet.* 128, 353-363. doi: 10.1007/s00122-014-2435-3

Thiel, T., Kota, R., Grosse, I., Stein, N., and Graner, A. (2004). SNP2CAPS: a SNP and indel analysis tool for CAPS marker development. *Nucl. Acids Res.* 32: e5. doi: 10.1093/nar/gnh006

Wang, S. C., Wong, D., Forrest, K., Allen, A. L., Chao, S., Huang, B. E., et al. (2014). Characterization of polyploid wheat genomic diversity using the high-density 90, 000 SNP array. *Plant Biotechnol. J.* 12, 787-796. doi: 10.1111/pbi.12183

Wei, T. M., Chang, X. P., Min, D. H., and Jing, R. L. (2010). Analysis of genetic diversity and tapping elite alleles for plant height in drought-tolerant wheat varieties. *Acta Agron. Sin.* 36, 895-904. doi: 10.3724/SP. J. 1006. 2010. 00895

Worland, A. J., Korzun, V., Röder, M. S., Ganal, M. W., and Law, C. N. (1998). Genetic analysis of the dwarfing gene *Rht8* in wheat. Part II. The distribution and adaptive significance of allelic variants at the *Rht8* locus of wheat as revealed by microsatellite screening. *Theor. Appl. Genet.* 96, 1110-1120. doi: 10.1007/ s001220050846

Wu, J., Kong, X., Wan, J., Liu, X., Zhang, X., Guo, X., et al. (2011). Dominant and pleiotropic effects of a *GAI* gene in wheat results from a lack of interaction between DELLA and GID1. *Plant Physiol.* 157, 2120-2130. doi: 10.1104/pp. 111. 185272

Wu, Q. H., Chen, Y. G., Fu, L., Zhou, S. H., Chen, J. J., Zhao, X. J., et al. (2016). QTL mapping of flag leaf traits in common wheat using an integrated high-density SSR and SNP genetic linkage map. Euphytica 208, 337-351. doi: 10.1007/s10681-015-1603-0

Würschum, T., Langer, S. M., and Longin, C. F. (2015). Genetic control of plant height in European winter wheat cultivars. Theor. Appl. Genet. 128, 865-874. doi: 10.1007/s00122-015-2476-2

Yang, T. Z., Zhang, X. K., Liu, H. W., and Wang, Z. H. (1993). Chromosomal arm location of a dominant dwarfing gene *Rht21* in common wheat variety—XN0004. *J. Northwest A F Univ.* 12, 13-17.

Zhai, S. N., He, Z. H., Wen, W. E., Jin, H., Liu, J. D., Zhang, Y., et al. (2016). Genome-wide linkage mapping of flour color-related traits and polyphenol oxidase activity in common wheat. *Theor. Appl. Genet.* 129, 377-394. doi: 10.1007/s00122-015-2634-6

Zhang, D. Q., Song, X. P., Feng, J., Ma, W. J., Wu, B. J., Zhang, C. L., et al. (2016). Detection of dwarf genes *Rht-B1b*, *Rht-D1b* and *Rht8* in Huang-Huai valley winter wheat areas and the influence on agronomic characteristics. *J. Triticeae Crops* 36, 975-981. doi: 10.7606/j. ssn. 1009-1041. 2016. 08. 01

Zhang, X. K., Yang, S. J., Zhou, Y., He, Z. H., and Xia, X. C. (2006). Distribution of the *Rht-B1b*, *Rht-D1b* and *Rht8* reduced height genes in autumn-sown Chinese wheats detected by molecular markers. *Euphytica* 152, 109-116. doi: 10.1007/ s10681-006-9184-6

Rht24b, an ancient variation of *TaGA2ox-A9*, reduces plant height without yield penalty in wheat

Xiuling Tian[1], Xianchun Xia[1], Dengan Xu[1], Yongqiang Liu[2], Li Xie[1], Muhammad Adeel Hassan[1], Jie Song[1], Faji Li[3], Desen Wang[1], Yong Zhang[1], Yuanfeng Hao[1], Genying Li[3], Chengcai Chu[2], Zhonghu He[1,4] and Shuanghe Cao[1]

[1] *Institute of Crop Sciences, National Wheat Improvement Center, Chinese Academy of Agricultural Sciences (CAAS), 12 Zhongguancun South Street, Beijing* 100081, *China*; [2] *State Key Laboratory of Plant Genomics, Institute of Genetics and Developmental Biology, The Innovative Academy for Seed Design, Chinese Academy of Sciences, 1 West Beichen Road, Beijing* 100101, *China*; [3] *Crop Research Institute, Shandong Academy of Agricultural Sciences, 202 Industry North Road, Jinan* 250100, *China*; [4] *International Maize and Wheat Improvement Center (CIMMYT) China Office, c/o CAAS, 12 Zhongguancun South Street, Beijing* 100081, *China*

Summary:

• *Rht-B1b* and *Rht-D1b*, the 'Green Revolution' (GR) genes, greatly improved yield potential of wheat under nitrogen fertilizer application, but reduced coleoptile length, seedling vigor and grain weight. Thus, mining alternative reduced plant height genes without adverse effects is urgently needed.

• We isolated the causal gene of *Rht24* through map-based cloning and characterized its function using transgenic, physiobiochemical and transcriptome assays. We confirmed genetic effects of the dwarfing allele *Rht24b* with an association analysis and also traced its origin and distribution.

• *Rht24* encodes a gibberellin (GA) 2-oxidase, TaGA2ox-A9. *Rht24b* conferred higher expression of *TaGA2ox-A9* in stems, leading to a reduction of bioactive GA in stems, but an elevation in leaves at the jointing stage. Strikingly, *Rht24b* reduced plant height, but had no yield penalty; it significantly increased nitrogen use efficiency, photosynthetic rate and the expression of related genes. Evolutionary analysis demonstrated that *Rht24b* first appeared in wild emmer and was detected in more than half of wild emmer and wheat accessions, suggesting that it underwent both natural and artificial selection.

• These findings uncover an important genetic resource for wheat breeding and also provide clues for dissecting the regulatory mechanisms underlying GA-mediated morphogenesis and yield formation.

Keywords: GA2-oxidase, grain yield, plant height, *Rht24*, wheat (*Triticum aestivum*)

Introduction

Wheat (*Triticum aestivum* L.) is one of the most important staple crops in the world. Reduced plant height can increase lodging resistance and consequently improve yield potential and stability, especially under conditions of high nitrogen fertilization (Evans, 1996). The 'Green Revolution' (GR) in wheat was initiated by the pervasive introduction of semi-dwarf

Published in New Phytologist, 2022, 233 (2): 738-750.

genes *Rht-B1b* and *Rht-D1b* that greatly improved capacity to achieve higher yield potential in favorable environments (Peng *et al.*, 1999; Silverstone & Sun, 2000). While reducing plant height, *Rht-B1b* and *Rht-D1b* also shorten coleoptile length and reduce seedling vigor, which is adverse to seedling emergence and establishment, especially in dryer environments (Rebetzke *et al.*, 1999, 2007). These *GR* genes also reduce grain weight and increase susceptibility to Fusarium head blight (FHB) (Li *et al.*, 2006; Lanning *et al.*, 2012; Herter *et al.*, 2018). Therefore, identification of new reduced height genes with little adverse effects could be beneficial for wheat breeding.

A huge effort has been made to identify genetic loci affecting plant height in recent decades (Gale & Youssefian, 1985; McIntosh *et al.*, 2017; Mo *et al.*, 2018). However, only a few reduced height genes have been cloned, mostly a variety of alleles at the orthologous *RHT-B1* and *RHT-D1* loci, such as *Rht-B1b* (syn. *Rht1*), *Rht-D1b* (*Rht2*), *Rht-B1c* (*Rht3*), *Rht-D1c* (*Rht10*), *Rht-B1e* (*Rht11*) and *Rht-B1p* (*Rht17*) (Peng *et al.*, 1999; Cao *et al.*, 2009; Pearce *et al.*, 2011; Li *et al.*, 2012; Bazhenov *et al.*, 2015). Genetic and transcriptome analyses showed that *GA2oxA14* was the downstream target of *RHT12* (Sun *et al.*, 2019), but later *GA2oxA14* was identified as the causal gene of *RHT12* through the MutChromSeq approach (Buss *et al.*, 2020). *GA2oxA9* was defined as the causal gene of *Rht18*, generated from artificial mutation, in durum wheat line Icaro by the same approach (Ford *et al.*, 2018).

Among several phytohormones regulating plant height, gibberellin (GA) was most documented and characterized. *GR* genes *Rht-B1b* and *Rht-D1b* encode DELLA proteins truncated in the N-terminal DELLA domain and can constitutively repress plant growth and development mediated by GA (Peng *et al.*, 1999; Van De Velde *et al.*, 2021). DELLA protein is an integrator of numerous development pathways and interacts with phytochrome interacting factors (PIFs), key regulators responsible for GA induction (Boccaccini *et al.*, 2014). The binding of DELLA and PIFs inhibits expression of GA- responsive genes (De Lucas *et al.*, 2008; Feng *et al.*, 2008). DELLA domain is responsible for the binding with GA receptor, GID1 and the GA-GID1-DELLA complex induces degradation of DELLA protein (Pearce *et al.*, 2011; Wu *et al.*, 2011). Disruption of the DELLA domain reduces DELLA protein binding with GID1 but not with PIFs, thus the truncated DELLA proteins encoded by *Rht-B1b* and *Rht-D1b* cannot be induced by GA for degradation but can still repress the promoting effects of GA on plant height (Peng *et al.*, 1999; Pearce *et al.*, 2011; Wu *et al.*, 2011).

We previously identified *RHT24* as a major locus controlling plant height on chromosome 6AL (Tian *et al.*, 2017). In contrast to plants carrying *GR* genes, *Rht-B1b* and *Rht-D1b*, those with the reduced height allele *Rht24b* were sensitive to exogenous GA and showed no negative effects in regard to coleoptile length and seedling vigor (Würschum *et al.*, 2017). Additionally, *Rht24b* has no adverse effects on Fusarium head blight infection (Herter *et al.*, 2018). Genotypic analyses using flanking markers indicated that *Rht24b* is widely present in wheat cultivars (Wü rschum *et al.*, 2017; Tian *et al.*, 2019), suggesting that *Rht24b* is a valuable allele in wheat breeding. Here *RHT24* was isolated through map-based cloning and the underlying regulatory mechanism governing plant height and yield was dissected according to com- prehensive investigation of GA isoforms, physiological measurement and transcriptome assays. Association analysis confirmed the genetic effects of *Rht24b* on plant height and yield using a natural population. The evolutionary origin and distribution of *Rht24b* were also investigated.

Materials and Methods

Plant materials and field trials

Two hundred and fifty-six recombinant inbred lines (RILs, $F_{2:6}$) were developed for genetic analysis using single seed descent from a cross between AK58 and JD8 and used for genetic analysis (Tian *et al.*, 2017). To develop secondary mapping populations for fine mapping of *Rht24*, we crossed JD8 (recurrent parent) with the RILs that have *Rht24b* allele from AK58 but similar background to JD8 based on single nucleotide polymorphism (SNP) chip-based genotyping. The population including 1245 BC_1F_3 plants was used for screening recombinant plants and grown at Gaoyi during the 2016-2017 cropping seasons. Each plot comprised three 2-m rows spaced 25 cm apart and 21 seeds were sown in each row. The derived population (BC_1F_4) of recombinant lines were used to measure plant height as a stable inheritance phenotype and planted with randomized complete blocks and three replications at Gaoyi and Beijing in the 2017- 2018 growing season. Each plot comprised three 1-m rows (10 seeds per row) spaced 25 cm apart.

A total of 284 diploid andtetraploid wheat accessions were employed to trace the origin of *Rht24b*, including 110 *T. urartu* accessions mainly from Turkey, Lebanon and Syria (kindly supplied by Dr Lingli Dong, Institute of Genetics and Developmental Biology, Chinese Academy of Sciences, Beijing, China), 50 *T. monococcum* accessions mostly from Turkey, Serbia and Euro- pean countries (kindly provided by Dr Dongcheng Liu, School of Chemistry and Biological Engineering, Beijing University of Science and Technology, China), 98 *T. dicoccoides* accessions mainly from Turkey, Lebanon and Syria (kindly provided by Dr Zhiyong Liu, Institute of Genetics and Developmental Biology, Chinese Academy of Sciences), 10 *T. dicoccum* accessions and 16 *T. durum* accessions from different countries (kindly provided by Dr Xinming Yang, Institute of Crop Sciences, Chinese Academy of Agricultural Sciences, Beijing, China) (Supporting Information Dataset S1).

Forty-two common wheat varieties were used to sequence the promoters and open reading frame (ORF) regions of *RHT24* for haplotype analysis (Dataset S2). One thousand common wheat varieties, including 486 Chinese elite varieties, 97 Chinese landraces, and 417 introduced varieties, were assembled to deter- mine the distribution of *Rht24b* (Datasets S3, S4).

Two sets of Chinese and introduced elite varieties were used to validate the genetic effects *of Rht24* allelic variants on plant height, grain yield (GY), and yield component traits (Datasets S5, S6). Set I included 137 elite cultivars from the Yellow andHuai Valley Winter Wheat Zone of China, which represent 60-70% of the total wheat production in China, and 17 accessions from other countries. These cultivars were planted at Anyang in Henan province and Suixi in Anhui province during the 2012- 2013 and 2013-2014 cropping seasons (F. *Li et al.*, 2018). Set II included 88 elite cultivars from the Northern China Winter Wheat Zone and were grown at Shijiazhuang in Hebei and Beijing, respectively, during the 2012-2013 and 2013-2014 cropping seasons (Tian *et al.*, 2017). All were grown in randomized complete blocks with three replications. Each plot consisted of four 2-m rows spaced 30 cm apart and c. 50 plants in each row. Field management was according to local practice.

Plant height was measured from the ground to spike (awns excluded) at grain fill stage. All plants in each plot were harvested at physiological maturity and GY as kg • ha^{-1} were measured when the moisture declined to c. 14%. Among yield components, spike number per m^2 (SN) was estimated by counting spikes at physio logical maturity in 1 m row section of each plot and expressed as m^{-2}; kernel number per spike (KNS) was measured by the mean kernel number of 20 randomly selected spikes in each plot at physiological maturity; and thousand kernel weight (TKW) was determined by the weight of 500 kernels from each plot after harvest (Tian *et al.*, 2017; F. Li *et al.*, 2018).

Fine mapping

RHT24 was initially mapped to a 1.85 cM interval flanked by markers *TaAP2* and *TaFAR* in the RIL population of JD8/AK58 (Tian *et al.*, 2017). To further narrow down the *Rht24* region, the secondary recombinant populations were generated and new markers were developed to map the *RHT24* region. To obtain the recombinant lines containing the *Rht24* region, JD8 was used as a recurrent parent to cross with the RIL harboring *Rht24b* allele and similar background to JD8 according to SNP chip based genotyping. The progenies were genotyped with *TaAP2* and *TaFAR* markers to screen recombinant lines. As mentioned earlier, 1245 BC_1F_3 plants were used for screening recombinant plants. Ten kinds of recombinant plants (eight heterozygous and two homozygous recombinant plants) were identified (Fig. 1b). The secondary mapping population (BC_1F_4) derived from the 10 kinds of recombinant plants includes *c*. 2000 plants. Plant heights were measured from the ground to spike (awns excluded) at the grain fill stage. Seven new polymorphic cleaved amplified polymorphic sequence (CAPS) markers were developed by a genome mining approach as described in Tian *et al*. (2017). The recombinant lines and their F_2 populations (secondary mapping populations) were genotyped by markers *TaCRY2*, *TaPAL*, *TaSR*, *TaRK*, *TaGA2ox-A9*, *TaRNT2* and *TaPL* (Dataset S7). The *RHT24* target region was then delimited to a smaller physical interval by genotyping and phenotyping the secondary mapping population.

Vector construction and genetic transformation

The full-length coding sequence (CDS) of *TaGA2ox-A9* (GenBank accession no. MH559347) from wheat variety AK58 (*Rht24b*) was first cloned into the entry vector pDONR207 and then recombined into the destination vector (pUbiGW) according to the handbook for Gateway cloning (12535-019; Invitro- gen, Carlsbad, CA, USA). The destination constructs were transformed into *Agrobacterium tumefaciens* strain EHA105 (BC303-1; Biomed) by the freezing-thawing method according to the product manual (BC303; Biomed, Beijing, China). Genetic transformation was performed by infecting immature embryos of wheat variety Fielder with *Agrobacterium tumefaciens* strain *EHA105* carrying the destination constructs. The detailed procedure for wheat transformation was reported in Ishida *et al*. (2015).

Nitrogen-15 (^{15}N) -ammonium and ^{15}N-nitrate accumulation assays

Nitrogen-15 (^{15}N) accumulation assays were performed as previously reported (Fang *et al*., 2019). Wheat seeds were surface sterilized with 2.5% (v/v) sodium hypochlorite (NaClO) for 15 min and germinated in sterile Petri dishes for 1 d. The seeds were immersed in sterile deionized water for germination until roots were 3-5 cm long (usually after 3 d). Then seedlings were transferred to Kimura B solution (0.37 mM $(NH_4)_2SO_4$, 0.18 mM KH_2PO_4, 0.18 mM KNO_3, 0.091 mM K_2SO_4, 0.37 mM $Ca(NO_3)_2$, 0.55 mM $MgSO_4 \cdot 7H_2O$, 1.6 mM $NaSiO_3 \cdot 9H_2O$, 40 μM Fe(II)-EDTA, μM H_3BO_3, 0.32 μM $CuSO_4 \cdot 5H_2O$, 0.76 μM $ZnSO_4 \cdot 7H_2O$, 9.14 μM $MnCl_2 \cdot 4H_2O$, 0.08 μM $(NH_4)_6Mo_7O_{24} \cdot 4H_2O$, pH 6.0) for hydroponic culture for 10 d, the solution was changed daily. For ^{15}N-ammonium and ^{15}N-nitrate labeling, seedlings were pretreated with the Kimura B solution for 2 h and then transferred to modified Kimura B solution containing 1 mM ^{15}N-NH_4Cl or 5 mM ^{15}N-KNO_3 for 3 or 20 h, after which the roots and shoots were collected and dried to constant weight at 70°C. Finally, the samples were ground to fine powder and analyzed by an isotope ratio mass spectrometer (Finnigan Delta Plus XP, Thermo Fisher Scientific, Shanghai, China) with an elemental analyzer (Flash EA 1112; Thermo Fisher Scientific).

Measurement of photosynthetic rate and chlorophyll content

A portable infrared gas analyzer system LI-6400XT (Li-Cor, Lincoln, NE, USA) was used to measure the maximum leaf photo- synthetic CO_2 uptake rate of flag leaves at the grain fill stage. Photosynthetic photon flux density (PPFD) was set at 1000 mol photo m^{-2} s^{-1} during measurement. Ambient temperature and CO_2 concentration during the measurement were 25-28° C and 407 ppm, respectively. Chlorophyll content was measured by a SPAD-502 chlorophyll meter (Minolta, Tokyo, Japan).

TaGA2ox-A9 promoter activity assays in *Nicotiana benthamiana*

The *TaGA2ox-A9* promoter (1.4 kb 5'upstream) from AK58 (GenBank accession no. MH559347) and JD8 (MH559346) were cloned into the pGreenII 0800-LUC vector following the manufacturer's recommendations (VT8124; Youbio Biological Technology, Co. Ltd, Beijing, China). The constructs were separately transferred into *Agrobacterium tumefaciens* strain *GV3101* (BC304-01; Biomed) and then infiltrated into *N. benthamiana* leaves. Luciferase activities in leaves were measured 50 h after infiltration by dual-luciferase reporter assay (E1910; Promega, Madison, WI, USA). Signal strength was determined by a SynergyH1 microplate reader (BioTek, Winooski, VT, USA).

RNA-sequencing, quantitative polymerase chain reaction, and quantitative assays of gibberellin isoforms

Three biological replicates of stem internodes and leaves at the jointing stage and grain fill stage were collected from *TaGA2ox-A9* overexpression lines (*TaGA2ox-A9*-OE) and transgenic null lines (TNLs) in Fielder background. RNA was extracted and sequenced by Novogene (Beijing, China). Briefly, 10 Gb of transcriptomic data for each sample was obtained from the Illumina HiSeq 4000 platform. Differentially expressed genes (DEGs) were defined with $|\text{fold change}| > 2$ and $P < 0.05$. A P-value (P_{adj}) < 0.05 was used as the threshold of significant enrichment for KEGG (Kyoto Encyclopedia of Genes and Genomes) pathway analysis. Gene annotation refers to Triticum _ aestivum. IWGSC. 44 (ftp: //ftp. ensemblgenomes. org/pub/release-44/ plants/gtf/triticum _ aestivum).

Fresh stems at the jointing stage were also harvested for quantitative polymerase chain reaction (qPCR) assays. Samples that had been quickly frozen by liquid nitrogen were used to extract total RNA with an EasyPure Plant RNA Kit (ER301; Transgene, Beijing, China). Complementary DNA (cDNA) was synthesized from total RNA using a PrimeScript RT Reagent Kit plus gDNA Eraser (RR047A; Takara, Shiga, Japan) and quantified by qPCR using iTaq Universal SYBR Green Supermix (1725124; Bio-Rad, Hercules, CA, USA) in a Bio-Rad CFX system following the manufacturer's recommendations. Primer pairs used in this study are listed in Dataset S8. Each sample was represented by three biological replicates. The *Actin* gene(GenBank accession no. AB181991) was used as an internal control to calibrate expression levels of genes of interest. Relative quantification gene expression was calculated by $2^{-\Delta\Delta C_t}$ method (Schmittgen & Livak, 2008).

Gibberellin contentsin *TaGA2ox-A9*-OE, TNL, Tall NIL, and Short NIL in stem internodes and leaves at the jointing and grain fill stages were determined as described by Ma *et al.* (2015). Three biological replicates were used for each line.

Measurement of cell size

Fourth stem internodes of *TaGA2ox-A9*-OE and TNL at the jointing stage were fixed in FAA solution (70% alcohol, 5% acetic acid, and 0.02% formaldehyde), embedded in paraffin, longitudinally sectioned, and stained with Safranin O Fast Green. Lengths and widths of *c*. 15 longitudinal stem cells were analyzed in each of three sections from one sample by IMAGEJ (https: //imagej. nih. gov/ij/).

Statistical analyses

Differences of genetic effects of two genotypes on plant height, KNS, TKW, SN, and GY in 242 elite cultivars, and between *TaGA2ox-A9*-OE and TNL were analyzed by *t*-tests (Duncan's method) in SAS 9.4 based on $P < 0.05$ for declaration of significance (https: //www. sas. com). Statistical analyses of GA isoforms and luciferase activity were performed using GraphPad PRISM 8 (https: //www. graphpad. com/).

Results

Map-based cloning of *RHT24*

In the previous study, *Rht24b* was mapped to a 1.85 cM chromosome interval flanked by markers *TaAP2* and *TaFAR* on chromosome 6AL in a RIL population of cross Jingdong 8 (JD8) /Aikang 58 (AK58) (Fig. 1a) (Tian *et al.*, 2017). To further narrow down the *RHT24* region, recombinant plants were identified using *TaAP2* and *TaFAR*, and seven gene-specific CAPS markers within the target interval were developed based on the IWGSC R_{EF} S_{EQ} v. 1.0 wheat genome sequence (IWGSC, International Wheat Genome Sequencing Consortium), 2018; http://www.wheatgenome.org/News/Latest-news/Annotation-RefSeq-v1.0- URGI)(Fig. 1b; Dataset S7). Genotypic and phenotypic anal yses of secondary populations derived from the earlier-mentioned recombinant plants delimited *RHT24* in an interval between CAPS markers *TaPL* and *TaSR*, spanning a *c*. 2.95 Mb physical region which contains six annotated genesaccording to R_{EF}Seq v. 1.0 (Fig. 1c; Dataset S9). Polymorphic and DEGs between short and tall plants in the secondary populations were analyzed through a bulked segregant RNA- sequencing(BSR-Seq) strategy to identify candidate genes. Two genes (*TraesCS6A02G221600* and *TraesCS6A02G222000*) each had a SNP in thecoding sequence, but neither difference caused an amino-acid change between parents (Dataset S9). Only one of two SNPs in *TraesCS6A02G221900* resulted in an amino-acid change between tall and short bulks (Fig. S1A; Dataset S9). Additionally, only *TraesCS6A02G221900* had different expression between two bulks in BSR-Seq assays (Dataset S10). qPCR assays confirmed that *TraesCS6A02G221900* was differentially expressed in stems between AK58 and JD8 (Fig. S1B; Dataset S8). *TraesCS6A02G221900* encodes a GA 2-oxidase, designated as *TaGA2ox-A9*. Previous studies showed that GA2-oxidases regulated plant height by inactivating endogenous bioactive GA isoforms (Lo *et al.*, 2008, 2017; Han & Zhu, 2011; Ford *et al.*, 2018; Liu *et al.*, 2018). We assumed that *TaGA2ox-A9* was the candidate gene for *RHT24*.

To validate the effect of *TaGA2ox-A9* on plant height, its full- length CDS from the dwarf parent AK58 was isolated and intro- duced into spring wheat variety Fielder by *Agrobacterium tumefaciens-mediated* transformation (Dataset S8). Thirty-one independent positive transgenic lines were generated, and three of them with different expression levels of *TaGA2ox-A9* were chosen for subsequent analyses (Fig. S2A-C). Phenotypic investigation demonstrated that *TaGA2ox-A9* overexpression lines (*TaGA2ox-A9*-OE) had significantly reduced plant height com- pared to TNL (Figs 1d, S2A, B; Dataset S11). We also generated short and tall near-isogenic lines (NILs) with *Rht24b* and *Rht24a*, designated as Short NIL and Tall NIL, respectively, and observed that Short NIL significantly reduced plant height compared to Tall NIL (Figs 1e, S2D, E). We further investigated dynamic effects of *RHT24* on plant height at different developmental phases and found that *RHT24* had the greatest effect on plant height at the jointing stage, the key spatio-temporal window of stem elongation (Fig. S3). These results suggested that *TaGA2ox-A9* was the causal gene of *Rht24b*.

Functions of *Rht24b* in gibberellin-mediated control of plant height

Considering that *Rht24* encoded a GA metabolic enzyme, *TaGA2ox-A9*, we measured GA isoforms in the internodes at the jointing stage. The result showed that quite a few GA isoforms had significantly different abundances between *TaGA2ox-A9*-OE and TNL. The levels of bioactive GA isoforms GA_1, GA_4 and GA_7 in *TaGA2ox-A9*-OE were significantly lower than those in TNL (Figs 1f, S4). Additionally, enhanced expression of *TaGA2ox-A9* led to reduced cell size (Fig. 1h). We also compared the contents of GA isoforms between Short and Tall NIL at the jointing stage and found that Short NIL had lower GA_4 but higher GA_1 than Tall NIL (Fig. 1g). Considering GA_4 is amajor bioactive GA isoform in the vegetative phase (Zhang *et al.*, 2007; Sun *et al.*, 2019), this confirmed

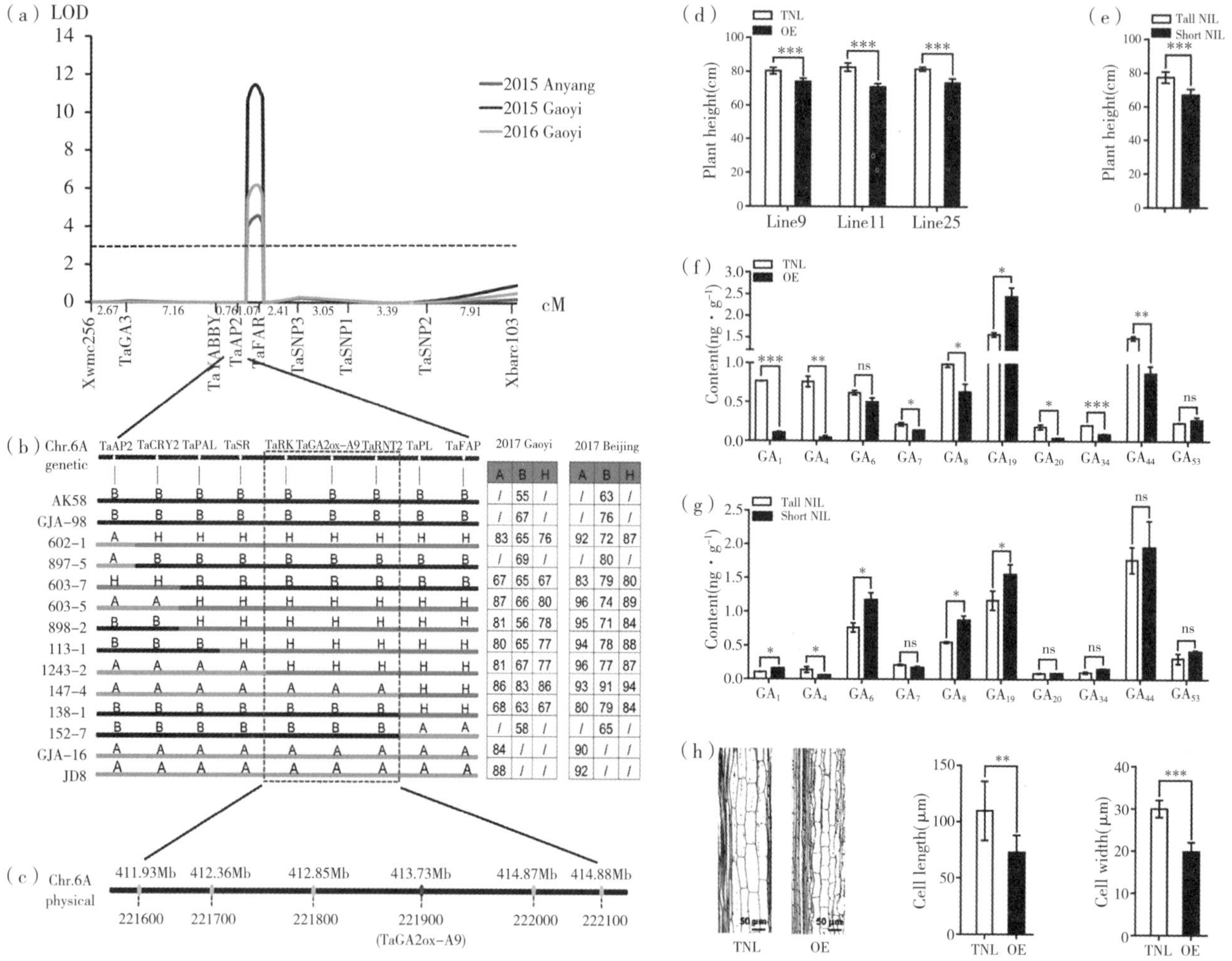

Fig. 1 Map-based cloning and functional validation of *RHT24* in wheat. (a) Linkage map of the *RHT24* locus for plant height in the JD8/AK58 recombinant inbred line (RIL) population. Horizontal dotted line represents a limit of detection (LOD) threshold value of 3.0 (Tian *et al.*, 2017). (b) Fine mapping of the *RHT24* locus based on critical crossovers. Left panel shows the genotyping of recombinant lines. Panels (a), (b), and (h) represent the JD8 allele, AK58 allele and heterozygosity, respectively, at the given marker locus. Right panel displays the genotype and corresponding plant height of secondary mapping populations derived from different recombinant lines. (c) Annotated genes in the target physical region of *RHT24*. *TraesCS6A02G221600*, *TraesCS6A02G221700*, *TraesCS6A02G221800*, *TraesCS6A02G221900*, *TraesCS6A02G222000*, and *TraesCS6A02G222100* are represented by 221 600, 221 700, 221 800, 221 900, 222 000, and 222 100, respectively. The candidate gene *TaGA2ox-A9* (*TraesCS221900*) for *RHT24* is highlighted in red. (d) Plant heights of *TaGA2ox-A9* overexpression positive lines (OEs) and their transgenic null lines (TNLs). (e) Plant heights of Short NIL and Tall NIL. Note: short and tall near-isogenic lines (NILs) with *Rht24b* and *Rht24a* were designated as Short NIL and Tall NIL, respectively. (f) Gibberellin (GA) isoform contents in stem internodes of TNL and OE transgenic plants at the jointing stage. Bioactive GA isoforms are shown in red. (g) GA isoforms in the stem internodes of Short NIL and Tall NIL plants at the jointing stage. Bioactive GA isoforms are shown in red. (h) Longitudinal sections (left panel), and measures of cell length (middle panel) and cell width (right panel) of stems from TNL and OE plants at the jointing stage. Note: error bars indicate SD among three biological replicates. *, $P < 0.05$; **, $P < 0.01$; ***, $P < 0.001$; ns, not significant.

that *TaGA2ox-A9* reduced plant height through decreasing bioactive GA levels. We further quantified the GA isoforms in stems at the grain fill stage, post determination of final plant height. Strikingly,

TaGA2ox-A9-OE still had lower levels of bioactive GA isoforms than TNL, suggesting that *TaGA2ox-A9* can also lower bioactive GA levels at the grain fill stage although its function in controlling plant height is completed at this stage (Fig. S5A). All bioactive GA isoforms had quite low abundance in the stem internodes of Short NIL and Tall NIL at the grain fill stage, and Short NIL showed little change in GA_4 but higher GA_1 and lower GA_7 compared with Tall NIL (Fig. S5B). The investigation of the native *TaGA2ox-A9* expression pattern in stems showed that it was little expressed in stem internodes at the grain fill stage (Fig. S2F). Thus, *TaGA2ox-A9* regulates plant height in a developmental stage-dependent way. Collectively, *Rht24b* reduces plant height by modulating bioactive GA levels and concomitant control of cell size.

To dissect the molecular mechanism of *Rht24b* in regulating plant height, we investigated DEGs in the stem internodes between *TaGA2ox-A9*-OE and TNL at the jointing stage. RNA-Seq assays identified 14 381 DEGs between *TaGA2ox-A9*-OE and TNL, including 6568 upregulated and 7813 downregulated DEGs involved in considerable number of pathways identified by KEGG. Quite a few GA metabolism-related genes, including *GA 2-oxidase*, *GA3-oxidase* and *GA 20-oxidase* family members, are DEGs according to KEGG analyses (Fig. S4A; Dataset S12). We also observed that a few genes putatively involving GA signal transduction, such as *PIF*-like genes, were differentially expressed (Fig. S4B; Dataset S12).

Effects of *Rht24b* on grain yield

To reveal the effects of *Rht24b* on GY, we compared agronomic traits between the transgenic groups (*TaGA2ox-A9*-OE vs TNL) and between NIL groups (short vs tall). Phenotypic analyses showed that there was no significant difference in TKW, KNS, SN and GY between *TaGA2ox-A9*-OE and TNL (Fig. 2a-d; Dataset S11). Likewise, Short NIL and Tall NIL showed no difference in TKW, KNS, SN, and GY (Fig. 2e-h). Thus, *Rht24b* reduces plant height without negative effects on yield.

To gain more insight into the physiological basis underlying *Rht24b* function in plant height reduction without yield penalty, we first investigated the uptake of nitrogen, one of the most vital macronutrients to plant growth. ^{15}N-nitrate and ^{15}N- ammonium feeding experiments were conducted on *TaGA2ox- A9*-OE, TNL, Short NIL, and Tall NIL. Nitrogen isotope analyses showed that uptake and assimilation of nitrate (NO_3^-), the major nitrogen form, was significantly increased in the roots and shoots of the *TaGA2ox-A9*-OE (Figs 3a, S6A), and the roots of the Short NIL (Figs 3g, S6B). We also investigated the expression of the genes putatively related to nitrogen use efficiency (NUE) in both roots and shoots of seedlings between contrasting genotypes using the RNA-Seq assays. Four high-affinity nitrate transporter-activating genes were upregulated by *TaGA2ox-A9* in roots, while *TaGA2ox-A9* had no significant effect on their expression in shoots (Fig. S6C, D; Dataset S13). By contrast, most of DEGs related to nitrate metabolism in shoots were upregulated by *TaGA2ox-A9* (Fig. S6D; Dataset S13). These data indicated that *TaGA2ox-A9* improves NUE by increasing nitrate uptake in roots and nitrogen metabolism in shoots. We further investigated the DEGs related to nitrogen metabolism in flag leaves at the grain fill stage and found that 11 of 15 DEGs were upregulated in *TaGA2ox-A9*-OE compared to the TNL (Fig. S7A; Dataset S12). It is well known that enhanced photosynthetic capacity and efficiency improves yield potential by increasing carbon assimilation (Zhu *et al.*, 2010; Parry *et al.*, 2011; Long *et al.*, 2015). We measured leaf width, chlorophyll content, and light-saturated photosynthetic rate, and found significantly higher values in *TaGA2ox-A9*-OE than in TNL (Fig. 3b-e). Likewise, leaf width, chlorophyll content, and light- saturated photosynthetic rate were higher in Short NIL than these in Tall NIL (Fig. 3h-k). Transcriptome analysis showed that 22 of 24 photosynthesis-related genes were upregulated in *TaGA2ox-A9*-OE compared to the counterparts in TNL (Fig. S7B; Dataset S12). Overexpression of *TaGA2ox-A9* down-regulated eight *GA 2-oxidases* and upregulated six *GA 20-oxidases* in leaves

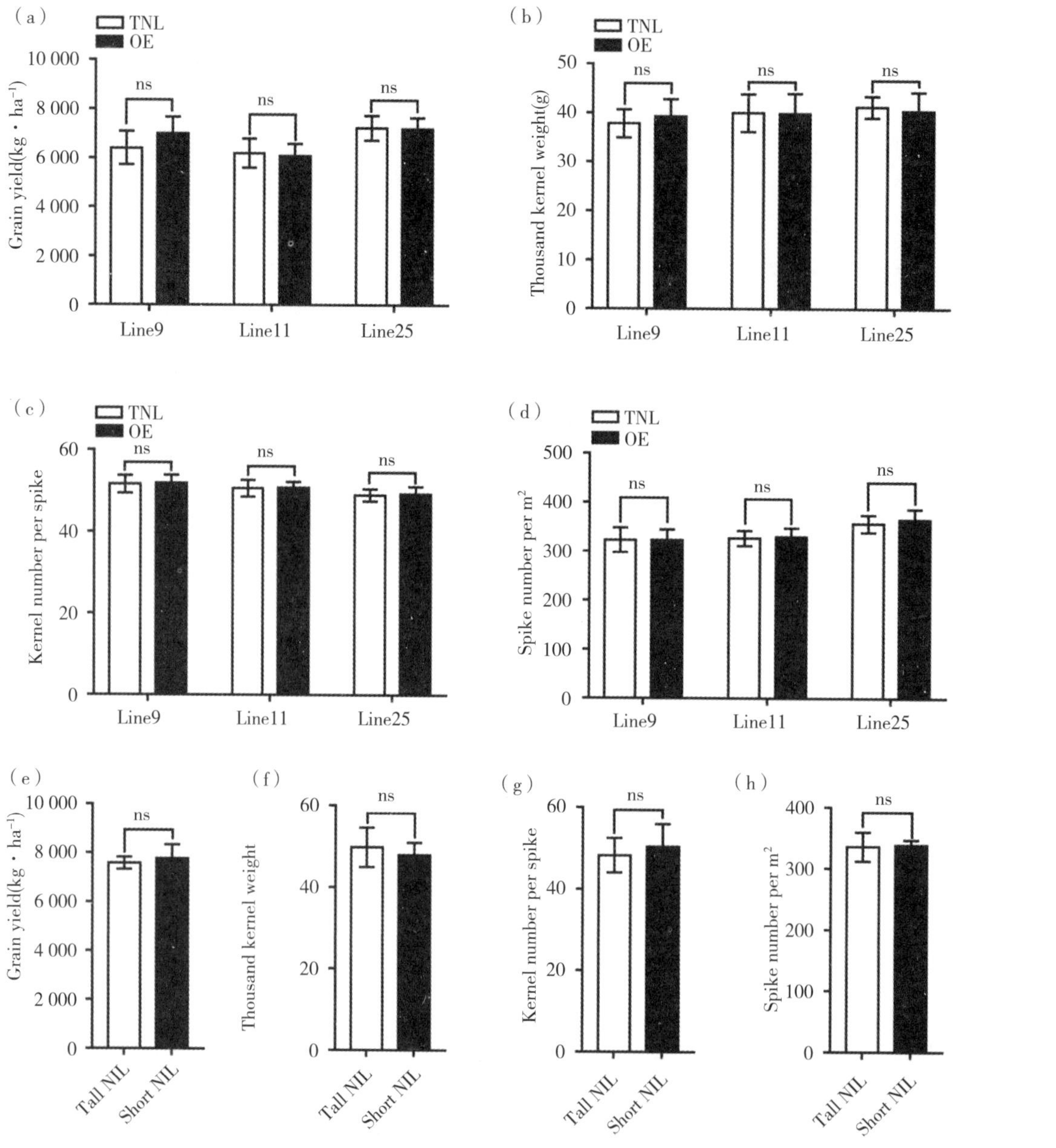

Fig. 2 Effects of *Rht24* on grain yield and yield component traits in wheat. Grain yield (a), thousand kernel weight (b), kernel number per spike (c) and spike number per m^2 (d) in TNL (transgenic null line) and *TaGA2ox-A9*-OE plants. Note: lines 9, 11, and 25 were derived from three independent transgenic plants. Grain yield (e), thousand kernel weight (f), kernel number per spike (g) and spike number per m^2 (h) in NIL (near- isogenic line) groups. Note: error bars indicate SD among three biological replicates. ns, not significant.

(Fig. S7C; Dataset S12). By contrast, both differentially expressed *GA2-oxidases* and *GA20-oxidases* were downregulated in the stems of *TaGA2ox-A9*-OE (Dataset S12). Considering that photosynthesis occurs mainly in leaves and provides the carbohydrates and energy for grain formation, we investigated GA levels in leaves at the jointing and grain fill stages. *TaGA2ox-A9*-OE had significantly higher GA_7 than TNL, whereas GA_1 and GA_4 contents were very low and not significantly different between *TaGA2ox-A9*-OE and TNL at the jointing stage (Fig. 3f). Moreover, all bioactive GA isoforms were significantly higher in the leaves of Short NIL than in Tall NIL at the jointing stage (Fig. 3l). GA_1, GA_3 and GA_4 at the grain fill stage had very low levels in flag leaves, whereas GA_7 contents showed no difference between *TaGA2ox-A9*-OE and TNL or between Short NIL and Tall NIL (Fig. S5C, D). Taken together, high NUE, chlorophyll content, and light-saturated photosynthetic rate combined with GA homeostasis in enhanced expression lines of *TaGA2ox-A9* addressed why *Rht24b* could reduce plant height without negative effects on yield.

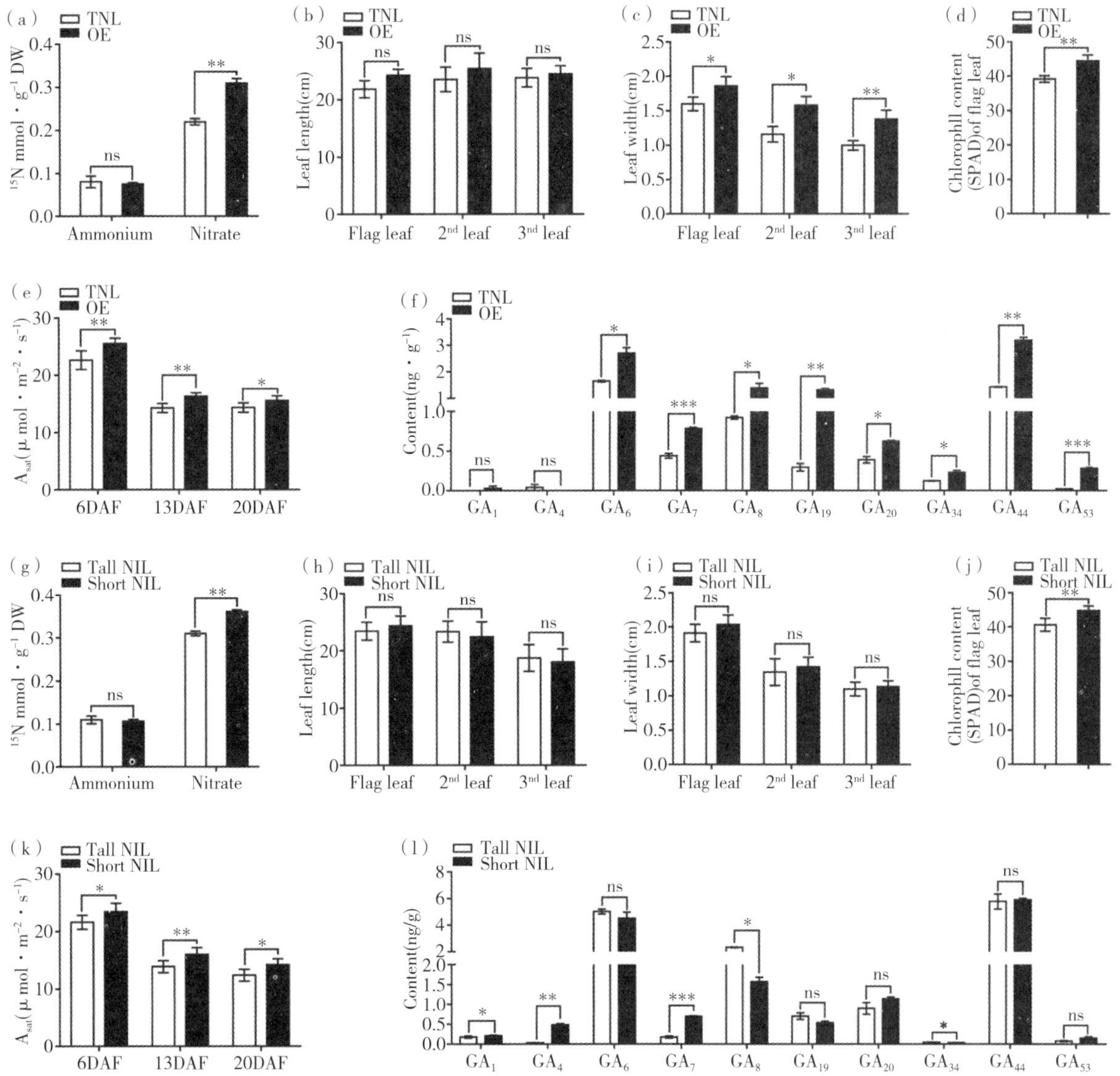

Fig. 3 Effects of *Rht24b* on nitrogen use efficiency, gibberellin (GA) content, and photosynthesis-related phenotypes in wheat. (a-f) Comparisons of TNL (transgenic null line) and *TaGA2ox-A9*-OE plants for nitrogen accumulation for 3 h in the roots at the seedling stage (a); length of upper three leaves (b); width of upper three leaves (c); chlorophyll content in flag leaves (d); light-saturated photosynthetic rate in flag leaves (e); and GA isoforms in flag leaves at the jointing stage (f). Bioactive GA isoforms are shown in red. (g^{-1}) Comparisons of Short NIL (near-isogenic line) and Tall NIL for nitrogen accumulation for 3 h in the roots at the seedling stage. Note: short and tall NILs with *Rht24b* and *Rht24a* were designated as Short NIL and Tall NIL, respectively (g); length of upper three leaves (h); width of upper three leaves (i); chlorophyll content of flag leaves (j); light-saturated photosynthetic rate of flag leaves (k); and GA isoforms of leaves at the jointing stage (l). Note: nitrate, nitrogen isotope labeled nitrate; ammonium, nitrogen isotope labeled ammonium. DAF, days after flowering. Error bars indicate SD among three biological replicates. *, $P < 0.05$; **, $P < 0.01$; ***, $P < 0.001$; ns, not significant.

Identification of the functional site of *RHT24*

We compared the full-length ORF of *TaGA2ox-A9* and its 1.4 kb upstream region in AK58 and JD8 to identify the functional site of *RHT24*. There were two SNPs in the ORF and one InDel, and one SNP in the promoter region

(Fig. 4a). The SNP in the first exon was a syno-nymous mutation, whereas the one in the third exon caused a non-synonymous JD8-proline/AK58-serine change (Fig. 4a, b). The haplotypes from JD8 and AK58 were confirmed as alleles *Rht24a* (tall) and *Rht24b* (dwarf), respectively (Fig. 4a). *TaGA2ox-A9* belongs to a gene family with two conserved domains, DIOX _ N and 2OG-FeII _ Oxy (Han & Zhu, 2011). The SNP causing the non-synonymous mutation was not within the conserved domains (Fig. 4b). Protein structure prediction also showed that the amino-acid transformation (proline to serine) was outside the key bioactive domains and did not change the three-dimensional structure of *TaGA2ox-A9* (Fig. 4c)(https: //swissmodel. expasy. org/interactive). Thus, the two SNPs in the ORF probably had no effect on the function of *TaGA2ox- A9*. The SNP at -841 bp did not cause *cis*-acting motif change according to promoter analysis by PlantCARE and PLACE. By contrast, the InDel (JD8-C/AK58-GT) at -182 bp of the pro- moter was close to the transcriptional start site (TSS) and generated two transcription-activated motifs (DOFCOREZM, S000265; CACTFTPPCA1, S000449) in the semi-dwarf parent AK58 (Fig. 4a) (Yanagisawa & Schmidt, 1999; Yanagisawa, 2000; Gowik *et al.*, 2004). Transgenic assays in *N. benthamiana* showed that the AK58 promoter of *TaGA2ox-A9* drove higher expression activity than its JD8 counterpart, in agreement with the differential transcription pattern of *TaGA2ox-A9* in the parents (Fig. 4d). Transcriptional analyses also showed that the varieties with *Rht24b* had obviously higher expression of *TaGA2ox-A9* than those with *Rht24a* (Fig. S8). We concluded that the InDel in the promoter of *TaGA2ox-A9* is a key functional poly-morphism conferring differential expression and concomitant change in plant height.

Evolutionary origin of *Rht24b*

Polyploidization events were responsible for the origin of wheat. It is well known that *T. urartu* (AA) was the progenitor of the A sub-genome of common wheat (AABBDD) (Dvorak *et al.*, 1993; Thomas *et al.*, 2014; Pont *et al.*, 2019). Allotetraploid wild emmer (*T. turgidum* ssp. *dicoccoides*, AABB) was generated by spontaneous chromosome doubling of a hybrid of *T. urartu* (AA) and an unknown progenitor species carrying the B sub- genome (possibly *Aegilops speltoides* or unknown extinct or extant *Sitopsis* species, https://www.sciencedirect.com/topics/agricultural-and-biological-sciences/aegilops-speltoides). Domesticated emmer (*T. turgidum* ssp. *dicoccum*), selected from wild emmer (*T. turgidum* ssp. *dicoccoides*), was the progenitor of modern durum. A second polyploidization event giving rise to common wheat involved domesticated tetraploid wheat and the D subgenome donor species *Ae. tauschii* (Pont *et al.*, 2019). *RHT24* was mapped on chromosome 6AL. To trace the evolutionary origin of the *Rht24b* allele, we assembled 284 accessions, comprising (Ⅰ) 110 *T. urartu* accessions mainly from Turkey, Lebanon, and Syria in the southwest 'Fertile Crescent', (Ⅱ) 50 *T. monococcum* accessions mostly from Turkey, Serbia, and European countries, (Ⅲ) 98 wild emmer accessions mainly from Turkey, Lebanon, and Syria, (Ⅳ) 10 cultivated emmer accessions from various countries, and (Ⅴ) 16 *T. durum* accessions from various geographic regions (Dataset S1). Genotyping-by-sequencing showed that all *T. urartu* accessions harbored the *Rht24a* allele (Dataset S1). A functional marker for *Rht24b* was developed to differentiate *Rht24a* and *Rht24b* according to the functional site identified earlier (Fig. 4a; Dataset S7). *Rht24b* was also not detected in *T. monococcum* accessions (Dataset S1). By contrast, wild emmer (54/98), cultivated emmer (3/10) and durum (13/16) accessions contained *Rht24b* (Dataset S1). Thus, the reduced height allele *Rht24b* originated from *Rht24a* and first arose in wild emmer. A time- line for the origin and spread of *Rht24b* is shown in Fig. 5 (a) based on evolutionary history of wheat (Thomas *et al.*, 2014; Pont *et al.*, 2019). Remarkably, *Rht24b* was present in more than half of the wild emmer accessions, suggesting that it has undergone natural selection (Dataset S1).

Allelic variation, genetic effect, and distribution of *RHT24* alleles

To identify more allelic variants of *RHT24* for wheat breeding, the 1.4-kb 5' upstream region and ORF of 42 elite wheat accessions were sequenced (Dataset S2). Only four polymorphic sites were detected, forming two haplotypes corresponding to *Rht24a* and

Rht24b (Fig. 4A). To further evaluate the genetic effects of the allelic variants, we performed an association analysis for plant height and yield component traits using a natural population including 242 elite wheat varieties (Datasets S5, S6). As identified in the JD8/AK58 RIL population, *Rht24b* significantly reduced plant height (Dataset S14). For yield component traits, *Rht24b* significantly decreased SN in Set I (137 elite varieties from the Yellow and Huai Valley Winter Wheat Zone of China and 17 accessions from other countries) but not in Set II varieties (88 elite varieties from the Northern China Winter Wheat Zone), whereas it increased KNS and TKW, albeit insignificantly, compared to *Rht24a* (Dataset S14). We also investigated the GY (in kg • ha^{-1}) and no significant difference in the GY was observed between the cultivars with *Rht24b* and those with *Rht24a* (Dataset S14).

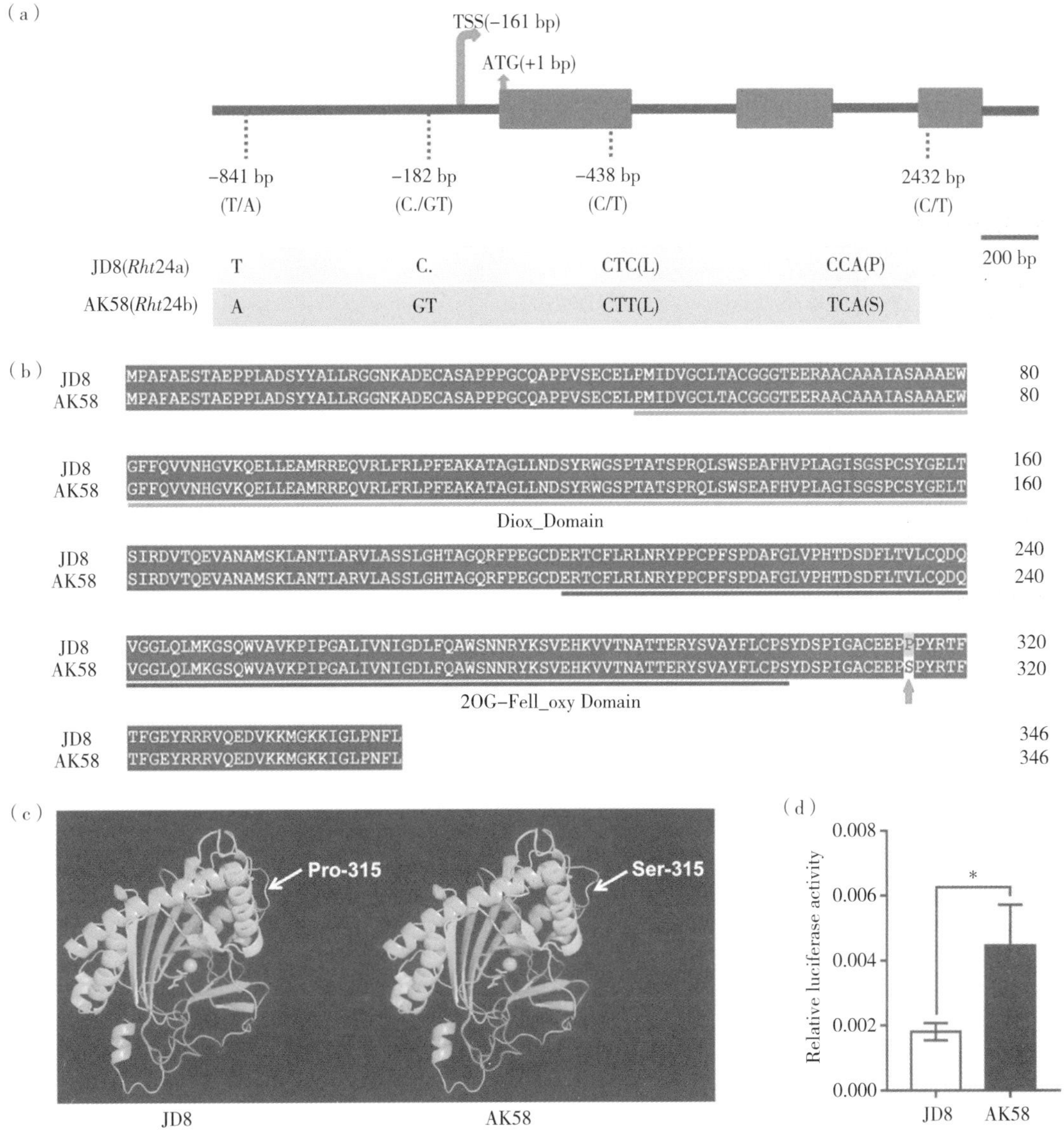

Fig. 4 Alignment and functional domains of *TaGA2ox-A9* amino-acid sequences from JD8 and AK58. Gray and black lines show the DIOX _ N and 2OG-FeII _ Oxy domains, respectively. The nonsynonymous change (P- S) in *TaGA2ox-A9* is labeled with an arrow. Predicted three-dimensional structures of *TaGA2ox-A9* from JD8 and AK58. Cyan and pink corkscrew spins display DIOX _ N and 2OG-FeII _ Oxy domains, respectively. Amino acid residues polymorphic between JD8 and AK58 are indicated by arrows. (d) Comparison of luciferase activities driven by *TaGA2ox-A9* promoters from JD8 and AK58. Note: error bars indicate SD among three biological replicates. *, $P < 0.05$.

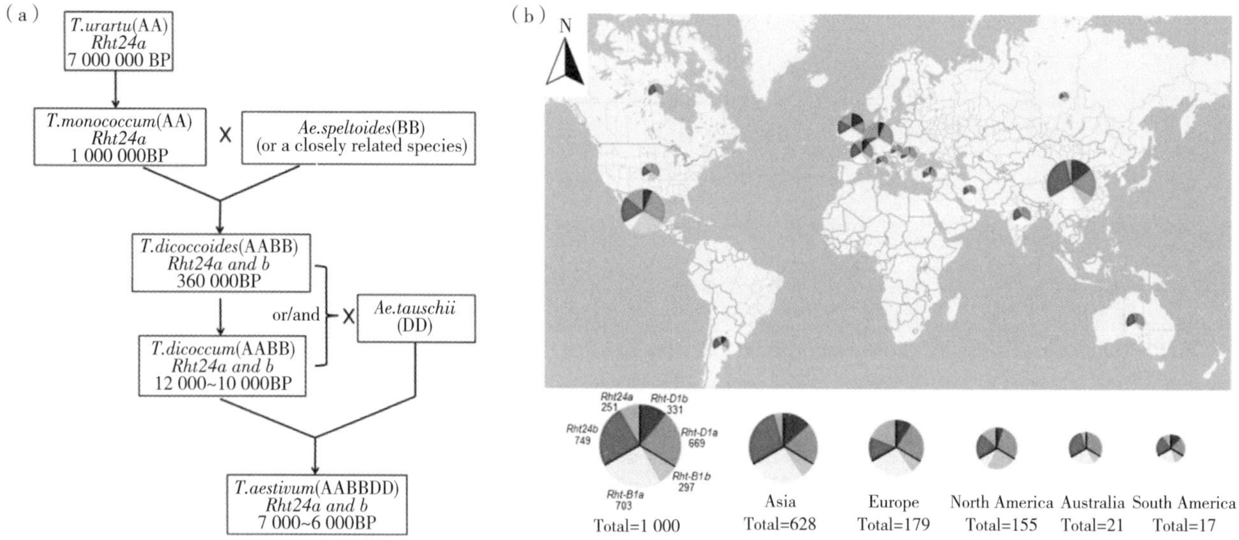

Fig. 5 Evolutionary history and distribution of alleles *Rht24a* and *Rht24b* in wheat and its relatives. (a) Evolutionary history of *Rht24*. BP, before present. The timeline for the origin and spread of *Rht24b* is calculated according to the origin time of the wheat progenitors (Thomas *et al.*, 2014; Pont *et al.*, 2019). (b) Worldwide distributions of major alleles of *RHT-B1*, *RHT-D1*, and *RHT24* in wheat.

To determine the distribution of *Rht24b* compared with *GR* genes *Rht-B1b* and *Rht-D1b*, we examined 1000 wheat accessions from various regions worldwide and identified 297 wheat varieties with *Rht-B1b*, 331 with *Rht-D1b*, and 749 with *Rht24b* (Fig. 5b; Datasets S3, S4). More than half of domesticated cultivars harbored *Rht24b*, suggesting that it had been subjected to positive selection in wheat breeding (Datasets S1, S4).

Discussion

Rht24b is an important reduced height gene for wheat breeding

The GR genes *Rht-B1b* and *Rht-D1b* greatly improved yield stability and potential due to increased lodging resistance and harvest index, especially under nitrogen fertilization, but they also led to shorter coleoptile length and reduced seedling vigor (Rebetzke *et al.*, 1999, 2007). *Rht24b* was identified as a major quantitative trait locus (QTL) for plant height without adverse effects on coleoptile length and seedling vigor (Würschum *et al.*, 2017). Here we identified *TaGA2ox-A9* as the causal gene of *RHT24*. The dwarfing allele *Rht24b* causes higher expression of *TaGA2ox-A9* and reduces plant height by lowering bioactive GA isoforms. Most importantly, *Rht24b* has no negative effect on GY. By contrast, the GR alleles *Rht-B1b* and *Rht-D1b* confer decreased grain weight, which is adverse to achievement of GY (Li *et al.*, 2006; Lanning *et al.*, 2012). *Rht24b* also promoted NUE and photosynthetic capacity, which is favorable for sustainable agriculture (Fig. 3a, e). Comprehensive genotyping analysis showed that more than half of wild emmer accessions (54/98) and 749 of 1000 wheat varieties examined in this study contain *Rht24b* (Datasets S1, S4). These results thus suggested that *Rht24b* had been subject to both natural and artificial selection. 104 of 486 elite Chinese wheat varieties harbor *Rht24b* but not *Rht-B1b* and *Rht-D1b*, and some of them, such as Jing 411, Jinnan 2, Taishan 1, Fan 6, Fan 7, Yannong 15, and Chuanmai 107, are major varieties in China (Dataset S15). The semi-dwarf high yielding wheat variety Zhongmai 175 containing *Rht24b* but not the GR genes, has been widely planted on a cumulative area of more than three million hectares. Overall, *Rht24b* is an important genetic resource with great potential for wheat breeding. Genotyping analysis of *Rht24* in wheat relatives showed that *Rht24b* originated in wild emmer

(Fig. 5a) that lacked *Rht-B1b* (Dataset S1). This suggests that *Rh24b* is an evolutionarily older reduced height gene than *Rht-B1b*. *Rht18* in durum accession Icaro, generated in variety Anhinga by fast neutron mutagenesis in 1980s, was also located in chromosome 6A (Konzak, 1987; Vikhe *et al.*, 2017) and *GA2oxA9* was identified as the causal gene (Ford *et al.*, 2018). However, no polymorphic sites were detected in the ORF and 7 kb region upstream of the transcriptional start site or 3' untranslated region-between Icaro (with *Rht18*, mutagenized from Anhinga by fast neutron treatment) and its tall parent Anhinga (without *Rht18*) (Ford *et al.*, 2018). We isolated and sequenced *GA2oxA9* (ORF and 1.4 kb 5' upstream promoter region) in Icaro and found that it had the same sequence as *Rht24b* (Fig. S9). In terms of genetic mapping, *Rht18* was close to the centromere of chromosome 6A and lack of recombination for further fine mapping (Ford *et al.*, 2018), whereas *Rht24* was mapped at the physical position of 413.7 Mb in chromosome 6AL, with a large distance from the centromere (283.3-288.7 Mb) according to $R_{EF}S_{EQ}$ v. 1.0. Therefore, *Rht24* is probably not equal to *Rht18*.

The gibberellin regulatory pathway underlying plant height and yield is implied by *Rht24*-mediated differential expression and gibberellin homeostasis

In the metabolic pathway of GA, decreased bioactive GA leads to accumulation of DELLA protein that inhibits expression of GA- responsive genes through interaction with PIFs, such as PIF3 and PIF4, which are central regulatory components integrating multiple internal and external signals (Zheng *et al.*, 2016). RNA-Seq analyses showed that three PIF genes (*TraesCS5D02G386500*, *TraesCS5D02G428400* and *TraesCS7A02G128300*) were differentially expressed in stems at the jointing stage between *TaGA2oxA9*-OE and TNL (Fig. S4B). Remarkably, phylogenetic analysis exhibited that *TraesCS5D02G428400* had high similarity with *OsPIL1* that promotes internode elongation in rice (Fig. S10) (Daisuke *et al.*, 2012). We speculated that upregulation of *TaGA2oxA9* reduced GA contents and protects DELLA proteins from degradation. PIFs are dissociated by highabundance DELLA from the promoters of downstream genes and concomitantly modulate plant height in *Arabidopsis* (De Lucas *et al.*, 2008; Feng *et al.*, 2008). These findings showed that a GA-DELLA-PIF module might be a conserved signal transduction pathway between wheat and model species.

Considering that *RHT24* encodes a GA2-oxidase, *TaGA2ox-A9*, we widely investigated effects of *TaGA2ox-A9* on the abundance of diverse GA isoforms in internodes and leaves at the jointing and grain fill stages, and found that its overproduction could result in significant difference in the spatio-temporal distribution of bioactive GA isoforms between contrasting lines (Figs 1f, g, 3f, l, S5). Strikingly, enhanced expression of *TaGA2ox-A9* caused a reduction of bioactive GA in stems but an elevation of bioactive GA in leaves at the jointing stage (Figs 1f, g, 3f, l). Additionally, *TaGA2ox-A9* overexpression upregulated *GA20-oxidase* genes and downregulates *GA2-oxidase* genes in leaves, suggesting that *TaGA2ox-A9* confers negative feed- back regulation of GA (Fig. S7C; Dataset S12). Some studies reported that GA can increase nitrogen content (Khan, 1996; Nagel & Lambers, 2002; Hedden, 2003; Gooding *et al.*, 2011; Bai *et al.*, 2016; S. Li *et al.*, 2018) and photosynthesis (Arteca & Dong, 1981; Sepehr, 1999; Yuan & Xu, 2001; Ashraf *et al.*, 2002; Biemelt *et al.*, 2004; Ouzounidou & Ilias, 2005; Iqbal & Ashraf, 2013; Hamdani *et al.*, 2019). To explore the underlying mechanisms of *Rht24b* in reducing plant height without yield penalty, we performed molecular and physiological analyses and observed that enhanced expression of *TaGA2ox-A9* significantly improved NUE, photosynthetic capacity as well as the expression of related genes (Figs 3a, e, S7A, B). *TaGA2ox-A9* also increased leaf width and chlorophyll content (Fig. 3c, d). Collectively, *TaGA2ox-A9* not only regulates bioactive GA metabolism but also rearranges the spatio-temporal distribution of bioactive GA, thus possibly accounting for reduced plant height without yield penalty. A working model of *TaGA2ox-A9* in modulating plant height and GY is proposed (Fig. S11). *TaGA2ox-A9* can trigger GA self-regulation and maintain its homeostasis while adjusting to conditions that are optimal for growth and development. However, more experimental work is required to explore how *Rht24b* enhances bioactive GA in leaves as well as NUE and photosynthetic efficiency, which is helpful to uncover the molecular mechanisms underlying the contribution of GA homeostasis to yield formation.

Acknowledgements

The authors are grateful to Prof Robert McIntosh, Plant Breeding Institute, University of Sydney, for revising this manuscript. The work was supported by the Natural Science Foundation of China (91935304), the National Key R&D Program of China (2016YFD0100502) and the Science and Technology Innovation Program of Chinese Academy of Agricultural Sciences (CAAS). The authors declare no conflicts of interest in regard to this manuscript. Ethical Standards: all experiments complied with the ethical standards in China.

References

Arteca R, Dong C. 1981. Increased photosynthetic rates following gibberellic acid treatments to the roots of tomato plants. *Photosynthesis Research* 2: 243- 249.

Ashraf M, Karim F, Rasul E. 2002. Interactive effects of gibberellic acid (GA3) and salt stress on growth, ion accumulation and photosynthetic capacity of two spring wheat (*Triticum aestivum* L.) cultivars differing in salt tolerance. *Plant Growth Regulation* 36: 49-59.

Bai L, Deng H, Zhang X, Yu X, Li Y. 2016. Gibberellin is involved in inhibition of cucumber growth and nitrogen uptake at suboptimal root-zone temperatures. *PLoS ONE* 11: e0156188.

Bazhenov MS, Divashuk MG, Amagai Y, Watanabe N, Karlov GI. 2015. Isolation of the dwarfing *Rht-B1p* (*Rht17*) gene from wheat and the development of an allele-specific PCR marker. *Molecular Breeding* 35: 213.

Biemelt S, Tschiersch H, Sonnewald U. 2004. Impact of altered gibberellin metabolism on biomass accumulation, lignin biosynthesis, and photosynthesis in transgenic tobacco plants. *Plant Physiology* 135: 254-265.

Boccaccini A, Santopolo S, Capauto D, Lorrai R, Minutello E, Serino G, Costantino P, Vittorioso P. 2014. The DOF protein DAG1 and the DELLA protein GAI cooperate in negatively regulating the *AtGA3ox1* gene. *Molecular Plant* 7: 1486-1489.

Buss W, Ford BA, Foo E, Schnippenkoetter W, Borrill P, Brooks B, Ashton AR, Chandler PM, Spielmeyer W. 2020. Overgrowth mutants determine the causal role of *GA2oxidaseA13* in *Rht12* dwarfism of wheat. *Journal of Experimental Botany* 47: 7171-7178.

Cao WG, Somers DJ, Fedak G. 2009. A molecular marker closely linked to the region of *Rht-Dlc* and *Ms2* genes in common wheat (*Triticum aestivum*). *Genome* 52: 95-99.

Daisuke T, Kazuo N, Kyonoshin M, Satoshi K, Yuriko O, Yusuke I, Satoko M, Yasunari F, Kyouko Y *et al*. 2012. Rice phytochrome-interacting factor-like protein OsPIL1 functions as a key regulator of internode elongation and induces a morphological response to drought stress. *Proceedings of the National Academy of Sciences, USA* 109: 15947-15952.

De Lucas M, Davière JM, Rodríguez-Falcón M, Pontin M, Iglesias-Pedraz JM, Lorrain S, Fankhauser C, Blázquez MA, Titarenko E, Prat S. 2008. A molecular framework for light and gibberellin control of cell elongation. *Nature* 451: 480-484.

Dvorak J, Terlizzi P, Zhang HB, Resta P. 1993. The evolution of polyploid wheats: identification of the A genome donor species. *Genome* 36: 21-31.

Evans LT. 1996. *Crop evolution, adaptation and yield*. Cambridge, UK: Cambridge University Press.

Fang J, Zhang FT, Wang HR, Wang W, Zhao F, Li ZJ, Sun CH, Chen FM, Xu F, Chang SQ *et al*. 2019. *Ef-cd* locus shortens rice maturity duration without yield penalty. *Proceedings of the National Academy of Sciences, USA* 116: 18717-18722.

Feng SH, Martinez C, Gusmaroli G, Wang Y, Zhou JL, Wang F, Chen LY, Yu L, Iglesias-Pedraz JM, Kircher S *et al*. 2008. Coordinated regulation of *Arabidopsis thaliana* development by light and gibberellins. *Nature* 451: 475-479.

Ford B, Foo E, Sharwood RE, Karafiatova M, Vrána J, Macmillan C, Nichols DS, Steuernagel B, Uauy C, Doležel J. 2018. *Rht18* semi-dwarfism in wheat is due to increased expression of *GA2-oxidaseA9* and lower GA content. *Plant Physiology* 177: 168-180.

Gale MD, Youssefian S. 1985. Dwarfing genes in wheat. In: Russell GE, ed. *Progress in plant breeding*. London, UK: Butterworths, 1-35.

Ghorbanli M, Kaveh S, Sepehr M. 1999. Effects of cadmium and gibberellin on growth and photosynthesis of *Glycine max*. *Photosynthetica* 37: 627-631.

Gooding MJ, Addisu M, Uppal RK, Snape JW, Jones HE. 2011. Effect of wheat dwarfing genes on nitrogen-use efficiency. Journal of Agricultural Science 150: 3-22.

Gowik U, Burscheidt J, Akyildiz M, Schlue U, Koczor M, Streubel M, Westhoff P. 2004. *cis*-Regulatory elements for mesophyll-specific gene expression in the C_4 plant *Flaveria trinervia*, the promoter of the C_4 phosphoenolpyruvate carboxylase gene. *Plant Cell* 16: 1077-1090.

Hamdani S, Wang H, Zheng G, Perveen S, Qu M, Khan N, Khan W, Jiang J, Li M, Liu X *et al*. 2019. Genome-wide association study identifies variation of glucosidase being linked to natural variation of the maximal quantum yield of photosystem II. *Physiologia Plantarum* 166: 105-119.

Han FM, Zhu BG. 2011. Evolutionary analysis of three gibberellin oxidase genes in rice, *Arabidopsis*, and soybean. *Gene* 473: 23-35.

Hedden P. 2003. Constructing dwarf rice. *Nature Biotechnology* 21: 873-874.

Herter CP, Ebmeyer E, Kollers S, Korzun V, Leiser WL, Würschum T, Miedaner T. 2018. *Rht24* reduces height in the winter wheat population 'Solitär × Bussard' without adverse effects on Fusarium head blight infection. *Theoretical and Applied Genetics* 131: 1263-1272.

Iqbal M, Ashraf M. 2013. Gibberellic acid mediated induction of salt tolerance in wheat plants: growth, ionic partitioning, photosynthesis, yield and hormonal homeostasis. *Environmental Experimental Botany* 86: 76-85.

Ishida Y, Tsunashima M, Hiei Y, Komari T. 2015. Wheat (*Triticum aestivum* L.) transformation using immature embryos. *Methods Molecular Biology* 1223: 189-198.

IWGSC (International Wheat Genome Sequencing Consortium). 2018. Shifting the limits in wheat research and breeding using a fully annotated reference genome. *Science* 361: eaar7191.

Khan NA. 1996. Effect of gibberellic acid on carbonic anhydrase, photosynthesis, growth and yield of mustard. *Biology Plant* 38: 145-147.

Konzak CF. 1987. Mutations and mutation breeding. In: Heyne EG, ed. *Wheat and wheat improvement*. Knoxville, TN, USA: The American Society of Agronomy Inc., 428-443.

Lanning SP, Martin JM, Stougaard RN, Guillen-Portal FR, Blake NK, Sherman JD, Robbins AM, Kephart KD, Lamb P, Carlson GR *et al*. 2012. Evaluation of near-isogenic lines for three height-reducing genes in hard red spring wheat. *Crop Science* 52: 1145-1152.

Li AX, Yang WL, Guo XL, Liu DC, Sun JZ, Zhang AM. 2012. Isolation of a gibberellin-insensitive dwarfing gene, *Rht-B1e*, and development of an allele- specific PCR marker. *Molecular Breeding* 30: 1443-1451.

Li FJ, Wen WE, He ZH, Liu JD, Jin H, Cao SH, Geng HW, Yan J, Zhang PZ, Wan YX *et al*. 2018. Genome-wide linkage mapping of yield-related traits in three Chinese bread wheat populations using high-density SNP markers. *Theoretical and Applied Genetics* 131: 1903-1924.

Li S, Tian YH, Wu K, Ye YF, Yu JP, Zhang JQ, Liu Q, Hu MY, Li H, Tong YP *et al*. 2018. Modulating plant growth-metabolism coordination for sustainable agriculture. *Nature* 560: 595-600.

Li XP, Lan SQ, Liu YP, Gale MD, Worland T. 2006. Effects of different *Rht-B1b*, *Rht-D1b* and *Rht-B1c* dwarfing genes on agronomic characteristics in wheat. *Cereal Research Communications* 34: 919-924.

Liu C, Zheng S, Gui JS, Fu CJ, Yu HS, Song DL, Shen JH, Qin P, Liu XM, Han B *et al*. 2018. *Shortened basal internodes* encodes a gibberellin 2-oxidase and contributes to lodging resistance in rice. *Molecular Plant* 11: 288-299.

Lo S-F, Ho T-H, Liu Y-L, Jiang M-J, Hsieh K-T, Chen K-T, Yu L-C, Lee M-H, Chen C-Y, Huang T-P *et al*. 2017. Ectopic expression of specific GA2 oxidase mutants promotes yield and stress tolerance in rice. *Plant Biotechnology Journal* 15: 850-864.

Lo SF, Yang SY, Chen KT, Hsing YI, Zeevaart JAD, Chen LJ, Yu SM. 2008. A novel class of gibberellin 2-oxidases control semidwarfism, tillering, and root development in rice. *Plant Cell* 20: 2603-2618.

Long SP, Marshall-Colon A, Zhu XG. 2015. Meeting the global food demand of the future by engineering crop photosynthesis and yield potential. *Cell* 161: 56-66.

Ma XD, Ma J, Zhai HH, Xin PY, Chu JF, Qiao YL, Han LZ, Chen ZH. 2015. CHR729 is a CHD3 protein that controls seedling development in rice. *PLoS ONE* 10: e0138934.

McIntosh RA, Dubcovsky J, Rogers WJ, Morris C, Xia XC. 2017. *Catalogue of gene symbols for wheat*: 2017 *supplement*, 137-143. [WWW document] URL https://shigen.nig.ac.jp/wheat/komugi/genes/macgene/supplement2017.pdf.

Mo Y, Vanzetti LS, Hale I, Spagnolo EJ, Guidobaldi F, Al-Oboudi J, Odle N, Pearce S, Helguera M, Dubcovsky J. 2018. Identification and characterization of *Rht25*, a locus on chromosome arm 6AS affecting wheat plant height, heading time, and spike development. *Theoretical and Applied Genetics* 131: 2021-2035.

Nagel OW, Lambers H. 2002. Changes in acquisition and partitioning of carbon and nitrogen in the gibberellin deficient mutants A70 and W335 of tomato (*Solanum lycopersicum* L.). *Plant, Cell & Environment* 25: 883-891.

Ouzounidou G, Ilias I. 2005. Hormone-induced protection of sunflower photosynthetic apparatus against copper toxicity. *Biologia Plantarum* 49: 223-228.

Parry MAJ, Reynolds M, Salvucci ME, Raines C, Andralojc PJ, Zhu XG, Price GD, Condon AG, Furbank RT. 2011. Raising yield potential of wheat. II. Increasing photosynthetic capacity and efficiency. *Journal of Experimental Botany* 62: 453-467.

Pearce S, Saville R, Vaughan SP, Chandler PM, Wilhelm EP, Sparks CA, Kaff NA, Korolev A, Boulton MI, Phillips AL *et al*. 2011. Molecular characterization of *Rht1* dwarfing genes in hexaploid wheat. *Plant Physiology* 157: 1820-1831.

Peng J, Richards DE, Hartley NM, Murphy GP, Devos KM, Flintham JE, Beales J, Fish LJ, Worland AJ, Pelica F *et al*. 1999. 'Green Revolution' genes encode mutant gibberellin response modulators. *Nature* 400: 256-261.

Pont C, Leroy T, Seidel M, Tondelli A, Duchemin W, Armisen D, Lang D, Bustos-Korts D, Goue' N, Balfourier F *et al*. 2019. Tracing the ancestry of modern bread wheats. *Nature Genetics* 51: 905-911.

Rebetzke GJ, Richards RA, Fettell NA, Long M, Condon AG, Forrester RI, Botwright TL. 2007. Genotypic increases in coleoptile length improves stand establishment, vigour and grain yield of deep-sown wheat. *Field Crops Research* 100: 10-23.

Rebetzke GJ, Richards RA, Fischer VM, Mickelson BJ. 1999. Breeding long coleoptile, reduced height wheats. *Euphytica* 106: 159-168.

Schmittgen TD, Livak KJ. 2008. Analyzing real-time PCR data by the comparative CT method. *Nature Protocols* 3: 1101-1108.

Silverstone AL, Sun T. 2000. Gibberellins and the green revolution. *Trends in Plant Science* 5: 1-2.

Sun LH, Yang WL, Li YF, Shan QQ, Ye XB, Wang DZ, Yu K, Lu WW, Xin PY, PeiZ *et al*. 2019. A wheat dominant dwarfing line with *Rht12*, which reduces stem cell length and affects gibberellic acid synthesis, is a 5AL terminal deletion line. *The Plant Journal* 97: 887-900.

Thomas M, Sandve SR, Heier LH, Spannagl M, Pfeifer M, IWGSC, Jakobsen KS, Wulff BBH, Steuernagel B, Mayer KFX *et al*. 2014. Ancient hybridizations among the ancestral genomes of bread wheat. *Science* 345: 1251788.

Tian XL, Wen WE, Xie L, Fu LP, Xu DA, Fu C, Wang DS, Chen XM, Xia XC, Chen QJ *et al*. 2017. Molecular mapping of reduced plant height gene *Rht24* in bread wheat. *Frontier in Plant Science* 8: 1397.

Tian XL, Zhu ZW, Xie L, Xu DG, Li JH, Fu C, Chen XM, Wang D, Xia XC, He ZH *et al*. 2019. Preliminary exploration of the source, spread, and distribution of reducing height ın bread wheat. *Crop Science* 59: 19-24.

Van DeVelde K, Thomas SG, Heyse F, Kaspar R, Van Der Straeten D, Rohde A. 2021. N-terminal truncated RHT-1 proteins generated by translational reinitiation cause semi-dwarfing of wheat green revolution alleles. *Molecular Plant* 14: 679-687.

Vikhe P, Patil R, Chavan A, Oak M, Tamhankar S. 2017. Mapping gibberellin-sensitive dwarfing locus *Rht18* in durum wheat and development of SSR and SNP markers for selection in breeding. *Molecular Breeding* 37: 28.

Wu J, Kong XY, Wan JM, Liu XY, Zhang X, Guo XP, Zhou RH, Zhao GY, Jing RL, Fu XD. 2011. Dominant and pleiotropic effects of a *GAI* gene in wheat results from a lack of interaction between DELLA and GID1. *Plant Physiology* 157: 2120-2130.

Würschum T, Langer SM, Longin C, Friedrich H, Tucker MR, Leiser WL. 2017. A modern green revolution gene for reduced height in wheat. *The Plant Journal* 92: 892-903.

Yanagisawa S. 2000. Dof1 and Dof2 transcription factors are associated with expression of multiple genes involved in carbon metabolism in maize. *The Plant Journal* 21: 281-288.

Yanagisawa S, Schmidt RJ. 1999. Diversity and similarity among recognition sequences of Dof transcription factors. *The Plant Journal* 17: 209-214.

Yuan L, Xu DQ. 2001. Stimulation effect of gibberellic acid

short-term treatment on leaf photosynthesis related to the increase in Rubisco content in broad bean and soybean. *Photosynthesis Research* 68: 39-47.

Zhang Y, Ni ZF, Yao YY, Nie XL, Sun QX. 2007. Gibberellins and heterosis of plant height in wheat (*Triticum aestivum* L.). *BMC Genetics* 8: 40.

Zheng YY, Gao ZP, Zhu ZQ. 2016. DELLA-PIF modules: old dogs learn new tricks. *Trends in Plant Science* 21: 813-815.

Zhu XG, Long SP, Ort DR. 2010. Improving photosynthetic efficiency for greater yield. Annual Review Plant Biology 61: 235-261.

A rapid monitoring of NDVI across the wheat growth cycle for grain yield prediction using a multi-spectral UAV platform☆

Muhammad Adeel Hassan[a], Mengjiao Yang[a,b], Awais Rasheed[a,e], Guijun Yang[c], Matthew Reynolds[d], Xianchun Xia[a], Yonggui Xiao[a,*], Zhonghu He[a,e,*]

[a] *Institute of Crop Sciences, National Wheat Improvement Centre, Chinese Academy of Agricultural Sciences (CAAS), Beijing 100081, China*

[b] *College of Agronomy, Xinjiang Agricultural University, Urumqi 830052, China*

[c] *Beijing Research Centre for Information Technology in Agriculture, Beijing Academy of Agricultural and Forestry Sciences, China*

[d] *Global Wheat Program, International Maize and Wheat Improvement Centre (CIMMYT), Apdo. Postal 6-641, 06600 Mexico DF, Mexico*

[e] *International Maize and Wheat Improvement Centre (CIMMYT) China Office, c/o CAAS, Beijing 100081, China*

☆ This article is part of a special issue entitled "Plant Phenotyping", published in the journal Plant Science 282, 2019.

* Corresponding authors at: Institute of Crop Sciences, National Wheat Improvement Centre, Chinese Academy of Agricultural Sciences (CAAS), Beijing 100081, China.

E-mail addresses: xiaoyonggui@caas.cn (Y. Xiao), zhhecaas@163.com (Z. He).

https://doi.org/10.1016/j.plantsci.2018.10.022

Keywords: High throughput phenotyping, Multi-spectral imaging, Normalized difference vegetation index, Unmanned aerial vehicle

Abbreviations: AtFFBM, at flowering time fresh biomass; AtFDBM, at flowering time dry biomass; AtMBM, at maturity time biomass; B, booting; EGF, early grain filling; F, flowering; GS, Greenseeker; H, heading; LGF, late grain filling; MGF, mid grain filling; NDVI, normalized difference vegetation index; NIR, near infra-red; SE, stem elongation; UAV, unmanned aerial vehicle

Abstract: Wheat improvement programs require rapid assessment of large numbers of individual plots across multiple environments. Vegetation indices (VIs) that are mainly associated with yield and yield-related physiological traits, and rapid evaluation of canopy normalized difference vegetation index (NDVI) can assist in-season selection. Multi-spectral imagery using unmanned aerial vehicles (UAV) can readily assess the Vis traits at various crop growth stages. Thirty-two wheat cultivars and breeding lines grown in limited irrigation and full irrigation treatments were investigated to monitor NDVI across the growth cycle using a Sequoia sensor mounted on a UAV. Significant correlations ranging from $R^2 =$ 0.38 to 0.90 were observed between NDVI detected from UAV and Greenseeker (GS) during stem e-

Published in Plant Science, 282 (2019): 95-103.

longation (SE) to late grain gilling (LGF) across the treatments. UAV-NDVI also had high heritabilities at SE ($h^2=0.91$), flowering (F) ($h^2=0.95$), EGF ($h^2=0.79$) and mid grain filling (MGF) ($h^2=0.71$) under the full irrigation treatment, and at booting (B) ($h^2=0.89$), EGF ($h^2=0.75$) in the limited irrigation treatment. UAV-NDVI explained significant variation in grain yield (GY) at EGF ($R^2=0.86$), MGF ($R^2=0.83$) and LGF ($R^2=0.89$) stages, and results were consistent with GS-NDVI. Higher correlations between UAV-NDVI and GY were observed under full irrigation at three different grain-filling stages ($R^2=0.40$, 0.49 and 0.45) than the limited irrigation treatment ($R^2=0.08$, 0.12 and 0.14) and GY was calculated to be 24.4% lower under limited irrigation conditions. Pearson correlations between UAV-NDVI and GY were also low ranging from $r=0.29$ to 0.37 during grain-filling under limited irrigation but higher than GS-NDVI data. A similar pattern was observed for normalized difference red-edge (NDRE) and normalized green-red difference index (NGRDI) when correlated with GY. Fresh biomass estimated at late flowering stage had significant correlations of $r=0.30$ to 0.51 with UAV-NDVI at EGF. Some genotypes Nongda 211, Nongda 5181, Zhongmai 175 and Zhongmai 12 were identified as high yielding genotypes using NDVI during grain-filling. In conclusion, a multispectral sensor mounted on a UAV is a reliable high-throughput platform for NDVI measurement to predict biomass and GY and grain-filling stage seems the best period for selection.

1 Introduction

Early field evaluation of large experimental plots based on secondary traits for selection is now a routine activity in plant breeding[1]. Monitoring of canopy traits provides a real-time assessment about the crop and assists in devising breeding strategies for yield improvement[2]. Previous findings have promoted vegetation indices especially normalized difference vegetation index (NDVI) as an important multi-spectral index to track the physiological dynamics of key traits such as biomass, nitrogen level and leaf area index[2-4]. Nowadays, several VIs has been reported using different reflectance bands ratios for the monitoring of crop plants, but breeders are not well aware and have less understanding about the application of these bands for their research. NDVI is the most used and consistent remote sensing vegetation index (VI), based on different ratios of reflected energy in the near infrared reflectance (NIR) and red proportion of the light (RED) spectrum[5]. Previous successful attempts involved the use of NDVI to predict the standing crop yield, increasing its importance in crop improvement studies[1,2,6-10]. The strong relationship of NDVI with yield has been demonstrated for several crop physiological attributes that influence yield under drought, heat, and biotic stresses[1,3,11,12]. Higher level of NDVI is associated with faster growth rate and higher biomass accumulation during the vegetative stage, and a longer grain filling period by delaying leaf senescence during the ripening phase thereby increasing yield[13].

Conventional breeding concepts are evolving with modern phenotyping strategies, which imply selecting the important secondary traits using advanced non-destructive remote sensing approaches[14]. Field measurement of NDVI for large experimental applications through destructive measurements even from a subsection of experimental plots may not perfectly characterize the entire attributes of the trait[15]. Ground based hand-held and auto-mobile based specialised platforms can allow finer band resolution collection using adequate spectral reflectance sensors for NDVI, but time and labour intensiveness remain major drawbacks[16,17]. Therefore, rapid, non-destructive and precise high-throughput phenotyping platforms could be useful for accurate selection[1,15,18,19]. Low altitude UAV based remote sensing has a high potential to do rapid and non-destructive phenotyping[20]. Near-infrared (NIR) cameras have been used in many studies[21-24], because plant leaves (or chlorophylls) strongly reflect NIR light[25] and some indices based on NIR reflectance rate, such as NDVI[26], are useful for assessing crop phenology at different developmental phases via remote sensing[6]. Previously, some studies were conducted to estimate plant height, biomass, and ground cover, senescence rate through UAV aerial imagery using RGB (imaging based on a red-green-blue additive colour model) and multi-spectral sensors in sorghum[20], barley[27] cotton[28] and wheat[22]. Daily operational advances in robotics and imagery have increased interest in these platforms[29]. Rapid estimation of NDVI's status of crops can provide useful information to breeders for timely selection of genotypes from large breeding populations regarding high grain yield. There is limited re-porting of the practical use of UAV-based NDVI in selection and yield prediction for wheat breeding[30]. Therefore, the aims of the current study were to (1) develop a rapid assessment platform to monitor variation in NDVI and its relationship with yield, (2) evaluate the full potential of UAV platform for selection using diverse wheat varieties and breeding lines from North China in two contrasting water treatments, (3) evaluate the environmental effect on NDVI and diversity among the germplasm using a UAV, and (4) identify the precise time points for UAV-based NDVI screening to predict grain yield.

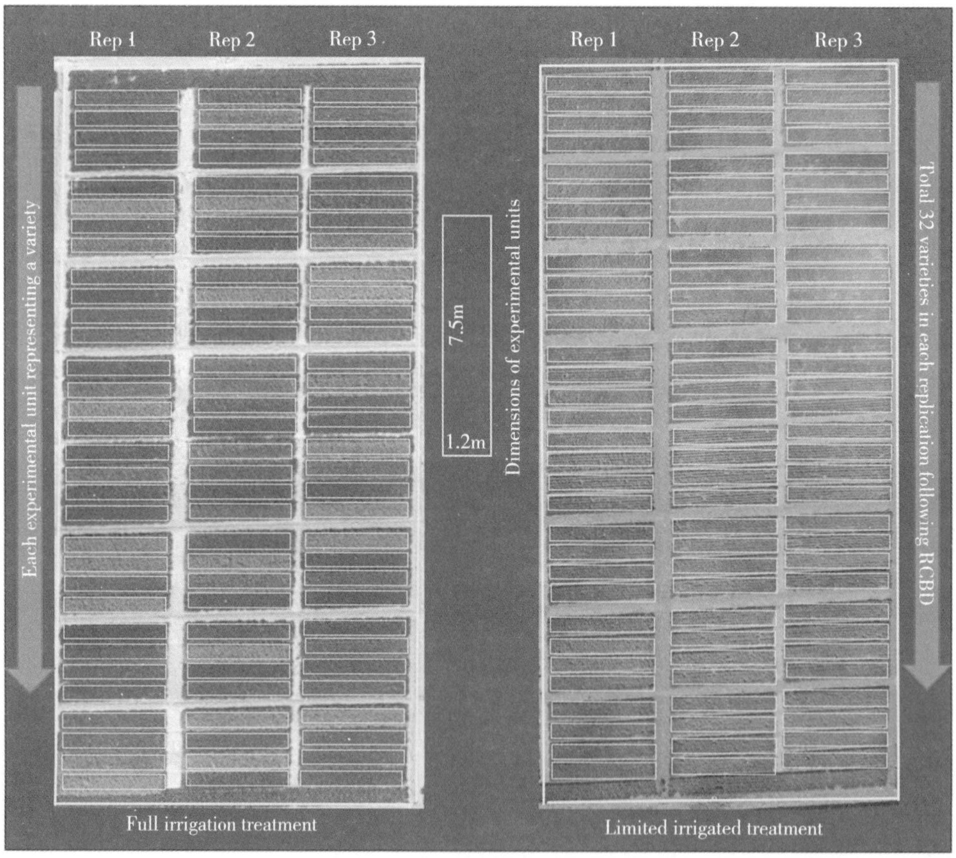

Fig. 1 Field management and design of the experiment to evaluate 32 winter wheat varieties and breeding lines subjected to two water treatments using UAV platform

2 Methodology

2.1 Germplasm and field trials

Thirty-two wheat varieties and breeding lines representing yield progress over the last 40 years in the North China Winter Wheat Region were used in remote sensing experiment. Information regarding the germplasm is given in Table A1 (Supplementary material). This study was conducted at the CAAS Shunyi experimental station in Beijing during the cropping season of 2016-2107 (Fig. A1). Germplasm was planted under two water treatments (fully irrigated and limitedly irrigated) to evaluate the full potential of cultivars and studied platform (UAV). The experiment was arranged in randomized complete blocks with three replications under both water treatments, and each plot was $9m^2$ with 7.5m×1.2m dimensions, representing one cultivar with six rows at 0.20m spacing (Fig. 1). Two irrigations at the seedling and tillering stages were applied for both treatments; the fully irrigated treatment was flooded at the stem elongation and early grain filling stages with $2250\text{-}2700m^3 \cdot ha^{-1}$ of water. Nutrient levels in both treatments were maintained at optimal level until the second irrigation. Seasonal precipitation was 128 mm, and the average seasonal temperature from March to July was 24 ℃ max and 12 ℃ min. Each plot was harvested with a combine harvester to measure grain yield.

2.2 Multi-spectral platform and imagery campaign

A Sequoia 4.0 multi-spectral sensor (Micasense Parrot, France) mounted on an advance auto-operational DJI inspires 1 model T600 (SZ DJI Technology Co., Shenzhen) was used for aerial imagery at seven-time points across the season (Table 1). The multi-channel Sequoia sensor consisted of four monochrome sensors (NIR, RED, RDG and GRE) of 1.4-megapixel definition fitted to a 16-megapixel RGB camera (https://www.micasense.com/parrotsequoia/). The sunshine sensor connected with Sequoia helped to calibrate the multispectral images according to the ambient sunlight to compare images over time, despite variation in ambient light during a mapping event. Flights were conducted during 12 pm to 1 pm under full sun light condition. The UAV was flown over the field through a fully automated flight pattern designed in the open source flight planning software Altizure DJI version 3.6.0 (https://www.altizure.com). To obtain a satisfactory image resolution each flight was taken at 30m and 40m height with 85% image overlapping (both forward and side). In total, three flights were taken at each time point; flight time was about 10 min with $2ms^{-1}-2.5ms^{-1}$ speed to cover the entire experimental area. Details of all flights are given in Table 2. Each monochrome sensor captured the same resolution (1280×960) image with a 10 nm bandwidth (half maximum bandwidth) for the red-edge band and 40 nm bandwidth for infra-red, red and green bands; details of band wavelengths are shown in Table 1. Pixel size was around 3 cm for images taken at 40m altitude and 2.5 cm for 30m flight elevation and images were captured at1.5 s intervals (Table 2).

Table 1 Specification of sensors used in the present study

Sensor	Band	Bandwidth	Wavelength	Definition	Picture resolution
Sequoia	NIR	40 nm	790 nm	1.4 mp	1280×960
	Red	40 nm	660 nm	1.4 mp	1280×960
	Green	40 nm	550 nm	1.4 mp	1280×960
	Red-edge	10 nm	735 nm	1.4 mp	1280×960
	RGB			16 mp	
Greenseeker	NIR	25 nm	770 nm		
	Red	25 nm	660 nm		

Acronyms: mp indicates megapixel.

Table 2 Flight details for the automated unmanned aerial vehicle imagery system during wheat growing season 2017

Time point	Zadok's stage	Flight attitude	Speed (ms^{-1})	Snap shoot Interval (s)	Pixel resolution (cm)
Stem elongation	ZS-39	40	3	2.5	3.5
Booting	ZS-49	40	3	2.5	3.5

(continued)

Time point	Zadok's stage	Flight attitude	Speed (ms^{-1})	Snap shoot Interval (s)	Pixel resolution (cm)
Heading	ZS-57	40	3	2.5	3.5
Flowering	ZS-65	30	2.5	1.5	2.5
Early grain filling	ZS-73	30	2.5	1.5	2.5
Mid grain filling	ZS-85	30	2.5	1.5	2.5
Late grain filling	ZS-91	30	2.5	1.5	2.5

Acronyms: ZS, Zadok's stage

2.3 Orthomosaic generation, segmentation and extraction of pixel values

Following detailed photo-shoots of experimental trials, raw images were processed for mosaic generation using a Pix4D mapper (https://pix4d.com/). The key steps in orthromosaic generation using the Pix4D mapper. Pix4D mapper comprised some important steps such as camera alignment, importing GCPs and geo-referencing, building dense point cloud, DSM and orthromosaics generation[22] (Fig. 2). World Geodetic System 1984 as a system coordinate was used for geo-referenced image generation.

To extract useful information foreach plot in the field, the orthomosaic images were segmented into 192 polygon shapes with assigned IDs defining the individual germplasms. Polygon shapes were generated in ArcMap scripting in ArcGIS (http://www.esri.com/arcgis/aboutarcgis)[23]. Vegetation analysis operated in IDL language plugin in ENVI software (https://esriaustralia.com.au/envi) was used to extract spectral band values for individual plots (in the form of average pixel values) from orthomosaic TIF images. Polygonal shapefile layer of plot boundaries with individual IDs generated in ArcMap and original mosaic image was uploaded to ENVI script for band extraction (Fig. 2).

2.4 Estimation of UAV-NDVI, NDRE and NGRDI

NDVI was calculated using following equation from the reflectance measurements in the near infrared (790 nm) and red (660 nm) (with 40 nm full width half full-maximum bandwidth) portion of the spectrum[31] and the band values ranging from −1.0 to 1.0[26,32].

$$NDVI = (R_{NIR} - R_{red}) / (R_{NIR} + R_{red})$$

Normalized difference red-edge (NDRE) and normalized green red difference index (NGRDI) were also calculated to compare the UAV-NDVI results for field-based selection of genotypes regarding their better performance. NDRE was estimated from near infrared (790 nm) and red-edge (735 nm) (with 10 nm full width half full-maximum bandwidth), while NGRDI was estimated from green (550 nm) and red (735 nm) bands as calculated in previous reports using following equations[33,34].

$$NDRE = (R_{NIR} - R_{red-edge}) / (R_{NIR} + R_{red-edge})$$

$$NGRDI = (R_{green} - R_{red}) / (R_{green} + R_{red})$$

2.5 Ground data collection and statistical analysis

Ground NDVI data were collected with ahandheld Greenseeker RT100® (N Tech Industries, Inc, Ukaiah, CA) at 1 m above the canopy; specifications of the ground sensor are given in Table 1. The wavelength of Greenseeker's bands were in the visible near infrared (770 nm) and red (660 nm) regions of the spectrum, and utilise active sensors. The full and half maximum bandwidth were approximately 25 nm (Table 1). Ground measurements were acquired at the same time point as the UAV passed over the field to minimize the environmental errors. The details of data acquiring time points are given in Table 2. Measurements of each experimental unit were taken at a height of 0.5-0.7 m above the canopy. Biomass were calculated at late flowering and maturity stages, while grain yield was measured according to the manual as described in Pask et al. [35]. Statistical analysis of the mean data estimated from 3 flights at each time point was conducted in R package. Linear regression was calculated to evaluate the relationship between UAV-NDVI and Greenseeker-based measurements. Whereas, Pearson correlation was calculated to indicate the relationship between secondary traits and grain yield at particular time points as well as between growth stages. Mixed linear model was used to test the significance variations at $P \leqslant 0.05$ among genotypes, treatments and their interactions.

To assess the accuracy of the UAV high throughput platform, we calculated broad-sense heritability for each time point of both treatments. Analysis of variance was conducted on mean values of in-dividual plots for each replication by considering entries as a random effect using the equation[36]:

$h^2 = \sigma_g^2 / (\sigma_g^2 + \sigma_\varepsilon^2 / r)$,

where σ_g^2 and σ_ε^2 represent genotypic and error variances, respectively.

3 Results

3.1 Assessment of UAV platform data accuracy

NDVI was recorded as a predictor of biomass at particular vegetative stages and yield advocator during the grain filling stages to identify critical time points for selection. Data from both platforms was normally distributed under full and limited irrigation treatments (Fig. A2). We scrutinised the UAV system as a high-throughput platform to explore the genetic diversity of germplasm for secondary traits. The accuracy of the UAV multi-spectral platform was evaluated though ground truth data using coefficient of determination (R^2) and root mean square error (RMSE) values for both data acquiring systems (Fig. 3). High correlations ranging from $R^2 = 79$ to 0.90 with low RMSE (0.02 to 0.03) between NDVI capture using the UAV-NDVI and GS-NDVI were observed at flowering to late grain filling I. Whereas, quite weak regression values ranging from $R^2 = 0.38$ to 0.42 at the first three time points (stem elongation, booting and heading stages) were obtained at the higher flight altitude and slightly increased image capture interval (Fig. 3 and Table 2). Pearson correlations between NDVI acquired from both platforms were also compared across the seven wheat developmental stages and significant correlations were observed at most of growth time points ranging from $r = 0.43$ to 0.93 under full irrigation, and $r = 0.26$ to 0.62 under limited irrigation condition (Fig. 4). Broad sense heritabilities specific for each platform were calculated and taken as a level of precision of spectral measurements. Overall heritabilities at all time points of both UAV and GS were high, and quite similar ranging from 0.68 to 0.96. Whereas, similar pattern of significant variations ($P \leqslant 0.05$) among germplasm and between water treatments were also observed from both platforms (Table 3).

3.2 Correlations between vegetative indices, biomass and grain yield

The correlation analysis of both platforms elaborated the relationship and influence of vegetation indices (VIs), especially NDVI on biomass and grain yield at seven different growth stages (Figs. 4 and A3). NDVI was strongly correlated with inseason biomass and GY for both platforms across the two water treatments. There were significant correlations between UAV-NDVI and fresh biomass (AtFFBM) taken at the late flowering stage ranging from $r = 0.33$ to $r = 0.55$ under full, and $r = 0.30$ to $r = 0.36$ under limited irrigation at maturity stages (Fig. 4). Whereas, AtFFBM also showed a quite similar trend in correlation with NDRE ($r = 0.19 - 0.41$) and NGRDI ($r = 0.2 - 0.45$) at maturity across the treatments. Very strong correlations were also esti-mated betwеen GY and UAV-NDVI, i.e. $r = 0.64$, 0.70, 0.67 under full, and $r = 0.29$, 0.36, 0.37 under limited irrigation at early, mid and late grain filling stages, respectively (Fig. 4). There were significant correlations ($r = 0.40$ in full irrigation and 0.24 in limited irrigation) between AtFFBM and GY, while also a strong correlation ($r = 0.55$) was observed between AtMBM and GY under limited-water condition. Similar trend for correlations of NDRE and NGRDI was also observed with biomass traits and grain yield across the treatments (Fig. A3).

3.3 Accuracy of UAV-platform for yield prediction

Early prediction of yield could be estimated through performance of yield related secondary traits at specific time-points. Regression analysis had estimated higher R^2 values between UAV-NDVI and GY at the early (0.40), mid (0.49) and late (0.45) grain filling stages under full water treatments compared to limited irrigated conditions where regression values (R^2 = 0.08, 0.12, 0.14 at early, mid and late grain filling stages, respectively) were relatively low (Fig. 5). Similar pattern of R^2 values was also observed in GS-NDVI and GY but slightly higher in full irrigation condition. On average, high UAV-NDVI during grain-filling period predicted greater GY in some genotypes such as Nongda 211, Nongda 5181, Zhongmai 175 and Zhongmai 12 under both treatments (Table A2).

3.4 Assessment of water treatments impact from UAV-NDVI

The NDVI trend over time for the two water regimes showed high resemblance both for UAV-NDVI and GS-NDVI (Fig. 6). There was significant variation between the two water treatments ($P \leqslant 0.0001$) at all growth stages (Fig. 6 and Table 3). A sudden decline in NDVI was observed after the booting stage, and germplasm in the limited irrigated treatment physiologically matured 10-15 days earlier than in the fully irrigated treatment. Significant differences between the two water treatments for AtFFBM certainly affect the level of NDVI at that time point. There was a 10% decrease in UAV-NDVI level at flowering time and a 29-75% decline during early to late grain filling stages, compared to a 16.5% reduction in fresh biomass, and 24.4% GY reduction in the limited water treatment (Figs. 6 and 7).

4 Discussion

4.1 Comparison of UAV and greenseeker

Image reflectance results were estimated according to previous literature available on application of the UAV system for high throughput phenotyping[23], and were based on a larger number of wheat genotypes than used in a recent report on dynamic monitoring of NDVI[6]. Our study demonstrated successful and rapid assessment of NDVI using a UAV platform that showed higher accuracy than a Greenseeker platform in predicting variation in biomass and GY (Table 3). The accuracy of the UAV platform was validated by ground truth data and proved a significant advantage of UAV over the hand-held data acqui-sition platform during the stem elongation to late grain filling stages, especially under limited water conditions. High correlations of UAV-NDVI with biomass and GY illustrated the efficiency of platform that had made this system useful for practical breeding. Previously, several reports were published on the benefits of hand-held and vehicle-based ground sensors to evaluate canopy traits in wheat, cotton and fruit crops[17,37-40]. Auto-vehicle based remote sensing only decreased labour costs, but provided no improvement in time and operational costs. Field design may also confer limitations in regard to on-going field conditions such as rainfall, timing of irrigation, and space[19]. UAV not only reduces the labor costs but also provides a means of phenotyping that overcomes limitations of large field trials. Low altitude flights with large image overlaps as demonstrated in this work permit high-resolution orthomosaic generation, allowing deep spectral extraction for important secondary traits. Estimations of Vis are mainly influenced by measurement time and sensor resolution[41], factors that can affect selection based on secondary traits. Sequoia sensor had the advantage of comprising sensors with high bandwidth which gave deeper information than the Greenseeker. UAV allows more data coverage across the growing season for the dynamic monitoring of NDVI and other VIs to screen large numbers of plots in a cost-effective way[42]. Our study suggested that continuous VIs data generation at critical time points through UAV could assist in prediction and selection for grain yield.

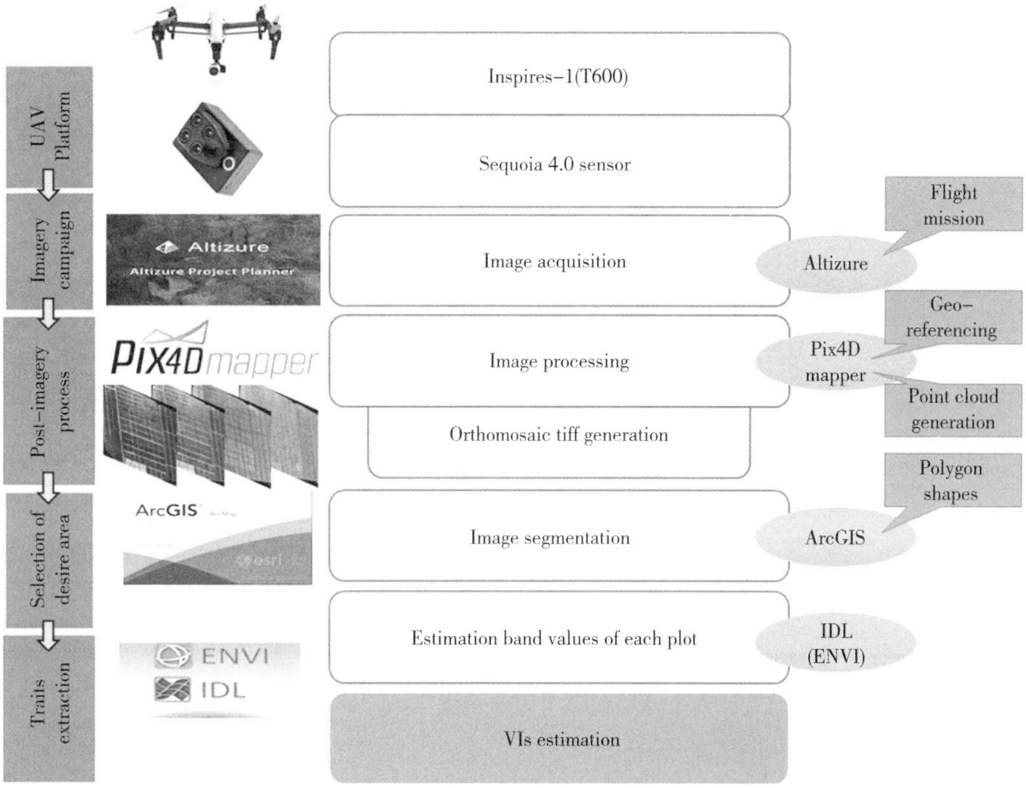

Fig. 2 A workflow diagram of UAV based multispectral trait extraction methodology

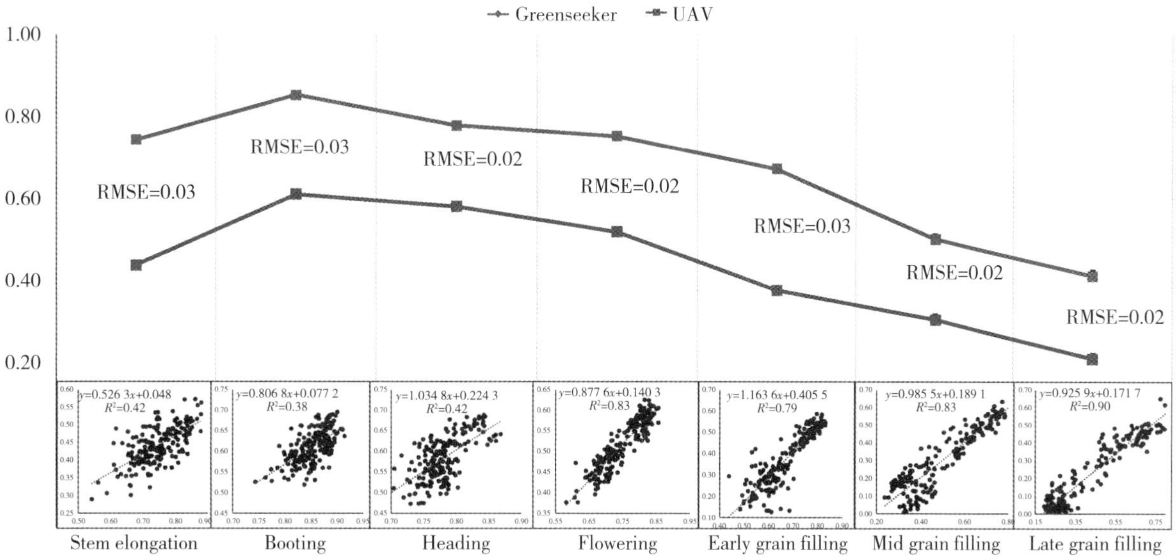

Fig. 3 Comparison of seasonal changes in NDVI from UAV-based multispectral imagery and Greenseeker. Root mean square error and coefficient of determination present the probability of noise at each developmental stage

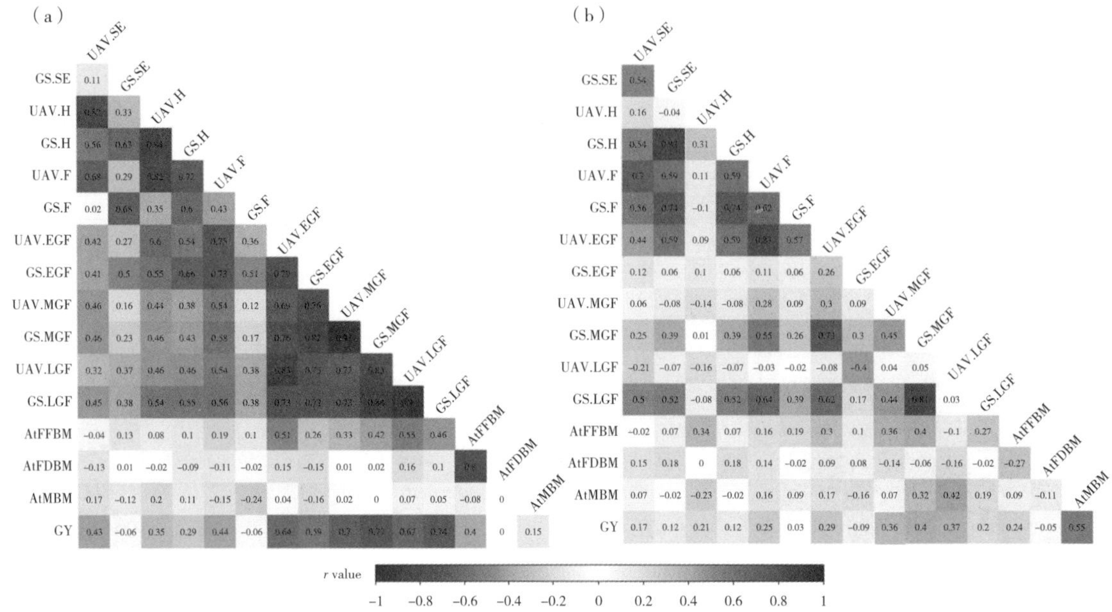

Fig. 4 Correlation coefficients of NDVI derived from UAV and GS with agronomic traits and grain yield under (a) full irrigation and (b) limited irrigation treatments at different growth stages (stem elongation to maturation). Color intensities indicate degrees of positive and negative significance at ($P\leqslant 0.05$).
Acronyms: B, booting; SE, stem elongation; EGF, early grain filling; F, flowering; GS, Greenseeker; H, heading; LGF, late grain filling; MGF, mid grain filling; UAV, unmanned aerial vehicle; AtFFBM, flowering time fresh biomass; AtFDBM, flowering time dry biomass; AtMMB, (maturity time biomass)

Table 3 Fitness test and heritabilities results of NDVI, agronomic and yield traits taken at different time points across wheat growing season

Trait	Cultivars F-value	Treatment F-value	C × T F-value	h^2	
				Full irrigation	Limited irrigatio
UAV. SE	13.02***	165.73***	1.82*	0.91	0.92
GS. SE	9.84***	83.35***	1.04	0.90	0.92
UAV. B	4.39***	230.56***	1.54	0.89	0.89
GS. B	5.99***	428.95***	0.82	0.93	0.86
UAV. H	3.00***	440.95***	2.76**	0.94	0.73
GS. H	15.24***	420.74***	5.32***	0.96	0.84
UAV. F	6.72***	1175.92***	1.05	0.95	0.87
GS. F	6.48**	1520.08***	1.29	0.92	0.88
UAV. EGF	2.63**	917.26***	0.85	0.79	0.75
GS. EGF	1.92*	663.84***	1.22	0.78	0.73
UAV. MGF	1.58*	360.91***	0.66	0.71	0.74
GS. MGF	2.02*	696.08***	0.94	0.68	0.83
UAV. LGF	2.25*	1379.34***	1.95*	0.76	0.72
GS. LGF	3.22***	1060.75***	2.24**	0.81	0.90
AtFFBM (kg)	2.14*	292.69***	1.25	0.86	0.79
AtFDBM (kg)	2.78**	10.18*	1.68*	0.79	0.72
AtMDBM (kg)	2.51*	11.20*	1.71*	0.80	0.72
GY (kg)	2.73**	574.02***	0.98	0.75	0.76

*, **, ***, significant at $P\leqslant 0.05$, $P\leqslant 0.001$ and $P\leqslant 0.0001$, respectively. Acronyms: B, booting; SE, stem elongation; EGF, early grain filling; F, flow-ering; GS, Greenseeker; GY, grain yield; H, heading; LGF, late grain filling; MGF, mid grain filling; UAV, unmanned aerial vehicle; AtFFBM, flowering time fresh biomass; AtFDBM, flowering time dry biomass; AtMMB, maturity time biomass.

4.2 UAV-data accuracy for biomass and yield prediction

Torres-Sánchez et al.[43] reported a soil background error during data extraction from aerial multispectral images that can affect data quality. Duan et al.[6] overcame the error using ground RGB images to improve the data quality by calculating ground cover adjusted NDVI that could be a time-consuming practice in the case of large breeding trials. Recent advancements in sensor technology are good enough to capture high resolution images with minimum data noise[44]. We expect this assumption to hold real for UAV high-throughput data collection while sensor and imaging pipelines continue to improve, further minimizing technical errors. The present report is based on original data generated through several flights at seven different growth stages. The accuracy of UAV data was higher with minimal RMSE as previously reported[6,23] (Fig. 3). High R^2 and correlation coefficients among the UAV and ground data from Greenseeker accorded with the recent study by Duan et al.[6], and were higher than reported by Haghighattalab et al.[23] (Fig. 3 and 4). Low correlation between the two methods at stem elongation, booting and heading stages are more likely due to low platform optimisation because of higher flight altitude, speed and image capture intervals. To get higher image resolution and sufficient image overlap for good quality orthomosaic generation, the UAV speed and altitude during imaging campaign was reduced at later growth stages. Higher R^2 with least RMSE at later growth stages indicated successful optimization of UAV-NDVI and enhanced the UAV data accuracy (Fig. 3). Another reason of low correlation at early stages might be due to operational and handling errors during data acquisition. As the UAV and Greenseeker measurements were taken in parallel at the same time, the environmental variance is assumed to be negligible. Broad sense heritability can be used to compare the level of precision of spectral measurements of both platforms, and the differ-ence could be attributed to bias in data acquisition[23]. Similar trends of high heritabilities for UAV and GS based NDVI were observed even higher for the UAV platform than for ground observations at SE (h^2 = 0.91), F (h^2 = 0.95), EGF (h^2 = 0.79) and MGF (h^2 = 0.71) under full irrigation, and at booting (h^2 = 0.89), EGF (h^2 = 0.75) in limited irrigation treatment demonstrating the higher repeatability and precision of UAV data (Table 3). Low heritabilities at LGF under full and at MGF to LGF under limited irrigation might be attributed to less UAV-image resolution taken from 30 m above the canopy compared with Greenseeker which was carried out at 1 m over the canopy to detect green proportion.

4.3 Significance of NDVI monitoring for yield prediction

Rapid estimation of vegetation indices through UAV platforms from the wheat canopy and its predictability for grain yield could speed up crop improvement programs[34]. Several vegetation indices have been introduced but most of the wheat physiology research and almost all breeding programs mainly rely on NDVI as most useful and consistent vegetation index[2,9]. The NDVI has been promoted as an indicator of chlorophyll level, biomass, and predictor of yield[2,7]. Its practical use in wheat breeding for selection and quantitative studies have increased the efforts towards rapid, precise and cost-effective estimation of NDVI throughout growth cycle. We monitored the fluctuating patterns of NDVI across the season from UAV-based system to validate its use in season selection and yield prediction using diverse genotypes in scenario of practical breeding. UAV-NDVI correlations with biomass and GY traits were consistent compared with NDRE and NGRDI (Fig. A3). The UAV-NDVI results also verified previous ground-based findings[2]. Highly significant correlations between prediction of yield at the grain filling stage and actual yield provided a strong evidence that UAV-NDVI can precisely explain yield variations among the genotypes and in two water treatments (Figs. 4 and 5). Regression between the UAV-NDVI and GY at early grain filling were similar, as recently demonstrated with ground cover-adjusted NDVI at near-flowering stages[6] (Fig. 5). The low association regarding prediction of yield under water-limited condition might be due to number of factors like low image resolution that could not capture the lowest fraction of green biomass under drought, low VI values due to high senesces rate and also some pre-harvest losses of grains. Therefore, more improvement in sensor resolution was required to measure minimum fraction of spectral bands to get accurate NDVI phenology. But UAV-NDVI had shown higher Pearson correlations with GY compared to GS-NDVI during grain-filling under limited irrigation treatment. This might be due to the fact that UAV-based NDVI averaged from whole plot compared to GS-NDVI. A strong association between fresh biomass estimated at late flowering time (AtFFBM) and UAV-NDVI in fully and limitedly irrigated treatments (r=0.33-0.51 and r=0.30-0.36, respectively) at EGF to MGF had supported previous findings indicating that NDVI can be used as a surrogate for biomass and leaf area index[2,6,9] (Fig. 4). Biomass as a secondary trait has been reported as direct associations with GY[34]. In this study the significant correlations (r=0.40 fully irrigated and 0.24 limitedly irrigated) between AtFFBM and GY also explained the im-portance of secondary traits in yield prediction (Fig. 4). Early season prediction of grain yield has been achieved based on optimum NDVI levels at two crucial time points i. e. heading (vegetative stage) and maturation (ripening stage) duration using ground platforms[2,9]. In the present study, optimum level of UAV-NDVI at vegetative stages was also just prior to heading and had shown significant correlation (r = 0.34) with AtFFBM under limited irrigation. Biomass estimated at maturity also had significant correlation with UAV-NDVI (r = 0.42) at LGF and with GY (r = 0.55) under the water-limited condition. Our results indicated that UAV platform can predict biomass status at both vegetative and maturity that could be useful information for breeders to select potential genotypes under drought conditions.

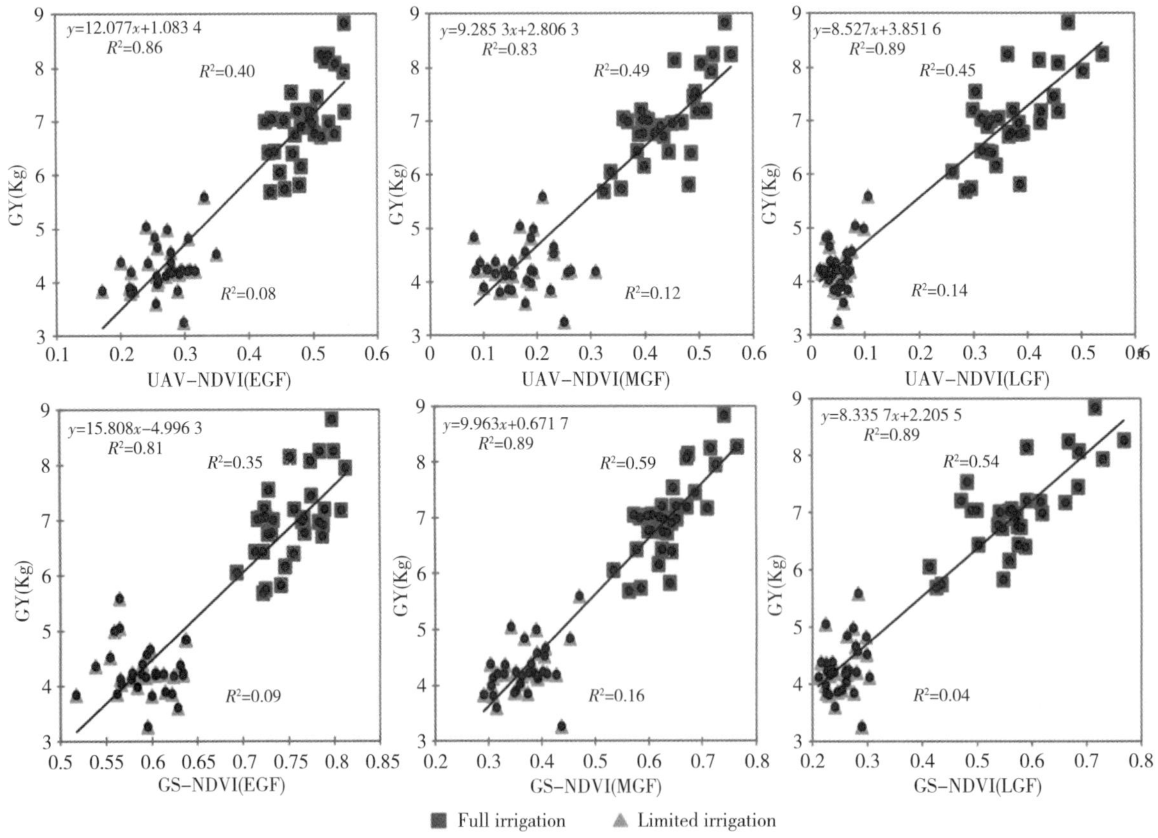

Fig. 5 Linear regression relationships (over all R^2 and separately for both treatments) between yield and vegetation index derived from UAV multispectral imagery and Greenseeker under two water treatments during three grain ripening time periods (early, mid, and late grain filling)

Acronyms: GS, Greenseeker; EGF, early grain filling; LGF, late grain filling; MGF, mid grain filing; UAV, unmanned aerial vehicle.

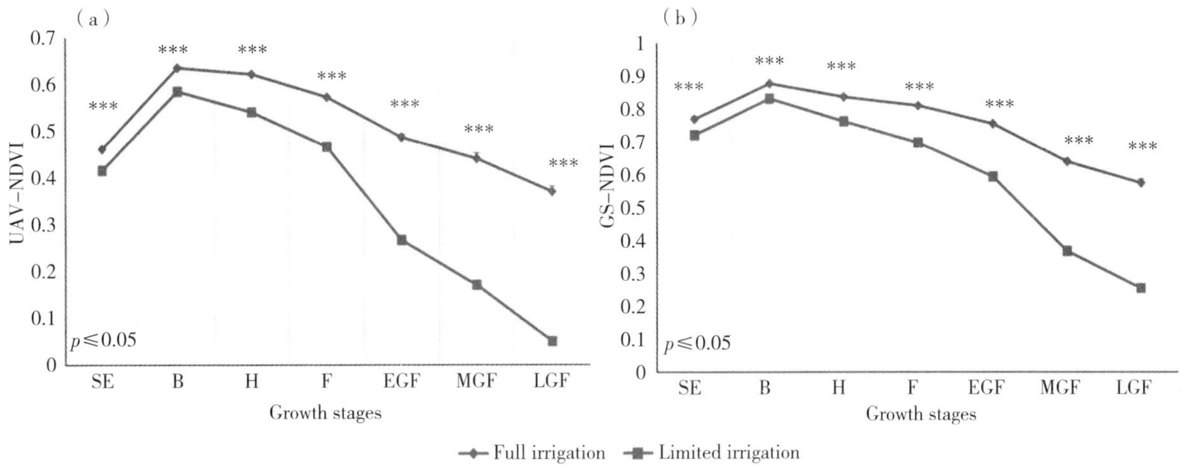

Fig. 6 Seasonal trend and variations in NDVI from UAV multispectral imagery (a) and Greenseeker (b) under different water treatments

Acronyms: B, booting; SE, stem elongation; EGF, early grain filling; F, flowering; GS, Greenseeker; H, heading; LGF, late grain filling; MGF, mid grain filling; UAV, unmanned aerial vehicle.

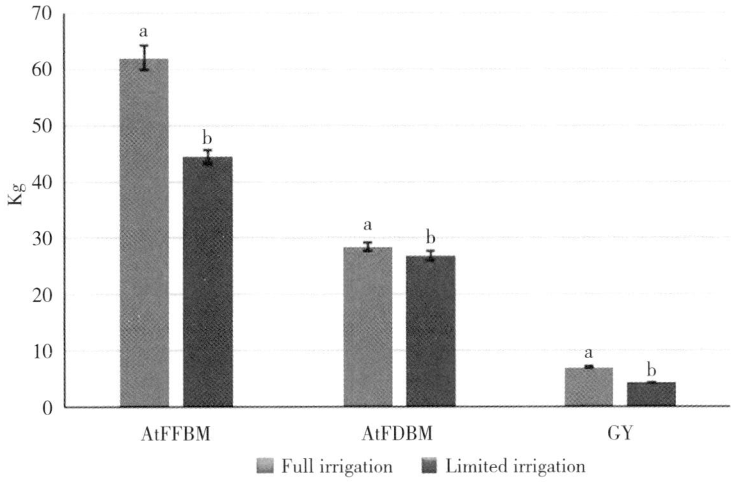

Fig. 7 In-season biomass and yield measurements under full irrigation and limited irrigation conditions. Error bars show standard deviation of mean data for 32 cultivars and breeding lines while alphabets indicate significant differ-ence between two treatments.

Acronyms: AtFFBM, flowering time fresh biomass; AtFDBM, flowering time dry biomass; GY, grain yield

4.4 Detection of drought effect on NDVI and yield

The maturation period also determines the grain quality because the rate of mobilization of nutrients into the grain is closely tied to the duration of senescence[34,45]. Water deficiency was generally attributed to the influence of rate of senescence that could affect the GY[34]. A rapid senescence had been linked to shorter grain filling time, resulting in low GY[46]. Our UAV remote sensing platform also demonstrated significant variation among the genotypes and difference between the treatments regarding NDVI status through significant relationships between UAV-NDVI and grain yield at three critical ripening time (Fig. 4, Tables 3 and A2). Lopes and Reynolds[9] used NDVI measured with a ground sensor to monitor wheat phenology and reported a negative correlation between yield and rapid senescence under drought conditions. In our results, a sudden decline in UAV-NDVI from 29 to 75% during the early to late grain filling stages under low water treatment, resulting in 24.4% of GY reduction (Figs. 6 and 7), was in line with previous reports[9,20,34]. A strong backup of previous findings confirmed our aerial data-acquiring approach to be a rapid and reliable phenotyping platform. Some genotypes such as Nongda 211, Nongda 5181, Zhongmai 175 and Zhongmai 12 were selected based on high UAV-NDVI and GY at grain-filling stages under both treatments. These genotypes can be used as parents in future crosses for variety development.

5 Conclusion

Ground truth validation demonstrated that NDVI measured using a UAV high-throughput phenotyping platform can increase the selection accuracy of breeding materials. The UAV-platform is not only a rapid data acquisition system but also reduces labor costs and problems associated with inclement weather. Therefore, high throughput phenotyping (HTP) of important multi-spectral secondary traits like NDVI appears to be a promising approach for in-season selection and yield prediction. We concluded that phenotypic data of NDVI generated through a UAV platform could be efficiently used for field-based selection and GY prediction. As the HTP phenotyping becomes more accessible and operational in the rapid estimation of secondary traits that explain yield it will be more reliably used in breeding. Machine learning approaches could make HTP more efficient by use of automated identification, classification, quantification and prediction for critical decision regarding selection of desired phenotypes in field conditions. Further improvement is also required in sensor resolution and data analysis for acquisition of precise HTP data.

Acknowledgments

We thank Prof. R. A. McIntosh, Plant Breeding Institute, University of Sydney, for review of this manuscript. This work was funded by the National Key Project (2016YFD0101804-6), the National Natural Science Foundation of China (31671691), the National Key Technology R&D Program of China (2014BAD01B05), and the International Science & Technology Cooperation Program of China (2016YFE0108600).

References

[1] J. Rutkoski, J. Poland, S. Mondal, E. Autrique, L. G. Pérez, J. Crossa, M. Reynolds, R. Singh, Canopy temperature and vegetation indices from high-throughput phenotyping improve accuracy of pedigree and genomic selection for grain yield in wheat, G3: Genes Genomes Genetics 6 (2016) 2799-2808.

[2] T. S. Magney, J. U. Eitel, D. R. Huggins, L. A. Vierling, Proximal NDVI derived phenology improves in-season predictions of wheat quantity and quality, Agric. For. Meteorol. 217 (2016) 46-60.

[3] A. Foster, V. Kakani, J. Mosali, Estimation of bioenergy crop yield and N status by hyperspectral canopy reflectance and partial least square regression, Precis. Agric. 18 (2017) 192-209.

[4] S. M. Samborski, D. Gozdowski, O. S. Walsh, D. Lamb, M. Stępień, E. S. Gacek, T. Drzazga, Winter wheat genotype effect on canopy reflectance: implications for using NDVI for in-season nitrogen topdressing recommendations, Agron. J. 107 (2015) 2097-2106.

[5] C. J. Tucker, Red and photographic infrared linear combinations for monitoring vegetation, Remote Sens. Environ. 8 (1979) 127-150.

[6] T. Duan, S. Chapman, Y. Guo, B. Zheng, Dynamic monitoring of NDVI in wheat agronomy and breeding trials using an unmanned aerial vehicle, Field Crops Res. 210 (2017) 71-80.

[7] K. Erdle, B. Mistele, U. Schmidhalter, Comparison of active and passive spectral sensors in discriminating biomass parameters and nitrogen status in wheat cultivars, Field Crops Res. 124 (2011) 74-84.

[8] L. Guo, N. An, K. Wang, Reconciling the discrepancy in ground-and satellite-observed trends in the spring phenology of winter wheat in China from 1993 to 2008, J. Geophys. Res. Atmos. 121 (2016) 1027-1042.

[9] M. S. Lopes, M. P. Reynolds, Stay-green in spring wheat can be determined by spectral reflectance measurements (normalized difference vegetation index) independently from phenology, J. Exp. Bot. 63 (2012) 3789-3798.

[10] J. Marti, J. Bort, G. Slafer, J. Araus, Can wheat yield be assessed by early mea-surements of Normalized

Difference Vegetation Index? Ann. Appl. Biol. 150 (2007) 253-257.

[11] S. Kumar, M. S. Röder, R. P. Singh, S. Kumar, R. Chand, A. K. Joshi, U. Kumar, Mapping of spot blotch disease resistance using NDVI as a substitute to visual ob-servation in wheat (*Triticum aestivum* L.), Mol. Breed. 36 (2016) 95.

[12] A. Kyratzis, D. Skarlatos, V. Fotopoulos, V. Vamvakousis, A. Katsiotis, Investigating correlation among NDVI index derived by unmanned aerial vehicle photography and grain yield under late drought stress conditions, Procedia Environ. Sci. 29 (2015) 225-226.

[13] M. Babar, M. Reynolds, M. Van Ginkel, A. Klatt, W. Raun, M. Stone, Spectral re-flectance indices as a potential indirect selection criteria for wheat yield under irrigation, Crop Sci. 46 (2006) 578-588.

[14] J. L. Araus, J. E. Cairns, Field high-throughput phenotyping: the new crop breeding frontier, Trends Plant Sci. 19 (2014) 52-61.

[15] D. Deery, J. Jimenez-Berni, H. Jones, X. Sirault, R. Furbank, Proximal remote sensing buggies and potential applications for field-based phenotyping, Agronomy 4 (2014) 349-379.

[16] T. S. Magney, L. A. Vierling, J. U. Eitel, D. R. Huggins, S. R. Garrity, Response of high frequency Photochemical Reflectance Index (PRI) measurements to environmental conditions in wheat, Remote Sens. Environ. 173 (2016) 84-97.

[17] M. Schirrmann, A. Hamdorf, A. Garz, A. Ustyuzhanin, K.-H. Dammer, Estimating wheat biomass by combining image clustering with crop height, Comput. Electron. Agric. 121 (2016) 374-384.

[18] S. C. Chapman, T. Merz, A. Chan, P. Jackway, S. Hrabar, M. F. Dreccer, E. Holland, B. Zheng, T. J. Ling, J. Jimenez-Berni, Pheno-copter: a low-altitude, autonomous remote-sensing robotic helicopter for high-throughput field-based phenotyping, Agronomy 4 (2014) 279-301.

[19] M. Tattaris, M. P. Reynolds, S. C. Chapman, A direct comparison of remote sensing approaches for high-throughput phenotyping in plant breeding, Front. Plant Sci. 7 (2016) 1131.

[20] K. Watanabe, W. Guo, K. Arai, H. Takanashi, H. Kajiya-Kanegae, M. Kobayashi, K. Yano, T. Tokunaga, T. Fujiwara, N. Tsutsumi, High-throughput phenotyping of sorghum plant height using an unmanned aerial vehicle and its application to genomic prediction modeling, Front. Plant Sci. 8 (2017) 421.

[21] B. Berger, B. Parent, M. Tester, High-throughput shoot imaging to study drought responses, J. Exp. Bot. 61 (2010) 3519-3528.

[22] M. A. Hassan, M. J. Yang, A. Rasheed, X. Jin, X. C. Xia, Y. G. Xiao, Z. H. He, Time-series multispectral indices from unmanned aerial vehicle imagery reveal senes-cence rate in bread wheat, Remote Sens. 10 (2018) 809.

[23] A. Haghighattalab, L. G. Pérez, S. Mondal, D. Singh, D. Schinstock, J. Rutkoski, Ortiz-Monasterio, R. P. Singh, D. Goodin, J. Poland, Application of unmanned aerial systems for high throughput phenotyping of large wheat breeding nurseries, Plant Methods 12 (2016) 35.

[24] J. Zhang, C. Yang, H. Song, W. C. Hoffmann, D. Zhang, G. Zhang, Evaluation of an airborne remote sensing platform consisting of two consumer-grade cameras for crop identification, Remote Sens. 8 (2016) 257.

[25] N. Fahlgren, M. A. Gehan, I. Baxter, Lights, camera, action: high-throughput plant phenotyping is ready for a close-up, Curr. Opin. Plant Biol. 24 (2015) 93-99.

[26] A. Kross, H. McNairn, D. Lapen, M. Sunohara, C. Champagne, Assessment of RapidEye vegetation indices for estimation of leaf area index and biomass in corn and soybean crops, Int. J. Appl. Earth Obs. Geoinf. 34 (2015) 235-248.

[27] J. Bendig, A. Bolten, S. Bennertz, J. Broscheit, S. Eichfuss, G. Bareth, Estimating biomass of barley using crop surface models (CSMs) derived from UAV-based RGB imaging, Remote Sens. 6 (2014) 10395-10412.

[28] T. Duan, B. Zheng, W. Guo, S. Ninomiya, Y. Guo, S. C. Chapman, Comparison of ground cover estimates from experiment plots in cotton, sorghum and sugarcane based on images and ortho-mosaics captured by UAV, Funct. Plant Biol. 44 (2017) 169-183.

[29] L. R. Khot, S. Sankaran, A. H. Carter, D. A. Johnson, T. F. Cummings, UAS imaging-based decision tools for arid winter wheat and irrigated potato production man-agement, Int. J. Remote Sens. 37 (2016) 125-137.

[30] M. Tattaris, M. P. Reynolds, S. C. Chapman, A direct comparison of remote sensing approaches for high-throughput phenotyping in plant breeding, Front. Plant Sci. 7 (2016).

[31] G. Alvarado, M. Suchismita, J. Rutkoski, L. González-

Pérez, J. Burgueño, Predicting grain yield using canopy hyperspectral reflectance in wheat breeding data, Plant Methods 13 (2017) 4.

[32] M. L. Gnyp, Y. Miao, F. Yuan, S. L. Ustin, K. Yu, Y. Yao, S. Huang, G. Bareth, Hyperspectral canopy sensing of paddy rice aboveground biomass at different growth stages, Field Crops Res. 155 (2014) 42-55.

[33] T. Ahamed, L. Tian, Y. Zhang, K. C. Ting, A review of remote sensing methods for biomass feedstock production, Biomass Bioenergy 35 (2011) 2455-2469.

[34] A. B. Potgieter, B. George-Jaeggli, S. C. Chapman, K. Laws, L. A. Suárez Cadavid, J. Wixted, J. Watson, M. Eldridge, D. R. Jordan, G. L. Hammer, Multispectral imaging from an unmanned aerial vehicle enables the assessment of seasonal leaf area dynamics of sorghum breeding lines, Front. Plant Sci. 8 (2017) 1532.

[35] A. J. D. Pask, J. Pietragalla, D. M. Mullan, M. P. Reynolds, Physiological breeding II: a field guide to wheat phenotyping, CIMMYT Mexico DF (Mexico) 4 (2012) 132.

[36] J. B. Holland, W. E. Nyquist, C. T. Cervantes-Martínez, Estimating and interpreting heritability for plant breeding: an update, in: Jules Janick (Ed.), Plant Breeding Reviews, John Wiley & Sons, Inc., 2010, pp. 9-112.

[37] S. Liu, F. Baret, B. Andrieu, P. Burger, M. Hemmerlé, Estimation of wheat plant density at early stages using high resolution imagery, Front. Plant Sci. 8(2017)739.

[38] M. Mora, F. Avila, M. Carrasco-Benavides, G. Maldonado, J. Olguín-Cáceres, proved image processing algorithms applied to canopy cover digital photograpies, Comput. Electron. Agric. 123 (2016) 195-202.

[39] B. Sharma, G. L. Ritchie, N. Rajan, Near-remote green: red perpendicular vegetation index ground cover fraction estimation in cotton, Crop Sci. 55 (2015) 2252-2261.

[40] M. P. Pound, A. P. French, E. H. Murchie, T. P. Pridmore, Automated recovery of three-dimensional models of plant shoots from multiple color images, Plant Physiol. 166 (2014) 1688-1698.

[41] L. G. T. Crusiol, Jd. F. C. Carvalho, R. N. R. Sibaldelli, W. Neiverth, A. do Rio, L. C. Ferreira, Sd. O. Procópio, L. M. Mertz-Henning, A. L. Nepomuceno, N. Neumaier, J. R. B. Farias, NDVI variation according to the time of measurement, sampling size, positioning of sensor and water regime in different soybean cultivars, Precis. Agric. 18 (2017) 470-490.

[42] J. Torres-Sánchez, F. López-Granados, N. Serrano, O. Arquero, J. M. Peña, High-throughput 3-D monitoring of agricultural-tree plantations with unmanned aerial vehicle (UAV) technology, PLoS One 10 (2015) e0130479.

[43] J. Torres-Sánchez, F. López-Granados, A. I. De Castro, J. M. Peña-Barragán, Configuration and specifications of an unmanned aerial vehicle (UAV) for early site specific weed management, PLoS One 8 (2013) e58210.

[44] G. Yang, J. Liu, C. Zhao, Z. Li, Y. Huang, H. Yu, B. Xu, X. Yang, D. Zhu, X. Zhang, R. Zhang, H. Feng, X. Zhao, Z. Li, H. Li, H. Yang, Unmanned aerial vehicle remote sensing for field-based crop phenotyping: current status and perspectives, Front. Plant Sci. 8 (2017) 1111.

[45] Z. Wang, J. Wang, C. Zhao, M. Zhao, W. Huang, C. Wang, Vertical distribution of nitrogen in different layers of leaf and stem and their relationship with grain quality of winter wheat, J. Plant Nutr. 28 (2005) 73-91.

[46] A. Madani, A. S. Rad, A. Pazoki, G. Nourmohammadi, R. Zarghami, Wheat (*Triticum aestivum* L.) grain filling and dry matter partitioning responses to source: sink modifications under postanthesis water and nitrogen deficiency, Acta Sci. Agron. 32 (2010) 145-151.

Genome-wide variation patterns between landraces and cultivars uncover divergent selection during modern wheat breeding

Jindong Liu[1,2] • Awais Rasheed[1,3,4] • Zhonghu He[1,3] • Muhammad Imtiaz[5] • Anjuman Arif[6] • Tariq Mahmood[4] • Abdul Ghafoor[7] • Sadar Uddin Siddiqui[7] • Muhammad Kashif Ilyas[7] • Weie Wen[1] • Fengmei Gao[8] • Chaojie Xie[2] • Xianchun Xia[1]

[1] *Institute of Crop Science, National Wheat Improvement Center, Chinese Academy of Agricultural Sciences (CAAS), 12 Zhongguancun South Street, Beijing* 100081, *China*

[2] *Department of Plant Genetics and Breeding/State Key Laboratory for Agrobiotechnology, China Agricultural University, 2 Yuanmingyuan West Road, Beijing* 100193, *China*

[3] *International Maize and Wheat Improvement Center (CIMMYT) China Office, c/o CAAS, 12 Zhongguancun South Street, Beijing* 100081, *China*

[4] *Quaid-i-Azam University, Islamabad, Pakistan*

[5] *International Maize and Wheat Improvement Center (CIMMYT) Pakistan Office, c/o National Agriculture Research Center (NARC), Islamabad, Pakistan*

[6] *National Institute of Agriculture and Biology (NIAB), Faisalabad, Pakistan*

[7] *Bio-resources Conservation Institute (BCI), National Agriculture Research Center (NARC), Islamabad, Pakistan*

[8] *Crop Research Institute, Heilongjiang Academy of Agricultural Sciences, Harbin* 150086, *Heilongjiang, China*

Key message: Genetic diversity, population structure, LD decay, and selective sweeps in 687 wheat accessions were analyzed, providing relevant guidelines to facilitate the use of the germplasm in wheat breeding.

Abstract: Common wheat (*Triticum aestivum* L.) is one of the most widely grown crops in the world. Landraces were subjected to strong human-mediated selection in developing high-yielding, good quality, and widely adapted cultivars. To investigate the genome-wide patterns of allelic variation, population structure and patterns of selective sweeps during modern wheat breeding, we tested 687 wheat accessions, including landraces (148) and cultivars (539) mainly from China and Pakistan in a wheat 90 K single nucleotide polymorphism array. Population structure analysis revealed that cultivars and landraces from China and Pakistan comprised three relatively independent genetic clusters. Cultivars displayed lower nucleotide diversity and a wider average LD decay across whole genome, indicating allelic erosion and a diversity bottleneck due to the modern breeding. Analysis of genetic differentiation between

Published in Theoretical and Applied Genetics, 2019, 132: 2509-2523.

landraces and cultivars from China and Pakistan identified allelic variants subjected to selection during modern breeding. In total, 477 unique genome regions showed sig-natures of selection, where 109 were identified in both China and Pakistan germplasm. The majority of genomic regions were located in the B genome (225), followed by the A genome (175), and only 77 regions were located in the D genome. EigenGWAS was further used to identify key selection loci in modern wheat cultivars from China and Pakistan by comparing with global winter wheat and spring wheat diversity panels, respectively. A few known functional genes or loci found within these genome regions corresponded to known phenotypes for disease resistance, vernalization, quality, adaptability and yield-related traits. This study uncovered molecular footprints of modern wheat breeding and explained the genetic basis of polygenic adaptation in wheat. The results will be useful for understanding targets of modern wheat breeding, and in devising future breeding strategies to target beneficial alleles currently not pursued.

Abbreviations

CL	Chinese landraces
CMC	Chinese modern cultivars
FMC	Foreign modern cultivars
F_{ST}	F-statistics
GWAS	Genome-wide association study
IWGSC	International Wheat Genome Sequencing Consortium
LD	Linkage disequilibrium
MAF	Minor allele frequency
MAS	Marker-assisted selection
	NJ Neighbor-joining
PCA	Principal components analysis
PL	Pakistan landraces
PMC	Pakistan modern cultivars
SNP	Single nucleotide polymorphism

Introduction

Hexaploid common wheat (genomes AABBDD) was formed~8000 years ago by hybridization of *Triticum turgidum* (AABB) with *Aegilops tauschii* (D) (Kihara 1944; McFadden and Sears 1946; Dvorak et al. 1993). It originated from the Middle East and expanded to China and Eastern Asia along the ancient Silk Road. The emergence of cultivars from landraces was achieved primarily by artificial selection aimed at developing high-yielding locally adapted varieties (Lopes et al. 2015). Although the landraces and modern cultivars have same genomes in both size and content, they have substantial genomic variations and exhibit substantial differences in agronomic traits, such as plant height, yield and disease resistance (Rasheed et al. 2018). Wheat genomic variations are determined by multiple factors, including polyploidization events (Choulet et al. 2014), domestication (Tanno and Willcox 2006), spread from origin sites to new geographic regions (Zhou et al. 2018), gene flow (Luo et al. 2007), and post-domestication selection (Cavanagh et al. 2013).

During the process of domestication and modern breeding, crops suffered from strong and extensive artificial selection for yield, stress tolerance, quality and environmental adaptability (Yamasaki et al. 2007; Cavanagh et al. 2013; Swarts et al. 2017). Evidence of the selection remains in the patterns of genetic variations and selection regions within cultivated genomes. Genes under artificial selection always associated with important and complex agronomic traits and reflect the main driving forces for each historical differentiation (Cavanagh et al. 2013; Zhou et al. 2018). Thus, details of genetic variations and identification of the selection regions during crop improvement can provide valuable guidelines for further crop improve-ment (Morrell et al. 2011; Cavanagh et al. 2013), as well as opportunities to improve genomic selection models (Calus et al. 2008; Heffner et al. 2009; Heslot et al. 2015). Although traditional linkage and association mapping are effective for identifying genetic loci, they are limited to identify genetic variations associated with domestication and improvement (Morrell et al. 2011). Investigation of genomic regions with patterns of genetic variations can identify loci subject to selection (Morrell et al. 2011; Akagi et al. 2016; Wei et al. 2017) and they do not require to measure phenotypes. Wheat has a complex and large genome size (~17.0 Gb) with high proportion (~80%) of repetitive sequences or transposons (Brenchley et al. 2012; Choulet et al. 2014; International Wheat Genome Sequencing Consortium (IWGSC) 2014; Marcussen et al. 2014; Pfeifer et al. 2014). Therefore, the use of high-throughput and high-density genotyping at low costs are suitable (Gupta et al. 2008; Wang et al. 2014; Edae et al. 2014; Allen et al. 2016). The high-throughput SNP genotyping has been used to better understand the impact of domestication and modern breeding on the genetic variations and identify genomic selection footprints in crops and vegetables, such as wheat (Cavanagh et al. 2013; Rasheed et al. 2017; Zhou et al. 2018), soybean (Li et al. 2013), rapeseed (Wei et al. 2017), pepper (Hill et al. 2013) and cotton (Hulse-Kemp et al. 2015). Cavanagh et al. (2013) identified selection regions during modern breeding progresses genotyped by the wheat 9 K SNP array in 2994 hexaploid wheat accessions, and suggested that most of selection regions were associated with vernalization, flowering time, disease resistance and plant height. Zhou et al. (2018) investigated 717 Chinese wheat landraces with 27, 933 DArT and 312, 831 SNP markers, and found 148 regions showing signatures of selection.

Some studieshave reported the analysis of genetic variations and the identification of selection regions in wheat worldwide (Cavanagh et al. 2013; Lopes et al. 2014, 2015). However, the impact of selection on the patterns of genetic variations and the selection footprints remained poorly understood in wheat adapted to major wheat growing countries. In the present study, 539 modern cultivars and 148 landraces mainly from China and Pakistan were collected. The cultivars represented the major breeding programs of China and Pakistan, whereas the landraces represented the genetic, geographic, and morphological diversity of wheat. We genotyped the 687 accessions with the wheat 90K SNP array to (1) investigate the population structure, (2) uncover the impact of selection on the genetic variations of wheat, and (3) gain insights into the genomic footprints of selection imposed by modern wheat breeding.

Materials and methods

Plant material

A diversepanel of 687 worldwide wheat accessions was used in this study, including landraces from China (57) and Pakistan (91), modern cultivars from China (351), Pakistan (105), France (23), Australia (11), America (11), Argentina (7), Romania (6), Russia (6), Italy (7), Britain (1), Japan (4), Canada (2), Hungary (2), Germany (1), Mexico (1), Norway (1) and Turkey (1) (Table S1, Fig. S1).

Genotyping and quality control

Genomic DNA was extracted from young leaf tissues according to Doyle and Doyle (1987). All the 687 accessions were genotyped using the wheat 90 K SNP array (Capital Bio Corporation, Beijing, China). SNP markers with missing data less than 20% and a minor allele frequency (MAF) exceeding 0.05 were used for further analysis. The physical positions of SNPs from the wheat 90 K SNP array were obtained from the

IWGSC (http://www.wheatgenome.org/) by local BLAST.

The 90 K SNP array data for 299 global winter wheat and 339 spring wheat accessions were retrieved from The Triticeae Toolbox (T3) maintained by Wheat Coordinated Agricultural Project (Wheat CAP). These datasets were used as reference to identify the selective sweeps in modern wheat cultivars from China and Pakistan.

Population structure

Population structure analysis was conducted using Admixture 1.3.0 (Alexander et al. 2009) with default cross-validation (K value ranged from 2 to 5). Neighbor-jointing (NJ) -tree construction was performed by PowerMarker v3.25 (Liu and Muse 2005). Principal component analysis (PCA) that summarizes the major patterns of variations was also used to reveal the relationships among 687 wheat accessions using Tassel v5.0 according to Bradbury et al. (2007). All the SNP markers with minor allele frequency<0.01 were excluded from the PCA analysis and population structure.

Linkage disequilibrium (LD) analysis

LD among markers was calculated by Tassel v5.0 as described by Bradbury et al. (2007). A subset of SNP markers with minor allele frequency>0.05 and the missing data less than 10% was used for LD analysis. The LD decays for the A, B, and D genomes were estimated following Breseghello and Sorrells (2006) for the whole population and the previous described clusters.

Genomic patterns of nucleotide variations

Phylogenetic and population structure analyses revealed spatial genetic structure among the Chinese landraces, Chinese cultivars, Pakistan landraces, Pakistan cultivars and other exotic cultivars. We divided the 687 accessions into 57 Chinese landraces (CL), 91 Pakistan landraces (PL), 351 Chinese modern cultivars (CMC), 105 Pakistan modern cultivars (PMC) and 84 worldwide or foreign modern cultivars (FMC).

The genetic diversity of CL, PL, CMC, PMC, FMC and the entire populations were analyzed with Variscan v2.0.3 (Hutter et al. 2006) for nucleotide diversity (θ_π) (Tajima 1983), Waterson's estimator of nucleotide diversity (θ_w) (Watterson 1975) and Fu and Li's D^* (Fu 1997). The 1000 kb sliding window with 100 kb step-size along genome was used to estimate these parameters. Furthermore, the MAF at each locus were also computed for each subset of genotypes using PowerMarker v3.25 (Liu and Muse 2005) to investigate the genetic diversity.

Identification of selective regions

F-statistics (F_{ST}) is the proportion of the variance in allele frequencies (Weir 1996; Holsinger and Weir 2009) and is used to measure the genetic differentiation and distance (Hudson et al. 1992) to estimate the selection footprints by the process of modern breeding. The F_{ST} was calculated via the VCFtools (Danecek et al. 2011). The genomic regions where the average F_{ST} fell in the top 5% of the empirical F_{ST} distribution were considered as the selection regions. Furthermore, to confirm the selection regions identified by F_{ST}, Tajiama's D was calculated by Variscan v2.0.3 (Hutter et al. 2006).

The selective sweeps were identified through genome-wide association studies of eigenvectors that were implemented by EigenGWAS (Chen et al. 2016). The selective sweeps in modern Chinese wheat cultivars were identified in an independent analysis using global winter wheat collection as the reference dataset. Similarly, the selective sweeps in modern wheat cultivars from Pakistan were identified using SNP data from global spring wheat collection as the reference dataset. The high-quality SNPs were used to generate genetic relationship matrix; the top 10 eigenvalues and their corresponding eigenvectors were calculated. The SNP effects, nearly equivalent to F_{ST}, could be estimated by regressing each SNP for a selected eigenvector. In principle, the estimated genetic effect for each locus is driven by genetic drift, which is random, and selection. To filter off the genetic drift, we adjusted the P-value with genomic control factor, and consequently the corrected P-value, P_{GC}, was used for detecting the loci under selection. To determine the cutoff of significance of selection loci, the first eigenvector was reshuffled 1000 times to evaluate the null distribution. The 95th quantile P-value of the 1000 most significant P-value across 1000 permutations was used as the threshold. After logarithm, the P-value threshold of 5.87 for experiment-wise type I error rate of 0.05 was used for the EigenGWAS analyses on the 10 eigenvectors.

Results

Marker density and allele frequency

In total, 80, 587 SNPs from the wheat 90 K SNP array were retrieved for data analysis and quality control. SNPs with MAF<5% and missing data>20% in the complete panel of 687 accessions were removed and finally 37, 856 SNPs were used for subsequent analysis (Table S2). However, the sub-populations still contained SNPs with minor allele frequency ranging from 0.01 to 0.05. These markers spanned a physical distance of 14,060.1 Mb, with an average density of 2.69 SNPs/Mb. The 13, 439 (35.5%), 15, 909 (42.0%) and 8508 (22.5%) markers were from A, B and D genomes, with corresponding map lengths of 4932.3, 5177.7 and 3950.1 Mb, respectively. The marker density for the D genome (2.154 SNPs/Mb) was lower than that for the A (2.723) and B (3.072) genomes (Table 1). The minor allele frequency(MAF) and heterozygosity were calculated for each of the four sub-populations along each chromosome (Table 1). The landraces collections from China (CL: Chinese landraces) had exceptionally higher heterozygosity of 0.24 compared to modern cultivars from China (CMC: Chinese modern cultivars). The heterozygosity differentiation between landraces (0.017) and cultivars (0.012) from Pakistan was much more minimal (Table 1).

Table 1 Basic statistics of the SNP markers distribution, allele frequency and heterozygosity in four different wheat collection from China and Pakistan

Chromosome	No. of markers	Length (Mb)	Density (Mb/marker)	MAF				Heterozygosity			
				CL	CMC	PL	PMC	CL	CMC	PL	PMC
1A	2 002	594.0	3.370	0.13	0.14	0.06	0.14	0.23	0.006	0.008	0.005
1B	2 591	689.4	3.758	0.17	0.18	0.06	0.15	0.29	0.01	0.02	0.021
1D	1 431	495.4	2.889	0.15	0.15	0.1	0.17	0.26	0.018	0.015	0.014
2A	2 114	780.7	2.708	0.18	0.15	0.08	0.13	0.3	0.011	0.015	0.01
2B	3 067	801.2	3.828	0.18	0.17	0.08	0.14	0.33	0.013	0.022	0.013
2D	1 759	651.6	2.700	0.12	0.17	0.07	0.15	0.23	0.008	0.005	0.006
3A	1 672	750.8	2.227	0.16	0.16	0.06	0.14	0.27	0.009	0.012	0.01
3B	2 230	830.3	2.686	0.16	0.18	0.08	0.14	0.27	0.015	0.013	0.012
3D	1 114	615.4	1.810	0.13	0.09	0.05	0.07	0.25	0.019	0.022	0.008
4A	1 531	744.5	2.056	0.18	0.15	0.09	0.15	0.33	0.009	0.028	0.02
4B	1 399	673.4	2.078	0.19	0.16	0.07	0.12	0.33	0.006	0.013	0.01
4D	717	509.6	1.407	0.12	0.16	0.07	0.17	0.23	0.017	0.018	0.005
5A	2 089	709.7	2.943	0.17	0.18	0.09	0.16	0.28	0.011	0.009	0.005
5B	2 530	713.0	3.548	0.14	0.18	0.08	0.15	0.24	0.006	0.012	0.009
5D	1 205	566.0	2.129	0.08	0.11	0.06	0.1	0.17	0.004	0.013	0.003

(continued)

Chromosome	No. of markers	Length (Mb)	Density (Mb/marker)	MAF				Heterozygosity			
				CL	CMC	PL	PMC	CL	CMC	PL	PMC
6A	1 869	618.0	3.024	0.21	0.15	0.08	0.14	0.33	0.003	0.024	0.014
6B	2 120	720.9	2.941	0.18	0.17	0.07	0.15	0.29	0.016	0.026	0.018
6D	996	473.5	2.103	0.17	0.13	0.1	0.13	0.31	0.019	0.009	0.01
7A	2 161	736.6	2.934	0.14	0.18	0.08	0.14	0.24	0.019	0.018	0.016
7B	1 972	750.6	2.627	0.18	0.17	0.08	0.12	0.31	0.016	0.023	0.015
7D	1 286	638.6	2.014	0.13	0.15	0.07	0.15	0.25	0.021	0.041	0.03
A genome	13 439	4 934.5	2.723	0.16	0.16	0.07	0.014	0.28	0.009	0.016	0.011
B genome	15 909	5 179.0	3.072	0.17	0.17	0.07	0.014	0.29	0.011	0.018	0.014
D genome	8 508	3 950.4	2.154	0.13	0.14	0.07	0.013	0.24	0.015	0.017	0.010
The whole genome	37 856	14 063.9	2.692	0.15	0.16	0.07	0.14	0.27	0.012	0.017	0.012

CL Chinese landraces, *CMC* Chinese modern cultivars, *PL* Pakistani landraces, *PMC* Pakistani modern cultivars

NJ-tree and PCA

The genetic relatedness and population structure within the collection were also determined using the NJ-tree analysis, which clearly divided the 687 wheat accessions into three groups: landraces from China assigned to subgroup 1 (SG 1), landraces from Pakistan to subgroup 2 (SG 2), and all cultivars to subgroup 3 (SG 3) (Fig. 1a). We also investigated the genetic structure and relatedness among the 687 accessions by PCA. The proportions of genotypic variance explained by the first three principal coordinates were 29.5%, 7.6% and 6.9%, respectively. PCA separated the 687 accessions into three major groups, which was in accordance with the NJ-tree results (Fig. 1b). The population structure of the 687 wheat accessions was inferred using Admixture 1.3.0. At $K=5$, and five distinct groups emerged (Fig. 1c), viz. two landrace groups referred to as Chinese landrace and Pakistan landrace, and three modern cultivar groups belonging to Chinese modern cultivar, Pakistan modern cultivar and Foreign modern cultivar. The relationship and population structure among all cultivars from China, Pakistan and other countries were also further analyzed by NJ-tree analysis. SG 3-1 comprised cultivars from China, SG 3-2 consisted of Pakistan cultivars, and SG 3-3 included cultivars from other countries (Fig. 2a). However, some counter-examples in which cultivars from different geographical origins grouped together were also observed. Sunstate, H45, Chara and Avocet'S originated from Australia, Norin 61 and Norin 67 from Japan, NSA 09-3645 from France, were clustered with Chinese modern cultivars into SG 3-1; Wheatear and Ac Barrie originated from Canada, Baxter, Drysadle, Livingston, Sunzell and Ellison from Australia, Aca 601, Aca 801, Klein Flecha, Klein Jabal 1 and Prointa Colibr 1 from Argentina, Festin and Insignia from France, and Pastor from Mexico were clustered with Pakistan modern cultivars into SG 3-2. Among all cultivars, the proportions of genotypic variance explained by the first three principal coordinates were 16.8%, 5.4% and 4.9%, respectively. Also, PCA separated all the cultivars into three major groups, which was consistent with assignments generated by NJ-tree analysis (Fig. 2b).

We calculated pair-wise genetic differentiation (F_{ST}) between the landraces and cultivars both in China and Pakistan to examine the population structure; F_{ST} among the four sub-populations ranged from 0.175 to 0.455 (Table S3). Compared with the landraces from China and Pakistan ($F_{ST}=0.402$), the cultivars from

China and Pakistan were least genetically differentiated ($F_{ST}=0.351$). Furthermore, the genetic difference between landraces and cultivars from China ($F_{ST}=0.288$) were higher than the landraces and cultivars from Pakistan ($F_{ST}=0.175$) (Table S3).

LD decay

The LD decay distances for the A, B, D and the whole genomes were estimated in CL, PL, CMC, PMC, FMC and entire populations, respectively (Fig. S2). According to Breseghello and Sorrells (2006), LD decay distances were about 7, 9, 16, 17, 16 and 15 Mb for the whole genome of CL, PL, CMC, PMC, FMC and entire populations, respectively (Fig. S2). Furthermore, the LD analysis revealed differences among genomes, and the D genome had the highest LD decay (11, 16, 19, 21, 20 and 19 Mb for the CL, PL, CMC, PMC, FMC and entire populations, respectively), followed by the A (6, 7, 13, 17, 14 and 13 Mb) and B genomes (5, 6, 12, 13, 13 and 13 Mb) (Fig. S2).

Genomic patterns of nucleotide variations

To investigate the nucleotide variations in wheat from landraces to cultivars, we quantified the genome-wide nucleotide diversity (θ_π) based on the wheat 90 K SNP array. The mean θ_π for all the 687 accessions were estimated at $2.2E^{-5}$. Genetic diversity was higher in the distal regions than in the proximal regions of all 21 chromosomes (Figs. 3 and 4). Furthermore, the B genome had the highest θ_π ($3.0E^{-5}$), followed by A ($2.5E^{-5}$) and D ($1.1E^{-5}$) genomes (Table S2). The Chinese landraces ($3.1E^{-5}$) showed the highest nucleotide diversity, followed by Pakistan landraces ($2.6E^{-5}$), Chinese cultivars ($1.9E^{-5}$) and foreign cultivars ($1.8E^{-5}$), whereas the Pakistan cultivars ($1.5E^{-5}$) possessed the smallest number of variants. These results revealed that landraces had slightly higher level of nucleotide diversity compared with cultivars at the whole genome level.

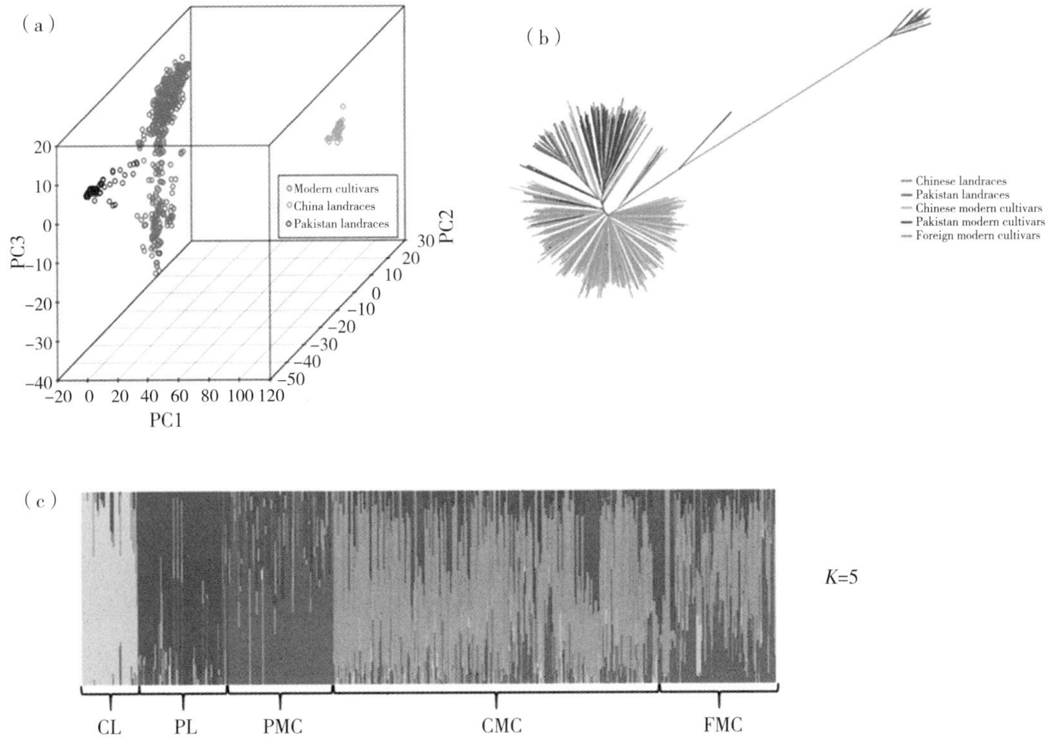

Fig. 1 Population structures of 687 wheat accessions. (a) Principal components analysis (PCA) plots; (b) neighbor-joining (NJ) tree; (c) population structure

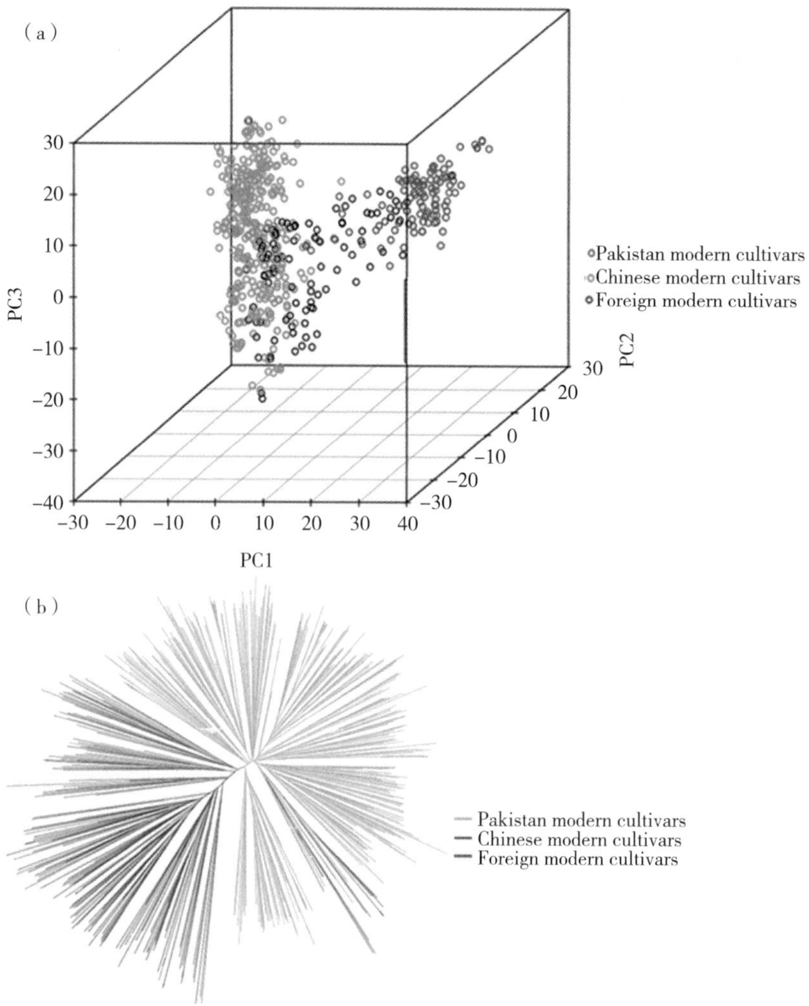

Fig. 2 Population structures of 539 common wheat cultivars. (a) Principal components analysis (PCA) plots; (b) neighbor-joining (NJ) tree

Selection regions identified in common wheat

To identify genomic regions affected by selection (so-called selective sweeps) during wheat improvement, we calculated F_{ST} across the 21 chromosomes (Fig. 5) with VCFtools. Foreign cultivars were not included in this analysis. The regions showed F_{ST} values that were in the top 5%, and thus were considered to be potentially positively selected. In total, we identified 268 (304.9 Mb) selection regions on all 21 chromosomes (Table S4; Fig. 5a) between Chinese landraces and cultivars. Similarly, 318 (372.6 Mb) genomic regions on all 21 chromosomes between Pakistan landraces and cultivars were detected by F_{ST} analysis. To examine the distribution of these selection regions between different wheat groups, we calculated the shared variants of 586 selection regions. A total of 109 out of 586 sweeps distributing on all chromosomes were shared between the China and Pakistan groups. Thus, 477 unique regions were identified in Chinese and Pakistan wheat germplasm. The majority of genomic regions were present in B genome (225), followed by A genome (175), whereas only 77 regions located in D genome. More than 30 shared selection regions were identified in chromosomes 1B, 2B, 3B, 6B, 7B, 4A, 5A and 7A, while only 6-10 in 4D, 5D, 6D or 7D (Table S4; Fig. 5). Furthermore, 159 selection regions were identified only between Chinese landraces and cultivars, whereas 209 selection regions were identified only between Pakistan landraces and cultivars. In addition, several selection regions were detected in presumably orthologous genes, and also in physical regions in homoeologous chromosomes, for example, the regions on chromosomes 2A at 704.7-706.8 Mb and 2B at 706.7-708.4 Mb.

For the statistical rigor, selective sweeps were further validated by EigenGWAS (Chen et al. 2016). EigenGWAS were conducted for the top 10 largest eigenvectors (i.e., Evl to Evl0) associated with the top 10 largest eigenvalues from Chinese and Pakistan modern wheat cultivars. In Chinese modern wheat cultivars, the largest eigenvalue was 72.12 explaining about 7.2% of the total genetic variation; the tenth largest eigenvalue was 16.2, explaining about 1.2% of the total genetic variation; and the top 10 largest eigenvalues covered 32.6% of the genetic variation. In modern wheat cultivars from Pakistan, the largest eigenvalue was 101.66 explaining about 8.1% of the total genetic variation; the tenth largest eigenvalue was 29.45, explaining about 2.3% of the total genetic variation; and the top 10 largest eigenvalues covered 41.1% of the genetic variation. The genomic inflation factor that is commonly used in adjusting population stratification for GWAS, namely λGC calculated from EigenGWAS, ranged from 23.2 to 4.09 (for Chinese germplasm) and 39.1-6.0 (for Pakistan germplasm) (Table S5). After correction by λGC, the SNPs with-log10 (PGC) exceeding the threshold of 6.01 were declared as the loci under selection at the genome-wide level. Upon positive or negative coordinates on the corresponding eigenvector that EigenGWAS was conducted, two subgroups were defined for the Chinese wheat materials (CMC versus global winter wheat diversity panel), Pakistani wheat material (PMC versus global spring wheat diversity panel), and their selection differentiations were quantified by FST. To facilitate the comparison, scanning results from-log10 (PGC) and FST were demonstrated together by Miami plot (Fig. 6). Generally, the peaks from-log10 (PGC) and FST were fairly mirrored each other, indicating reasonable grouping as defined by EigenGWAS.

In CMC, 138 significant SNPs were identified in these 10EigenGWAS analyses which were distributed over 13 different loci (Table S5; Fig. 6a). Significant hits were only identified on Evl, Ev2, Ev8, Ev9 and Evl0. The most hits (i.e., 103 SNPs over 3 loci) were on Ev8. The distribution of the identified SNPs across different chromosomes varies considerably, with chromosome 2A (i.e., 102 SNPs) having the highest numbers of SNPs and chromosome 3A (i.e., 1 SNP) the lowest. The genomic region on chromosome 4A between 103 Mb and 504 Mb consistently had several selective sweeps with more than 98 SNPs. All the selective sweep identified by EigenGWAS were shared by the F_{ST} scan. In PMC, 426 significant SNPs were identified in these 10 EigenGWAS analyses which were distributed over 30 loci (Table S5; Fig. 6b). Three genomic regions over group 1 homeologous chromosomes were consistently identified as selective sweeps (Fig. 6b). Significant hits were only identified on Evl, Ev2, Ev4, Ev8 and Evl0. The highest number of 373 hits were on Ev4, while only 8 hits were identified on Evl0. The maximum number of hits were identified on chromosome 1B, where 263 SNPs were distributed over 3 loci, whereas only one significant hit was identified on chromosomes 3A, 3D, 4D and 7D.

Although selection regions have been identified, the functions of these genes remained unclear. To investigate the function of these selection regions, we compared the selection regions with previous reported genes and found those regions associated with important agronomic traits and diverse biological pathways, including drought tolerance, early flowering, grain size, lodging, plant height, pre-harvest sprouting, end-use quality, disease resistance, thousand-grain weight andvernalization (Table S6).

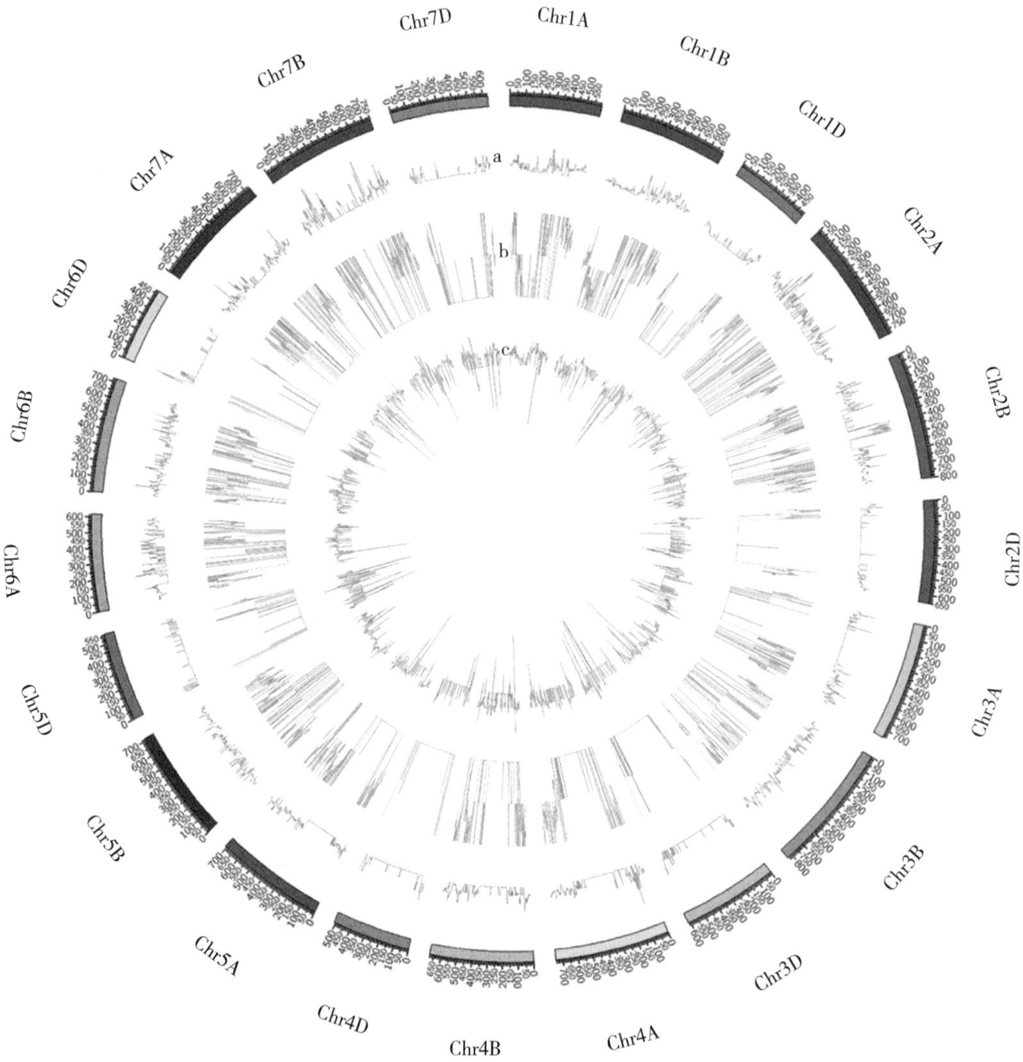

Fig. 3 Summary of SNP markers from the wheat 90 K SNP array between landraces and cultivars in China. a The ratio of the nucleotide diversity (θ_{π}) in the landraces versus modern cultivars; b ratio of the Water-son's estimator of nucleotide diversity (θ_{w}) in the landraces versus modern cultivars, and c ratio of the Fu and Li's D^{*} values in the landraces versus modern cultivars. The 21 chromosomes are portrayed along the perimeter of each circle. Physical positions of SNP markers from the wheat 90 K SNP array are based on IWGSC v1.0 (http://www.wheatgenome.org/)

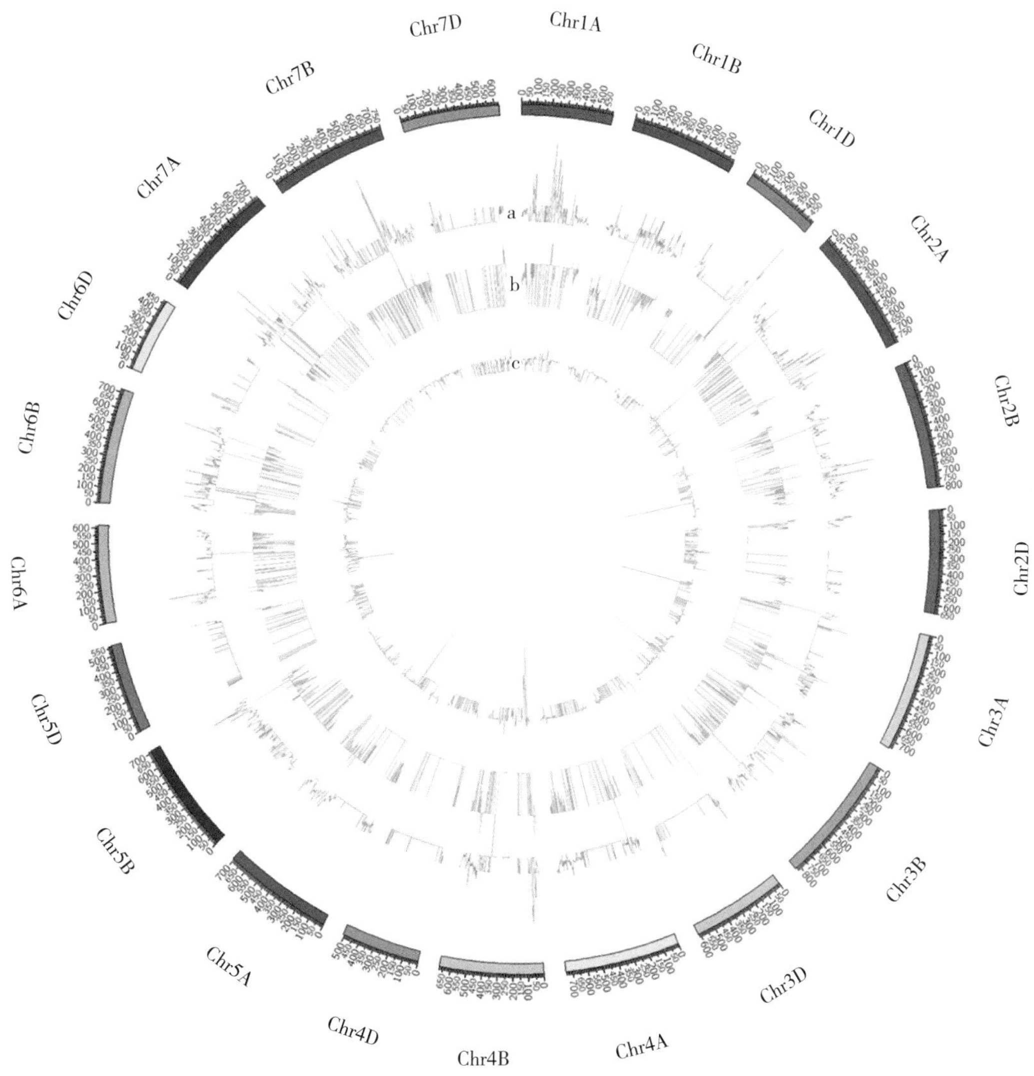

Fig. 4 Summary of SNP mark-ers from the wheat 90 K SNP array between landraces and cultivars in Pakistan. a The ratio of the nucleotide diversity (θ_π) in the landraces versus modern cultivars; b ratio of the Water-son's estimator of nucleotide diversity (θ_w) in the landraces versus modern cultivars, and c ratio of the Fu and Li's D^* values in the landraces versus modern cultivars. The 21 chromosomes are portrayed along the perimeter of each circle. Physical positions of SNP markers from the wheat 90 K SNP array are based on IWGSC v1.0 (http://www.wheatgenome.org/)

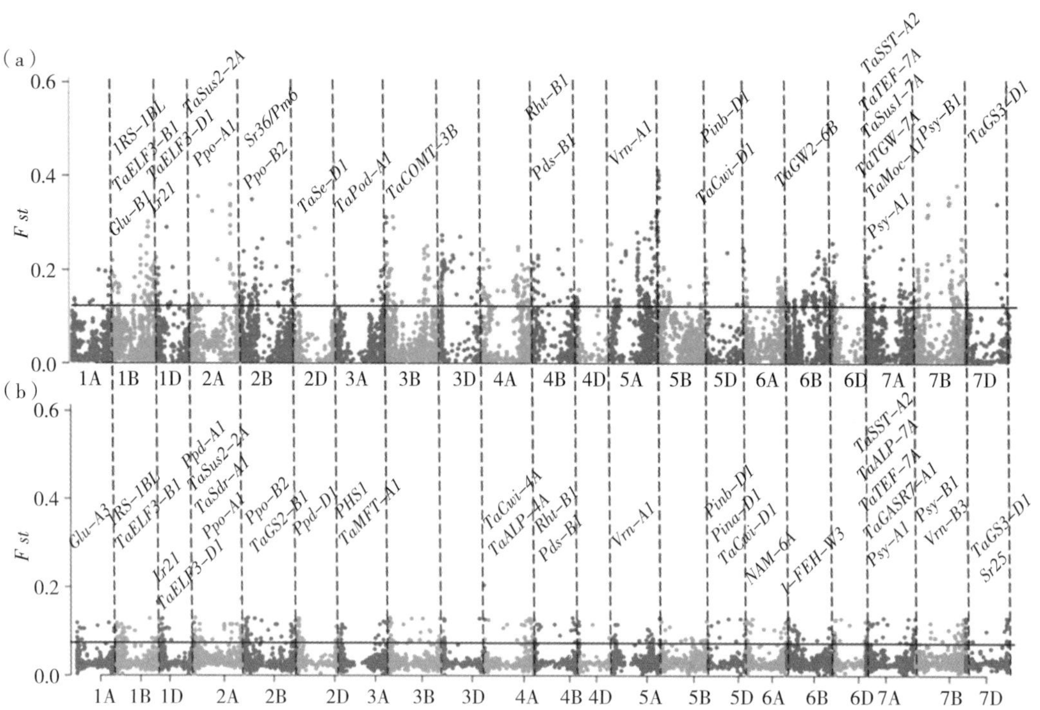

Fig. 5 Functional genes identified in the selection regions by F_{ST} statistics for (a) China and (b) Pakistan

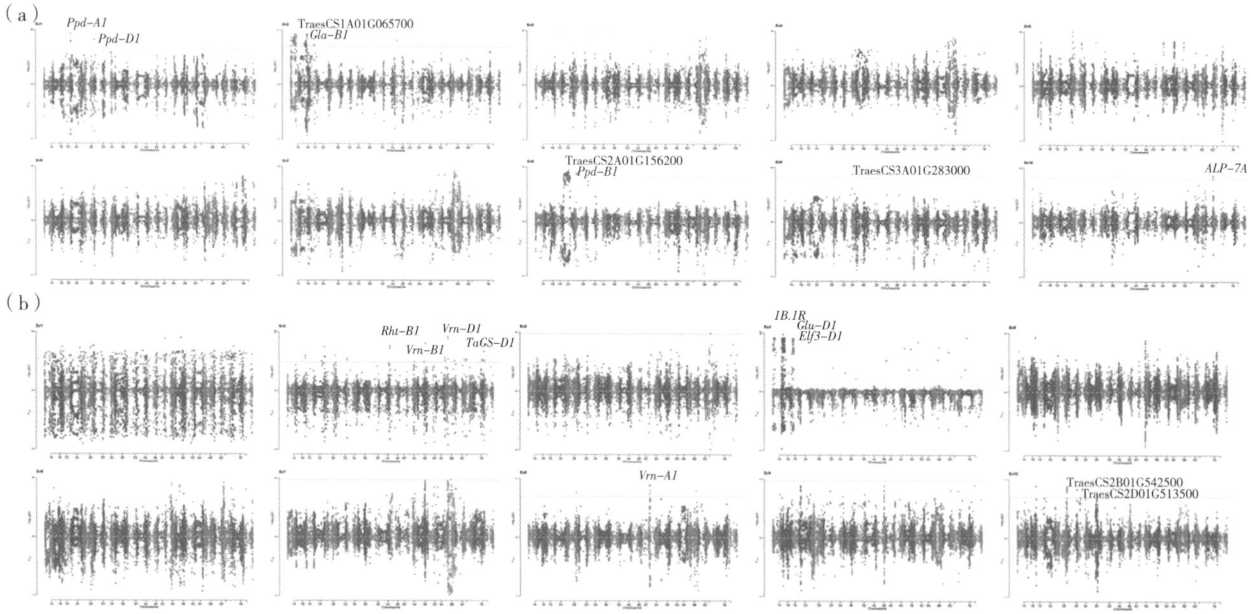

Fig. 6 Miami plots showing selective sweeps identified by EigenG-WAS: (a) showing selective sweeps on top 10 eigenvectors in modern Chinese wheat cultivars versus global winter wheat collection and (b) showing selective sweeps on top 10 eigenvectors in modern Pakistani wheat cultivars versus global winter wheat collection

Discussion

Better understanding the patterns of nucleotide variations in landraces and cultivars allows breeders to identify novel alleles to improve productivity, adaptation, end-use quality and disease resistance in modern cultivars (Yamasaki et al. 2007; Morrell et al. 2011). Furthermore, knowledge about the selection regions is an important step to exploit markers associated with important agronomic traits (Heslot et al. 2015). Most of the landraces have yet to be employed in modern breeding. Therefore, this study investigated nucleotide variations, population structures and identified selection regions between landraces and cultivars in China and Pakistan to facilitate wheat breeding and the use of germplasm.

Population structure and genetic relatedness

The population structure, PCA and NJ-tree analysis separated the 687 accessions into three main clusters: Chinese landraces, Pakistan landraces and modern cultivars (Fig. 1), confirming that artificial selection is a major source for population structure in the wheat germplasm, in accordance with previous reports (Cavanagh et al. 2013; Bonman et al. 2015; Hao et al. 2017). The distance between modern cultivars and landraces from China was much larger than that between modern cultivars and landraces from Pakistan, indicating the relatively frequent use of landraces in developing elite cultivars in Pakistan (Fig. 1). Furthermore, all modern cultivars were separated into three subgroups, which is consistent with geographic origins, and in agreement with previous reports (Cavanagh et al. 2013; Hao et al. 2017; Zhou et al. 2018).

Compared with landraces, a higher level of admixture was detected in modern cultivars. Clustering of cultivars from the Pakistan with varieties from China and wheat accessions from other countries suggests the extensive use of lines sharing common ancestry in the development of these cultivars. The PCA and NJ-tree results indicated that Chinese cultivars influenced by foreign cultivars heavily, particularly the cultivars from Japan, French, Italy and Romania. For example, Abbondanza, Funo, Pascal and Libellula from Italy, Lovrin 10 and Lovrin 13 from Romania, have exerted important influence on wheat breeding in China. Likewise, the Pakistan modern cultivars were also influenced by foreign cultivars, such as Australia, Argentina and Canada (Fig. 2). It was also reported that most of the released cultivars of Pakistan have been developed by crossing exotic parents introduced from Mexico and USA (Rasheed et al. 2016; Ain et al. 2015).

Although China is the largest producer and consumer of wheat in the world, wheat breeding in China initiated until 1930s. Before 1949, most farmers used landraces and only a few introductions were used for production. Most landraces cannot be applied for wheat breeding due to its higher plant height, lower yield potential and disease tolerance. Thus, several outstanding modern cultivars from Europe and America were widely adopted and became the core parents of breeding programs, such as Nanda 2419 (i. e., Mentana, Italy), Ardito (Italy), Lovrin 10 and Lovrin 13 (Romania), Songhuajiang 2 (i. e., Minn 2761, USA), Funo (Italy), Abbondanza (Italy), Orofen (Chile) and Gansu 96 (i. e., CI12203, USA). Furthermore, semi-dwarf materials from the International Maize and Wheat Improvement Center (CIMMYT) have been widely used in China wheat breeding and many excellent varieties have been bred since the 1960s. Wheat breeding in Pakistan started in the early 1930s and was accelerated after the Green Revolution. The early breeding efforts focused on the selection of landraces with higher yield and disease tolerance. After the 1970s, the emphasis of wheat breeding was focused on the introduction and application of outstanding cultivars from Europe and America, and a number of modern cultivars with better yield potential as well as resistance to biotic and abiotic stresses were developed. Based on knowledge of pedigrees of modern cultivars that were included in this study, the PCA and NJ-tree obtained a number of consistent relationships. For example, Jingdong 17, derived from Jindong 8//RHT3//931, positioned close to Jindong 8; Zhongmai 871, Zhongmai 875 and Zhongmai 895 derived from Zhoumai 16, positioned close to Zhoumai 16 in SG 3-1; Lumai 9, derived from a cross of Lovrin 13/71 (17) 6-1, close to Lovrin 13. In addition, some modern cultivars with different backgrounds were clustered together, such as Zhongyou 9507 from China and Sunstate from Australia (Fig. 1). Knowledge about population structure and estimation of the degree of differentiation among germplasm accessions can assist the selection of crossing parents and maximize potential for genetic gain in the progeny (Flint-Garcia et al. 2003; Winfield et al. 2017). The introduction of parental lines from distinct populations has been proposed as a potential path to wheat improvement (Lopes et al. 2015). Therefore, these results provide relevant guidelines to improve wheat by select suitable parents.

Linkage disequilibrium (LD) decay

LD decay is one of the main factors to evaluate the density of markers to conduct a high-resolution association mapping and marker-assisted selection (MAS) (Yu and Buckler 2006; Hao et al. 2011; Sukumaran et al. 2015). LD decay analysis is crucial for species with a large and complex genome, such as wheat (~17.0 Gb) (IWGSC 2014) and barley (~5.1 Gb) (Mascher et al. 2017). In the present study, the marker density was of sufficient resolution for evolution analyses and association mapping because the distance between markers is less than the LD decay distances (Breseghello and Sorrells 2006). In addition, the distances of LD decays in the landraces were shorter than those in the cultivars in both China and Pakistan due to the cultivars germplasm under higher selective pressure than the landraces germplasm (Figure S2). These differences are to be expected, in accordance with previous reports (Han et al. 2016; Hao et al. 2017).

Furthermore, the slower LD decay of A genome than B genome was mainly caused by the stronger selection pressure to A genome than B genome (Marcussen et al. 2014). Our analysis was hampered by the rather low number of mapped SNPs for the D genome. The availability of a SNP array with a much higher number of SNPs in D genome, such as 660 K (Zhou et al. 2018) and 820 K (Winfield et al. 2015) SNP arrays, will likely alleviate this problem.

Patterns of genetic diversity and population structure

The increasing reliance on relatively few varieties and selection in modern breeding programs would have resulted in a progressive reduction in genetic diversity throughout the genome (Tanksley and McCouch 1997; Wright et al. 2005; Doebley et al. 2006; Meyer and Purugganan 2013; Lopes et al. 2015; Han et al. 2016). In the present study, nucleotide diversity (θ_π) was evaluated for CL, PL, CMC, PMC, FMC and all 687 accessions separately. We observed a reduction of 15.0% of θ_π from the landraces to cultivars in China, and 20.8% in Pakistan. These results revealed higher genetic diversity in landraces compared with cultivars, consistent with previous studies (Doebley et al. 2006; Meyer and Purugganan 2013; Cavanagh et al. 2013). Besides, the genome-wide allele frequency has dramatically altered for landraces and cultivars of China and Pakistan ($P < 0.05$). Furthermore, the genetic diversity of Chinese landraces was higher than Pakistan landraces, suggesting a substantial level of genetic variations presented within Chinese landraces, which mainly due to the various eco-geographical regions of China.

Modern breeding suffers from low genetic diversity seriously (Tanksley and McCouch 1997; Doebley et al. 2006; Meyer and Purugganan 2013; Lopes et al. 2015). Previous studies have demonstrated that hybrids between different wheat ecotypes exhibit higher heterosis and enhance the increase in genetic diversity (Betran et al. 2003; Yao et al. 2013). The population structure analysis between accessions from China and Pakistan indicated that the genetic exchange between China and Pakistan are lower. The germplasm introduction and exchanges among populations plays a vital role in improving genetic diversity (Cavanagh et al. 2013; Lopes et al. 2015). Therefore, Chinese and Pakistan germplasm can serve as a valuable source to broaden the genetic basis each other. Landraces can serve as a valuable source to broaden the genetic basis since it harbors favorable alleles, which are not present in the current gene pool (Baenziger and DePauw2009; Cavanagh et al. 2013; Lopes et al. 2015). Furthermore, allelic variations of genes originally found in the landraces but gradually lost through breeding have been recovered only by going back to landraces (Lopes et al. 2015; Hao et al. 2017). Therefore, it is urgent to measure landraces diversity to conserve genetic variability for selection. All of the landraces from China and Pakistan were grouped into two separate clades, showcasing their integrity and diversity. Such diversity could be used to extend the wheat gene pool and promotion wheat improvement.

The present results also revealed different genetic diversities among the three wheat genomes. The highest genetic diversity was identified in the B genome, followed by the A and D genomes, in accordance with previous studies (Chen et al. 2012; Cavanagh et al. 2013; Gao et al. 2017; Hao et al. 2017). Compared with the A genome, the higher genetic diversity of the B genome is mainly attributed to the fact that the A genome carries more agronomic trait-related genes than the B genome and the selection pressure to the A genome was stronger than the B genome (Peng et al. 2003; Marcussen et al. 2014; Hao et al. 2017). The low genetic diversity in the D genome was resulted from the bottleneck caused when hexaploid wheat arose from extremely few hybridizations of tetraploid wheat with *Aegilops tauschii* around 10, 000 years ago (Lupton 1987; Petersen et al. 2006). Previous reports indicated that the D genome made less contributions in modern breeding (Gao et al. 2017) due to the lower genetic diversity. However, *Aegilops tauschii*, the D genome contributor, encompasses huge diversity for various important agronomic traits, such as yield, stress tolerance and quality (Gianibelli et al. 2002; Yan et al. 2004). Therefore, the D genome should have abundant, important, but unexploited alleles, and has a high potential for wheat breeding (Chao et al. 2009; Gao et al. 2017; Rasheed et al. 2018). *Aegilops tauschii* has been used to introgress various traits of economic importance such as biotic stress resistance (Miranda et al. 2006; Trethowan and Mujeeb-Kazi 2008; Olson et al. 2013; Li et al. 2014) and yield traits (Watanabe et al. 2006) into bread wheat. These indicated that the *Aegiolops tauschii* should be more widely exploited in future breeding, and exploiting agronomical important genes from the D genome will be important in future wheat improvement.

Footprints for selection in the wheat genome

During crop breeding, artificial selection results in the reduction or elimination of variation in the selective sweeps. The identification of selection regions is a basic step in under-standing the breeding history (Helyar et al. 2011; Narum and Hess 2011). The genes/SNPs located in these regions should have been under selection during wheat breeding and will be valuable for the MAS of important traits and facilitate the use of germplasm. In this study, selective regions between landraces and cultivars in China and Pakistan were deter-mined among genomes using F_{ST} values. A total of 109 selection regions (28.8-40.7%) were found in both China and Pakistan. In addition, the Chinese cultivars harbored more private variants than the Pakistan cultivars. Selective sweeps of wheat during modern breeding progresses have been detected in several previous studies (Cavanagh et al. 2013; Zhou et al. 2018). Cavanagh et al. (2013) identified selection regions by the wheat 9 K SNP array in 2994 hexaploid wheat accessions, and found that most selection regions associated with yield potential, vernalization, plant height, biotic and abiotic stress tolerance. Zhou et al. (2018) investigated 717 Chinese wheat landraces with 27, 933 DArT and 312, 831 SNP markers, and found 148 selection regions associated with yield and disease tolerance. The limited overlap among various populations were also observed in maize (Pickrell et al. 2009) and rice (Jiao et al. 2012), suggesting that selection may occur on different loci. The reasons for the significant variation among populations could be that (a) selection pressures are likely to vary temporally and spatially, and strongly influenced by environmental conditions, such as climatic regime, biotic and abiotic stresses, altitude, annual sunlight, temperature and precipitation; (b) genotypes underwent different selection pressures to adapt to local agricultural conditions; and (c) selection may be acting on multiple functionally equivalent mutations in different populations (Ralph and Coop 2010;

Cavanagh et al. 2013). The majority of selective sweeps in wheat were in the B genome (47.2%), followed by A (28.8%) and D genomes (16.2%), in contrast to the previous reports, also suggesting that selection pressure to the A genome was stronger than to the B genome (Peng et al. 2003).

We further explored the selective sweeps by EigenGWAS in improved germplasm from Pakistan and China by comparing the allele frequency differentiation in global spring and winter wheat diversity panels, respectively. Chen et al. (2016) proposed a single-marker regression approach based on principal component analysis (or eigen-analysis), called EigenGWAS. This method is similar to genome-wide association studies (GWAS); the analysis procedure of EigenGWAS is nearly identical to the conventional simple regression analysis for GWAS but the phenotype is an eigenvector capturing genetic variation of the studied population. An eigenvalue represents the average genetic variation captured, whereas λGC of the EigenGWAS on the corresponding eigenvector indicates the median of the variation. In an analogue, the difference between eigenvalue and λGC is equivalent to the difference between the mean and a median of a population, implicating the existence of strong selection, or selection sweep by nature or artificial selection by domestication. The regression coefficient of EigenGWAS was approximately to FST. In recent studies, EigenGWAS has been success fully deployed to identify selection signals in species, such as human (Parolo et al. 2017) and wild birds (Bosse et al. 2017; Kim et al. 2017). They identified the genes under selection or adaption, and illustrated how genetic signatures of selection translate into variation in fitness and phenotypes. The number of selection hits in EigenGWAS analyses are relative low because the global diversity panels were used as a reference sets which could compensate the allele frequency at most of the loci. However, several blocks of selective sweeps were identified like genomic region with 1B.1R translocation in wheats from Pakistan and ALP-7A gene in wheats from China. These results also validated the selective sweeps identified by FST scan because all the selection loci by EigenGWAS were shared by FST scan.

A number of selective regions identified in this study are associated with important agronomic traits, including yield-related traits, end-use quality, plant height, biotic and abiotic resistance and vernalization (Table S6). Similar studies reported in maize, rice, soybean, rapeseed, pepper and cotton, also suggested that selective regions are associated with important agronomic traits and physiological and biochemical process (van Heerwaarden et al. 2012; Li et al. 2014; Qin et al. 2014; Han et al. 2016). Furthermore, 11 selective regions were only identified in China (Table S6), whereas 18 selections were only identified in Pakistan, indicating that selection is not uniform in both germplasms due to various environments and breeding targets. Here, this study verified that the selection regions identified between landraces and cultivars is an efficient strategy for exploring complex agronomic trait-related genes or loci on a large scale. Although the function of many selection regions remains unclear, the selection regions observed in this study provide an opportunity for the characterization of genes associated with agronomic traits and facilitate wheat breeding.

Modern breeding has effectively improved wheat adaptation to human preferences and various environments. This study illustrates the nature and extent of genetic differentiation between landraces and cultivated wheat species in China and Pakistan. Furthermore, 198 selection regions were identified in China and Pakistan, which harbored a number of known genes related with environmental adaptability, biotic and abiotic resistance and yield-related traits. Comprehensive knowledge of genetic diversity, population structure, LD decay and the identification selection regions with new genomics tools would provide relevant guidelines to improve wheat by targeting those selected loci and facilitate the use of the germplasm.

Acknowledgements

We thank Prof. R. A. McIntosh, at Plant Breeding Institute, University of Sydney, for reviewing this manuscript. We are grateful to Dr. Huihui Li, at CAAS, for the help in statistical analysis. We acknowledge the Triticeae Toolbox (http://triticeaetoolbox.org; T3) for 90K data availability used in this manuscript. This work was funded by the National Natural Science Foundation of China (31461143021, 31550110212), the National Key Research and Development Program of China (2016YFD0101802, 2016YFE0108600, 2014BAD01B05), and CAAS Science and Technology Innovation Program.

References

Ain QU, Rasheed A, Anwar A, Mahmood T, Imtiaz M, He ZH, Quraishi UM (2015) Genome-wide association for grain yield under rainfed conditions in historical wheat cultivars from Pakistan. Front Plant Sci 6: 743

Akagi T, Hanada T, Yaegaki H, Gradziel TM, Tao R (2016) Genome-wide view of genetic diversity reveals paths of selection and cultivar differentiation in peach domestication. DNA Res 23: 271-282

Alexander DH, Novembre J, Lange K (2009) Fast model-based estimation of ancestry in unrelated individuals. Genome Res 19: 1655-1664

Allen AM, Winfield MO, Burridge AJ, Downie RC, Benbow HR, Barker GL, Scopes G (2016) Characterization of a wheat breeders, array suitable for high-throughput SNP genotyping of global accessions of hexaploid bread wheat (*Triticum aestivum*). Plant Biotechnol J 15: 390-401

Baenziger PS, DePauw RM (2009) Wheat breeding: procedures and strategies. In: Carver BF (ed) Wheat: science and trade. Wiley-Blackwell Publishing, Ames, pp 275-308

Betran FJ, Ribaut JM, Beck D, De Leon DG (2003) Genetic diversity, specific combining ability, and heterosis in tropical maize under stress and nonstress environments. Crop Sci 43: 797-806

Bonman JM, Babiker EM, Cuesta-Marcos A, Esvelt-Klos K, Brown-Guedira G, Chao S, See D, Chen JL, Akhunov E, Zhang JL, Bockelman HE, Gordon TC (2015) Genetic diversity among wheat accessions from the USDA National Small Grains Collection. Crop Sci 55: 1243-1253

Bosse M, Spurgin LG, Laine VN, Cole EF, Firth JA, Gienapp P, Gosler AG, McMahon K, Poissant J, Verhagen I, Groenen MA (2017) Recent natural selection causes adaptive evolution of an avian polygenic trait. Science 358: 365-368

Bradbury PJ, Zhang Z, Kroon DE, Casstevens TM, Ramdoss Y, Buckler ES (2007) TASSEL: software for association mapping of complex traits in diverse samples. Bioinformatics 23: 2633-2635

Brenchley R, Spannagl M, Pfeifer M, Barker GLA, D'Amore R, Allen AM, McKenzie N, Kramer M, Kerhornou A, Bolser D, Kay S, Waite D, Trick M, Bancroft I, Gu Y, Huo NX, Luo MC, Sehgal S, Gill B, Kianian S, Anderson O, Kersey P, Dvorak J, McCombie WR, Hall A, Mayer KFX, Edwards KJ, Bevan MW, Hall N (2012) Analysis of the bread wheat genome using whole-genome shotgun sequencing. Nature 491: 705-710

Breseghello F, Sorrells ME (2006) Association mapping of kernel size and milling quality in wheat (Triticum aestivum L.) cultivars. Genetics 172: 1165-1177

Calus MPL, De Roos APW, Veerkamp RF (2008) Accuracy of genomic selection using different methods to define haplotypes. Genetics 178: 553-561

Cavanagh CR, Chao S, Wang S, Huang BE, Stephen S, Kiani S, For-rest K, Saintenac C, Brown-Guedira GL, Akhunova A, See D, Bai G, Pumphrey M, Tomar L, Wong D, Kong S, Reynolds M, da Silva ML, Bockelman H, Talbert L, Anderson JA, Dreisigacker S, Baenziger S, Carter A, Korzun V, Morrell PL, Dubcovsky J, Morell MK, Sorrells ME, Hayden MJ, Akhunov E (2013) Genome-wide comparative diversity uncovers multiple targets of selection for improvement in hexaploid wheat landraces and cultivars. Proc Natl Acad Sci USA 110: 8057-8062

Chao SM, Zhang WJ, Akhunov E, Sherman J, Ma YQ, Luo MC, Dubcovsky J (2009) Analysis of gene derived SNP marker polymor-phism in US wheat (Triticum aestivum L.) cultivars. Mol Breed 23: 23-33

Chen X, Min D, Yasir TA, Hu YG (2012) Genetic diversity, population structure and linkage disequilibrium in elite Chinese winter wheat investigated with SSR markers. PLoS ONE 7: e44510

Chen GB, Lee SH, Zhu ZX, Benyamin B, Robinson MR (2016) EigenGWAS: finding loci under selection through genome-wide association studies of eigenvectors in structured populations. Heredity 117: 51-61

Choulet F, Alberti A, Theil S, Glover N, Barbe V,

Daron J, Pingault L, Sourdille P, Couloux A, Paux E, Leroy P, Mangenot S, Guilhot N, Gouis JL, Balfourier F, Alaux M, Jamilloux V, Poulain J, Durand C, Bellec A, Gaspin C, Safar J, Dolezel J, Rogers J, Vandepoele K, Aury JM, Mayer K, Berges H, Quesneville H, Wincker P, Feuillet C (2014) Structural and functional partitioning of bread wheat chromosome 3B. Science 345: 1249721

Danecek P, Auton A, Abecasis G, Albers CA, Banks E, DePristo MA, Handsaker RE, Lunter G, Marth GT, McVean G, Durbin R (2011) The variant call format and VCFtools. Bioinformatics 27: 2156-2158

Doebley JF, Gaut BS, Smith BD (2006) The molecular genetics of domestication. Cell 127: 1309-1321

Doyle JJ, Doyle JL (1987) A rapid DNA isolation procedure from smallquantities of fresh leaf tissues. Phytochem Bull 19: 11-15

Dvorak J, Terlizzi PD, Zhang HB, Resta P (1993) The evolution of polyploid wheats: identification of the A genome donor species. Genome 36: 21-31

Edae EA, Byrne PF, Haley SD, Lopes MS, Reynolds MP (2014) Genome-wide association mapping of yield and yield components of spring wheat under contrasting moisture regimes. Theor Appl Genet 127: 791-807

Flint-Garcia SA, Thornsberry JM, Buckler ES (2003) Structure of link-age disequilibrium in plants. Annu Rev Plant Biol 54: 357-374

Fu YX (1997) Statistical tests of neutrality of mutations against population growth, hitckhiking and background selection. Genetics 147: 915-925

Gao LF, Zhao GY, Huang DW, Jia JZ (2017) Candidate loci involved in domestication and improvement detected by a published 90 K wheat SNP array. Sci Rep-UK 7: 44530

Gianibelli MC, Echaide M, Larroque OR, Carrillo JM, Dubcovsky J (2002) Biochemical and molecular characterization of Glu-1 loci in Argentinean wheat cultivars. Euphytica 128: 61-73

Gupta PK, Rustgi S, Mir RR (2008) Array-based high-throughput DNA markers for crop improvement. Heredity 101: 5-18

Han YP, Zhao X, Liu DY, Li YH, Lightfoot DA, Yang ZJ, Zhao L, Zhou G, Wang ZK, Huang L, ZW Zhang, Qiu LJ, Zheng HK, Li WB (2016) Domestication footprints anchor genomic regions of agronomic importance in soybeans. New Phytol 209: 871-884

Hao CY, Wang LF, Ge HM, Dong YC, Zhang XY (2011) Genetic diversity and linkage disequilibrium in Chinese bread wheat (*Triticum aestivum* L.) revealed by SSR markers. PLoS ONE 6: e17279

Hao CY, Wang YQ, Chao SAM, Li T, Liu HX, Wang LF, Zhang XY (2017) The iSelect 9 K SNP analysis revealed polyploidization induced revolutionary changes and intense human selection causing strong haplotype blocks in wheat. Sci Rep-UK 7: 41247

Heffner EL, Sorrells ME, Jannink JL (2009) Genomic selection for crop improvement. Crop Sci 49: 1-12

Helyar SJ, Hemmer-Hansen J, Bekkevold D, Taylor MI, Ogden R, Limborg MT, Cariani A, Maes JE, Diopere E, Carvalho GR, Nielsen EE (2011) Application of SNPs for population genetics of non-model organisms: new opportunities and challenges. Mol Ecol Resour 11: 123-136

Heslot N, Jannink JL, Sorrells ME (2015) Perspectives for genomic selection applications and research in plants. Crop Sci 55: 1-12

Hill TA, Ashrafi H, Reyes-Chin-Wo S, Yao J, Stoffel K, Truco MJ, Kozik A, Michelmore RW, Deynze AV, Van Deynze A (2013) Characterization of Capsicum annuum genetic diversity and population structure based on parallel polymorphism discovery with a 30 K unigene pepper gene chip. PLoS ONE 8: e56200

Holsinger KE, Weir BS (2009) Genetics in geographically structured populations: defining, estimating and interpreting FST. Nat Rev Genet 10: 639-650

Hudson RR, Slatkin M, Maddison WP (1992) Estimation of levels of gene flow from DNA sequence data. Genetics 132: 583-589

Hulse-Kemp AM, Lemm J, Plieske J, Ashrafi H, Buyyarapu R, Fang DD et al (2015) Development of a 63 K SNP array for cotton and high-density mapping of intra-and inter-specific populations of Gossypium spp. G3-Genes Genom Genet 5: 1187-1209

Hutter S, Vilella AJ, Rozas J (2006) Genome-wide DNA polymorphism analyses using VariScan. BMC Bioinform 7: 1-10

International Wheat Genome Sequencing Consortium (2014) A chromosome-based draft sequence of the hexaploid bread wheat (*Triticum aestivum*) genome. Science 345: 1251788

Jiao YP, Zhao HN, Ren LH, Song WB, Zeng B, Guo JJ, Wang BB, Liu ZP, Chen J, Zhang M, Lai JS (2012) Genome-wide genetic changes during modern breeding of maize. Nat Genet 44: 812

Kihara H (1944) Discovery of the DD-analyser, one of the ancestors of Triticum vulgare. Agric Hortic 19: 889-890

Kim KW, Bennison C, Hemmings N, Brookes L, Hurley LL, Griffith SC, Burke T, Birkhead TR, Slate J (2017) A sex-linked supergene controls sperm morphology and swimming speed in a songbird. Nat Ecol Evol 1: 1168-1176

Li YH, Zhao SC, Ma JX, Li D, Yan L, Li J, Chang RZ et al (2013) Molecular footprints of domestication and improvement in soy-bean revealed by whole genome re-sequencing. BMCGenom 14: 579

Li J, Wan HS, Yang WY (2014) Synthetichexaploid wheat enhances variation and adaptive evolution of bread wheat in breeding processes. J Syst Evol 52: 735-742

Liu K, Muse SV (2005) PowerMarker, an integrated analysis environment for genetic marker analysis. Bioinformatics 21: 2128-2129 Lopes M, Dreisigacker S, Peña R, Sukumaran S, Reynolds M (2014) Genetic characterization of the Wheat Association Mapping Initiative (WAMI) panel for dissection of complex traits in spring wheat. Theor Appl Genet 128: 453-464

Lopes MS, El-Basyoni I, Baenziger PS, Singh S, Royo C, Ozbek K, Aktas H, Ozer E, Ozdemir F, Manickavelu A, Ban T, Vikram P (2015) Exploiting genetic diversity from landraces in wheat breeding for adaptation to climate change. J Exp Bot 66: 3477-3486

Luo MC, Yang ZL, You FM, Kawahara T, Waines JG, Dvorak J (2007) The structure of wild and domesticated emmer wheat populations, gene flow between them, and the site of emmer domestication. Theor Appl Genet 114: 947-959 Lupton FGH (ed) (1987) Wheat breeding: its scientific basis. Chapman and Hall Ltd, London

Marcussen T, Sandve SR, Heier L, Spannagl M, Pfeifer M, Jakobsen KS, Jakobsen KS, The International Wheat Genome Sequencing Consortium, Wulff BBH, Steuernage B, Mayer KFX, Olsen OA (2014) Ancient hybridizations among the ancestral genomes of bread wheat. Science 345: 1250092

Mascher M, Gundlach H, Himmelbach A, Beier S, Twardziok SO, Wicker T et al (2017) A chromosome conformation capture ordered sequence of the barley genome. Nature 544: 427-433

McFadden ES, Sears ER (1946) The origin of Triticum spelta and its free-threshing hexaploid relatives. J Hered 37: 107-116

Meyer RS, Purugganan MD (2013) Evolution of crop species: genetics of domestication and diversification. Nat Rev Genet 14: 840-852 Miranda LM, Murphy JP, Marshall D, Leath S (2006) Pm34: a new powdery mildew resistance gene transferred from *Aegilops tauschii* Coss. to common wheat (*Triticum aestivum* L.). Theor Appl Genet 113: 1497-1504

Morrell PL, Buckler ES, Ross-Ibarra J (2011) Crop genomics: advancesand applications. Nat Rev Genet 13: 85-96

Narum SR, Hess JE (2011) Comparison of FST outlier tests for SNP loci under selection. Mol Ecol Resour 11: 184-194

Olson EL, Rouse MN, Pumphrey MO, Bowden RL, Gill BS, Poland JA (2013) Introgression of stem rust resistance genes SrTA10187 and SrTA10171 from Aegilops tauschii to wheat. Theor Appl Genet 126: 2477-2484

Parolo S, Lacroix S, Kaput J, Scott-Boyer MP (2017) Ancestors' dietary patterns and environments could drive positive selection in genes involved in micronutrient metabolism-the case of cofactor transporters. Gene Nutr 12: 28

Peng JX, Ronin Y, Fahima T, Röder MS, Li YC, Nevo E, Korol A (2003) Domestication quantitative trait loci in Triticum dicoccoides, the progenitor of wheat. Proc Natl Acad Sci USA 100: 2489-2494

Petersen G, Seberg O, Yde M, Berthelsen K (2006) Phylogenetic relationships of Triticum and Aegilops and evidence for the origin of the A, B, and D genomes of common wheat (Triticum aestivum). Mol Phylogenet Evol 39: 70-82

Pfeifer M, Kugler KG, Sandve SR, Zhan B, Rudi H, Hvidsten TR, International Wheat Genome Sequencing Consortium, Mayer KFX, Olsen OA (2014) Genome interplay in the grain transcrip-tome of hexaploid bread wheat. Science 345: 1250091

Pickrell JK, Coop G, Novembre J, Kudaravalli S, Li JZ, Absher D, Srinivasan BS, Barsh GS, Myers RM, Feldman MW, Pritchard JK (2009) Signals of recent positive selection in a worldwide sample of human populations. Genome Res 19: 826-837

Qin L, Hao CY, Hou J, Wang YQ, Li T, Wang LF, Ma ZQ, Zhang XY (2014) Homologous haplotypes, expression, genetic effects and geographic distribution of the wheat yield gene TaGW2. BMC Plant Biol 14: 107

Ralph P, Coop G (2010) Parallel adaptation: one or many waves ofadvance of an advantageous allele? Genetics 186: 647-668

Rasheed A, Xia XC, Mahmood T, Quraishi UM, Aziz A,

Bux H, Mahmood Z, Mirza JI, Mujeeb-Kazi A, He ZH (2016) Comparison of economically important loci in landraces and improved wheat cultivars from Pakistan. Crop Sci 56: 1-15

Rasheed A, Mujeeb-Kazi A, Ogbonnaya FC, He ZH, Rajaram S (2017) Wheat genetic resources in the post-genomics era: promise and challenges. Ann Bot-Lond 121: 603-616

Rasheed A, Ogbonnaya FC, Lagudah E, Appels R, He ZH (2018) The goat grass genome's role in wheat improvement. Nat Plants 4: 56

Sukumaran S, Dreisigacker S, Lopes M, Chavez P, Reynolds MP (2015) Genome-wide association study for grain yield and related traits in an elite spring wheat population grown in temperate irrigated environments. Theor Appl Genet 128: 353-363

Swarts K, Gutaker RM, Benz B, Blake M, Bukowski R, Holland J, Kruse-Peeples M, Lepak N, Prim L, Romay MC, Ross-Ibarra J, Gonzalez JJS, Schmidt C, Schuenemann VJ, Krause J, Matson RG, Weige D, Buckler ES, Burbano HA (2017) Genomic estimation of complex traits reveals ancient maize adaptation to temperate North America. Science 57: 512-515

Tajima F (1983) Evolutionary relationship of DNA sequences in finitepopulations. Genetics 105: 437-460

Tanksley SD, McCouch SR (1997) Seed banks and molecular maps: unlocking genetic potential from the wild. Science 277: 1063-1066 Tanno KI, Willcox G (2006) How fast was wild wheat domesticated. Science 311: 1886

Trethowan RM, Mujeeb-Kazi A (2008) Novel germplasm resources for improving environmental stress tolerance of hexaploid wheat. Crop Sci 48: 1255-1265

van Heerwaarden J, Hufford MB, Ross-Ibarra J (2012) Historical genomics of North American maize. Proc Natl Acad Sci USA 109: 12420-12425

Wang S, Wong D, Forrest K, Allen A, Chao S, Huang BE, Maccaferri M, Salvi S, Milner SG, Cattivelli L, Mastrangelo AM, Stephen S, Barker G, Wieseke R, Plieske J, International Wheat Genome Sequencing Consortium, Lillemo M, Mather D, Appels R, Dulferos R, Brown-Guedira G, Korol A, Akhunova AR, Feuillet C, Salse J, Morgante M, Pozniak C, Luo MC, Dvorak J, Morell M, Dubcovsky J, Ganal M, Tuberosa R, Lawley C, Mikoulitch I, Cavanagh C, Edwards KJ, Hayden M, Akhunov E (2014) Characterization of polyploid wheat genomic diversity using the high-density 90, 000 SNP array. Plant Biotech J 12: 787-796

Watanabe N, Fujii Y, Takesada N, Martinek P (2006) Cytological and microsatellite mapping of genes for brittle rachis in a *Triticum aestivum-Aegilops tauschii* introgression line. Euphytica 151: 63-69 Watterson GA (1975) On the number of segregating sites in genetical models without recombination. Theor Popul Biol 7: 256-276

Wei DY, Cui YX, He YJ, Xiong Q, Qian LW, Tong CB, Lu GY, Ding YJ, Li JN, Jung C, Qian W (2017) A genome-wide survey with different rapeseed ecotypes uncovers footprints of domestication and breeding. J Exp Bot 68: 4791-4801

Weir BS (1996) Genetic data analysis II, 2nd edn. Sinauer, Sunderland Winfield MO, Allen AM, Burridge AJ, Barker GL, Benbow HR, Wilkinson PA, Coghill J (2015) High-density SNP genotyping array for hexaploid wheat and its secondary and tertiary gene pool. Plant Biotechnol J 14: 1195-1206

Winfield MO, Allen AM, Wilkinson PA, Burridge AJ, Barker GL, Coghill J, Waterfall C, Wingen LU, Griffiths S, Edwards KJ (2017) High density genotyping of the AE Watkins Collection of hexaploid landraces identifies a large molecular diversity compared to elite bread wheat. Plant Biotechnol J. https://doi.org/10.1111/pbi.12757

Wright SI, Bi IV, Schroeder SG, Yamasaki M, Doebley JF, McMullen MD, Brandon S, Gaut BS (2005) The effects of artificial selection on the maize genome. Science 308: 1310-1314

Yamasaki M, Wright SI, McMullen MD (2007) Genomic screening for artificial selection during domestication and improvement inmaize. Ann Bot-Lond 100: 967-973

Yan Y, Zheng J, Xiao Y, Yu J, Hu Y, Cai M, Li Y, Hsam SLK, Zeller FJ (2004) Identification and molecular characterization of a novel y-type *Glu-Dt1* glutenin gene of *Aegilops tauschii*. Theor Appl Genet 108: 1349-1358

Yao H, Gray AD, Auger DL, Birchler JA (2013) Genomic dosage effects on heterosis in triploid maize. Proc Natl Acad Sci USA 110: 2665-2669

Yu J, Buckler ES (2006) Genetic association mapping and genome organization of maize. Curr Opin Biotechnol 17: 155-160

Zhou Y, Chen, ZX, Cheng MP, Chen J, Zhu TT, Wang R, Liu YX, Qi PF, Chen GY, Jiang QT, Wei YM, Luo MC, Nevo E, Allaby RG, Liu DC, Wang JR, Dvorak J, Zheng YL (2018) Uncovering the dispersion history, adaptive evolution and selection of wheat in China. Plant Biotechnol J 16: 280-291

节水节肥与抗病性

从产量和品质性状的变化分析北方冬麦区小麦品种抗热性

韩利明[1,6]，张　勇[1]，彭惠茹[2]，乔文臣[3]，何明琦[4]，王洪刚[5]，
曲延英[6]，刘春来[7]，何中虎[1,8]

([1]中国农业科学院作物科学研究所/国家小麦改良中心/农作物基因资源与基因改良国家重大科学工程，北京 100081；[2]中国农业大学农学与生物技术学院，北京 100193；[3]河北省农林科学院旱作农业研究所，河北衡水 053000；[4]石家庄市农林科学院，河北石家庄 050000；[5]山东农业大学农学院，山东泰安 271018；[6]新疆农业大学农学院，新疆乌鲁木齐 830052；[7]国营保定农场，河北保定 072550；[8]国际玉米小麦改良中心（CMMYT）中国办事处，北京 100081)

摘要：培育抗热性强的品种对应对气候变化、保障小麦稳产性具有重要意义。选用北方冬麦区53份主栽品种和苗头品系，于2008—2009年度分别种植在北京、石家庄、衡水、安阳和泰安5点，各点设正常温度和塑料棚热胁迫处理。结果表明，千粒重可作为抗热性筛选的简易指标。农大189、CA0518和京冬8号的产量和千粒重在正常和热处理环境中均较高，抗热性好；衡观33和CA0736的产量在正常和热处理环境中均较高，但千粒重均表现中等，抗热性较好；农大211、石麦15、济麦22、农大3432和山农2149在正常环境中的产量和千粒重均较高，但在热处理环境中产量和千粒重均较低，抗热性差。53份品种按和面时间和峰值曲线面积可分为强筋、中强筋、中筋、中弱筋和弱筋5大类。热处理使所有类型材料的蛋白质含量和籽粒硬度增加，峰值带高、8min带高和8min带宽降低，并降低强筋、中强筋和中筋类型材料的和面时间与峰值曲线面积，增加中弱筋和弱筋类型材料的和面时间与峰值曲线面积。

关键词：普通小麦；热胁迫；产量；千粒重；和面仪参数

Analysis of Heat Resistance for Cultivars from North China Winter Wheat Region by Yield and Quality Traits

HANLi-Ming[1,6], ZHANG Yong[1], PENG Hui-Ru[2], QIAO Wen-Chen[3], HE Ming-Qi[4], WANG Hong-Gang[5], QU Yan-Ying[6], LIU Chun-Lai[7], and HE Zhong-Hu[1,8]

([1]Institute of Crop Sciences/National Wheat Improvement Center/National Key Facility for Crop Gene Resources and Genetic Improvement, Chinese Academy of Agricultural Sciences, Beijing 100081, China; [2]Agronomy and Biotechnology College, China Agricultural University, Beijing 100193, China; [3]Dryland Agricultural Research Institute, Hebei Academy of Agricultural and Forestry Sciences, Hengshui 053000, China; [4]Shijiazhuang Academy of Agricultural and forest

---

原文发表在《作物学报》，2010，36（9）：1538-1540.

Sciences, Shijiazhuang 050000, China;[5] *Agronomy College, Shandong Agricultural University, Tai'an 271018, China*;[6] *Agronomy College, Xinjiang Agricultural University, Urumqi 830052, China*;[7] *Hebei Baoding State-owned Farms, Baoding 072550, China*;[8] *CIMMYT-China Office, c/o CAAS, Beijing 100081, China*)

Abstract: Selection of heat-resistance cultivars is an important approach for coping climate changes as well as ensuring stable production. Fifty-three wheat (*Triticum aestivum* L.) cultivars and advanced lines from the North China Winter Wheat Region, planted in five locations including Beijing, Shijiazhuang, Hengshui, Anyang, and Tai'an in 2008-2009 wheat season, were used to analyze the effect of heat stress on yield and quality traits. The result indicated that thousand-kernel weight (TKW) can be used as a simple criterion for heat-resistance selection; Nongda 189, CA0518, and Jingdong 8 performed high yield and high TKW under both normal and heat-stress environments, and were characterized with high resistance to heat stress; Hengguan 33 and CA0736 performed high yield but medium TKW in all environments, and were characterized with good resistance to heat stress; Nongda 211, Shimai 15, Jimai 22, Nongda 3432, and Shannong 2149 performed high yield and high TKW in normal environments, but low yield and low TKW in heat stress environments, and were characterized with poor resistance to heat stress. The 53 cultivars and lines were classified into five groups including strong, medium strong, medium, medium weak, and weak gluten strength based on Mixograph Midline peak time and peak integral. The grain protein content and grain hardness were increased, and Midline peak value, time x value, and time x width were decreased for all materials under heat stress when comparing with that of normal environment, and the Midline peak time and peak integral were increased for materials with medium weak and weak gluten strength whereas decreased for materials with strong, medium strong, and medium gluten strength.

Keywords: Common wheat (*Triticum aestivum* L.); Heat stress; Yield; Thousand kernel weight; Mixograph parameters

应对全球气候变化已成为国际小麦育种的重要方向，气候变化对小麦生产的影响方向和程度因地区而异，但主要表现为干旱和热害，因此提高水分利用效率和品种抗热性至关重要[1]。包括北京、河北、河南和山东等省市在内的北方冬麦区约占全国年播种面积和总产量的65%和75%，在小麦生产中占举足轻重的地位[2]。IPCC资料表明，过去50年中气温以每10年0.13℃的速率增加[3]，小麦灌浆中后期的热害胁迫越来越严重。邓振镛等[4]表明高温往往伴随着干旱和大风而形成干热风，减慢籽粒灌浆速度，甚至停止灌浆，形成瘦秕粒，严重影响产量。干热风在我国小麦主产区基本是十年七遇，约使其减产10%～20%[5]，因此培育灌浆速率快的抗热性品种是北方冬麦区的重要目标。在北方冬麦区，灌浆期高温对北部冬麦区和黄淮北片冬麦区的影响更大，而对黄淮南片冬麦区的影响相对小一些。

国外早在20世纪90年代就开始小麦抗热性研究，其中美国、澳大利亚和国际玉米小麦改良中心的工作较为系统，国际影响大。对春麦一般通过晚播的方法进行全生育期抗热性筛选；对冬麦则通过各种人工措施在灌浆期进行热处理，高温使灌浆期缩短，灌浆速率减小，所以育种工作以提高灌浆期的抗热性为主。He等[6]认为，春小麦发生热害时，株高、穗粒数和产量所受影响比穗数和千粒重大，灌浆速率比灌浆持续期对热害更敏感。Dias等[7]表明，千粒重和灌浆速率对最终产量贡献较大，Keling等[8]和Rijven[9]认为灌浆期热害抑制了淀粉合成相关酶的活性而影响灌浆质量，Altenbach等[10]表明热害使淀粉积累提前结束而降低产量。Lawlor和Mitchell[11]发现，灌浆期温度每上升1℃，灌浆期缩短5%，并降低收获指数和产量。Blum[12]和Rintamaki等[13]认为灌浆期温度的异常升高和强光辐射可对小麦造成双重胁迫，使叶片光合反应中心异常，从而影响光合作用的进程和籽粒的灌浆质量，降低产量。Randall和

Moss[14]研究表明，开花后高温会加快胚乳中淀粉积累速度，但这不足以弥补因积累时间缩短所造成的粒重损失。郑飞等[15]发现，花后高温影响小麦灌浆期的物质运输与分配过程，从而影响籽粒存储光合物质的能力。热害还严重影响加工品质。Spiertz 等[16]表明，热害使籽粒硬度增加，籽粒变小，蛋白质相对含量提高，磨粉时间延长，出粉率降低。灌浆期热害对淀粉的影响比对蛋白质含量的影响大[17]，热敏感品种蛋白质相对含量增幅比抗热性品种大[18-19]。Gupta 等[20]发现灌浆期热害可增加可溶性谷蛋白聚合体、醇溶蛋白等的相对含量，降低不溶性谷蛋白聚合体相对含量，从而影响面团加工品质。

国内外对抗热性的分子机制也进行了一些探讨。Jane 等[21]发现，增加转录因子、去氧化损伤、对蛋白质起保护作用的分子伴侣等相关基因的表达，减少细胞程序性死亡和生物应激反应基因的表达，是植物获得抗热性的分子机制。Qin 等[22]认为，热应答基因包含热休克蛋白、转录因子、植物激素、钙/糖信号途径、RNA 代谢、核糖体蛋白激酶等，抗热性品种在热害发生时基因表达量变化比热敏感性品种小，控制新陈代谢和蛋白质合成的基因只有在长期热胁迫下才受到影响。Sairam 等[23]认为，在热害发生时，抗热性品种的超氧化物歧化酶和过氧化氢酶表达量比热敏感性品种高。国内一些学者就热害对蔗糖合成酶和可溶性淀粉合成酶的影响进行了研究[24-26]。李浩等[27]认为，高温胁迫总体上诱导品质不稳定品种的基因表达，而抑制品质稳定品种的基因表达。虽然对抗热性分子机制的研究取得了一定进展，但尚不能用于品种抗热性筛选，因此常规方法仍是鉴定抗热性的主要途径。

国内对北方冬麦区主要品种和品系的抗热性还缺乏系统研究，本研究采用塑料大棚升温方法进行抗热性鉴定，分析热胁迫和正常环境下北方冬麦区代表性品种（系）的产量和主要品质性状的变化，旨在为培育抗热性品种提供参考依据。

1 材料与方法

1.1 试验材料

选用来自我国北方冬麦区北京、河北、河南和山东的 53 份代表性主栽品种和新育成品系（表 1），其中 28 份为审定品种，25 份为参加区域试验的新品系，2008—2009 年度分别种植在北京、河北石家庄、河北衡水、河南安阳和山东泰安，小区面积 1.2m^2，3 行区，2m 行长。

1.2 热处理

开花后 10d（灌浆期）进行塑料大棚升温，用 TRS1S 型温度纪录仪记载棚内热处理和棚外正常处理离地 90cm 左右（小麦冠层）处的温度。开花后 10d 到成熟期高温处理比正常环境的温度平均高 1.6℃，棚内热处理平均最高温度为 35.6℃，棚外正常处理平均最高温度为 28.7℃（表 2）。北京点设高温处理和正常处理各 3 次重复；石家庄、安阳和泰安点各设正常处理 1 次重复和高温处理 2 次重复；衡水点仅设高温处理 3 次重复，其结果在分析时仅做参考。

1.3 产量性状调查方法

田间和室内分别记载抽穗期、开花期、株高、穗粒数（30 穗）、千粒重和产量。由于穗数在灌浆前已经确定，故没有对其进行分析。灌浆期为开花至成熟的天数，灌浆速率为千粒重除以灌浆天数（g·d^{-1}）。

1.4 品质分析方法

利用单籽粒谷物特性系统（Single Kernel Characterization System 4100，SKCS 4100，USA）测定籽粒硬度和水分。参照 AACC 26-20 方法，按籽粒硬度和水分调整加水量，用德国 Brabender 实验磨（Junior D-47055 型）制粉[28]，出粉率 60%左右。参照 AACC 39-10A 方法，用近红外分析仪（Foss 1241，Sweden）测定面粉蛋白质和水分含量。参照 AACC 54-40A 方法，用 10g 和面仪（National MFG Mixograph）分析和面特性，选用和面时间（Midline peak time，MPT）、峰值带高（Midline peak value，MPV）、峰值曲线面积（Midline peak integral，MPI）、8 min 带高（Midline time x value，MTxV）和 8 min 带宽（Midline time x width，MTxW）等和面仪参数。

1.5 统计分析

产量和千粒重热感指数 $S=(1-YD/YP)/(1-\overline{YD}/\overline{YP})$[29]，$YD$ 为某品种在热胁迫下的千粒重或产量，YP 为某品种在正常环境下的千粒重或产量，$\overline{YD}$ 为所有品种在热胁迫处理下千粒重或产量的平均值，$\overline{YP}$ 为所有品种在正常环境下千粒重或产量的平均值。$S<1$ 为抗热性品种，$S\geqslant 1$ 为热敏感品种[30]。

采用Statistical Analysis System（SAS Institute，2000）统计分析软件进行基本统计量和方差分析。采用欧氏距离按类平均法进行聚类分析。在分析高温处理和正常环境对各类品种品质参数的影响时，采用混合线性模型进行计算，其中类为固定效应，类内品种为随机效应。

2 结果

2.1 热胁迫对产量性状的影响

2.1.1 方差分析 方差分析（表3）表明，穗粒数、灌浆速率、千粒重和产量的基因型效应均达0.1%显著水平，灌浆期的基因型效应达1%显著水平，地点对穗粒数、灌浆期、灌浆速率、千粒重和产量的影响分别达5%、1%或0.1%显著水平，处理对灌浆速率、千粒重和产量的影响达0.1%显著水平，对穗粒数和灌浆期的影响达5%显著水平，基因型和地点互作对穗粒数、灌浆期、灌浆速率、千粒重和产量的影响分别达5%、1%或0.1%显著水平，地点和处理互作对灌浆期、千粒重和产量的影响达5%或0.1%显著水平。综合以上分析可知，热处理主要通过影响千粒重来影响产量，而千粒重受灌浆期和灌浆速率的共同影响。比较两种处理下产量的变化（表4），热处理显著降低了灌浆速率、千粒重和产量，而穗粒数和灌浆期变化不显著。

2.1.2 产量热感指数 046402等9份材料产量热感指数2.0～3.1，067257等15份材料产量热感指数1.1～1.8，CA0629等4份材料产量热感指数为1.0，这28份材料为产量热敏感性品种（表5）。其中中麦349在正常和热处理环境下产量都较高，分别为8.24t·hm^{-2}和7.64t·hm^{-2}，农大211、农大318、08CA190、石麦15、济麦22、农大3432、济麦19、山农2149、衡4422、中麦875和石麦18在正常环境中产量较高（7.53～8.26t·hm^{-2}）。在产量较高的品种中，农大211的产量热敏感性较高，其次为08CA190、农大318和石麦15、济麦22、农大3432、济麦19、山农2149、衡4422、中麦875和石麦18。农大189等19份材料产量热感指数0～0.9，CA0736等6份材料产量热感指数−2.2～−0.1，这25份材料为产量抗热性品种，其中农大189、京冬8号、农大3659、衡观33、CA0415、农大3634、DH155和CA0736在正常（7.53～8.33 t·hm^{-2}）和热处理环境中（7.56～8.13t·hm^{-2}）产量都较高，衡观115（7.62t·hm^{-2}）和CA0518（7.74t·hm^{-2}）在正常环境中产量较高。在这部分材料中，CA0736和DH155抗热性好，其次为农大3634、CA0415、衡观33、农大3659、京冬8号和农大189。正常环境中产量最高的是农大189（8.33t·hm^{-2}），热处理环境中产量最高的是农大3659（8.13t·hm^{-2}）。

2.1.3 千粒重热感指数 农大3432等3份材料千粒重热感指数2.0～2.7，济麦22等18份材料千粒重热感指数1.1～1.7，07CA010等4份材料千粒重热感指数为1.0，这25份材料为千粒重热敏感性品种（表5）。其中CA0629、农大3492、中麦875、农大3634、07CA010和济麦19在正常（44.1～47.7g）和热处理环境中（40.8～43.1g）千粒重都较高，农大3432、农大211、中麦175、济麦22、DH155、石麦15、山农2149、046402、烟农19和035037在正常环境中千粒重较高（40.3～47.7g）。在千粒重较高的品种中，农大3432、农大211和中麦175的千粒重热敏感性高。石07-6023等27份材料千粒重热感指数0.1～0.9，衡4399千粒重热感指数为−0.4。这28份材料的抗热性较好，其中农大413、农大189、CA0548、055319、京冬8号、邯6228、石家庄8号、CA0518、中麦306和08CA190在正常（41.7～46.3g）和热处理环境中（40.3～44.6g）千粒重都较高，农大408和农大318在正常环境中千粒重较高，分别为40.7g和40.4g。在这部分材料中，08CA190、中麦306、石家庄8号、邯6228、京冬8号、CA0548、055319、CA0518和农大189的抗热性较好，热处理下千粒重均在40g以上。正常环境中千粒重最高的是中麦875（47.7g），热处理环境中千粒重最高的是京冬8号（44.6g）。

综合产量和千粒重热感指数分析可知，农大189、CA0518和京冬8号的产量和千粒重在正常和热处理环境中均较高，抗热性好；衡观33和CA0736的产量在正常和热处理环境中均较高，但千粒重在正常和热处理环境中均一般，抗热性较好；农大211、石麦15、济麦22、农大3432和山农2149在正常环境中的产量和千粒重均较高，但在热处理环境中产量和千粒重均较低，抗热性较差。

2.2 热胁迫对品质性状的影响

2.2.1 聚类分析 在正常环境下，将53份品种和品系按和面时间和峰值曲线面积进行聚类分析，在

0.65 欧氏距离上 53 份材料可分为强筋、中强筋、中筋、中弱筋和弱筋 5 大类（图略）。强筋品种包括 037042 和中优 206，中强筋品种包括山农 2149、石 6207、石 07-6023、农大 3492、济麦 20 和石 B07-4056 等 6 份材料，中筋品种包括 CA0736、035037、石优 17、CA0518、中麦 175、中麦 306、07CA010 和 067257 等 8 份材料，中弱筋品种包括 CA0548、中麦 875、京冬 8 号、08CA190、055319、烟农 19、衡观 111、衡 4399、济麦 22、衡 6607、邯 6228、CA0415、石 06-6051、DH155、046402、石新 733、农大 413、中麦 349、CA0629 和济麦 19 等 20 份材料，弱筋品种包括农大 3432、石 8 号、衡观 216、农大 189、衡 4422、农大 211、农大 212、石麦 15、石麦 18、农大 3659、农大 3634、衡观 33、农大 408、农大 318、衡观 115、石 4185 和衡 6632 等 17 份材料。

2.2.2 热处理对不同筋力类型品种主要品质性状的影响 热处理使强筋材料蛋白质含量显著提高（表 6），而正常和热处理间籽粒硬度、和面时间和峰值曲线面积等品质参数差异未达显著水平，但热处理环境中和面时间、峰值带高、峰值曲线面积、8min 带高和 8min 带宽稍低于正常环境，籽粒硬度则稍高于正常环境。

热处理使中强筋材料蛋白质含量显著提高，8min 带高显著降低，而正常和热处理间籽粒硬度、和面时间和峰值曲线面积等品质参数差异未达显著水平，但热处理环境下和面时间、峰值带高、峰值曲线面积和 8min 带宽等均稍低于正常环境，籽粒硬度则稍高于正常环境。

对于中筋材料来说，热处理使 8min 带高和 8min 带宽显著降低，而正常和热处理间蛋白质含量、籽粒硬度、和面时间等品质参数差异未达显著水平，但热处理环境下峰值带高和峰值曲线面积稍低于正常环境，蛋白质含量和籽粒硬度稍高于正常环境，和面时间则基本没有变化。

对于中弱筋材料来说，热处理使和面时间显著提高，峰值带高、8min 带高和 8min 带宽显著降低，而正常和热处理间蛋白质含量、籽粒硬度和峰值曲线面积等品质参数差异未达显著水平，但热处理环境下蛋白质含量、籽粒硬度和峰值曲线面积均稍高于正常环境。

对于弱筋材料来说，热处理使和面时间和峰值曲线面积显著提高，而正常和热处理间其他品质参数差异未达显著水平，但热处理环境下蛋白质含量和籽粒硬度稍高于正常环境，峰值带高、8min 带高和 8min 带宽则稍低于正常环境。

总之，热处理使所有筋力类型材料的蛋白质含量和籽粒硬度提高，峰值带高、8min 带高和 8min 带宽降低，强筋、中强筋和中筋类型材料的和面时间和峰值曲线面积降低，中弱筋和弱筋类型材料的和面时间与峰值曲线面积增加。

表 1 参试品种代号和系谱

Table 1 Code and pedigree of 53 cultivars and advanced lines

编号 Code	品种 Cultivar	类型 Type	来源 Origin	系谱 Pedigree
1	农大 189	品种	北京	S915/T208
2	农大 211	品种	北京	农大 3291（农大 3338/S180）系选
3	农大 212	品种	北京	农大 211 系选
4	农大 318	品种	北京	农大 9516//N 早/贵 411/3/BL193/4/农大 189 * 2
5	农大 408	品种	北京	农大 9516//N 早/贵 411/3/BL193/4/农大 189 * 2
6	农大 413	品种	北京	农大 9516//N 早/贵农 411/3/京 4112
7	农大 3432	品种	北京	农大 9136/F390
8	农大 3492	品种	北京	农大 3251/中优 9507
9	农大 3634	品种	北京	农大 179/农大 3251
10	农大 3659	品种	北京	农大 2911/黑小麦 76
11	中优 206	品种	北京	CA9614/中优 9507

（续）

编号 Code	品种 Cultivar	类型 Type	来源 Origin	系谱 Pedigree
12	中麦 175	品种	北京	BPM27/京 411
13	CA0415	品系	北京	京 411/贵农 11//京 411
14	CA0518	品系	北京	30095/中优 9701
15	CA0548	品系	北京	CA9722/中优 9507//CA9722
16	CA0629	品系	北京	30095/中优 9701
17	CA0736	品系	北京	CA9722/中优 9507//CA9722
18	京冬 8 号	品种	北京	[（阿夫乐尔/5238-016）F_1/红良 4 号] F_4/（有芒红 7 号/洛夫林 10 号）F_7
19	石 4185	品种	河北	Ta 不育株（Ta 不育株/豫麦 2 号）/冀麦 26
20	石家庄 8 号	品种	河北	石 84-7111/21124//冀麦 26///冀麦 38
21	石麦 15	品种	河北	冀麦 38^3/92R137
22	石优 17	品种	河北	冀 935-352/鲁麦 21
23	石麦 18	品种	河北	（92 鉴 3/T447）F_2/冀麦 38//石 4185 * 3
24	石 6207	品系	河北	冀 935-352/鲁麦 21
25	石 06-6051	品系	河北	88119-19-3-5/石 4185
26	石 07-6023	品系	河北	周麦 11/石 01-5070
27	石 B07-4056	品系	河北	冀师 02-1/烟优 361
28	石新 733	品种	河北	大拇指矮异型株/石新 163
29	衡观 33	品系	河北	冀 84-5148/衡 4041//冀 84-5418/运 85-24
30	衡观 111	品系	河北	衡 4119/石家庄 1 号
31	衡观 115	品系	河北	衡 4119/石家庄 1 号
32	衡观 216	品种	河北	冀麦 36/鲁麦 21//邯 5316
33	衡 4399	品种	河北	邯 6172/衡穗 28（冀 5418/运 85-24）
34	衡 4422	品系	河北	衡 7228/曲农 96-1
35	衡 6607	品系	河北	邯 6172/衡穗 28（冀 5418/运 85-24）
36	衡 6632	品系	河北	黑小麦/衡 8116
37	邯 6228	品种	河北	山农太 91136/冀麦 36
38	中麦 306	品系	河南	济麦 19/豫麦 47
39	中麦 349	品种	河南	陕优 225-2/98 中 443
40	中麦 875	品系	河南	周麦 16/荔垦 4 号
41	07CA010	品系	河南	周麦 14/藁城 8901
42	08CA190	品系	河南	中育 5 号/邯 3475
43	山农 2149	品种	河南	莱州 137/L126
44	烟农 19	品种	山东	烟 1933/陕 82-29
45	济麦 19	品种	山东	鲁麦 13/临汾 5064
46	济麦 20	品种	山东	鲁麦 14/84187（80Q16-22-4-5/Lancota）
47	济麦 22	品种	山东	935024/935106
48	DH155	品系	山东	济麦 19/鲁麦 21

（续）

编号 Code	品种 Cultivar	类型 Type	来源 Origin	系谱 Pedigree
49	035037	品系	山东	莱州 95021/运丰早 18
50	037042	品系	山东	957069/烟 861602
51	046402	品系	山东	90（4）015/邯 5136
52	055319	品系	山东	965261/烟农 19
53	067257	品系	山东	965261/济麦 19

表 2　开花后 10d 至成熟期热处理和正常处理温度比较

Table 2　Comparison of temperatures in normal and heat-stressenvironment，10 d after anthesis to maturity（℃）

处理 Treatment	均值 Mean	最高温均值 Mean of highest temperature	变幅 Range
热处理 Heat	24.7	35.6	7.4～49.5
正常处理 Normal	23.1	28.7	6.7～37.5

表 3　产量因子方差分析

Table 3　Sum square of analysis of variance for yield component

变异来源 Source of variance	自由度 *df*	穗粒数 KNS	灌浆期 GFP	灌浆速率 GFR	千粒重 TKW	产量 Yield
基因型 Genotype（G）	52	3 341***	219**	5***	460***	506 285***
地点 Location（L）	3	6 245***	8 825***	43***	374***	13 334**
处理 Treat（T）	1	26*	11*	1***	148***	50 413***
基因型×地点（G×L）	156	1 136***	287*	2*	144***	385 543***
基因型×处理（G×T）	52	159	86	1	376	72 521
地点×处理（L×T）	3	17	38**	0	220***	102 267***
基因型×地点×处理（G×L×T）	156	380	173	9	555	158 156
误差 Error	156	682	202	1	890	205 200

*5%显著水平，**1%显著水平，***0.1%显著水平。

*，**，and*** indicates significance at 5%，1%，and 0.1% probability levels，respectively. KNS：kernel number per spike；GFP：grain filling period；GFR：grain filling rate；TKW：1000-kernel weight.

表 4　两种处理的产量性状比较结果

Table 4　Comparison of yield parameters in normal and heat-stress treatments

性状 Trait	均值 Mean		变幅 Range	
	正常 Normal	热处理 Heat	正常 Normal	热处理 Heat
穗粒数 Kernel number per spike	29a	28a	10～40	8～40
灌浆期 Grain filling period（d）	28a	28a	21～37	21～36
灌浆速率 Grain filling rate（$g \cdot d^{-1}$）	1.5a	1.4b	0.8～2.5	0.7～2.3
千粒重 1000-kernel weight（g）	40a	37b	26～52	25～49
产量 Yield（$t \cdot hm^{-2}$）	6.75a	6.33b	4.22～9.12	4.05～9.27

同列内标以不同字母的值表示 5%水平上显著差异。

Values followed by different letters in the same column were significantly different at 5% probability level.

表5　53份小麦材料依产量和千粒重热感指数的分类

Table 5　Classification of 53 wheat cultivars based on S_{yield} and S_{TKW}

品种 Cultivar	产量 Yield (t·hm⁻²) 热处理 Heat	正常 Normal	热感指数 S_{yield}	分类 Group	品种 Cultivar	产量 Yield (t·hm⁻²) 热处理 Heat	正常 Normal	热感指数 S_{yield}	分类 Group	品种 Cultivar	千粒重 TKW (g) 热处理 Heat	正常 Normal	热感指数 S_{TKW}	分类 Group	品种 Cultivar	千粒重 TKW (g) 热处理 Heat	正常 Normal	热感指数 S_{TKW}	分类 Group
046402	4.52	5.62	3.1	S	烟农19	6.78	7.26	1.0	S	046402	39.3	43.2	1.3	S	烟农19	38.5	41.5	1.1	S
农大211	6.76	8.26	2.8	S	农大189	7.83	8.33	0.9	R	农大211	36.4	43.2	2.3	S	农大189	40.8	43.0	0.8	R
农大3492	6.17	7.43	2.6	S	邯6228	6.28	6.66	0.9	R	农大3492	41.6	46.3	1.5	S	邯6228	40.3	41.7	0.5	R
08CA190	6.61	7.82	2.4	S	衡6632	6.70	7.09	0.9	R	08CA190	40.9	41.7	0.3	R	衡6632	32.8	37.1	1.7	S
衡6607	6.03	7.02	2.2	S	CA0518	7.31	7.74	0.9	R	衡6607	35.9	36.0	0.1	R	CA0518	41.5	43.6	0.7	R
石新733	5.36	6.21	2.1	S	京冬8号	7.79	8.24	0.8	R	石新733	37.7	39.8	0.8	R	京冬8号	44.6	46.3	0.5	R
石06-6051	6.23	7.18	2.1	S	衡4399	7.13	7.49	0.8	R	石06-6051	33.7	35.2	0.6	R	衡4399	37.3	36.3	−0.4	R
农大318	7.07	8.14	2.1	S	石优17	5.90	6.18	0.7	R	农大318	39.2	40.4	0.4	R	石优17	37.6	39.2	0.6	R
石麦15	6.83	7.85	2.0	S	农大212	6.71	7.02	0.7	R	石麦15	39.7	44.0	1.4	S	农大212	37.0	39.8	1.1	S
067257	6.13	6.93	1.8	S	衡观115	7.35	7.62	0.5	R	067257	35.7	37.3	0.6	R	衡观115	35.1	36.2	0.5	R
济麦22	7.01	7.91	1.8	S	中麦306	6.52	6.70	0.4	R	济麦22	38.1	43.2	1.7	S	中麦306	40.7	42.0	0.5	R
衡观111	6.29	7.07	1.7	S	衡观216	6.69	6.84	0.3	R	衡观111	33.7	35.6	0.8	R	衡观216	32.4	36.7	1.7	S
农大3432	6.77	7.60	1.7	S	石B07-4056	7.15	7.29	0.3	R	农大3432	38.3	47.1	2.7	S	石B07-4056	33.4	35.8	1.0	R
济麦20	6.45	7.21	1.6	S	农大3659	8.13	8.25	0.2	R	济麦20	35.0	38.1	1.2	S	农大3659	32.9	37.2	1.7	S
中麦175	6.58	7.35	1.6	S	衡观33	7.79	7.85	0.1	R	中麦175	37.8	43.8	2.0	S	衡观33	36.1	37.9	0.7	R
济麦19	6.88	7.64	1.6	S	石4185	7.32	7.37	0.1	R	济麦19	41.4	44.4	1.0	S	石4185	33.9	37.4	1.4	S
中优206	6.60	7.27	1.4	S	农大413	7.30	7.34	0.1	R	中优206	36.2	38.3	0.8	R	农大413	41.7	44.0	0.8	R
山农2149	6.86	7.53	1.4	S	CA0415	7.75	7.79	0.1	R	山农2149	36.9	40.6	1.4	S	CA0415	32.6	35.4	1.1	S
石6207	6.55	7.15	1.3	S	农大3634	7.82	7.83	0	R	石6207	37.4	39.2	0.7	R	农大3634	40.8	44.7	1.3	S
石07-6023	6.29	6.85	1.3	S	DH155	7.64	7.62	0	R	石07-6023	31.2	33.2	0.9	R	DH155	35.9	40.3	1.6	S
石家庄8号	6.43	6.97	1.2	S	CA0736	7.56	7.53	−0.1	R	石家庄8号	41.4	42.8	0.5	R	CA0736	35.8	36.6	0.3	R
衡4422	7.13	7.69	1.1	S	农大408	6.69	6.61	−0.2	R	衡4422	35.4	37.5	0.8	R	农大408	38.3	40.7	0.9	R
中麦349	7.64	8.24	1.1	S	07CA010	7.35	7.22	−0.3	R	中麦349	33.5	37.5	1.5	S	07CA010	41.0	44.1	1.0	S
中麦875	7.04	7.55	1.1	S	037042	6.55	6.42	−0.3	R	中麦875	43.1	47.7	1.4	S	037042	35.7	36.9	0.5	R
CA0629	6.55	7.02	1.0	S	055319	7.15	6.93	−0.5	R	CA0629	41.4	47.0	1.7	S	055319	41.1	42.7	0.6	R

（续）

品种 Cultivar	产量 Yield (t·hm^{-2}) 热处理 Heat	正常 Normal	热感指数 S_{yield}	分类 Group	品种 Cultivar	产量 Yield (t·hm^{-2}) 热处理 Heat	正常 Normal	热感指数 S_{yield}	分类 Group	品种 Cultivar	千粒重 TKW (g) 热处理 Heat	正常 Normal	热感指数 S_{TKW}	分类 Group	品种 Cultivar	千粒重 TKW (g) 热处理 Heat	正常 Normal	热感指数 S_{TKW}	分类 Group
石麦 18	7.19	7.70	1.0	S	CA0548	6.69	5.86	−2.2	R	石麦 18	37.5	38.2	0.3	R	CA0548	41.7	43.1	0.5	R
035037	6.70	7.17	1.0	S						035037	38.4	41.3	1.0	S					

R：抗；S：感；S_{yield}：产量热感指数；S_{TKW}：千粒重热感指数。

R：resistance；S：susceptibility；S_{yield}：thermal index of yield；S_{TKW}：thermal index of 1000-kernel weight.

表 6　5 类品种在两种处理环境中品质性状比较结果

Table 6　Comparison of mean values of quality parameters among five groups of cultivars in two environments

性状 Trait	处理 Treatment	强筋 Strong 均值 Mean	强筋 Strong 变幅 Range	中强筋 Medium strong 均值 Mean	中强筋 Medium strong 变幅 Range	中筋 Medium 均值 Mean	中筋 Medium 变幅 Range	中弱筋 Medium weak 均值 Mean	中弱筋 Medium weak 变幅 Range	弱筋 Weak 均值 Mean	弱筋 Weak 变幅 Range
蛋白质含量 Protein content	热处理 Heat	15.1a	15.0～15.3	14.4a	13.2～15.6	15.0a	13.6～16.9	14.5a	13.1～15.5	14.6a	13.2～16.4
	正常 Normal	14.6b	14.6～14.7	13.9b	12.8～15.7	14.8a	13.6～16.2	14.4a	12.4～15.8	14.5a	13.0～16.5
硬度 Hardness	热处理 Heat	62.5a	59.8～65.2	56.9a	16.7～68.8	51.6a	24.9～64.2	54.2a	22.5～66.6	59.0a	50.4～67.3
	正常 Normal	60.3a	50.4～70.2	55.3a	13.8～70.5	50.0a	25.4～67.4	53.4a	22.2～66.7	58.7a	49.1～65.3
和面时间 MPT	热处理 Heat	4.7a	4.5～5.2	3.3a	2.8～3.6	2.8a	2.0～3.6	2.0a	1.7～2.5	1.6a	1.0～2.5
	正常 Normal	4.9a	4.4～5.0	3.6a	3.3～3.9	2.8a	2.5～3.1	1.9b	1.6～2.3	1.2b	0.9～1.5
峰值带高 MPV	热处理 Heat	46.4a	45.8～48.2	44.2a	38.1～48.8	46.3a	41.1～51.2	47.2b	36.9～56.2	46.0a	39.6～56.4
	正常 Normal	47.0a	43.2～49.6	45.4a	40.1～48.7	46.8a	40.6～50.6	49.0a	40.1～56.6	46.4a	39.4～59.1
峰值曲线面积 MPI	热处理 Heat	177a	170～183	121a	106～142	104a	69～145	74a	60～98	53.8a	29.2～86.4
	正常 Normal	178a	177～179	133a	119～151	105a	89～127	73a	55～103	42.9b	27.3～56.0
8min 带高 MTxV	热处理 Heat	41.5a	42.5～43.8	35.7b	31.6～39.1	35.4b	29.2～41.7	33.2b	27.2～40.2	28.6a	23.5～35.9
	正常 Normal	43.2a	39.1～43.9	38.2a	34.4～40.8	36.9a	30.7～41.0	34.7a	27.8～41.7	29.1a	23.8～38.7
8min 带宽 MTxW	热处理 Heat	14.2a	12.8～18.8	10.3a	7.4～15.0	7.5b	3.6～14.5	6.0b	3.3～12.6	3.2a	2.3～5.8
	正常 Normal	15.8a	14.1～14.3	12.1a	8.5～15.7	9.2a	5.1～12.5	6.8a	3.1～12.3	3.5a	2.3～5.3

同列内标以不同字母的值在表示 5%水平上显著差异。

Values followed by different letters in the same column were significantly different at 5% probability level. MPT：midline peak time；MPV：midline peak value；MPI：midline peak integral；MT×V：midline time × value；MT×W：midline time × width.

3 讨论

本研究表明，灌浆期热害胁迫主要通过影响粒重即籽粒灌浆速率来影响产量，且不同材料的产量和千粒重热敏感性明显不同，这与前人研究结果一致[4,6-11,31]。说明在抗热性品种选育中，可以通过考察千粒重来进行大规模品种筛选。需要说明的是，本研究试验小区面积偏小，因此穗粒数和产量等指标仅供参考，而千粒重基本不受小区面积的影响，可靠性较高。

产量和千粒重热感指数分析表明，农大189、CA0518和京冬8号的产量和千粒重在正常和热处理环境中均较高，说明这些品种抗热性好，这与我们多年观察结果一致。京冬8号已经20多年考验，其抗热性表现突出，这是它能大面积长时间推广的主要原因之一。新育成的CA0518和其姊妹系CA0627（已参加区试）不仅高产优质，而且抗热性突出，籽粒红色，抗穗发芽能力强，有望成为北部冬麦区的苗头品种。农大189的抗热性也已经过不同地点的验证。衡观33和CA0736的产量在正常和热处理环境中均较高，千粒重在正常和热处理环境中均较低，但变化较小，品种抗热性也较好。在抗热性育种中，上述5份材料可作为杂交亲本，用于培育高产抗热新品种。农大211、石麦15、济麦22、农大3432和山农2149在正常环境中的产量和千粒重均较高，但在热处理环境产量和千粒重均较低，品种抗热性差，应注重改良后加以利用。同时热感指数分析表明，所选材料的产量和千粒重热感指数变幅均较大（－2.2～3.1）和（－0.4～2.7），其中中麦349产量热感指数为1.1，但其产量在正常和热处理环境中都较高；济麦19的千粒重热感指数为1.0，但其千粒重在正常和热处理环境中均较高，表明传统的热感指数以1.0为分类的标准尚需进一步研究，所反映品种的抗热性指标只表示其在所有参试品种中的相对表现。根据实际育种和生产表现，结合本研究结果，建议今后在进行品种抗热性鉴定时，以京冬8号为抗热性对照品种，石麦15为中感对照品种，农大211为高感对照品种。CA0736等6份材料产量热感指数（－2.2～－0.1）和衡4399千粒重热感指数（－0.4）小于0，原因可能在于本研究小区面积较小，存在取样误差。除此之外，产量和千粒重热感指数分析结果不一致的原因还可能在于花后10d即开始覆膜进行热处理，从而部分影响了穗粒数，建议今后的抗热性试验中在开花后15d进行塑料大棚增温处理。

Hurkman等[32]发现，高温可导致蛋白质含量增加，面筋强度降低，但二者之间无因果关系。本文通过对参试品种的主要品质性状进行聚类分析，发现强筋和中强筋材料037042、中优206、山农2149、石6207、石07-6023、农大3492、济麦20和石B07-4056在正常和热胁迫环境下均有较好和面特性，在今后的品质育种中应注重利用它们作为亲本，培育加工品质稳定的品种。热处理可不同程度地增加所有筋力类型材料的蛋白质含量和籽粒硬度，降低峰值带高、8min带高和8min带宽，并不同程度地降低强筋、中强筋和中筋类型材料的和面时间和峰值曲线面积，而增加中弱筋和弱筋类型材料的和面时间与峰值曲线面积，这可能与高温对麦谷蛋白的影响有关[20]。有关热处理对和面特性和面筋强度的影响还有待进一步深入研究。

4 结论

千粒重可作为抗热性筛选的简易指标。农大189、CA0518和京冬8号的产量和千粒重在正常和热处理环境中均较高，抗热性好；衡观33和CA0736的产量在正常和热处理环境中均较高，但千粒重表现中等，抗热性较好。热处理使所有类型材料的蛋白质含量和籽粒硬度增加，峰值带高、8min带高和8min带宽降低。

References

[1] Dixon J, Braun H J, KosinaP, Crouch J, eds. Wheat Facts and Futures 2009. Mexico, D. F.: CIMMYT, 2009

[2] Zhao J-Y（赵俊晔）, Yu Z-W（于振文）. Production status and development of productive capacity of wheat in China. *Res Agric Modernization*（现代农业研究）, 2005, 26 (5): 344-348 (in Chinese with English abstract)

[3] IPCC. IPCC fourth assessment report-climate change 2007, 2007 [2007-05-30]. http://www.ipcc.ch/

[4] Deng Z-Y（邓振镛）, Xu J-F（徐金芳）, Huang L-N（黄蕾诺）, Zhang S-Y（张树誉）. Research summary of the damage characteristics of the wheat dry hot wind in northern China. *J Anhui Agric Sci*（安徽农业科学）,

2009, 37 (20): 9575-9577 (in Chinese with English abstract)

[5] Zhang P-P (张平平), He Z-H (何中虎), Xia X-C (夏先春), Wang D-S (王德森), Zhang Y (张勇). Effect of heat stress on wheat protein and starch quality. *J Triticeae Crops* (麦类作物学报), 2005, 25 (5): 129-132 (in Chinese with English abstract)

[6] He Z H, Rajaram S. Differential responses of bread wheat characters to high temperature. *Euphytica*, 1994, 72: 197-203

[7] Dias A S, Lidon F C. Evaluation of grain filling rate and duration in bread and durum wheat under heat stress after anthesis. *J Agron Crop Sci*, 2009, 195: 137-147

[8] Keling P L, Bacon P J, Holt D C. Elevated temperature reduced starch deposition in wheat endosperm by reducing the activity of soluble starch synthase. *Planta*, 1993, 191: 342-348

[9] Rijven A H G C. Heat inactivation of starch synthase in wheat endosperm tissue. *Plant Physiol*, 1986, 81: 448-453

[10] Altenbach S B, Dupont F M, Kothari K M. Temperature, water and fertilizer influence the timing of key events during grain development in a US spring wheat. *J Cereal Sci*, 2003, 37: 9-20

[11] Lawlor DW, Mitchell R A C. Wheat. In: Reddy K R, Hodges H F, eds. Climate Change and Global Crop Productivity. Wallingford, UK: CAB International, 2000. pp 81-106

[12] Blum A. The effect of heat stress on wheat leaf and ear photosynthesis. *J Exp Bot*, 1986, 37: 111-118

[13] Rintamaki E, Kettunen R, Aro E M. Differential dephosphorylation in functional and photodamaged photo system II centers. *J Biol Chem*, 1996, 271: 14870-14875

[14] Randall P J, Moss H J. Some effect of temperature regimeduring grain filling on wheat quality. *Aust J Agric Res*, 1990, 41: 602-617

[15] Zheng F (郑飞), Shao Y-H (邵运辉), He Z-P (何钟佩). The effects of high-temperture stress on the component of the phloem exudates of flag leaf and peduncle of winter wheat. *Acta Agric Boreali-Sin* (华北农学报), 2003, 18 (3): 12-14 (in Chinese with English abstract)

[16] Spiertz J H J, Hamer R J, Xu H, Primo-Martin C, Don C, van der Putten P E L. Heat stress in wheat (*Triticum aestivum* L.): effects on grain growth and quality traits. *Eur J Agron*, 2006, 25: 89-95

[17] Sofield I, Evans LT, Cook M G, Wardlaw I F. Factors influencing the rate and duration of grain filling in wheat. *Aust J Plant Physiol*, 1977, 4: 785-797

[18] Stone P J, Nicolas M E. Wheat cultivars vary widely in their responses of grain yield and quality to short periods of post-anthesis heat stress. *Aust J Plant Physiol*, 1994, 21: 887-900

[19] Stone P J, Nicolas M E. The effect of duration of heat stress during grain filling on two wheat varieties differing in heat tolerance: grain growth and fractional protein accumulation. *Aust J Plant Physiol*, 1998, 25: 13-20

[20] Gupta R B, Masci S, Lafiandra D. Accumulation of protein subunits and their polymers in developing grains of hexaploid wheats. *J Exp Bot*, 1996, 47: 1377-1385

[21] Larkindale J, Vierling E. Core genome responses involved in acclomation to high temperature. *Plant Physiol*, 2008, 146: 748-761

[22] Qin D D, Wu H, Peng H R, Yao Y Y, Ni Z F, Li Z X, Zhou C L, Sun Q X. Heat stress-responsive transcriptome analysis in heat susceptible and tolerant wheat (*Triticum aestivum* L.) by using wheat genome array. *BMC Genomics*, 2008, 9: 432

[23] Sairam R K, Srivastava GG, Saxena D C. Increased antioxidant activity under elevated temperatures: a mechanism of heat stress. *Biol Plant*, 2000, 43: 245-251

[24] Yan S-H (闫素辉), Yin Y-P (尹燕枰), Li W-Y (李文阳), Li Y (李勇), Liang T-B (梁太波), Wu Y-H (邬云海), Geng Q-H (耿庆辉), Wang Z-L (王振林). Effect of high temperature after anthesis on starch formation of two wheat cultivars differing in heat tolerance. *Acta Ecol Sin* (生态学报), 2008, 28 (12): 6138-6147 (in Chinese with English abstract)

[25] Zhang Y-H (张英华), Wang Z-M (王志敏). The effect of heat stress om grain growth in the growing period of wheat. *Chin J Eco-Agric* (中国生态农业学报), 2006, 3 (14): 8-11 (in Chinese with English abstract)

[26] Liu P (刘萍), Guo W-S (郭文善), Pu H-C (浦汉春), Feng C-N (封超年), Zhu X-K (朱新开), Peng Y-X (彭永欣). Effects of transient high temperature during grain filling period on starch formation in wheat. *Acta Agron Sin* (作物学报), 2006, 32 (2): 182-188 (in Chinese with English abstract)

[27] Li H（李浩），Zhang P-P（张平平），Zha X-D（查向东），Xia X-C（夏先春），He Z-H（何中虎）. Isolation of differentially expressed genes from wheat cultivars Jinan 17 and Yumai 34 with good bread quality under heat stress during grain filling stage. *Acta Agron Sin*（作物学报），2007，33（10）：1644-1653（in Chinese with English abstract）

[28] AACC. Approved methods of the American Association of Cereal Chemists，9th edition，St. Paul，MN，USA，1995

[29] Bruker P L，Frohberg R. Stress tolerance and adaptation in spring wheat. *Crop Sci*，1987，27：31-36

[30] Chen X-Y（陈希勇），Sun Q-X（孙其信），Sun C-Z（孙长征）. Performance and evaluation of spring wheat heat tolerance. *J China Agric Univ*（中国农业大学学报），2000，5（1）：43-49（in Chinese with English abstract）

[31] Lillemo M，van Ginkel M，Trethowan R M，Hernandez E，Crossa J. Differential adaptation of CIMMYT bread wheat to global high temperatures. *Crop Sci*，2005，45：2443-2453

[32] Hurkman W J，DuPont F M，Altenbach S B. BiP，HSP70，NDK and PDI in wheat endosperm：II. Effects of high temperature on protein and mRNA accumulation. *Physiol Plant*，1998，103：80-90

小麦骨干亲本京411及衍生品种苗期根部性状的遗传

肖永贵[1]，路亚明[1]，闻伟锷[1]，陈新民[1]，夏先春[1]，王德森[1]，李思敏[1]，童依平[2]，何中虎[1,3]

（[1]中国农业科学院作物科学研究所/国家小麦改良中心，北京 100081；[2]中国科学院遗传与发育生物学研究所/植物细胞与染色体工程国家重点实验室，北京 100101；[3]CIMMYT 中国办事处，北京 100081）

摘要：【目的】利用高密度SNP标记解析骨干亲本京411的遗传结构，并研究其根系性状遗传特征，为培育高产广适性品种奠定基础。【方法】选用京411及其14个衍生品种（系），包括衍生一代6份，衍生二代8份，每品种选取饱满且大小一致的种子在发苗网上生长6d，每份材料选取大小一致的幼苗转移至培养盘，每个培养盘种植3次重复，连续培养15d后测定苗期根部性状的最大根长、侧根长、主根长、总根长、侧根表面积、主根表面积、总根表面积、总根尖数和根系干重。利用90K SNP芯片分析京411及其衍生后代群体的遗传结构和遗传区段传递，结合逐步回归分析定位苗期根部性状基因，探讨京411携带的优异根系基因在衍生后代中的分布。【结果】京411衍生群体平均遗传相似性为57.9%，该亲本与其衍生一代和衍生二代相同的等位变异频率分别为63.9%和67.9%，显著高于理论值。在A、B和D基因组间的相同等位变异频率分别为62.2%、61.3%和74.3%。京411的侧根和根尖数均优于其衍生品种，衍生品种的主根长和根干物质重等性状的改良较为显著。以已有定位信息的SNP标记与根系性状进行逐步回归分析，共发掘出35个根系性状位点，3DL和5BL上携带控制根长性状的主效位点，分别与SNP标记*wsnp _ Ex _ c1032 _ 1972861*和*BS00100708 _ 51*关联，可用于京411衍生群体中分子辅助选择。来自京411的26个位点对根部性状起正向效应，中麦175和CA0958携带正向效应区段最多，占正向效应位点总数的73.1%。【结论】传统育种对地上部农艺性状的选择，可有效促进地下根系的改良，相当于对优异性状遗传片段的聚合。在京411衍生群体中，中麦175是针对骨干亲本遗传改良最成功的品种，不仅携带较多的京411优异根系基因，而且其主要根系性状、产量表现及广适性均优于京411。

关键词：普通小麦；骨干亲本；根部性状；SNP标记；新基因位点

Genetic Contribution of Seedling Root Traits Among Elite Wheat Parent Jing 411 to Its Derivatives

XIAO Yong-gui[1], LU Ya-ming[1], WEN Wei-e[1], CHEN Xin-min[1], XIA Xian-chun[1], WANG De-sen[1], LI Si-min[1], TONG Yi-ping[2], HE Zhong-hu[1,3]

(*[1]Institute of Crop Sciences, Chinese Academy of Agricultural Sciences (CAAS) /National*

原文发表在《中国农业科学》，2014，47（15）：2916-2926.

基金项目：国家自然科学基金青年科学基金（31201207）、“948”计划［2011-G3（4）］、作物生物学国家重点实验室开放课题基金（2014KF02）

联系方式：肖永贵，E-mail：xiaoyonggui@caas.cn

通信作者：童依平，010-64806556；E-mail：yptong@genetics.ac.cn；何中虎，Tel：010-82108547；E-mail：zhhecaas@gmail.com

Wheat Improvement Center, Beijing 100081;
[2] *Institute of Genetics and Developmental Biology, Chinese Academy of Sciences/The State Key Laboratory of Plant Cell and Chromosome Engineering, Beijing 100101*; [3] *International Maize and Wheat Improvement Center (CIMMYT) China Office, c/o CAAS, Beijing 100081*)

Abstract: 【Objective】 The breeding program in crop root plays an important role in determining yield potential and broad adaptation. Jing 411 as one of the most important elite parents has been widely used in the Northern Winter Wheat Region. Understanding the principal components and detecting the root traits loci in Jing 411 will provide important information for future genetic improvement by high density SNP markers. 【Method】 The core parent Jing 411 and its 14 derivative varieties and lines including six derivative and eight varieties in the first and second generations were used in present study. The uniform grains of each variety were selected for germinating in seedling plate for 6 days, and then six healthy and consistent plants were selected to culture 15 days in cultivating plates. Every variety with three replicates in each cultivating plate was planted in nutrient solution. Nine root parameters including maximum length root, length of branch root, length of main root, total length of roots, surface area of branch roots, surface area of main root, surface area of total roots, number of total root tips, and root dry weight were measured after 15 days of culture. The genetic structure and components of Jing 411 and its derivative populations of different generations were compared. The whole-genome association mapping was employed to identify the chromosome region controlling seedling root traits loci using 90K SNP markers. 【Result】 The averaged genetic similarity index was 57.9% among Jing 411 and its derivatives, and the clustering results for these genotypes were generally in consistent with their pedigree. Jing 411 had the same allele ratios of 63.9% and 67.9% to its firstand second generation derivatives, respectively. Most of Jing 411 derivatives were obtained from the second generation by backcrossing with the elite parent or its derivatives, and the contribution of Jing 411 was higher than theoretical expectation. On A, B and D genomes, the same allele ratios from the elite parent were 62.2%, 61.3% and 74.3%, respectively. Jing 411 exhibited more branch root and total root tips than its derivatives, however, the maximum length root and root dry weight has been improved significantly in the derivatives. Thirty-five loci showed significant association with root traits using stepwise model selection with Proc GLMselect in SAS software. Two new loci for root traits on the long arm of chromosomes 3D and 5B were linked with the SNP markers *wsnp_Ex_c1032_1972861* and *BS00100708_51*, respectively. The gene and its linked SNP marker could be used for selecting root traits in the derivatives of Jing 411. Alleles were positively related with root traits from Jing 411 at 26 loci, and Zhongmai 175 and CA0958 remained more positive root loci delivered from the core parents, accounted for 73.1%. 【Conclusion】 Conventional breeding program could effectively collect available genetic components, and promoted the root characters through selected outstanding above-agronomic parameters. A notable example in the improvement of Jing 411 was the variety Zhongmai 175, which carried more loci from the elite parent than other derivatives, and had superior root traits, agronomic traits and wide adaptation than its parent Jing 411.

Keywords: common wheat; elite parent; root traits; SNP marker; new locus

0 引言

【研究意义】作物发达的根系系统和良好的根系活力可有效增加对土壤营养的吸收面积和吸收效率，进而提高品种的适应能力[1]。培育稳产、高产、广适、水肥高效利用型小麦新品种，不仅需要优良的群体结构和协调的产量构成，而且需要稳固的根系锚定、发达的根系系统和持久的根系活力[2]。目前，作物育种工作中尚未建立起高效的根系改良指标，育种家主要依据地上部生物量等来判断品种的根系活力，并在小麦品种根系性状改良及水肥利用效率方面取得一定进展。因此，结合育种实际工作需求，发掘骨干亲本携带的优良根系性状基因，探讨根系基因在衍生品种的遗传传递，对深入了解骨干亲本遗传机制及优异品种遗传结构具有重要意义。【前人研究进展】半矮秆品种的根系生物量较农家种呈下降趋势，但品种的根系活力和水肥利用效率显著增加[3]。矮秆基因 *Rht－B1b* 和 *Rht－8c* 均有利于提高根系总长、表面积、体积和干物质重[4-5]。在拔节至开花期，*Rht－B1b*＋*Rht－D1b* 和 *Rht－D1b*＋*Rht－B1c* 基因型品种的根系生长迅速，其根系生物量较 *Rht－D1b* 半矮秆品种提高5%[6]。1RS易位系的根系生物量显著提高，最高达31%[7]。可见，育种家在提高小麦产量和适应性方面取得显著进展的同时，根系生物量和根系活力也得到一定改良。尽管栽培环境影响作物根系性状，但品种基因型是决定根系特性的关键[8]。水稻根系性状的主效基因位点分布在第1、2、3、7和9染色体上，位于第3染色体短臂上的水稻同源基因还对玉米根系性状具有调控作用[9-10]。已克隆的水稻根系相关基因 *OsGnom1*，可能调控茎内维管束外侧邻近的细胞分生组织，产生新根冠原基，并与根冠大小、侧根数、根尖向地下和根部激素运输有关[11]。此外，激素和细胞分裂素相关的基因 *OsWOX11* 也参与并调节水稻根顶端分生区、冠根和侧根原基的发育[12]。小麦的根长度、表面积、体积和干物质重等均由多基因控制，主要分布于1D、2A、2D、3A、5A、6A和7D染色体[5,13-14]，并且通常呈簇存在，部分基因存在上位效应[15-16]。多数根部性状基因与农艺性状、产量性状相关联[5]。【本研究切入点】由于小麦根系所处土壤环境的复杂性及其研究方法和技术的局限性[4,16-17]，相对地上部组织而言根系的研究严重滞后，以致长期以来对不同基因型根系特征的遗传研究不够深入和全面，制约了高产广适品种的培育。因此，研究小麦品种间根系性状遗传差异，筛选根系优异的品种，对培育优良品种具有重要价值。京411是20世纪90年代初育成的半矮秆小麦品种，具有抗寒性强、分蘖力强、成穗率高、抗病性好、高产稳产、适应性广等优点。从育成至21世纪初，京411一直是北部冬麦区高产广适育种的骨干亲本。中国农业科学院作物科学研究所育成14个新品种（系），其中，中麦175、北京0045等成为冬麦区的主栽品种。【拟解决的关键问题】本研究以冬小麦骨干亲本京411及其衍生后代为材料，在水培环境下研究不同基因型苗期根部性状，利用全基因组SNP标记分析京411及其衍生品种携带的根系性状遗传区段，解析其遗传传递的规律，为今后利用骨干亲本的优异基因，培育更优良的新品种奠定理论基础。

表1　小麦骨干亲本京411及其衍生品种（系）

Table 1　Elite wheat parent Jing 411 and its derivatives

编号 No.	品种（系） Variety (Line)	系谱 Pedigree	世代 Generation	审定年份 Released year
1	京411 Jing 411	丰抗2号/长丰1号 Fengkang 2/Changfeng 1	骨干亲本 Elite parent	1991
2	CA9722	京411/贵农11//京411 Jing 411/Guinong 11//Jing 411	衍生一代 1st generation	2003
3	北京0045 Beijing 0045	中麦9号/京411 Zhongmai 9/Jing 411	衍生一代 1st generation	2004
4	中麦175 Zhongmai 175	BPM27/京411 BPM27/Jing 411	衍生一代 1st generation	2008
5	中麦415 Zhongmai 415	京411/贵农11//京411 Jing 411/Guinong 11//Jing 411	衍生一代 1st generation	2010
6	新冬37 Xindong 37	京411/贵农11//京411 Jing 411/Guinong 11//Jing 411	衍生一代 1st generation	2012
7	CA0958	CA9722/皖麦33//京411 CA9722/Wanmai 33//Jing 411	衍生一代 1st generation	—
8	CA1055	CA9722/肖试4//CA9722 CA9722/Xiaoshi 4//CA9722	衍生二代 2nd generation	—
9	CA0548	CA9722/中优9507//CA9722 CA9722/Zhongyou 9507//CA9722	衍生二代 2nd generation	—
10	CA1090	CA9722/中优9507//CA9722 CA9722/Zhongyou 9507//CA9722	衍生二代 2nd generation	—
11	中麦818 Zhongmai 818	CA9722/轮选987 CA9722/Lunxuan 987	衍生二代 2nd generation	—
12	中麦816 Zhongmai 816	CA9722/轮选987 CA9722/Lunxuan 987	衍生二代 2nd generation	2013
13	09抗1027 09 Kang 1027	（YW243/Pm13//Pm97033）/2＊M23-4//2＊CA9722	衍生二代 2nd generation	—

（续）

编号 No.	品种（系）Variety (Line)	系谱 Pedigree	世代 Generation	审定年份 Released year
14	CA1119	中麦 175/CA0493 Zhongmai 175/CA0493	衍生二代 2nd generation	—
15	CA1133	中麦 175/CA0477 Zhongmai 175/CA0477	衍生二代 2nd generation	—

1 材料与方法

1.1 试验材料

选用京 411 及其 14 个衍生品种（系），包括衍生一代 6 份，衍生二代 8 份，材料名称及系谱见表 1。试验材料来自 2011—2012 年度在中国农业科学院作物科学研究所昌平试验基地收获的种子，没有穗发芽发生。

1.2 根部性状测定

根部试验在中国科学院遗传与发育生物学研究所进行，采用水培法对不同材料进行培养。发苗和培养的方法如下：每品种选取饱满且大小一致的种子 50 粒，用 10%的 H_2O_2 处理 20～30min，无菌水冲洗 5～6 次。选取处理过的种子置于铺有滤纸的培养皿中，在培养箱暗室催芽 18～24h。每个材料挑选发芽一致的种子 25 粒置于发苗网上生长 6d，每份材料选取大小发育一致的幼苗 6 株转移至培养盘，每个培养盘种植 3 次重复，在可控温室的营养液中培养，营养液配置参照文献 [17]。幼苗培养条件为 (22±1)℃，相对湿度在 50%～60%。每 3d 换一次营养液，连续培养 15d 后对苗期根部性状进行调查。调查性状包括最大根长（maximum length root，MLR）、侧根长（length of branch root，LBR）、主根长（Length of main root，LMR）、总根长（total length of roots，TLR）、侧根表面积（surface area of branch roots，SABR）、主根表面积（surface area of main root，SAMR）、总根表面积（surface area of total roots，SATR）、总根尖数（number of total root tips，NTRT）和根系干重（root dry weight，RDW）。

1.3 SNP 标记分析

每份材料选取 10 个大小和色泽均匀一致的籽粒在培养皿中发苗 2 周，混合剪取幼苗，采用澳大利亚 Triticarte 公司提供的 CTAB 法提取小麦全基因组 DNA（http：//www.diversityarrays.com）。利用 UVS-99 微量分光光度计（ACTGene USA）检测 DNA 质量和浓度，DNA 终浓度调为 300ng·μL^{-1}。

利用 Illumina SNP 基因分型研究平台对京 411 及其衍生品种（系）DNA 进行 90K SNP 芯片分型，包括 BS、BobWhite、CAP、D_contig 等系列标记，共计 81 587 个，其中 38 819 个 SNP 标记在京 411 衍生群体内存在差异。

1.4 数据处理和图谱绘制

利用 SAS9.2 软件[18]进行基本统计量、多重比较分析，并结合 SAS 的 Glmselect 程序对 SNP 数据和根系性状进行逐步回归，根据 P 值（$P<0.01$）判断关联位点。采用 PowerMarker V3.25 软件对 SNP 标记的等位变异、多态性信息含量、等位变异频率和遗传距离进行分析。利用 Mega 5.10 软件进行 UPGMA 聚类分析和绘制聚类图[19]。

2 结果

2.1 京 411 与其衍生品种的遗传关系

38 819 个 SNP 标记在 15 个品种（系）中基因分型频率为 0.07～0.93，平均为 0.40。UPGMA 聚类分析（图 1）表明，京 411 及其衍生品种（系）的平均遗传相似性为 57.9%，最高为 89.5%，最低 37.4%。京 411 与衍生一代的遗传相似性为 65.4%，与衍生二代的遗传相似性为 63.9%。中麦 415、CA0958、中麦 175 和 CA1119 与京 411 聚为第一亚群，遗传相似性为 72.9%，主要是京 411 的第一代衍生系。第二亚群主要是京 411 的第二代衍生系，也是 CA9722 的后代品种，与 CA9722 的遗传相似性为 72.2%，与京 411 的遗传相似性为 63.1%。值得说明的是，该亚群中 CA1133 虽然是中麦 175/CA0477 组合的后代，但是其父本 CA0477 来自于 CA9722/中优 9507//CA9722 的后代，因此与 CA9722 聚为一类。北京 0045 和新冬 37 各自独立成为一个亚群，与京 411 的遗传相似性分别为 65.6% 和 43.0%。

2.2 基因型对苗期根部性状的影响

方差分析表明（表略），不同基因型对苗期根系性状具有极显著的影响（$P<0.001$），重复间差异影响较小（$P>0.05$）。骨干亲本京 411 较其子代在侧根长、主根长、总根长、侧根表面积、主根表面积、总根表面积和总根尖数等根部性状上具有显著的优势（$P<0.05$），衍生一代的最大根长和根干物质重显著高于骨干亲本和衍生二代（表 2）。中麦 175 的最大根长、主根长、主根表面积和总根表面积显著优于其他品种（系）（$P<0.05$，图 2），京 411 的侧根长、总根长和侧根表面积最大，显著高于其他品种（系），北京 0045 和中麦 818 具有最高的总根尖数和根系干重。综合而言，骨干亲本京 411 仍具有较好的侧根性状，其后代衍生种中以主根长和根干物质重等性状改良较为显著，其中，衍生一代品种中麦 175 改良最为明显（图 2）。

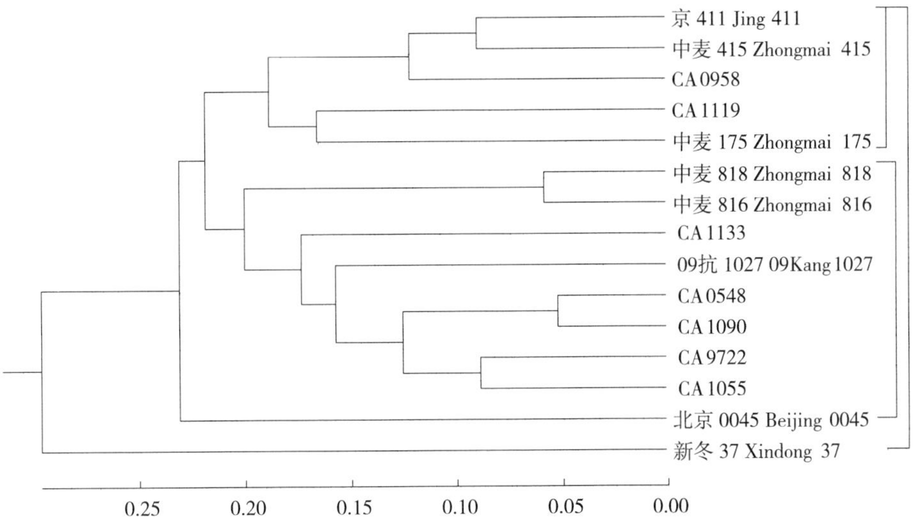

图 1 供试材料的 UPGMA 聚类树分析

Fig. 1 UPGMA dendrogram of 15 varieties studied based on Nei's distance

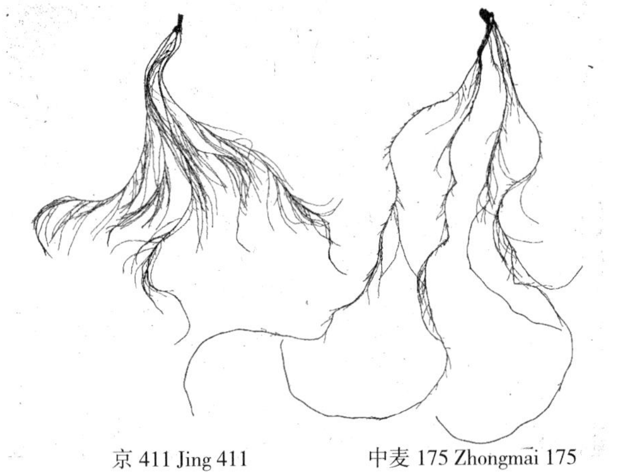

图 2 京 411 和中麦 175 的苗期根系图

Fig. 2 The seedling root graph of elite wheat parent Jing 411 and its derivative Zhongmai 175

2.3 京 411 遗传信息对其衍生后代的影响

38 819 个标记在 A、B 和 D 基因组上的分布频率分别为 38.4%、46.4%和 15.2%。在 A、B、D 基因组，衍生一代与京 411 相同等位变异的频率分别为 61.1%、56.6%和 74.0%，衍生二代与京 411 相同等位变异的频率分别为 63.3%、66.1%和 74.5%。该频率均高于理论遗传贡献率 50.0%和 25.0%，主要原因是衍生后代多来自骨干亲本京 411 或衍生一代材料的二次回交，所有后代材料中骨干亲本的遗传比例均≥50%，尽管如此，骨干亲本的遗传频率仍高于每个后代的理论遗传贡献。

衍生世代间遗传贡献率表现为衍生二代>衍生一代，基因组间贡献率为 D>A>B，可见骨干亲本京 411 的遗传贡献率随子代增加呈逐渐下降趋势，其 D 基因组的遗传对衍生后代贡献较大（表 3）。在 21 条染色体中，京 411 与其衍生一代品种相同等位变异的频率变幅为 92.3%（7D）～47.7%（5D），与其衍生二代品种相同等位变异的频率变幅为 83.3%（4D）～51.9%（1A），表明在对京 411 遗传改良过程中，不同染色体间所受的选择压力存在差异，D 基因组受到选择压力较小，A 和 B 基因组受到压力较大，其中衍生群体中 1A 和 2A 染色体受到选择压力较大，而 4D 和 7D 染色体受到选择压力较小，保留较高的遗传传递。

表 2 供试材料根部性状的多重比较分析

Table 2 Comparison of root phenotypic traits of Jing 411 and its derivatives

品种 Variety	最大根长 MLR (cm)	侧根长 LBR (cm)	主根长 LMR (cm)	总根长 TLR (cm)	侧根表面积 SABR (cm^2)	主根表面积 SAMR (cm^2)	总根表面积 SATR (cm^2)	总根尖数 NTRT	根系干重 RDW (mg)
京 411 Jing 411	19.5b	365.8a	88.0a	453.8a	15.5a	11.0a	26.5a	1 026.3a	8.1b
衍生一代 1st generation	21.7a	256.3b	72.1b	328.4b	11.2b	9.0b	20.2b	966.8ab	14.0a
衍生二代 2nd generation	17.9b	243.3b	56.2c	299.4b	11.1b	6.8c	18.0b	857.8b	12.2a
京 411 Jing 411	19.5bcd	365.8a	88.0bc	453.8a	15.5a	11.0b	26.5ab	1026.3b	8.1fg
CA9722	17.5defg	186.6f	41.4fg	227.9ef	8.9ef	4.9fg	13.7f	598.5e	11.9def
北京 0045 Beijing 0045	22.1b	333.5ab	72.9bcd	406.4ab	13.4ab	8.8bcd	22.2bc	1254.8a	15.5cd
中麦 175 Zhongmai 175	33.7a	235.1def	138.7a	373.8bc	11.1bcde	18.7a	29.7a	985.2bc	12.7de
中麦 415 Zhongmai 415	21.6bc	253.9cde	62.4def	316.3cd	11.4bcde	7.6cdef	18.9cde	812.2cd	6.9g
新冬 37 Xindong 37	16.5defg	228.2def	47.7efg	275.9def	9.5def	5.5efg	15.0def	913.3bcd	19.8ab
CA0958	18.6cdef	300.5bc	69.4bcde	370.0bc	12.9bc	8.6bcd	21.5cd	1236.5a	17.1bc
CA1055	22.2b	216.5ef	90.9b	307.4cd	11.3bcde	10.8b	22.1bc	892.0bcd	15.0cd
CA0548	15.0g	258.5cde	67.8cde	326.3cd	12.3bc	8.3bcde	20.6cd	887.8bcd	17.3abc
CA1090	15.2fg	259.5cde	25.8g	285.4def	12.0bcd	3.2g	15.2def	592.3e	9.3efg
中麦 818 Zhongmai 818	18.5cdefg	188.9f	32.7g	221.5f	7.9f	3.9g	11.8f	735.3de	21.5a
中麦 816 Zhongmai 816	18.3cdefg	185.9f	34.7g	231.5ef	7.8f	4.1g	12.0f	699.4de	19.9ab
09 抗 1027 09 Kang 1027	18.8bcde	259.1cde	36.3g	295.4de	10.6cde	4.2g	14.8ef	856.5bcd	6.1g
CA1119	19.8bcd	241.4def	79.2bcd	320.6cd	11.4bcde	10.2bc	21.6cd	999.2bc	8.4fg
CA1133	15.5efg	279.0bcd	60.5def	339.5bcd	12.5bc	7.2def	19.7cd	1 041.8b	7.6fg
平均值 Mean	19.5	252.8	63.2	316.8	11.2	7.8	19.0	902.1	13.1
变异系数 *C.V*%	24.1	19.5	45.0	19.8	16.8	49.8	25.9	21.6	48.6

数据后字母不同表示品种间差异显著（$P<0.05$）。

Values within a column followed by different letters are significantly different at $P<0.05$.

表 3 不同染色体上骨干亲本京 411 的 SNP 位点对其衍生品种的遗传贡献率

Table 3 Genomic contribution of different chromosomes of the elite parent Jing 411 in two generations

A 基因组 A Genome	SNPs	相同等位变异比例 Same allele ratio (%)		B 基因组 B Genome	SNPs	相同等位变异比例 Same allele ratio (%)		D 基因组 D Genome	SNPs	相同等位变异比例 Same allele ratio (%)	
		衍生一代 1st generation	衍生二代 2nd generation			衍生一代 1st generation	衍生二代 2nd generation			衍生一代 1st generation	衍生二代 2nd generation
1A	2 072	51.7	51.9	1B	3 396	50.2	80.0	1D	908	68.6	63.4
2A	2 085	55.4	54.9	2B	3 156	58.4	66.1	2D	1 336	50.1	65.0
3A	1 782	58.6	63.0	3B	2 414	60.7	70.7	3D	781	76.1	79.3
4A	1 909	66.5	68.9	4B	1 411	54.4	64.8	4D	288	85.8	83.3
5A	2 431	65.3	74.7	5B	2 885	47.7	53.9	5D	959	81.5	82.7
6A	2 094	66.1	57.9	6B	2 397	63.6	62.6	6D	497	63.8	68.2
7A	2 544	64.2	71.4	7B	2 347	60.9	64.4	7D	1 127	92.3	79.5
平均 Average	2131	61.1	63.3	平均 Average	2572	56.6	66.1	平均 Average	842	74.0	74.5

2.4 根部性状遗传区段解析

利用已定位的 SNP 标记对根部性状进行遗传区段检测分析[20]，共检测出 35 个 SNP 标记与根系性状相关（表 4），分别位于 1B、2A、2D、3D、4A、4B、5A、5B、5D、6A、6B、6D、7A 和 7B 等 14 条染色体上（表 4 和图 3）。4B、5A 和 5B 染色体上携带根系位点最多，其中，4B 染色体短臂 6cM 区段有控制总根表面积和侧根长的位点，该染色体长臂 70cM 附近携带有控制总根长和侧根长的位点；5AL 上 80cM 区段存在控制侧根长、总根长和总根尖数的位点；5BL 上 130cM 区段存在侧根长和总根长相关位点。此外，位于 3DL 的 *wsnp_Ex_c1032_1972861* 区段同时参与主根长和主根表面积调控，5BL 上 *BS00100708_51* 区段同时参与调控侧根长和总根长。在发掘的根部性状关联区段中，衍生一代得到遗传比例较高，为 55.2%；衍生二代相对较低，为 45.0%。不同品种间携带与骨干亲本一致的位点存在明显差异（图 3），中麦 175、北京 0045 和 CA0958 携带与京 411 相同位点最多，均为 23 个，占京 411 总位点数的 65.7%；中麦 816 和中麦 818 与骨干亲本的相同位点最少，为 10 个，仅占京 411 位点数的 28.6%。京 411 群体发掘的 35 个遗传区段中，对根部性状起正向效应的区段为 26 个，占总位点数的 74.3%，其中，中麦 175 和 CA0958 携带正向效应区段最多，均为 19 个，占正向效应位点总数的 73.1%。通过对全基因组根系性状遗传分析，说明衍生后代获得的骨干亲本的遗传区段存在差异，衍生一代具有较高的遗传比例，中麦 175 根部性状的基因主要来自于骨干亲本遗传。

3 讨论

3.1 骨干亲本及衍生品种根部性状遗传分析

作物根系大小和吸收效率对地上部生长和产量形成具有重要影响。传统育种方法塑造小麦地上部生长发育形态和结构，为适应地上部发展，根系形态和活力也得到相应改良。由于根系所处土壤环境的复杂性，使品种的根系遗传研究滞后，营养培养法可简便、有效地反映作物大田条件下根系特征[21]。京 411 衍生品种的苗期最长根长和根干物质重较京 411 有显著改良，而京 411 在侧根和主根特征性状上仍优于各个衍生世代的均值（表 2）。就单个基因型而言，其衍生品种的一些根部性状遗传改良显著，如中麦 175 的最大根长、主根长、主根表面积和根干物质重均优于骨干亲本（表 2，图 2）。发达的根系系统，特别是根系深层及阔度，不仅抵御土壤干旱胁迫，而且与肥料利用效率有密切关系[17,21]。最大根长及根干物质重、根表面积是判断品种水肥利用效率的主要指标[22]。由此可以推测，中麦 175 的节水和广适性应超过其亲本。实际生产中的表现有力佐证了上述推理。

表 4 根系性状基因定位分析

Table 4 Analysis of association mapping for rooting traits

性状 Trait	标记 Marker	染色体 Chromosome	位置 Position(cM)	京 411 对应变异 Jing411's allele	遗传效应 Genetic effect	*P* 值 *P-value*	总 R^2 TotalR^2
最大根长 MRL	*RAC875_rep_c74595_301*	1B	11.0	A:G	−2.72	0.0002	
	Kukri_c4213_2363	2A	102.0	A:A	0.41	0.0036	
	wsnp_Ex_c45468_51254832	2AL	135.6	A:A	1.38	0.0011	
	Ku_c68144_972	2AL	167.9	A:A	0.53	0.0019	
	Tdurum_contig100733_89	6A	37.0	A:G	5.17	<.0001	0.99
侧根长 LBR	*BobWhite_c9589_56*	4BS	6.0	A:G	77.82	0.0002	
	Ra_c27465_569	4BL	68.5	A:G	111.83	0.001	
	wsnp_Ex_rep_c68117_66883366	5AS	43.3	G:G	1.00	0.0042	
	BS00062729_51	5A	82.7	C:C	−7.89	0.0002	
	CAP11_c6746_112	5BL	117.9	C:C	2.26	0.0016	
	BS00100708_51	5BL	132.3	A:C	−36.87	0.0053	0.99
主根长 LMR	*wsnp_Ex_c1032_1972861*	3DL	149.8	A:A	−63.07	<.0001	0.74
总根长 TLR	*Ex_c32540_659*	4BL	72.5	A:A	144.49	0.0004	
	Excalibur_rep_c103747_193	5AL	89.6	A:A	−37.70	0.0009	
	BS00100708_51	5BL	132.3	A:C	85.49	0.0006	0.97
侧根表面积 SABR	*tplb0040j04_1007*	4A	126.7	A:A	2.38	0.0008	
	BS00037003_51	6B	0.4	A:A	1.88	0.0053	
	IACX10982	6DL	83.4	G:G	0.75	0.0007	
	Kukri_c25145_332	7B	163.2	G:G	2.83	0.0003	0.98
主根表面积 SAMR	*wsnp_Ex_c1032_1972861*	3DL	149.8	A:A	−9.41	0.0008	0.78
总根表面积 SATR	*BS00021701_51*	4B	6.0	A:G	−3.03	0.0008	
	RAC875_c18002_58	6DS	82.1	A:A	11.59	0.0007	
	BS00064367_51	7B	116.5	A:A	7.26	0.002	0.99
总根尖数 NTRT	*Kukri_c18971_279*	2DL	103.3	G:G	175.02	0.0004	
	wsnp_Ku_c3237_6024936	4A	70.0	G:G	1.98	0.0035	
	BS00003618_51	5AL	56.5	A:A	21.01	<.0001	

（续）

性状 Trait	标记 Marker	染色体 Chromosome	位置 Position(cM)	京411对应变异 Jing411's allele	遗传效应 Genetic effect	P值 *P-value*	总 R^2 TotalR^2
	wsnp_Ex_c15342_23592789	5AL	88.0	A：C	6.12	0.0026	
	Kukri_rep_c101981_260	5BL	40.8	C：C	70.77	0.0003	
	TA001996-0818	5BL	55.5	G：G	66.87	0.0044	0.99
根系干物重 RDW	*tplb0040d04_873*	5D	75.6	A：G	1.38	0.0004	
	wsnp_JD_c7795_8868122	6A	138.3	A：A	1.61	0.0035	
	D_GBB4FNX02FV5R9_171	6BS	0.4	A：A	−5.18	<.0001	
	BS00066884_51	6BL	91.5	A：G	6.28	0.0026	
	BS00097659_51	7AS	113.3	A：A	0.25	0.0003	
	BS00110894_51	7A	241.4	A：G	−14.93	0.0044	0.99

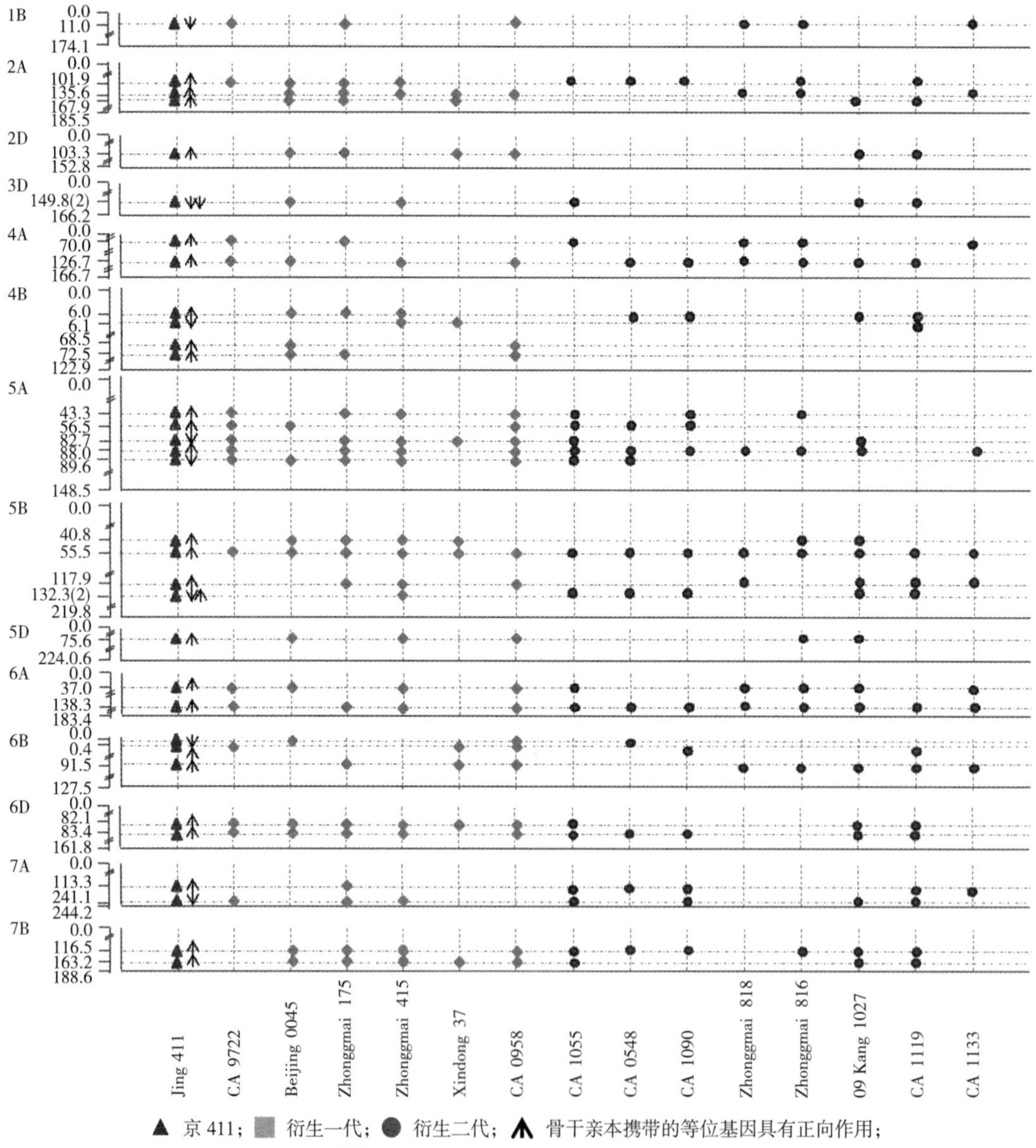

图3 根系性状的位点在京411及其衍生品种（系）中的分布图

Fig. 3 The distribution graph of QTL regions of root traits in elite wheat parent Jing 411 and its derivatives

3.2 京 411 对高产稳产广适小麦品种中麦 175 的遗传贡献分析

京 411 具有适应性广、丰产性好等突出特点，在北部冬麦区推广年限较长，1989—2010 年累计推广 74.3 万 hm^2，其中 1993—1995 年为北部麦区第一大品种。针对京 411 植株偏高、抗病丧失、耐高温差等缺点，中国农业科学院作物科学研究所利用 BPM27 的抗病、矮秆等特点配制组合，选育出中麦 175。该品种实现了高产稳产、抗寒、节水节肥与早熟性的良好结合，且面条品质优良，2007—2012 年分别通过北京市、山西省、国家北部水地与国家黄淮旱肥地及河北省、青海省和甘肃省审定。由于其株型紧凑，成穗率高，产量水平显著提高，株高在水地和旱地变幅相对较小，适合水浇地、补充灌溉地及旱肥地种植，已成为北部冬麦区及甘肃、青海春改冬的主栽品种，并在黄淮旱肥地大面积推广。2010 年北部冬麦区遭遇罕见低温危害，而中麦 175 在天津武清区 6.7hm^2 高产田平均产量 7.89$kg \cdot hm^{-2}$；2012 年陕西省旱肥地实产 8.75$kg \cdot hm^{-2}$，创当地高产纪录，充分说明其产量潜力高，适应性广，已成为中国北部冬麦区第 7 轮更新换代品种。

中麦 175 是京 411 的第一代衍生品种，在 81 590 个 SNP 标记中检测到 72 669 个位点的等位变异相同，占全基因组遗传信息的 89.1%，远超出理论遗传贡献。骨干亲本本身具备遗传传递力强、有利基因多等特点[23-24]，在遗传和驯化过程中，其携带的对物种本身或人类有益的优势等位基因很容易被高频保留，使衍生品种中骨干亲本遗传成分增多。本研究发掘的 35 个根部性状关联区段中，中麦 175 携带骨干亲本贡献的等位基因最多，达 23 个，且多为正向遗传效应基因。充分说明传统育种针对优异地上部分性状为目标进行高强度选择，可使优异目标性状基因有效聚集，使骨干亲本携带的正向遗传位点在衍生材料中传递率增加，新创制品种与亲本共性遗传成分提高显著。

3.3 京 411 根部性状优异等位基因解析

本研究利用 90K SNP 标记分析京 411 衍生群体发现，在 3DL 和 5BL 均存在一个共性位点，分别与标记 *wsnp _ Ex _ c1032 _ 1972861* 和 *BS00100708 _ 51* 紧密关联，并参与根长性状的调控（表 4 和图 3），与已报道的根部共性位点不一致[13-14]，可能属于京 411 特有遗传位点。前人曾在 1D、2A、2BS、2D、3B、4B、6A、7A 和 7B 染色体上发掘一些根部性状共性位点，但不同研究群体所定位的共性遗传位点存在差异[13-17]，说明品种背景对根系有重要影响，而且根部性状基因存在成簇分布现象。本研究在 3DL 和 5BL 上发掘的根系性状共性位点，可能在京 411 衍生群中对根部性状遗传具有一定价值，该位点的 SNP 标记可用于鉴定此群体。

此外，本研究还检测到控制单个根系性状的遗传区段，分布在 1B、2A、2D、3D、4A、4B、5A、5B、5D、6A、6B、6D、7A 和 7B 上，在京 411 衍生种群体中可能这些位点各自参与单个性状的特异调控。然而，在连锁定位研究中，位于这些染色体上根部性状位点多数参与调控其他重要农艺性状。例如，Ibrahim 等[14]认为 2A 上携带有控制根部性状的基因位点，在灌溉和干旱胁迫下均能够有效表达，不仅对总根长、总根尖数、总根体积和表面积有正向遗传效应，且有利于提高收获指数、穗数[25]和产量[26]。2DS 着丝点附近同样存在有既参与调控根系性状又作用于产量性状的遗传区段[14,27]。任永哲等[28]在 3A 上定位 3 个控制苗期最大根长、根干物质重和地上物质干重的主效 QTL 位点，但解释表型变异 10%以上的位点均来自京 411。另外，多数根部性状基因位点直接参与调控水肥利用效率[21,29]。

4 结论

骨干亲本京 411 仍具有较好的侧根性状，但主根长和根干物质重等性状在其后代衍生种中改良较为显著。京 411 的遗传成分对其衍生后代的贡献比例高于理论值，其衍生子一代品种中麦 175 是针对骨干亲本遗传改良的成功范例，不仅高频保留骨干亲本的遗传成分，且携带京 411 的优异根部性状基因最多。在京 411 衍生群体中发现 35 个 SNP 标记与根系性状相关位点，其中 2 个京 411 特有的遗传位点参与多个根长性状调控，分别位于在 3DL 和 5BL 染色体上，与标记 *wsnp _ Ex _ c1032 _ 1972861* 和 *BS00100708 _ 51* 紧密关联，可用于鉴定京 411 衍生群体的根部性状。

致谢：感谢中国农业科学院作物科学研究所李慧慧副研究员在基因定位中给予的帮助与支持。

参考文献

References

[1] 程建峰，戴廷波，荆奇，姜东，潘晓云，曹卫星．不

同水稻基因型的根系形态生理特性与高效氮素吸收．土壤学报，2007，44：266-272.
Cheng J F，Dai T B，Jin Q，Jiang D，Pan X Y，Cao W X. Root morphological and physiological characteristics in relation to nitrogen absorption efficiency in different rice genotypes. *Acta Pedologica Sinica*，2007，44：266-272.（in Chinese）

[2] Palta J A，Chen X，Milroy P S，Rebetzke G J，Dreccer M F，Watt M. Large root systems：Are they useful in adapting wheat to dry environments? *Functional Plant Biology*，2011，38：347-354.

[3] Waines J G，Ehdaie B. Domestication and crop physiology：Roots of green-revolution wheat. *Annals of Botany*，2007，100：991-998.

[4] Wojciechowski T，Gooding M J，Ramsay L，Gregory P J. The effects of dwarfing genes on seedling root growth of wheat. *Journal of Experimental Botany*，2009，60：2565-2573.

[5] Bai C，Liang Y，Hawkesford M J. Identification of QTLs associated with seedling root traitsand their correlation with plant height in wheat. *Journal of Experimental Botany*，2013，64：1745-1753.

[6] Miralles D J，Slafer G A，Lynch V. Rooting patterns in near-isogenic lines of spring wheat for dwarfism. *Plant and Soil*，1997，197：79-86.

[7] Waines J G，Ehdaie B. Optimizing root characters and grain yield in wheat//*Proceedings of the 5th International Triticeae Symposium*. Prague，Czech Republic. *Czech Journal of Genetics and Breeding*，2005，41：326-330.

[8] Jia Y，Yang X，Feng Y，Ghulam J. Differential response of root morphology to potassium deficient stress among rice genotypes varying in potassium efficiency. *Journal of Zhejiang University：Science B*，2008，9：427-434.

[9] Hochholdinger F. The maize root system：Morphology，anatomy and genetics//Bennetzen J，Hake S，eds. *The Handbook of Maize*. New York：Springer，2009：145-160.

[10] Coudert Y，Périn C，Courtois B，Khong N G，Gantet P. Genetic control of root development in rice，the model cereal. *Trends in Plant Science*，2010，15：219-226.

[11] Liu S，Wang J，Wang L，Wang X，Xue Y，Wu P，Shou H. Adventitious root formation in rice requires *OsGNOM1* and is mediated by the *OsPINs* family. *Cell Research*，2009，19：1110-1119.

[12] Zhao Y，Hu Y，Dai M，Huang L，Zhou D. The WUSCHEL-related homeobox gene *WOX11* is required to activate shoot-borne crown root development in rice. *The Plant Cell*，2009，21：736-748.

[13] Petrarulo M，Marone D，De Vita P，Sillero J C，Ferragonio P，Giovanniello V，Blanco A，Cattivelli L，Rubiales D，Mastrangelo A M. Mapping QTLs for root morphological traits in durum wheat. //*International Symposium "Root Research and Applications"*，Root RAP，2-4 September 2009，Boku-Vienna，Austria.

[14] Ibrahim S E，Schubert A，Pillen K，Léon J. QTL analysis of drought tolerance for seedling root morphological traits in an advanced backcross population of spring wheat. *International Journal of Agricultural Science*，2012，2：619-629.

[15] 周晓果，景蕊莲，郝转芳，昌小平，张正斌．小麦幼苗根系性状的QTL分析．中国农业科学，2005，38：1951-1957.
Zhou X G，Jing R L，Hao Z F，Chang X P，Zhang Z B. Mapping QTL for seedling root traits in common wheat. *Scientia Agricultura Sinica*，2005，38：1951-1957.（in Chinese）

[16] 刘秀林，昌小平，李润植，景蕊莲．小麦种子根结构及胚芽鞘长度的QTL分析．作物学报，2011，37：381-388.
Liu X L，Chang X P，Li R Z，Jing R L. Mapping QTLs for seminal root architecture and coleoptile length in wheat. *Acta Agronomica Sinica*，2011，37：381-388.（in Chinese）

[17] Ren Y，He X，Liu D，Li J，Zhao X，Li B，Tong Y，Zhang A，Li Z. Major quantitative trait loci for seminal root morphology of wheat seedlings. *Molecular Breeding*，2011，30：139-148.

[18] SAS Institute. SAS user's guide：Statistics. SAS Inst.，Cary，NC，2000.

[19] Liu K，Muse S V. PowerMarker：Integrated analysis environment for genetic marker data. *Bioinformatics*，2005，21：2128-2129.

[20] Wang S，Wong D，Forrest K，Allen A，Chao S，Huang E，Maccaferri M，Salvi S，Milner S，Cattivelli L，Mastrangelo A M，Whan A，Stephen S，Barker G，Wieseke R，Plieske J，International Wheat Genome Sequencing Consortium，Lillemo M，Mather D，Appels R，Dolferus R，Brown-Guedira G，Korol A，Akhunova A R，Feuillet C，Salse J，Morgante M，Pozniak C，Luo M，Dvorak J，Morell

M, Dubcovsky J, Ganal M, Tuberosa R, Lawley C, Mikoulitch I, Cavanagh C, Edwards K J, Hayden M, Akhunov E. Characterization of polyploid wheat genomic diversity using a high-density 90,000 SNP array. *Plant Biotechnology Journal*, 2014, 13: 1-10.

[21] An D G, Su J Y, Liu Q Y, Zhu Y G, Tong Y P, Li J M, Jing R L, Li B, Li Z S. Mapping QTLs for nitrogen uptake in relation to the early growth of wheat (*Triticum aestivum* L.). *Plant and Soil*, 2006, 284: 73-84.

[22] 徐吉臣，李晶昭，郑先武，邹亮星，朱立煌. 苗期水稻根部性状的 QTL 定位. 遗传学报，2001，28: 433-438.
Xu J C, Li J Z, Zheng X W, Zou L X, Zhu L H. QTL mapping of the root traits in rice seedling. *Acta Genetica Sinica*, 2001, 28: 433-438. (in Chinese)

[23] 盖红梅，李玉刚，王瑞英，李振清，王圣健，高峻岭，张学勇. 鲁麦 14 对山东新选育小麦品种的遗传贡献. 作物学报，2012，38: 954-961.
Ge H M, Li Y G, Wang R Y, Li Z Q, Wang S J, Gao J L, Zhang X Y. Genetic contribution of Lumai 14 to novel wheat varieties developed in Shandong province. *Acta Agronomica Sinica*, 2012, 38: 954-961. (in Chinese)

[24] 肖永贵，殷贵鸿，李慧慧，夏先春，阎俊，郑天存，吉万全，何中虎. 小麦骨干亲本"周 8425B"及其衍生品种的遗传解析和抗条锈病基因定位. 中国农业科学，2011，44: 3919-3929.
Xiao Y G, Yin G H, Li H H, Xia X C, Yan J, Zheng T C, Ji W Q, He Z H. Genetic diversity and genome-wide association analysis of stripe rust resistance among the core wheat parent Zhou 8425B and its derivatives. *Scientia Agricultura Sinica*, 2011, 44: 3919-3929. (in Chinese)

[25] Ibrahim S E, Schubert A, Pillen K, Léon J. Quantitative trait loci analysis for drought tolerance in an advanced backcross population of spring wheat. *Sudan Journal of Agricultural Research*, 2010, 15: 1-18.

[26] Huang X Q, Cöster H, Ganal M W, Röder M S. Advanced backcross QTL analysis for the identification of quantitative trait loci alleles from wild relatives of wheat (*Triticum aestivum* L.). *Theoretical and Applied Genetics*, 2003, 106: 1379-1389.

[27] Li W L, Nelson J C, Chu C Y, Shi L H, Huang S H, Liu D J. Chromosomal locations and genetic relationships of tiller and spike characters in wheat. *Euphytica*, 2002, 125: 357-366.

[28] 任永哲，徐艳花，贵祥卫，王素平，丁锦平，张庆琛，马原松，裴冬丽. 盐胁迫下调控小麦苗期性状的 QTL 分析. 中国农业科学，2012，45: 2793-2800.
Ren Y Z, Xu Y H, Gui X W, Wang S P, Ding J P, Zhang Q C, Ma Y S, Pei D L. QTLs analysis of wheat seedling traits under salt stress. *Acta Agricultura Sinica*, 2012, 45: 2793-2800. (in Chinese)

[29] Xu Y, Wang R, Tong Y, Zhao H, Xie Q, Liu D, Zhang A, Li B, Xu H, An D. Mapping QTLs for yield and nitrogen-related traits in wheat: Influence of nitrogen and phosphorus fertilization on QTL expression. *Theoretical and Applied Genetics*, 2014, 127: 59-72.

不同水分条件下广适性小麦品种中麦175的农艺和生理特性解析

李兴茂，倪胜利

（甘肃省农业科学院旱地农业研究所，兰州 730070）

摘要：【目的】探讨小麦广适性品种在不同水分条件下的农艺和生理特征，为广适性新品种培育提供参考依据。【方法】2011—2012年和2012—2013年，选用节水品种京冬8号、高产抗倒品种矮抗58与广适性品种中麦175，在干旱和充分供水2个水分处理下，分别在河北省高邑原种场和北京中国农业科学院作物科学研究所中圃场种植，小区间随机区组设计。测定株高、抽穗期、籽粒产量和产量三要素等农艺学性状，以及春季苗期植被覆盖度（GC）、抽穗前植被归一化指数（NDVIv）、灌浆中期植被归一化指数（NDVIg）、抽穗前叶绿素含量（SPADv）、灌浆中期叶绿素含量（SPADg）、灌浆中期叶片衰老等级评分（LSS）、抽穗前叶面积指数（LAIv）、灌浆中期叶面积指数（LAIg）、叶片衰老速率（LSR）和叶片卷曲度（LRS）等生理性状，比较研究不同类型品种间的性状差异。【结果】穗粒数、LAIg和SPADv品种间差异不显著，其他性状品种间存在显著差异。与干旱条件相比，充分灌水条件下中麦175、京冬8号和矮抗58的籽粒产量分别增加1 072.6、1 274和991.7kg·hm^{-2}，穗数分别增加104、15和47个/m^2，千粒重分别增加0.9、1.8和1.7g，株高分别增加7.2、9和6.2cm，抽穗分别推迟3、2和3d。中麦175产量和穗数较其他品种高647.3～1 505.8kg·hm^{-2}和38～68.8个/m^2。中麦175的平均株高、GC、NDVIv、NDVIg、SPADg、LSR和干旱环境下LRS分别是75.8cm、13.2%、0.64、0.55、56.5、14.10和4.0；其产量、千粒重、株高、SPADg和LAIv的水分敏感指数分别为0.86、0.64、0.94、0.61和0.57。【结论】广适性品种中麦175穗数多，苗期繁茂性好，株高中等，株型紧凑，抽穗后植被指数低，叶片衰老速度慢；干旱条件下叶片发生明显卷曲，但叶绿素含量较高，叶功能好；产量、千粒重和株高的水分敏感性低，而穗数和穗粒数的水分敏感性较高，灌浆中期的叶绿素含量和抽穗前叶面积指数对水分不敏感。以上这些特性使其在水旱条件下均表现出高产。

关键词：普通小麦；广适性；中麦175；水分利用效率；生理特性

Agronomic and Physiological Characterization of the Wide Adaptable Wheat CultivarZhongmai 175 Under Two Different Irrigation Conditions

LI Xing-mao，NI Sheng-li

（Dryland Farming Institute of Gansu Academy of Agricultural Sciences，Lanzhou 730070）

Abstract：【Objective】This study aims to differentiate the agronomic and physiologic characteristics

---

原文发表在《中国农业科学》，2015，48（21）：4374-4380.

基金项目：国家自然科学基金（31460349）、甘肃省农业科学院科技创新专项（2013GAAS14，2014GAAS06）

联系方式：李兴茂，Tel：09317614854；E-mail：lxm759@163.com

among three winter wheat cultivars under water-limited and full irrigated conditions, and to identify the traits closely associated with the adaptability of Zhongmai 175, which will provide the reference in a breeding program for improving wheat adaptability. 【Method】 Three cultivars were grown in the 2011-2012 and 2012-2013 cropping seasons at the CAAS, Beijing Experimental Station, and at Gaoyi in Hebei province. Randomized complete blocks were used to incorporate water in two levels with two replications. Heading date, plant height, grain yield and components, ground cover after green recover stages (GC), normalized difference vegetation index before heading (NDVIv) and at mid-filling (NDVIg), SPAD before heading (SPADv) and at mid-filling (SPADg), leaf senescence score at mid-filling (LSS), leaf area index before heading (LAIv) and at mid-filling (LAIg), leaf senescence rate (LSR), and leaf rolling score (LRS) were measured. The differences of traits were analyzed among three varieties. 【Result】 All traits except kernel number per spike, LAIg, and SPADv showed significant genetic variation. The values of all traits under full irrigation were increased more than those under limited irrigation. Grain yield, spike number per square meter, thousand kernel weight, plant height, and heading date of Zhongmai 175 were increased 1 072. 6kg • hm^{-2}, 104, 0. 9 g, 7. 2 cm, and 3 days, while those of Jingdong 8 were 1 274kg • hm^{-2}, 15, 1. 8 g, 9 cm, and 2 days, and Aikang 58 were 991. 7kg • hm^{-2}, 47, 1. 7 g, 6. 2 cm, and 3 days. Grain yield and spike number per square meter of Zhongmai 175 were 647. 3-1 505. 8kg • hm^{-2} and 38-68. 8 higher than other cultivars. Plant height, GC, NDVIv, NDVIg, SPADg, LSR, and LRS of Zhongmai 175 were 75. 8 cm, 13. 2%, 0. 64, 0. 55, 56. 5, 14. 10, and 4. 0, respectively. The water sensitive index of Zhongmai 175 was 0. 86, 0. 64, 0. 94, 0. 61, and 0. 57 for yield, thousand kernel weight, plant height, SPADg, and LAIv, respectively. 【Conclusion】 Zhongmai 175 showed the highest yield and spike number per unit area, early maturity, and semi-dwarf in limited irrigation and full irrigation areas. It was characterized by higher ground cover after the green recover stages, but lower normalized difference vegetation index in the heading and mid-filling stages, higher SPADg, and higher leaf rolling score. The grain yield, thousand kernel weight, plant height, SPADg, and LAIv were insensitive to water, but the spike number and kernel number of Zhongmai 175 were sensitive.

Keywords: common wheat; adaptability; Zhongmai 175; water use efficiency; physiological trait

0 前言

【研究意义】广泛适应性是新品种大面积推广的基本前提，选育广适性品种是育种工作的基本目标。研究不同类型品种的农艺和生理特征，有助于理解品种适应环境的优势性状，为广适性品种选育提供依据。

【前人研究进展】中国育成并推广了鲁麦 21、济麦 22、邯优 6172 和矮抗 58 等一批有影响力的广适性品种[1]，并初步明确了其半矮秆且株高稳定、抗逆性好、生长势强、不早衰、结实率高、有效穗数多而稳定、千粒重稳定等基本特征[2]。不同试验条件下，也有一些研究认为小麦幼苗早期活力、抽穗前光谱指数、灌浆期叶片衰老、冠层温度和茎秆可溶性糖等性状与产量和适应性有关[3-8]。借助田间高通量数据采集手段，上述一些指标可用于高世代育种选择[3,9]。国际玉米小麦改良中心（CIMMYT）在相关生理性状应用于小麦育种方面的研究已取得显著进展，出版了小麦生理育种的性状鉴定手册[10-11]，但针对广适性品种的生理特征研究相对较少，制约了高产广适性品种的选育。中麦 175 通过了北部冬麦区水地组和黄淮旱肥组 2 次国家审定以及北京、山西、河北、青海和甘肃 5 省（市）品种审定，是近 30 年唯一同时通过水地和旱地国家审定的品种，是国家北部冬麦区区试、北京市、山西省和河北省冀中北区试对照品种。大面积推广和研究表明，中麦 175 不仅产量潜力高，而且水肥利用效率高，是水地和旱肥地兼用型品种，具有广泛的适应性[12-14]。京冬 8 号有稳产、节水等特点，曾是北部冬麦区国家区试和河北省冀中北区试对照品种，因产量潜力偏低和植株偏高、容易倒伏，逐渐被中麦175所替代，但至今仍是北京市节水区试对照品种。矮抗58有矮秆、抗倒伏、高产、抗病等特点，是黄淮南片尤其是河南省的第一大品种。

【本研究切入点】在不同水分条件下，比较京冬8号、矮抗58和中麦175的产量构成因素及生理特征。

【拟解决的关键问题】通过比较不同类型品种在不同水分条件下的特征表现，明确中麦175高产、广适性的农艺和生理特征，为广适性品种选育提供理论依据。

1　材料与方法

1.1　试验设计

试验于2011—2012和2012—2013年在河北省高邑原种场（北纬38.02°、东经114.30°）和北京中国农业科学院作物科学研究所中圃场（北纬39.55°、东经116.24°）种植。尽管北京点位于北部冬麦区，但由于受城市小环境影响，冬季气温与河北高邑基本一致，对品种的抗寒性要求低，其生育期与石家庄接近。参试品种中麦175、京冬8号和矮抗58在北京和高邑两点种植，均能正常生长和越冬，基本可以排除品种间抗寒性差异对产量等性状的影响。试验以品种为主处理，水分胁迫和正常灌水两个水平为副处理，采取裂区设计，2个重复，小区长4m，行距30cm，小区面积6m²。水分胁迫处理：仅播种前和越冬前各灌溉1次，保证出苗和安全越冬；正常灌溉处理：在播种前、越冬前、返青期、拔节期和灌浆期各灌溉1次，共灌水5次，每次灌水量约70mm。施肥、除草等田间管理依据大田生产进行。2011—2012年小麦生育期内北京气温正常，灌浆期高邑点遭遇持续高温天气影响。

2012—2013年小麦生育前期温度均偏低，小麦生长发育在两个试验点都推迟一周。2012—2013年北京春季有效降雨少，干旱严重，干旱处理小麦生长发育速度较快，生长抑制明显。由于小麦苗期没有水分处理，因此苗期植被覆盖度按照4个重复的平均值进行分析。灌浆中期叶绿素含量仅测定了2011—2012年北京试验点的2个水分处理下的数据；叶片卷曲度仅在干旱环境下测定。

1.2　调查测定项目与方法

1.2.1　苗期植被覆盖度（GC）测定　春季返青期，利用佳能数码相机G12，在高度一致条件下，采用3万像素，垂直拍摄数码照片，然后依据肖永贵等[15]提供的方法，用Photoshop软件的扩展功能，计算平均灰度值，根据公式（1）计算植被覆盖度。计算公式：

$$GC（\%）=灰度值\times 100/255 \quad (1)$$

1.2.2　植被归一化指数（NDVI）　用便携式光谱仪GreenSeeker（Ntech Industries，USA），分别测定抽穗前归一化植被指数（NDVIv）和灌浆中期植被指数（NDVIg）。植株衰老速率（LSR）估算公式：

$$LSR=（NDVIv-NDVIg）/NDVIv \quad (2)$$

1.2.3　叶片衰老等级（LSS）　在灌浆中期，按照CIMMYT提供的田间小麦表型测定手册[11]划定的0—10级标准纪录数据，其中0：无枯叶；1：10%枯叶；2：20%枯叶；3：30%枯叶；4：40%枯叶；5：50%枯叶；6：60%枯叶；7：70%枯叶；8：80%枯叶；9：90%枯叶；10：100%枯叶。

1.2.4　叶面积指数（LAI）　用植物冠层分析仪AccuPAR LP-80（Decagon，USA），分别测定抽穗前叶面积指数（LAIv）和灌浆中期叶面积指数（LAIg）。

1.2.5　叶绿素含量（SPAD）　用叶绿素含量仪，测定旗叶中部的SPAD值。每小区选取10片叶，取平均值。在抽穗前和灌浆中期测定相应的SPAD值，分别记为SPADv和SPADg。

1.2.6　叶片卷曲度（LRS）　依据O'toole等[16]的方法，在灌浆中期晴天午后2点，按照5个等级对旗叶卷曲度进行评价，1～5级分别代表旗叶从无卷曲到完全卷曲。

1.2.7　产量及农艺性状　全区收获后，风干计产，折算成产量（$kg\cdot hm^{-2}$）。千粒重：取2组500粒的平均值折算为千粒重（g）。小区约2/3穗抽出旗叶时记载抽穗时间，计算从播种到抽穗的间隔天数作为抽穗时间进行统计分析。收获前每小区随机取3组共30个穗计算平均穗粒数；选取2个50cm样段统计穗数，平均折合成每平米穗数。成熟期测定株高。

1.2.8　水分敏感指数　按照Fischer[17]的方法，水分敏感指数的函数表达式：

$$WSI=（1-Xi/Yi）/（1-Xij/Yij） \quad (3)$$

式中Xi为品种在干旱条件下的产量或生理性状，Yi为品种在充分灌水条件下的产量或生理性状，Xij为所有品种在干旱条件下产量或生理性状的均值；Yij为所有品种在充分灌水条件下产量或生理性状的均值。$WSI<1$表明品种对水分不敏感，$WSI\geq$

1 表明品种对水分敏感。

1.2.9　计算与统计方法　使用 Excell 2007 进行数据整理，SAS 软件进行方差分析。

2　结果

2.1　农艺和生理性状方差分析

表 1 中方差分析表明，穗粒数、灌浆中期叶面积指数和抽穗前叶绿素含量 3 个品种间差异不显著，其他性状品种间存在显著差异。灌浆中期叶绿素含量品种、年份与处理间互作显著，其他性状年份、处理和品种的互作均不显著。穗粒数和抽穗前叶面积指数在不同水分处理间差异不显著，而其他性状差异显著。

2.2　产量及其构成因素的水分敏感性

从表 2 可知，与干旱处理相比，充分灌水条件下，中麦 175、京冬 8 号和矮抗 58 的籽粒产量分别增加 1 072.6、1 274 和 991.7kg • hm^{-2}，穗数增加 104、15、47 个/m^2，千粒重增加 0.9、1.8、1.7g，株高增加 7.2、9、6.2cm，抽穗时间分别推迟 3d、2d 和 3d。中麦 175 产量的水分敏感指数小于 1，而京冬 8 号和矮抗 58 大于 1，分别为 1.13 和 1.03；中麦 175 千粒重和株高的水分敏感指数分别为 0.64 和 0.94，低于京冬 8 号和矮抗 58；而中麦 175 的水分敏感指数为 1.67，高于京冬 8 号和矮抗 58 穗数的 0.30 和 0.87。干旱环境下，中麦 175 籽粒产量比京冬 8 号和矮抗 58 增产 848.7 和 1 424.9kg • hm^{-2}，而在充分灌水环境下，增产 647.3 和 1 505.8kg • hm^{-2}。

中麦 175 穗数比京冬 8 号和矮抗 58 分别多 68.8 和 38.0 个/m^2；京冬 8 号千粒重显著高于中麦 175 和矮抗 58；3 个品种穗粒数差异不显著。株高从高到低，依次是京冬 8 号、中麦 175 和矮抗 58；中麦 175 抽穗最早，京冬 8 号其次，矮抗 58 最晚。因此，中麦 175 的显著特征是早熟、半矮秆、穗数多，其穗数对水分敏感，而千粒重、株高和产量的水分敏感性较低。

2.3　生理性状与水分敏感性

从表 3 可知，与干旱处理相比，充分灌水条件下，所有品种的衰老速度明显减缓，其生理特征值 *NDVI*、*SPAD* 和 *LAI* 均增大。分析各项指标的水分敏感程度可知，中麦 175 的 *NDVIv* 和 *NDVIg* 水分敏感指数大于 1，显著高于京冬 8 号和矮抗 58，与其穗数的敏感性相一致，说明干旱对中麦 175 生物量有较大影响。然而，中麦 175 的 *LAIv* 水分敏感系数为 0.57，低于京冬 8 号和矮抗 58 的 1.43 和 0.87，对水分反应敏感性小。中麦 175 的 *LSR* 和 *LSS* 水分敏感指数都大于 1，且介于京冬 8 号和矮抗 58 之间。品种间衰老速率的敏感性差异与株高有关（表 2），表现为株高越高，衰老速率的敏感性越小。

中麦 175 和京冬 8 号的春季苗期覆盖度显著大于矮抗 58，苗期繁茂性较好，在干旱环境下有利于抑制大气蒸腾。在抽穗前和灌浆中期，中麦 175 均保持最低的植被指数。中麦 175 和矮抗 58 的 LSR 显著低于京冬 8 号，灌浆中期中麦 175 叶片的 LSS 低于京冬 8 号，其衰老速率较慢。抽穗前中麦 175 和矮抗 58 的 LAI 差异不显著，但均显著低于高秆品种京冬 8 号；灌浆中期 LAI 品种间差异不显著，说明中麦 175 灌浆中后期的光合面积没有显著优势。抽穗前品种间 SPAD 差异不显著，但灌浆中期中麦 175 和矮抗 58 显著高于京冬 8 号，说明灌浆期中麦 175 的 SPAD 保持相对较高水平。灌浆中期中麦 175 SPAD 水分敏感系数仅为 0.61，表现为水分敏感性低。品种间 LRS 差异显著，在干旱环境下，中麦 175 叶片发生明显卷曲，与高产品种矮抗 58 表现一致，而节水型品种京冬 8 号的叶片轻微卷曲。

表 1　性状方差分析表

Table 1　Analysis of variance for traits

性状 Trait	环境数 Environment number	处理间 Treatment	年份间 Year	品种间 Line	重复 Rep	品种×年份×处理 Line×Year×Treatment	误差 Error
籽粒产量 Grain yield	8	79 544.3**	11 642.4	37 495.5**	2 249.5	5 570.2	3 518.7
穗数 Spike number	8	42 742.7**	37 807.5**	16 869.1*	2 675.9	2 049.3	3 151.4
穗粒数 Kernel number per spike	8	22.3	1 023.8**	27.6	6.3	21.1	9.3
千粒重 Thousand kernel weight	8	24.2**	1 013.2**	65.7**	0.9	3.8	2.9
株高 Plant height	8	262.2**	49.4*	3 760.8**	0.8	14.2	6.9

（续）

性状 Trait	环境数 Environment number	处理间 Treatment	年份间 Year	品种间 Line	重复 Rep	品种×年份×处理 Line×Year×Treatment	误差 Error
抽穗 Heading date	8	382.4**	6 021.1**	19.2**	6.8*	1.2	1.5
植被覆盖度 GC	4	158.1**	13.4	19.3*	7.9	5.5	5.8
抽穗前植被归一化指数 NDVIv	8	0.03**	0.15**	0.05**	0.004	0.002	0.002
灌浆中期植被归一化指数 NDVIg	8	0.11**	0.095**	0.017**	0.003	0.004	0.002
抽穗前叶面积指数 LAIv	4	1.3	18.1**	1.43*	0.24	0.47	0.31
灌浆中期叶面积指数 LAIg	4	5.17**	0.004	0.51	0.005	0.21	0.44
抽穗前叶绿素含量 SPADv	6	108.9**	0.03	2.1	2.4	10.1	16.7
灌浆中期叶绿素含量 SPADg	2	122.7**		31.1**	0.26**	6.2**	0.016
叶片衰老速率 LSR	8	92.03*	738.4**	153.54**	31.36	51.72	10.36
叶片衰老等级 LSS	4	75.10**	1.33	47.10**	0.56	4.98	1.77
叶片卷曲度 LRS	3	1.33*	0.81	18.1**	0.51	0.22	1.25

* 表示显著（$P<0.05$）；** 表示极显著（$P<0.01$） * Significant at $P<0.05$; ** Significant at $P<0.01$.

表 2 不同水分条件下 3 个品种的产量及主要农艺性状的比较

Table 2 Grain yield and major traits of three cultivars in two environments

品种 Cultivar	处理 Treatment	籽粒产量 Grain yield ($kg \cdot hm^{-2}$)	穗数 Panicle ($panicle/m^2$)	千粒重 1000-grain weight (g)	穗粒数 Kernel number per spike	株高 Plant height (cm)	抽穗天数 Heading date (d)
中麦 175 Zhongmai175	干旱 Drought	5 662.2	447.9	40.5	29.7	72.2	203.2
	灌水 Irrigation	6 734.8	551.8	41.4	29.2	79.4	206
	平均 Mean	6 198.5a	499.9a	41.0b	29.5	75.8b	204.6c
	WSI	0.86	1.67	0.64	2.3	0.94	1.08
京冬 8 号 Jingdong 8	干旱 Drought	4 813.5	423.6	43.7	30.5	85.2	204.8
	灌水 Irrigation	6 087.5	438.5	45.5	30.5	94.2	206.8
	平均 Mean	5 450.5b	431.1b	44.6a	30.5	89.7a	205.8b
	WSI	1.13	0.3	1.16	0.05	1	0.77
矮抗 58 号 Aikang 58	干旱 Drought	4 237.3	438.2	40.3	32.5	53.6	205.3
	灌水 Irrigation	5 229.0	485.6	42.0	32.4	59.8	208.3
	平均 Mean	4 733.2c	461.9ab	41.2b	32.5	56.7c	206.8a
	WSI	1.03	0.87	1.19	0.72	1.08	1.15

同列不同小写字母表示 5%差异显著。下同。

Values followed by different small letters within the same column are significantly different at 5% probability level. The same as below.

表 3 不同水分条件下 3 个品种的生理性状比较

Table 3 The physiological traits of three cultivars in two environments

品种 Cultivar	处理 Treatment	植被覆盖度 *GC*	抽穗前植被归一化指数 *NDVIv*	灌浆中期植被归一化指 *NDVIg*	抽穗前叶面积指数 *LAIv*	灌浆中期叶绿素含量 *SPADg*	叶片衰老速率 *LSR*	叶片衰老等级 *LSS*	叶片卷曲度 *LRS*
中麦 175 Zhongmai 175	干旱 Drought		0.60	0.44	2.88	56.20	19.10	7.5	4.0a
	灌水 Irrigation		0.68	0.66	3.21	56.80	9.10	3.5	
	平均 Mean	13.2a	0.64b	0.55b	3.05b	56.50a	14.10b	5.5b	
	WSI		1.48	1.17	0.57	0.61	1.09		

（续）

品种 Cultivar	处理 Treatment	植被覆盖度 *GC*	抽穗前植被归一化指数 *NDVIv*	灌浆中期植被归一化指数 *NDVIg*	抽穗前叶面积指数 *LAIv*	灌浆中期叶绿素含量 *SPADg*	叶片衰老速率 *LSR*	叶片衰老等级 *LSS*	叶片卷曲度 *LRS*
京冬 8 号 Jingdong 8	干旱 Drought		0.72	0.48	3.07	50.20	31.40	10	1.3b
	灌水 Irrigation		0.79	0.67	4.12	52.20	16.00	1.4	
	平均 Mean	13.0a	0.76a	0.58a	3.59a	51.20c	23.70a	6.2a	
	WSI		1.02	0.99	1.43	1.00	0.87		
矮抗 58 Aikang 58	干旱 Drought		0.72	0.55	2.53	55.40	20.28	4.0	5.0a
	灌水 Irrigation		0.75	0.71	2.99	55.80	9.44	2.2	
	平均 Mean	11.2b	0.74a	0.63a	2.76b	55.60b	14.86b	3.1c	
	WSI		0.53	0.84	0.87	1.41	1.14		

3 讨论

3.1 中麦 175 农艺学特征对适应性的影响

成熟期是决定品种对环境适应性的一个重要时期，直接或间接影响产量、抗病性及抗逆性。中麦 175 成熟期早，能够避免灌浆后期高温等不利环境的影响；同时其抽穗期对水分敏感，表现为干旱条件下抽穗早，充分供水条件下与京冬 8 号抽穗时间基本一致。这种生长发育速率受水分变化影响较大的特征，使中麦 175 在不同的水分环境都能正常成熟，从而适应更广的环境。

具有高产潜力和水分不敏感的品种，在有限灌溉和充分供水条件下都能获得高产。中麦 175 不但在有限灌溉环境下和充分供水环境下都能高产，而且产量对水分敏感性较低。从产量三要素来看，中麦 175 穗数和穗粒数对水分敏感性较大，而千粒重敏感性较低。中麦 175 分蘖多、成穗率高，在有限灌水条件下，其穗数也多于一般品种（如本试验中京冬 8 号和矮抗 58），因此穗数多是中麦 175 高产的重要基础，而株型紧凑，叶片小又是穗数多不倒伏的重要保障。中麦 175 的千粒重较低，株高中等，但这两个指标的水分敏感指数都低于京冬 8 号和矮抗 58。因此，中麦 175 的高产和广适性主要表现为穗数多，株高和千粒重水分敏感性低，穗数水分敏感性较高，这些特征与前人对广适性品种的研究结果基本一致[2,18]。中麦 175 的育成和推广也表明，培育水旱兼用的半矮秆品种是可行的，其关键指标是穗数多、株高适中，这与耐密性好的玉米品种产量高有相似之处。根据中麦 175 亲本来源，推测其分蘖力强、成穗率高的特性来自于父本京 411，抗病性来自于母本 BMP27，BMP27 是中国农业大学培育的抗条锈、抗白粉病亲本，因此中麦 175 兼具抗病与高产基因，是其广适性的重要遗传基础。

3.2 生理特征对适应性的影响

植被覆盖度与产量和千粒重呈正相关[19]。苗期植被覆盖率高，植株竞争能力强，能提高小麦苗期抗逆能力。快速的地表覆盖能增大地表遮荫，从而有效减少土壤水分蒸发，同时增加营养吸收和生物量积累，提高小麦抵抗水分胁迫的能力[10,20]，在水分限制环境下有明显的节水效果[21-22]。中麦 175 尽管千粒重较低，但春季表现快速的植被覆盖度，之前研究表明该品种苗期最大根长、根鲜重具有明显优势，具有较高的肥料吸收效率[13-14]，因此中麦 175 较好的苗期繁茂性可能对提高其肥料和水分利用效率有利，是适应不同环境的优势性状之一。

不同水分条件下 NDVI 的增加与冬小麦产量的变化密切相关[23]。中麦 175 在抽穗前到灌浆中期，都保持相对低的 NDVI，且灌浆中期的叶面积指数没有优势（可能与其紧凑型株型有关），但中麦 175 仍然在不同环境下取得了高产。干旱条件下，耐旱性较强的品种叶绿素含量变化较小，且叶绿素含量高的品种表现出抗衰老的特征[3]，中麦 175 在灌浆中期叶绿素含量显著高于京冬 8 号，有利于抵抗后期衰老。中麦 175 与矮抗 58 的后期叶片衰老速率都显著慢于京冬 8 号，这与广适性品种鲁麦 21 的特征[18]有相似之处。因此中麦 175 叶片叶绿素含量高，叶功能较好，可能是其发挥光合效率的优势性状之一，是获得高产的基础。中麦 175 叶绿素含量水分敏感性低，说明其群体光合能力比较稳定，能在水、旱环境下取得较高产量。

叶片卷曲是植物水分胁迫适应性反应之一。尽管叶片卷曲的光合作用机制仍然不十分明确，但是干旱胁迫下小麦叶片卷曲，减少了有效叶面积和蒸腾作用，有利于节约水分，使光系统功能免受损害[24]，对产量和穗粒数有正效应[25]，因此叶片卷曲有助于小麦抵抗干旱等胁迫，提高品种适应性。中麦175在干旱环境下叶片发生明显卷曲，可能对其抗旱节水有重要贡献。

中麦175表现出矮抗58叶片卷曲、衰老缓慢及京冬8号早熟、灌浆速率快等特点，并且其株型紧凑、半矮秆。控制这些性状的基因聚合在一起，可能对小麦抗旱节水与高产的结合有利，从而使品种表现广适性特征。培育广适性品种可借鉴中麦175的选育方法，在多个环境下进行熟期、株高、千粒重及生长发育相关特征的稳定性比较，注意穗数、灌浆期对干旱和耐热性等逆境的表现，借助生理性状开展辅助选择，可提高广适性品种选择效率和准确性。

4 结论

广适性品种中麦175具有株型直立、叶片小、高产、穗数多、早熟、半矮秆、苗期繁茂性好及较慢的衰老速率等特点。干旱环境下，灌浆期叶片发生明显卷曲，但叶绿素含量较高，叶功能好。中麦175叶面积指数和叶绿素含量的水分敏感性低，抽穗后植被指数维持较低水平，株高和千粒重在不同水分条件下表现稳定，产量水分敏感性较低，穗数对水分反应敏感。对这些性状开展基因聚合与表型选择，有助于提高培育广适性品种的效率。

参考文献

References

[1] 何中虎，夏先春，陈新民，庄巧生．中国小麦育种进展与展望．作物学报，2011，37：202-215.

He Z H，Xia X C，Chen X M，Zhuang Q S. Progress of wheat breeding in China and future perspective. *Acta Agronomica Sinica*，2011，37：202-215.（in Chinese）

[2] 孙道杰，王辉，闵东红，李学军，冯毅．广适性小麦品种的重要性状指标．中国农学通报，2002，18：83-85.

Sun D J，Wang H，Min D H，Li X J，Feng Y. The important characteristics index to wide adaptation of wheat cultivars. *Chinese Agricultural Science Bulletin*，2002，18：83-85.（in Chinese）

[3] Babar M A，vanGinkel M，Klatt A R，Prasad B，Reynolds M P. The potential of using spectral reflectance indices to estimate yield in wheat grown under reduced irrigation. *Euphytica*，2006，150：155-172.

[4] Babar M A，Reynolds MP，van Ginkel M，Klatt A R，Raun W R，Stone M L. Spectral reflectance indices as a potential indirect selection criteria for wheat yield under irrigation. *Crop Science*，2006，46：578-588.

[5] Chen J，Liang Y，Hu X，Wang X，Tan F，Zhang H，Ren Z，Luo P. Physiological characterization of stay green wheat cultivars during the grain filling stage under field growing conditions. *Acta Physiologiae Plantarum*，2010，32：875-882.

[6] Christopher J T，Manschadi A M，Hammer G，Borrell A K. Developmental and physiological traits associated with high yield and stay-green phenotype in wheat. *Australian Joural of Agricultural Research*，2008，59：354-364.

[7] Gupta A K，Kaur K，Kaur N. Stem reserve mobilization and sink activity in wheat under drought conditions. *American Journal of Plant Sciences*，2011，2：70-77.

[8] 张嵩午，王长发，冯佰利，苗芳，周春菊，刘党校．冷型小麦对干旱和阴雨的双重适应性．生态学报，2004，24（4）：680-685.

Zhang S W，Wang C F，Feng B L，Miao F，Zhou C J，Liu D X. Double adaptability of cold type wheat to drought and rainy weather. *Acta Ecologica Sinica*，2004，24（4）：680-685.（in Chinese）

[9] FanT，Balta M，Rudd J，Payne W A. Canopy temperature depression as a potential selection criterion for drought resistance in wheat. *Agricultural Sciences in China*，2005，4：793-800.

[10] Reynolds MP，Ortiz-Monasterio J I，McNab A. *Application of Physiology in Wheat Breeding*. Mexico：CIMMYT，2001，88-100.

[11] Pask A，Pietragalla J，Mullan D M，Reynolds M P. *Physiological Breeding II：A Field Guide to Wheat Phenotyping*. Mexico：CIMMYT，2012，60-62.

[12] 张国庆．广适高产小麦品种中麦175在晋城市的推广应用．中国种业，2013，10：24-25.

Zhang G Q. Application ofZhongmai 175 with wide adaptability and high yield in Jincheng. *China Seed Industry*, 2013, 10: 24-25. (in Chinese)

[13] 肖永贵，路亚明，闻伟锷，陈新民，夏先春，王德森，李思敏，童依平，何中虎．小麦骨干亲本京 411 及衍生品种苗期根部性状的遗传．中国农业科学，2014，47 (15)：2916-2926.
Xiao Y G, Lu Y M, Wen W E, Chen X M, Xia X C, Wang D S, Li S M, Tong Y P, He Z H. Genetic contribution of seedling root traits among elite wheat parent Jing 411 to its derivatives. *Scientia Agricultura Sinica*, 2014, 47 (15): 2916-2926. (in Chinese)

[14] 肖永贵，李思敏，李法计，张宏燕，陈新民，王德森，夏先春，何中虎．两种施肥环境下冬小麦京 411 衍生系的产量和生理性状分析．作物学报，2015，41 (9)：1333-1342.
Xiao Y G, Li S M, Li F J, Zhang H Y, Chen X M, Wang D S, Xia X C, He Z H. Genetic analysis of yield and physiological traits among genotypes derived from elite parent Jing 411 under two fertilizer environments. *Acta Agronomica Sinica*, 2015, 41 (9): 1333-1342. (in Chinese)

[15] 肖永贵，刘建军，夏先春，陈新民，Matthew Reynolds，何中虎．基于图像处理的冬小麦植被覆盖率测定及其遗传解析．作物学报，2013，39：1935-1943.
Xiao Y G, Liu J J, Xia X C, Chen X M, Matthew Reynolds, He Z H. Genetic analysis of vegetative ground cover rate in winter wheat using digital imaging. *Acta Agronomica Sinica*, 2013, 39: 1935-1943. (in Chinese)

[16] O'toole J C, Cruz R T. Response of leaf water potential, stomatal resistance, and leaf rolling to water stress. *Plant Physiology*, 1980, 65: 428-432.

[17] Fischer R A, Maurer R. Drought resistance in spring wheat cultivars 1. Grain yield responses. *Australian Journal of Agricultural Research*, 1978, 29: 897-912.

[18] 胡延吉．两个不同适应性冬小麦品种的竞争能力．作物学报，2003，29：175-180.
Hu Y J. Competitive ability of two winter wheat cultivars with different adaptability. *Acta Agronomica Sinica*, 2003, 29: 175-180. (in Chinese)

[19] Li X M, Xiao Y G, Cheng X M, Xia X C, Wang D S, He Z H, Wang H J. Identification of QTLs for seedling vigor in winter wheat. *Euphytica*, 2014, 198: 199-209.

[20] Mullan D J, Reynolds M P. Quantifying genetic effects of ground cover on soil water evaporation using digital imaging. *Functional Plant Biology*, 2010, 37: 703-712.

[21] Casadesús J, Kaya Y, Bort J, Nachit M M, Araus J L, Amor S, Ferrazzano G, Maalouf F, Maccaferri M, Martos V, Ouabbou H, Villegas D. Using vegetation indices derived from conventional digital cameras as selection criteria for wheat breeding in water-limited environments. *Annals of Applied Biology*, 2007, 150: 227-236.

[22] Araus J L, Bort J, Steduto P, Villegas D, Royo C. Breeding cereals for Mediterranean conditions: Ecophysiological clues for biotechnology application. *Annals of Applied Biology*, 2003, 142: 129-141.

[23] 丛建鸥，李宁，许映军，顾卫，乐章燕，黄树青，席宾，雷镲．干旱胁迫下冬小麦产量结构与生长、生理、光谱指标的关系．中国生态农业学报，2010，18：67-71.
Cong J O, Lin N, Xu Y J, Gu W, Le Z Y, Huang S Q, Xi B, Lei C. Relationship between indices of growth, physiology and reflectivity and yield of winter wheat under water stress. *Chinese Journal of Eco-Agriculture*, 2010, 18: 67-71. (in Chinese)

[24] Nar H, Saglam A, Terzi R, Várkonyi Z, Kadioglu A. Leaf rolling and photosystem II efficiency in Ctenanthe setosa exposed to drought stress. *Photosynthetica*, 2009, 47 (3): 429-436.

[25] Bogale A, Tesfaye K, Geleto T. Morphological and physiological attributes associated to drought tolerance of Ethiopian durum wheat genotypes under water deficit. *Journal of Biodiversity and Environmental Sciences*, 2011, 1: 22-36.

两种施肥环境下冬小麦京411及其衍生系产量和生理性状的遗传分析

肖永贵[1]，李思敏[1]，李法计[1]，张宏燕[2]，陈新民[1]，王德森[1]，夏先春[1]，何中虎[1,3]

([1]中国农业科学院作物科学研究所/国家小麦改良中心，北京 100081；
[2]天津市武清区种子管理站，天津 301700；[3]CIMMYT 中国办事处，北京 100081)

摘要：明确高产广适性与肥料高效利用相关性状的遗传特性对培育优良新品种具有重要意义。本研究以京411及其衍生后代共15个品种（系）为材料，在4个正常施肥和1个常年不施肥环境下研究品种（系）的产量构成因素和生理性状，结合90K SNP芯片，解析骨干亲本携带的产量和生理性状等位基因信息，探讨优异基因对高产品种的贡献。结果表明，正常施肥环境下，京411衍生后代的产量和收获指数均随世代增加呈逐渐上升趋势，其中收获指数增加较为显著（$P<0.05$）。冠层温度对籽粒产量有重要贡献（$P<0.05$），而叶面积指数和光合速率对收获指数有显著贡献（$P<0.05$）。两种施肥条件下中麦175具有较高且稳定的产量和生物学产量，主要与其较高的肥料吸收效率有关。控制产量和生理性状的遗传区段主要分布在A和B基因组上，2B、3A和5A染色体分别携带控制穗粒数、叶面积指数和光合速率的位点。京411携带31个对产量和生理性状为正向效应的等位基因，衍生品种CA0958和中麦175携带的正向效应区段最多，分别占正向效应位点总数的53.85%和51.35%。中麦175的高产潜力和广适性可能与其携带较多有利等位基因有关。

关键词：普通小麦；衍生品种；肥料利用效率；生理性状；SNP标记

Genetic Analysis of Yield and Physiological Traits in Elite Parent Jing 411 and Its Derivatives under Two Fertilization Environments

XIAO Yong-Gui[1], LI Si-Min[1], LI Fa-Ji[1], ZHANG Hong-Yan[2], CHEN Xin-Min[1], WANG De-Sen[1], XIA Xian-Chun[1], and HE Zhong-Hu[1,3,*]

([1]Institute of Crop Science, Chinese Academy of Agricultural Sciences (CAAS) /National Wheat Improvement Center, Beijing 100081, China; [2]Wu-qing Seed Management Station, Tianjin 301700, China; [3]International Maize and Wheat Improvement Center (CIMMYT) China Office, c/o CAAS, Beijing 100081, China)

原文发表在《作物学报》，2015，41（9）：1333-1342.
本研究由国家自然科学基金项目（31161140346）和作物生物学国家重点实验室开放课题基金（2014KF02）资助。
通讯作者（Corresponding author）：何中虎，E-mail：zhhecaas@163.com
第一作者联系方式：E-mail：xiaoyonggui@caas.cn

Abstract: Understanding the genetics of the traits related to yield potential, stability and high fertilizer-use efficiency is important for breeding new varieties. The elite parent Jing 411 has been widely used in wheat breeding programs in Northern Winter Wheat Region. In this study, Jing 411 and its 14 derivatives were sown in normal fertilized and no-fertilized environments to assess the changes in yield stability and fertilizer-use efficiency. Grain yield and harvest index increased following the generation advance, particularly harvest index increased significantly ($P<0.05$). Zhongmai 175 showed high and stable grain yield and above ground biomass under two fertilized environments, mainly due to its high fertilizer-uptake efficiency. Canopy temperature was signifi-cantly associated with grain yield ($P<0.05$), and leaf area index, and photosynthetic rate were significantly correlated with harvest index ($P<0.05$). Several important loci associated with yield and physiological traits were found in A and B genomes. Three new loci for grains per spike, leaf area index, and photosynthetic rate were detected on chromosomes 2B, 3A, and 5A, respectively. Furthermore, 31 positive genetic loci for yield and physiological traits were detected in elite parent Jing 411, and two derivatives, CA0958 and Zhongmai 175, had the most positive loci than others, accounting for 53.85% and 51.35%, respectively. Zhongmai 175 had higher yield potential and stability under different fertilizer environments, possibly due to the presence of positive loci for yield and physiological traits.

Keywords: Common wheat; Derivative genotypes; Fertilizer-use efficiency; Physiological traits; SNP markers

提高品种的产量潜力及肥料利用效率是作物育种的重要发展方向，其优势主要体现在提高种植效益，节约资源投入，减少环境污染[1-2]。不同基因型在有效利用土壤潜在养分特性方面存在显著差异[3]，高肥效品种在营养缺乏的条件下，也可以通过一系列形态、生理和生化机制活化土壤中的营养元素，从而提高土壤中潜在肥料的生物有效性[2]。小麦高肥效特性主要表现为植株对土壤固有营养的高效利用，营养元素在植株体内的合理分配和高效转运，并使机体保持最优的光合势和碳同化效率，从而获得较高的生物产量和籽粒产量[4]。因此，品种的高肥效特性可进一步区分为高效吸收型和高效转化型。不同小麦品种的肥料吸收效率和转化效率存在显著差异，在低肥环境下，墨西哥和芬兰的春小麦具有较高的肥料吸收效率，而法国和英国的冬小麦具有较高的肥料转化效率；在高肥环境下，英国、墨西哥和芬兰小麦的吸收和转化效率基本一致[3,5-6]。Gaju等[2]认为，土壤肥料供给量与小麦品种存在显著互作关系，在低肥条件下更能有效鉴定品种的肥料吸收效率。

有关产量性状基因定位的报道很多，几乎每个遗传背景下都会定位到数十个产量相关性状的基因位点，而且多数位点不尽相同。已发掘的产量性状基因位点的遗传效应相对较小（2%～10%），对表型变异的解释率（R^2）大于15%的主效基因很少[7-10]。主要原因有二：一是环境因素诱导产量及其相关性状之间存在相互协调或此消彼长的关系[11-12]；二是影响产量及其相关性状的基因不仅具有主效效应，而且还受基因×基因、基因×环境等复杂网络关系的共同影响[7]。目前，关于产量性状的遗传研究不足以解析其性状和基因间的关系，仍需要在多种试验环境下发掘产量性状基因及其协同效应[9-10]。另外，理想的基因定位研究应在一个能够造成产量相关性状差异较大的环境（如干旱胁迫）下进行，尤其是遗传力相对较低的目标性状（如生理性状），这样可降低基因与环境的互作效应，更容易鉴定出目标性状的主基因效应及基因间的互作关系[8]。在多种环境下均稳定表达的产量相关性状基因更受育种家的青睐[12]，因为它决定品种的广适性和稳产性。

京411于1991年通过北京市品种审定，具有高产稳产、适应性广等突出特点，在北部冬麦区推广近20年，累计种植74.3万 hm^2[13]。从该品种育成至21世纪初，京411一直是北部冬麦区高产广适育种的骨干亲本，已育成16个衍生品种（系），包括中麦175、北京0045等本麦区的主栽品种[14]。本研究以冬小麦骨干亲本京411及其14个衍生后代为材料，在正常施肥环境和常年不施肥环境下研究不同基因型的产量构成因素及相关生理性状的变化，并结合全基因组 SNP 标记分析该衍生群体的产量、生理性状遗传区段，为今后利用骨干亲本培育更优良的新品种提供理论基础。

1 材料与方法

1.1 材料及试验设计

京 411 是 20 世纪 90 年代初育成的半矮秆小麦品种，具有抗寒性强、分蘖力强、成穗率高、高产稳产、抗病性好、适应性广等特点，本课题组针对其植株偏高、灌浆较慢、落黄差等缺点进行改造，育成的衍生品种（系）多数具有高产广适的特性，其中北京 0045、中麦 175 等品种已在北部冬麦区生产中发挥重要作用。本研究选用骨干亲本京 411 及其 14 个衍生品种（系）为材料（表 1），包括衍生子一代 6 份、子二代 8 份，这些材料基本上反映了当前北部冬麦区小麦品种的产量现状。

2012—2013 和 2013—2014 年度，将 15 份品种（系）种植于中国农业科学院作物科学研究所顺义试验站（40.13°N，116.57°E）和天津市武清区种子管理站农场（39.50°N，117.00°E），北京试验点的播期为 2012 年 9 月 28 日和 2013 年 9 月 29 日，天津试验点的播期为 2012 年 10 月 20 日（因前茬玉米收获晚，导致小麦迟播）和 2013 年 9 月 26 日。仅 2013—2014 年度顺义试验站设置正常施肥和全生育期不施肥两组试验。各试验点播种时施纯 N 180kg·hm^{-2}、P_2O_5 120kg·hm^{-2} 和 K_2O 60kg·hm^{-2}，拔节期追施氮肥约 150kg·hm^{-2}。均采取随机区组设计，3 次重复，小区行距 20cm，行长 4m，面积为 5.6m^2，种植密度为 300 株·m^{-2}。全生育期浇越冬水、拔节水和灌浆水共 3 次，并控制病虫草害。

1.2 产量和生理性状测定

测定了两种试验条件下产量及其相关性状。越冬前，在小区内出苗均匀且长势旺盛的非边行区域选定 2 个 1m 样段，调查基本苗。春季拔节前调查样段的最大分蘖，收获前纪录样段穗数，并折算成单位面积穗数。成熟期在小区内随机选取 3 点，量取株高，计算平均株高；在非边行区随机选取 20 穗，计算穗粒数，折算成每穗粒数。收获前从根部拔取收获每小区的样段，剪去根部，置于 80℃烘箱烘干至恒重，测定生物学产量，样品穗单独脱粒计算收获指数。田间记载抽穗期和成熟期。小区收获后晾晒干燥至含水量为 12%时称重，测定千粒重。根据小区收获产量和样段的籽粒重量的总合计算供试品种的公顷产量。

本试验仅在正常施肥条件下测定生理性状。采用 GreenSeeker 冠层光谱仪（Trimble，美国）分别在越冬前、起身期、拔节期和灌浆期采集植被归一化指数（normalized difference vegetation index，NDVI），测定时传感器与小麦冠层 60cm 高度上方保持平衡，沿播种方向测定，在小区非边行区域往返测定 2 次。用 REYTEK ST20XB 型手持式红外测温仪（Reytek Corporation，美国）测定冠层温度，光谱通带为 8～14μm，灰度值 ε=0.95，选择抽穗期晴朗无风少云的天气，于午后 13：30—15：00 测定一次[15]。开花期和灌浆中后期在小区非边行区域随机选择 10 片旗叶，利用 SPAD 502Plus 叶绿素仪（Konica Minolta Inc.，日本）测定叶绿素相对含量，10 片旗叶的均值作为被测群体的叶绿素相对含量。开花期在每个小区非边行区随机选择 10 片旗叶，用 YMJ-B 叶面积测定仪（浙江托普仪器有限公司）测定叶面积，其均值作为被测群体的旗叶面积。利用 AccuPAR 叶面积指数仪（Decagon Devices Inc.，美国）测定群体冠层上下的入射光和透射光强度，并计算群体的冠层 LAI 值。另外，灌浆初期利用 Li-6400XT 光合仪（LI-COR，美国），在恒定光源 1 200μmol·m^{-2}·s^{-1}和二氧化碳 400μmol·mol^{-1}条件下，测定每小区 5 片旗叶的光合速率，其均值视为被测群体的旗叶光合速率。

1.3 SNP 标记分析

从每份材料选取 10 个大小和色泽均匀一致的籽粒在培养皿中发芽 2 周，混合剪取幼苗，用澳大利亚 Triticarte 公司提供的 CTAB 法提取小麦全基因组 DNA（http://www.diversityarrays.com/）。利用 UVS-99 微量分光光度计（ACTGene，美国）检测 DNA 质量和浓度，将 DNA 终浓度调为 300ng·μL^{-1}。

利用 Illumina SNP 基因分型研究平台对京 411 及其衍生品种（系）DNA 进行 90 K SNP 芯片分型，包括 BS、BobWhite、CAP、D_contig 等系列标记，共计 81 587 个，其中 38 819 个 SNP 标记在京 411 衍生群体内存在差异，已有染色体定位信息的标记为 18 011个[16]。本研究仅选取有染色体定位信息的标记进行分析，并去除掉重复标记，剩余 4 846 个标记对产量和生理性状进行遗传区段检测分析。

1.4 数据处理和图谱绘制

按 Bruker 和 Frohberg[17] 的方法计算肥料敏感指数 FSI。

$$FSI=\left(1-\frac{0F}{NF}\right)\bigg/\left(1-\frac{\overline{0F}}{\overline{NF}}\right)$$

其中，0F 为某品种在不施肥条件下的产量或生理性状，NF 为某品种在正常施肥条件下的产量或生理性状，$\overline{0F}$为所有品种在不施肥条件下产量或生理性状的均值；$\overline{NF}$为所有品种在正常施肥条件下产量或生理性状的均值。FSI＜1 为肥料不敏感型品种，FSI≥1 为肥料敏感型品种。

利用 SAS 9.3 软件对产量和生理性状进行基本统计量、方差和相关分析，结合 SAS 的 Glmselect 程序对 SNP 数据和产量相关性状进行逐步回归，当 $P<0.001$ 时认为标记区段与性状相关联[14]。

2 结果与分析

2.1 方差分析与试验总体表现

对 15 个品种的产量和生理性状的方差分析表明（表 2），除越冬前和灌浆期的 NDVI 之外，其他性状均受基因型、环境和基因型×环境的显著影响（$P<0.05$）。由于越冬前和灌浆期的 NDVI 在品种间差异不显著，不再进行后续分析。冠层温度和旗叶光合速率在相同环境下的重复间存在显著差异（$P<0.05$），可能是这两个性状在测定时受环境的温度和 CO_2 浓度变化影响所致。

2012—2013 年度，供试材料在北京点产量平均为 5 671.5kg • hm^{-2}，在天津点的平均产量为 5 164.8 kg • hm^{-2}；2013—2014 年度，两点产量介于二者之间，全生育期不施肥处理的平均产量为 2 840.7kg • hm^{-2}。2012—2013 年度北部麦区冬季降水量达 108mm，播种期土壤墒情好，气温偏高，出苗均匀，有利于分蘖形成；春季持续低温（≤5℃）直至 3 月中下旬，抽穗后低温时间长，生育期较往年推迟 5～7d，灌浆期相对较长，因而产量较高。2012—2013 年度天津武清试验点产量偏低的主要原因是播期偏晚。2013—2014 年度的有效积温和降雨量与历年相近，冬前温度适宜，春季气温回升较往年快，但灌浆后期（5 月 26 日至 31 日）遇到高温天气，气温高达 42℃，成熟期较往年提早 7d 左右。

表 1 供试品种（系）的矮秆基因型、抽穗期和株高

Table 1 *Rht-B1a*/*Rht-B1b* and *Rht-D1a*/*Rht-D1b* genotype, heading date, and plant height of varieties/lines used in this study

品种 Variety	矮秆基因型 Dwarf genotype		抽穗期 Heading date (month/day)	株高 Plant height (cm)
	Rht-B1a/*Rht-B1b*	*Rht-D1a*/*Rht-D1b*		
京 411 Jing 411	*Rht-B1a*	*Rht-D1a*	5/4	89.7
CA9722	*Rht-B1b*	*Rht-D1a*	5/2	79.5
北京 0045 Beijing 0045	*Rht-B1a*	*Rht-D1b*	5/5	79.0
中麦 175 Zhongmai 175	*Rht-B1a*	*Rht-D1a*	5/3	81.8
中麦 415 Zhongmai 415	*Rht-B1b*	*Rht-D1a*	5/5	78.6
新冬 37 Xindong 37	*Rht-B1b*	*Rht-D1a*	5/5	74.7
CA0958	*Rht-B1a*	*Rht-D1a*	5/6	89.3
CA1055	*Rht-B1b*	*Rht-D1a*	5/2	83.5
CA0548	*Rht-B1b*	*Rht-D1a*	4/30	79.3
CA1090	*Rht-B1b*	*Rht-D1a*	5/2	83.6
中麦 818 Zhongmai 818	*Rht-B1b*	*Rht-D1a*	5/4	87.3
中麦 816 Zhongmai 816	*Rht-B1b*	*Rht-D1a*	5/4	87.0

（续）

品种 Variety	矮秆基因型 Dwarf genotype		抽穗期 Heading date (month/day)	株高 Plant height (cm)
	Rht-B1a/Rht-B1b	*Rht-D1a/Rht-D1b*		
09抗1027 09-Kang 1027	*Rht-B1b*	*Rht-D1a*	5/6	79.2
CA1119	*Rht-B1a*	*Rht-D1a*	5/5	84.7
CA1133	*Rht-B1b*	*Rht-D1a*	5/1	77.8

15份供试材料的光周期均为*Ppd-A1b*、*Ppd-B1b*和*Ppd-D1a*类型；春化基因均呈隐性，即*vrn-A1*、*vrn-B1*和*vrn-D1*。

Fifteen varieties tested have the same photoperiod (*Ppd-A1b*, *Ppd-B1b*, *and Ppd-D1a*) and vernalization (*vrn-A1*, *vrn-B1*, *and vrn-D1*) types.

表2 正常施肥条件下京411及其衍生品种的产量和生理性状方差分析

Table 2 Analysis of variance for yield and physiological traits in elite parent Jing 411 and its derivatives under normal fertilizer environments

来源 Source		基因型 Genotype (G)	环境 Environment (E)	重复 Replicate	基因型×环境 G×E	误差 Error
自由度 *df*		14	3	2	56	148
最大分蘖 Maximum seedling		3 985 621.87***	26 066 869.37***	82 173.60	32 022 951.64**	3 361 165.08
植被归一化指数 NDVI	越冬前 Pre-winter	0.02	1.10***	0.002	0.09***	0.09
	返青期 Green-up	0.02***	0.13***	0.005*	0.06***	0.06
	拔节期 Jointing	0.05***	1.34**	0.002	0.03***	0.02
	灌浆期 Grain-filling	0.02	0.21***	0.000 5	0.05	0.18
叶绿素含量 SPAD value	开花期 Anthesis	631.83***	1 603.22***	9.10	164.62*	315.04
	灌浆期 Grain-filling	78.62*	863.91***	12.87	97.34**	61.12
旗叶面积 Flag leaf area		168.58*	822.07***	14.40	180.48***	296.60
叶面积指数 Leaf area index		7.45***	51.76***	0.24	17.70***	12.63
冠层温度 Canopy temperature		39.25***	362.01***	5.44*	29.62***	31.82
旗叶光合速率 Photosynthetic rate		184.32***	28.80***	33.51*	213.72***	110.47
每平方米穗数 Spike number per square meter		341 819.92***	1 719 981.49***	25 517.29	675 756.94*	1 159 083.17
穗粒数 Grain number per spike		355.14**	2 708.16***	28.70	2 625.75***	2 798.44
千粒重 Thousand-grain weight		2 124.62**	6 576.44***	4.77	1 215.09***	1 397.87
株高 Plant height		2 072.69**	4 900.36***	7.36	829.79***	872.95
地上部生物量 Above-ground biomass		49 706.74***	38 277.45***	3 328.00	92 835.91***	95 558.42
收获指数 Harvest index		0.04*	0.26***	0.002	0.09***	0.13
产量 Yield		11 958 435.47**	6 332 302.28***	115 897.72	30 837 903.06***	30 919 802.30

*、**和***分别表示在 $P<0.05$、$P<0.01$ 和 $P<0.001$ 水平显著。

*, **, and *** indicate significance at $P<0.05$, $P<0.01$, and $P<0.001$, respectively.

2.2 衍生品种的产量和生理性状及肥料敏感性分析

正常施肥条件下衍生子代的每平方米穗数与京411接近，穗粒数随世代更迭略呈增加趋势，千粒重和株高有所降低（表3），收获指数和产量随世代演变均呈递增趋势，其中衍生二代的收获指数和产量显著高于骨干亲本京411（$P<0.05$）。说明在对京411的产量性状改良过程中，基本保持了单位面积有效穗数，提高穗粒数，进而提高收获指数和籽粒产量。在正常施肥条件下，中麦175、中麦818、CA1055、中麦816和CA1119产量较高（5 741.9～5 882.6kg·hm^{-2}）；在不施肥条件下，CA1119和中麦175同样具有较高的产量，分别为3 237.8kg·hm^{-2}和3 135.1kg·hm^{-2}。在产量较高的品种中，CA1055的产量、穗粒数、千粒重和株高均对肥料的敏感性较高，敏感指数在1.05～1.57之间，中麦818和中麦816的产量和穗数对肥料敏感性较高，敏感指数在1.04～1.34之间（表3）。虽然CA1119的产量对肥料表现不敏感（0.90），但穗数（1.22）和穗粒数（1.06）对肥料敏感性较高。而中麦175的产量、产量构成因素和株高均对肥料表现不敏感，敏感指数在0.70～0.97之间，说明中麦175的产量及其相关性状在不同肥料环境下均具有较好的稳定性。

在正常施肥条件下，中麦175在苗期和成熟期均具有较高的生物学产量（拔节期NDVI值为0.50，成熟期生物量为16.94kg·m^{-2}，穗数为741.25个·m^{-2}），而旗叶面积（17.28cm^2）、叶绿素含量（SPAD值49.04）和冠层温度（19.72℃）其他品种较低。在不施肥环境下，中麦175仍具有较高的生物量（8.80kg·m^{-2}），仅次于中麦415（9.80kg·m^{-2}）。可见，中麦175在不同肥料条件下均具有较高的肥料吸收效率，能保持较高的生物量和籽粒产量，这可能与其株型直立、有效穗数较多、群体蒸腾效率高有关。

2.3 正常施肥条件下生理性状和产量构成因素的相关分析

在正常施肥条件下，每平方米穗数与起身期群体最大分蘖数（$r=0.72$，$P<0.01$）和拔节期NDVI（$r=0.53$，$P<0.05$）呈显著正相关（表4），而与叶绿素含量（$r=-0.54$，$P<0.05$）、旗叶面积（$r=-0.52$，$P<0.05$）和光合速率（$r=-0.54$，$P<0.05$）呈显著负相关，说明苗期群体量对发育后期的单位面积穗数有重要影响，而随着群体量的增加，旗叶光合作用能力呈下降趋势。灌浆期旗叶光合速率与穗粒数（$r=0.56$，$P<0.05$）和收获指数（$r=0.59$，$P<0.05$）呈显著正相关，灌浆期旗叶的光合产物主要作用于籽粒灌浆，进而提高收获指数。此外，生物量与产量呈极显著正相关（$r=0.70$，$P<0.01$），灌浆初期的冠层温度与产量呈显著负相关（$r=-0.61$，$P<0.05$），冠层温度低意味着植物群体的水分吸收和蒸腾系统高效，生理代谢旺盛，具有群体高光效能力，利于光合产物的积累和转化，进而影响籽粒产量。

2.4 京411及其衍生品种产量和生理性状的遗传区段解析

在4 846个SNP标记中筛选出70个与产量和生理性状相关的标记，A、B基因组携带的位点较多，分别为29个和36个，D基因组仅有5个位点。2B和5A染色体分布最多，分别为10个和9个，其中2B携带的位点主要与穗粒数、千粒重和产量等性状相关，而5A携带的位点多数与生理性状相关，如旗叶叶绿素含量、光合速率和株高等（表5）。控制小麦NDVI的位点主要与1B、4A、4B和5B染色体有关，而控制叶绿素含量的位点主要位于4A、4D、5A和5B染色体。在3A染色体的80cM区段附近和5A染色体的300cM区段附近分别发掘到3个与叶面积指数和光合速率相关联的SNP位点，这两个遗传区段可能参与小麦生理性状调控。另外，在2B染色体发掘到4个与穗粒数相关的SNP位点（表5），位于染色体短臂近着丝点附近的240cM区域，该区域可能携带穗粒数相关基因。

在发掘的产量和生理性状关联区段中，衍生一代获得骨干亲本的遗传位点较高，为51.43%，衍生二代相对较低，为43.75%，正向效应遗传位点也在衍生一代和二代之间存在差异，分别为46.12%和40.50%（表6）。不同品种间携带与骨干亲本一致的位点存在较大差异，如中麦415、CA0958和中麦175携带与京411相同位点较多，所占比例均在50%以上；CA1090与骨干亲本的相同位点最少，占京411位点数的37.14%。在发掘的70个遗传区段中，对产量和生理性状起正向效应的区段为31个，占总区段数的44.29%，其中CA0958和中麦175携带正向效应区段最多，分别占骨干亲本贡献位点总数的53.85%和51.35%（表6），说明这两份基因型获得较多的骨干亲本优异产量和生理性状遗传成分。

表3　京411衍生品种的产量及其构成因素、株高和肥料敏感性分析

Table 3　Analysis of yield and its components, plant height and fertilizer susceptibility in Jing 411 and its derivatives under two treatments

品种/衍生世代 Variety/derived generation	穗数 Spikes mumber（m^{-2}）			穗粒数 Grain number per spike			千粒重 Thousand-grain weight(g)			株高 Plant height（cm）			产量 Yield（$kg \cdot hm^{-2}$）		
	NF	0F	FSI	NF	0F	FSI	NF	0F	FSI	NF	0F	FSI	NF	0F	FSI
京411 Jing 411	674.6	413.3	0.81	28.8	22.7	0.87	32.2	44.3	0.78	89.7	67.8	1.12	4 956.7	2 863.7	0.88
CA9722	642.3	336.7	0.99	28.8	24.3	0.64	27.2	41.2	1.07	79.5	61.8	1.02	5 375.7	3 071.2	0.89
北京0045 Beijing 0045	520.6	328.3	0.77	32.1	20.2	1.52	31.1	51.2	1.35	79.0	63.0	0.93	5 121.5	2 401.1	1.10
中麦175 Zhongmai 175	741.3	386.7	0.99	29.4	24.4	0.70	30.6	44.1	0.92	81.8	67.0	0.83	5 882.6	3 135.1	0.97
中麦415 Zhongmai 415	648.4	343.3	0.98	32.7	28.6	0.51	27.6	41.4	1.04	78.6	59.8	1.09	5 518.4	2 938.5	0.97
新冬37 Xindong 37	663.0	368.3	0.92	30.8	23.5	0.97	23.6	42.8	1.70	74.7	59.7	0.92	4 711.7	3 078.0	0.72
CA0958	745.8	343.3	1.12	28.3	19.3	1.30	33.0	44.3	0.71	89.3	64.9	1.25	5 313.2	2 475.9	1.11
CA1055	640.1	473.3	0.54	31.0	19.1	1.57	28.5	42.8	1.05	83.5	60.1	1.28	5 814.8	2 595.0	1.15
CA0548	631.4	365.0	0.88	29.4	22.9	0.90	33.1	50.4	1.09	79.3	62.1	0.99	5 455.0	2 873.9	0.98
CA1090	609.2	326.7	0.96	30.5	20.2	1.38	33.3	48.7	0.96	83.6	63.6	1.09	5 476.8	2 666.4	1.06
中麦818 Zhongmai 818	686.9	243.3	1.34	30.5	27.1	0.50	28.9	40.1	0.83	87.3	67.3	1.01	5 821.0	2 897.7	1.04
中麦816 Zhongmai 816	668.3	241.1	1.33	30.9	26.3	0.61	29.3	40.4	0.79	87.0	68.1	0.99	5 812.2	2 886.7	1.04
09抗1027 09-kang 1027	724.9	338.3	1.11	29.2	20.4	1.23	23.9	34.7	0.94	79.2	62.4	0.97	5 586.2	2 642.6	1.09
CA1119	727.0	301.7	1.22	27.1	20.1	1.06	34.0	45.7	0.72	84.7	69.1	0.84	5 741.9	3 237.8	0.90
CA1133	661.6	373.3	0.91	30.2	21.0	1.25	31.4	49.8	1.22	77.8	67.7	0.59	5 660.8	2 836.5	1.04
亲本 Parent	674.6	413.3	0.81	28.8	22.7	0.87	32.1a	44.3	0.79	89.7a	67.8	1.12	4 956.7b	2 863.7	0.88
衍生一代 First generation	660.2	351.1	0.97	30.4	23.4	0.94	28.8b	44.2	1.11	80.5c	62.7	1.01	5 302.5ab	2 850.0	0.96
衍生二代 Second generation	668.7	332.8	1.04	29.9	22.1	1.06	30.3ab*	44.1	0.95	82.8b	65.2	0.97	5 671.1b	2 830.9	1.04
均值 Mean	665.7	345.5		30.0	22.7		29.8	44.1		82.3	64.3		5 483.2	2 840.7	
标准差 *SD*	44.2	59.6		1.2	5.4		1.9	2.7		3.5	2.7		251.4	327.3	

NF：正常施肥；0F：不施肥；FSI：肥料敏感指数。部分性状数据后不同字母表示亲本及衍生世代间差异（$P<0.05$）。

NF：normal fertilizer；0F：no fertilizer；FSI：fertilizer susceptibility index；Diffenert letters after partial values indieme significant different among the parent and its decived generation at $P<0.05$.

表4 正常施肥条件下生理性状与产量及其构成因素的相关分析

Table 4 Correlation coefficients of grain yield and its components with physiological traits under normal fertilizer environments

生理性状 Physiological trait		穗数 SN	穗粒数 GNS	千粒重 TGW	株高 PH	生物量 Biomass	收获指数 HI	产量 Yield
最大分蘖 Maximum seedlings		0.72**	−0.16	−0.20	0.45	0.12	−0.62*	−0.36
植被归一化指数 NDVI	返青期 Green-up	0.41	−0.22	−0.16	0.37	−0.26	−0.31	0.04
	拔节期 Jointing	0.53*	−0.25	−0.37	0.14	−0.01	−0.38	0.01
叶绿素含量 SPAD value	开花期 Anthesis	−0.50*	0.36	−0.26	−0.69**	−0.05	0.44	0.21
	灌浆期 Grain-filling	−0.54*	0.31	−0.46	−0.67**	−0.26	0.14	0.07
旗叶面积 Flag leaf area		−0.52*	−0.07	0.06	−0.23	−0.26	0.08	−0.17
叶面积指数 Leaf area index		0.38	−0.20	−0.28	0.35	−0.46	−0.55*	−0.22
冠层温度 Canopy temperature		−0.22	0.22	0.34	0.62*	−0.40	−0.36	−0.61*
旗叶光合速率 Photosynthetic rate		−0.54*	0.56*	0.35	−0.16	0.22	0.59*	0.01

穗数为每平方米穗数。*和**分别表示在 $P<0.05$ 和 $P<0.01$ 水平显著。

SN: spike number per square meter; GNS: grain number per spike; TGW: thousand-grain weight; PH: plant height; HI: harvest index. * and **indicate significance at $P<0.05$ and $P<0.01$, respectively.

表5 京411及其衍生品种的产量和生理性状遗传区段分析

Table 5 Genetic loci for grain yield and physiological traits in Jing 411 and its derivatives

产量和生理性状 Yield and physiological trait		位点总数 Total loci	A基因组 A genome	B基因组 B genome	D基因组 D genome
最大分蘖 Maximum seedlings		2		1B (1); 4B (1)	
植被归一化指数 NDVI	返青期 Green-up	3	4A (1)	4B (1); 5B (1)	
	拔节期 Jointing	1		1B (1)	
叶绿素含量 SPAD value	开花期 Anthesis	5	4A (2); 5A (1)	5B (1)	4D (1)
	灌浆期 Grain-filling	6	4A (2); 5A (2)	5B (2)	
旗叶面积 Flag leaf area		2	7A (1)	2B (1)	
叶面积指数 Leaf area index		5	3A (3)	2B (1)	3D (1)
冠层温度 Canopy temperature		5	2A (1); 7A (1)	2B (1); 6B (1); 7B (1)	
旗叶光合速率 Photosynthetic rate		6	4A (1); 5A (3)	4B (1); 5B (1)	
每平方米穗数 Spike number per square meter		6	6A (1)	6B (2); 7B (1)	2D (1); 7D (1)
穗粒数 Grain number per spike		6	1A (1)	1B (1); 2B (4)	
千粒重 Thousand-grain weight		5	3A (1); 6A (1)	2B (1); 3B (1); 4B (1)	
株 高 Plant height		5	5A (2)	1B (1); 4B (1)	2D (1)
地上部生物量 Above-ground biomass		7	6A (2)	3B (1); 6B (2); 7B (2)	
收获指数 Harvest index		3	4A (1); 5A (1)	5B (1)	
产 量 Yield		3	1A (1)	2B (2)	

括号中数字表示该染色体上相关位点数。The number in a pair of parenthese indicates the quanlity of loci on the chromosome.

3 讨论

3.1 京 411 衍生品种的产量和肥料利用效率分析

理想的小麦基因型应在多种环境下具有较高且稳定的产量，并表现较高的水肥利用效率[1]。育种过程中通常将世代材料和稳定品系在肥水充足的环境下进行产量鉴定，但由于基因型与肥料用量存在互作关系，很难鉴定出品种间肥料利用效率的差异[3,6]。本研究表明，生物量对籽粒产量有重要贡献，尤其在低肥条件下，生物量高的品种一般籽粒产量较高（$r=0.70$，$P<0.01$）。在京 411 衍生后代中，中麦 175 在两种肥料条件下均具有较高的肥料吸收能力，表现较高的生物量和产量，而骨干亲本京 411 具有较高的肥料转化能力，低肥环境下具有最高的收获指数（0.46）。本研究还发现，低肥环境下的千粒重显著高于正常施肥条件下的，可能原因是低肥条件下单位面积穗数少，群体内竞争力小，光温通透，植株光合产物能够充分向籽粒转化，进而提高粒重。

中麦 175 是骨干亲本京 411 最成功的改造品种，具有较好的丰产性、稳产性和广适性，分别通过北部麦区水地和黄淮旱肥地品种审定，也是北部冬麦区国家区域试验的对照品种，近几年来一直是北部冬麦区种植面积最大的品种[14]。该品种不仅具有较强的耐低磷[18]和氮肥利用效率（童依平，个人交流），而且水分利用效率高。2012—2013 和 2013—2014 年对 100 个国家审定的冬小麦品种在无灌水、灌 1 水和灌 2 水条件下进行比较，中麦 175 在灌 1 水的条件下产量最高（郭进考，个人交流）。该品种的节水性可能是其在河南、陕西和甘肃陇东等旱肥地区大面积推广种植的主要原因。

3.2 京 411 产量和生理性状的优异等位基因解析

在京 411 衍生群体中共发掘 70 个控制产量和生理性状的遗传区段，其中 2B 染色体携带较多控制穗粒数、千粒重和产量的遗传区段（表 6），5A 染色体上控制叶绿素含量和光合速率等生理性状位点较多。已有研究证实，5A 染色体携带较多控制小麦环境适应性、抗病性和驯化相关的基因[19-20]，最重要的 2 个基因是春化基因（*Vrn1*）和易脱粒基因（*q*），*Vrn1* 基因除参与调控小麦春化作用之外，也可能参与株高、穗数和穗粒数等产量相关性状的调控[20]；*q* 基因同时具有控制穗部形态和易脱粒性的作用[19]。Graziani 等[21]发现，2B 染色体携带一个重要的控制千粒重、穗下节长、叶绿素含量和产量的基因，且在 8 个试验环境下均稳定表达。另外，在京 411 衍生群体中发现 3A 和 5A 短臂分别存在控制叶面积指数和光合速率相关基因位点，这些影响生理性状的位点也可能参与植物发育[22]、叶片衰老相关性状[23]的调控。

此外，本研究还检测到控制单个产量和生理性状的遗传区段，分布在 1A、1B、2A、2D、3A、3B、3D、4A、4B、4D、5B、6A、6B、7A、7B 和 7D，可能这些区段的等位基因仅属于京 411 衍生种群体特有的遗传变异，也可能是因群体数量小造成了等位变异的伪关联。然而，利用全基因组关联分析时，经常遇到单标记负责调控单一性状的结果，可能是高密度的标记信息更能精准评估种质的单倍型模式，并能够准确发掘出单倍型与性状表型的关系[24]。连锁定位表明，分布在这些遗传区段的多数位点具有多效性，同时参与产量和生理相关性状的调控。例如，5B 染色体 NDVI 位点，不仅在灌溉和干旱胁迫下能够有效作用于苗期 NDVI 的表达，而且对植株发育后期的叶片衰老机制有一定的影响[22]。3B 染色携带的基因位点同时参与调控千粒重、生物学产量、叶片叶绿素含量和籽粒产量[21]。

3.3 骨干亲本的遗传解析及其对衍生品种的遗传贡献

骨干亲本京 411 具有突出的适应性和丰产性，可能与其携带较多正向遗传效应基因有关（表 6）[14,18]。本研究表明，京 411 携带 31 个产量和生理性状正向关联区段，其衍生后代 CA0958 和中麦 175 携带骨干亲本贡献的正向基因位点最多。通常骨干亲本具备有利基因多、遗传传递力强等特点[14,25-26]，育种过程主要是保留其携带的有益等位基因，并使其在多个环境下稳定表达。这也充分说明传统育种针对优异农艺性状为目标进行高强度选择，可使优异目标性状基因有效聚集，使亲本携带的正向遗传位点在衍生材料中增加。对京 411 衍生群体研究发现，CA0958 是携带骨干亲本的根系性状[14]、肥料利用效率[18]、产量和生理性状正向遗传区段（表 6）最多的品种，然而，在本试验和实际生产情况下，该品系的产量均不突出，且稳定性较差。而另外一个携带正向遗传区段较多的品种中麦 175，其产量和适应性明显优于其他品种。由此说明，正向等位基因的富集是构成优良品种的遗传基础，而优势等位基因的协同表达可能是提高品种高产和稳产性的重要因素之一。

表 6 京 411 及其衍生品种的多效性标记遗传位点差异

Table 6 Difference of alleles linked with pleiotropic SNP markers among Jing 411 and its derivatives

品种 Variety	位点数 Loci	与京 411 相同位点 Same allele as that of Jing 411		正向效应位点 Positive allele	
		数目 Number	比例 Percentage（%）	数目 Number	比例 Percentage（%）
亲本 Parent					
京 411 Jing 411	70	—	—	31	44.29
衍生一代 First generation					
CA9722	70	32	45.71	14	43.75
北京 0045 Beijing 0045	70	31	44.29	12	38.71
中麦 175 Zhongmai 175	70	37	52.86	19	51.35
中麦 415 Zhongmai 415	70	42	60.00	17	40.48
新冬 37 Xindong 37	70	35	50.00	17	48.57
CA0958	70	39	55.71	21	53.85
衍生二代 Second generation					
CA1055	70	35	50.00	15	42.86
CA0548	70	30	42.86	15	50.00
CA1090	70	26	37.14	12	46.15
中麦 818 Zhongmai 818	70	32	45.71	11	34.38
中麦 816 Zhongmai 816	70	31	44.29	11	35.48
09 抗 1027 09-Kang 1027	70	32	45.71	14	43.75
CA1119	70	27	38.57	10	37.04
CA1133	70	32	45.71	11	34.38

4 结论

在京 411 衍生群体中，中麦 175 是改造骨干亲本较成功的范例。该品种在越冬前后幼苗旺盛，生长势较强，分蘖成穗率多，单位面积穗容量多，株型好，叶面积指数较高（光截获面积大），群体光能利用率高，具有较高的生物学产量。此外，高效肥料吸收效率和水分利用效率为该品种的稳产和广适性奠定了基础。这些优良的农艺性状表现可能与其携带较多正向遗传位点有关。

参考文献

References

[1] Chloupek O, HrstkovaP, Schweigert P. Yield and its stability, crop diversity, adaptability and response to climate change, weather and fertilisation over 75 years in the Czech Republic in comparison to some European countries. *Field Crops Res*, 2004, 85: 167-190.

[2] Gaju O, AllardV, Martre P, Snape J W, Heumez E, LeGouis J, Moreau D, Bogard M, Griffiths S, Orford S, Hubbart S, Foulkes M J. Identification of traits to improve the nitrogen-use efficiency of wheat genotypes. *Field Crops Res*, 2011, 123: 139-152.

[3] Foulkes M J, Sylvester-Bradley R, Scott R K. Evidence for dif-ferences between winter wheat cultivars in acquisition of soil mineral nitrogen and uptake and utilization of applied fertilizer nitrogen. *J Agric Sci* (Camb), 1998, 130: 29-44.

[4] Barraclough P B, Lopez-Bellido R, Hawkesford M J. Genotypic variation in the uptake, partitioning and remobilisation of nitro-gen during grain-filling in wheat. *Field Crops Res*, 2014, 156: 242-248.

[5] Ortiz-Monasterio R J I, Sayre K D, Rajaram S, Mc-Mahom M. Genetic progress in wheat yield and nitrogen

use efficiency under four nitrogen rates. *Crop Sci*, 1997, 37: 898-904.

[6] Muurinen S, Slafer G A, Peltonen-SainioP. Breeding effects on nitrogen use efficiency of spring cereals under northern condi-tions. *Crop Sci*, 2006, 46: 561-568.

[7] Pushpendra KG, Harindra S B, Pawan L K, Neeraj K, Ajay K, Reyazul R M, Amita M, Jitendra K. QTL analysis for some quan-titative traits in bread wheat. *J Zhejiang Univ Sci B*, 2007, 8: 807-814.

[8] Kuchel H, Williams K J, LangridgeP, Eagles H A, Jefferies S P. Genetic dissection of grain yield in bread wheat: I. QTL analysis. *Theor Appl Genet*, 2007, 115: 1029-1041.

[9] Bennett D, Reynolds M, Mullan D, Izanloo A, Kuchel H, Lan-gridgeP, Schnurbusch T. Detection of two major grain yield QTL in bread wheat (*Triticum aestivum* L.) under heat, drought and high yield potential environments. *Theor Appl Genet*, 2012, 125: 1473-1485.

[10] Maphosa L, Langridge P, Taylor H, Parent B, Emebiri L C, Ku-chel H, Reynolds M P, Chalmers K J, Okada A, Edwards J, Mather D E. Genetic control of grain yield and grain physical characteristics in a bread wheat population grown under a range of environmental conditions. *Theor Appl Genet*, 2014, 127: 1607-1624.

[11] Reynolds MP, Trethowan R, Crossa J, Vargas M, Sayre K D. Physiological factors associated with genotype by environment interaction in wheat. *Field Crops Res*, 2002, 75: 139-160.

[12] Maccaferri M, Sanguineti M C, Corneti S, Ortega J L A, Salem M B, Bort J, DeAmbrogio E, del Moral L F G, Demontis A, El-Ahmed A, Maalouf F, Machlab H, Martos V, Moragues M, Motawaj J, Nachit M, Nserallah N, Ouabbou H, Royo C, Slama A, Tuberosa R. Quantitative trait loci for grain yield and adapta-tion of durum wheat (*Triticum durum* Desf.) across a wide range of water availability. *Genetics*, 2008, 178: 489-511.

[13] 陈新民，何中虎，王德森，庄巧生，张运宏，张艳，张勇，夏先春．利用京 411 为骨干亲本培育高产小麦新品种．作物杂志，2009，(4)：1-5.
Chen X M, He Z H, Wang D S, Zhuang Q S, Zhang Y H, Zhang Y, Zhang Y, Xia X C. Developing high yielding wheat varieties from core parent "Jing 411". *Crops*, 2009, (4): 1-5. (in Chinese with English abstract)

[14] 肖永贵，路亚明，闻伟锷，陈新民，夏先春，王德森，李思敏，童依平，何中虎．小麦骨干亲本京 411 及衍生品种苗期根部性状的遗传．中国农业科学，2014，47：2916-2926.
Xiao Y G, Lu Y M, Wen W E, Chen X M, Xia X C, Wang D S, Li S M, Tong Y P, He Z H. Genetic contribution of seedling root traits among elite wheat parent Jing 411 to its derivatives. *Sci Agric Sin*, 2014, 47: 2916-2926. (in Chinese with English abstract)

[15] Reynolds MP, Nagarajan S, Razzaque M A, Ageeb O A A. Heat tolerance. In: Reynolds M P, Ortiz-Monasterio J I, McNab A, eds. Application of Physiology in Wheat Breeding. Mexico: CIMMYT Press, 2001. pp 124-135.

[16] Allen A M, Barker G L, Berry S T, Coghill J A, Gwilliam R, Kirby S, Robinson P, Brenchley R C, D'Amore R, McKenzie N, Waite D, Hall A, Bevan M, Hall N, Edwards K J. Transcript-specifie, single-nucleotide polymorphism discovery and linkage analysis in hexaploid bread wheat (*Triticum aestivum* L.). *Plant Biotechnol J*, 2011, 9: 1086-1099.

[17] Brucker P L, Frohberg R C. Stress tolerance and adaptation in spring wheat. *Crop Sci*, 1987, 27: 31-36.

[18] 李法计，肖永贵，金松灿，夏先春，陈新民，汪洪，何中虎．京 411 衍生系苗期氮和磷利用效率相关性状的遗传解析．麦类作物学报，2015，35：737-749.
Li F, Xiao Y G, Jin S C, Xia X C, Chen X M, Wang H, He Z H. Genetic analysis of nitrogen and phosphorus utilization efficiency related traits at seeding stage of Jing 411 and its derivatives. *J Tritical Crops*, 2015, 35: 737-749. (in Chinese with English abstract)

[19] Kato K, Miura H, Akiyama M, Kuroshima M, Sawada S. RFLP mapping of the three major genes, *Vrn1*, *Q* and *B1*, on the long arm of chromosome 5A of wheat. *Euphytica*, 1998, 101: 91-95.

[20] Vitulo N, Albiero A, Forcato C, Campagna D, Pero F D, Bag-naresi P, Colaiacovo M, Faccioli P, Lamontanara A, Šimková H, Kubaláková M, Perrotta G, Facella P, Lopez L, Pictrella M, Gianese G, Doležel J, Giuliano G, Cattivelli L, Valle G, Stanca A M. First survey of the wheat chromosome 5A composition through a next generation sequencing approach. *PLoS One*, 2011, 6: e26421.

[21] Graziani M, Maccaferri M, Royo C, Salvatorelli F, Tuberosa R. QTL dissection of yield components and

morpho-physiological traits in a durum wheat elite population tested in contrasting themo-pluviometric conditions. *Crop & Past Sci*, 2014, 65: 80-95.

[22] Li X M, Chen X M, Xiao YG, Xia X C, Wang D S, He Z H, Wang H J. Identification of QTLs for seedling vigor in winter wheat. *Euphytica*, 2014, 198: 199-209.

[23] Vijayalakshmi K, Fritz A K, Paulsen G M, Bai G, Pandravada S, Gill B S. Modeling and mapping QTL for senescence-related traits in winter wheat under high temperature. *Mol Breed*, 2010, 26: 163-175.

[24] Maccaferri M, Sanguineti M C, Demontis A, El-Ahmed A, del Moral LG, Maalouf F, Nachit M, Nserallah N, Ouabbou H, Rhouma S, Royo C, Villegas D, Tuberosa R. Association map-ping in durum wheat grown across a broad range of water regimes. *J Exp Bot*, 2011, 62: 409-438.

[25] 韩俊，张连松，李静婷，石丽娟，解超杰，尤明山，杨作民，刘广田，孙其信，刘志勇．小麦骨干亲本“胜利麦/燕大 1817”杂交组合后代衍生品种遗传构成解析．作物学报，2009，35：1395-1404.

Han J, Zhang L S, Li J T, Shi L J, Xie C J, You M S, Yang Z M, Liu G T, Sun Q X, Liu Z Y. Molecular dissection of core parental cross "Triumph/Yanda 1817" and its derivatives in wheat breeding program. *Acta Agron Sin*, 2009, 35: 1395-1404. (in Chinese with English abstract)

[26] 肖永贵，殷贵鸿，李慧慧，夏先春，阎俊，郑天存，吉万全，何中虎．小麦骨干亲本“周 8425B”及其衍生品种的遗传解析和抗条锈病基因定位．中国农业科学，2011，44：3919-3929.

Xiao Y G, Yin G H, Li H H, Xia X C, Yan J, Zheng T C, Ji W Q, He Z H. Genetic diversity and genome-wide association analysis of stripe rust resistance among the core wheat parent Zhou 8425B and its derivatives. *Sci Agric Sin*, 2011, 44: 3919-3929. (in Chinese with English abstract)

不同氮素处理对中麦175和京冬17产量相关性状和氮素利用效率的影响

李法计[1]　徐学欣[2]　肖永贵[1]　何中虎[1,3]　王志敏[2]

（[1]中国农业科学院作物科学研究所/国家小麦改良中心，北京 100081；
[2]中国农业大学农学院，北京　100193；[3]CIMMYT 中国办事处，北京 100081）

摘要：本研究旨在了解我国黄淮和北部冬麦区不同施氮量和施氮模式对氮高效吸收和利用的影响，以及中麦175和京冬17产量对不同施氮处理的响应。2013—2014和2014—2015连续两年在河北吴桥和北京顺义两地种植两品种，观测不同施氮量和基追比处理下，冬小麦的群体特性、产量相关性状，以及氮素吸收效率（NUpE）和氮素利用效率（NUtE）。在吴桥点设0、60＋0、120＋0、120＋60、120＋120、120＋180kg·hm^{-2}（基肥＋拔节肥）6个处理，在顺义点仅设前5个处理。在总施氮量0～240kg·hm^{-2}（吴桥）和0～180kg·hm^{-2}（顺义）范围内，随施氮量增加，归一化植被指数（NDVI）和气冠温差（CTD）提高，群体总粒数和成熟期生物量增加，进而产量提高；但继续增加施氮量会导致粒重、开花前干物质向籽粒转运量、转运率、对籽粒贡献率、收获指数、氮肥偏生产力、氮素吸收和利用效率降低。在不同施氮水平下，中麦175的产量和稳定性均优于京冬17，表现出穗数多、穗粒重稳定性好、群体活力持久、生物量和收获指数高、花前干物质积累量高和花后干物质转运能力强、氮素吸收效率高，这可能是其高产高效的重要基础。考虑到产量回报和经济效益，推荐中麦175和京冬17在黄淮麦区（北片）施氮量为180～240kg·hm^{-2}，在北部冬麦区施氮量为120～180kg·hm^{-2}。灌浆中后期，NDVI和CTD与穗数、产量和生物量相关性高，可作为快速评价品种氮肥敏感性的指标。

关键词：冬小麦；氮素利用效率；干物质；高产；稳产

Effect of Nitrogen on Yield Related Traits and Nitrogen Utilization Efficiency in Zhongmai 175 and Jingdong 17

LI Fa-Ji[1]，XU Xue-Xin[2]，XIAO Yong-Gui[1]，HE Zhong-Hu[1,3]，and WANG Zhi-Min[2]

（[1] *Institute of Crop Science*，*Chinese Academy of Agricultural Sciences*（*CAAS*）/ *National Wheat Improvement Center*，*Beijing* 100081，*China*；[2] *College of Agronomy and Biotechnology*，*China Agricultural University*，*Beijing* 100193，*China*；[3] *CIMMYT-China Office*，*c/o CAAS*，*Bei-*

原文发表在《作物学报》，2016，42（12）：1853-1863.

本研究由中国农业科学院科技创新工程，国家公益性行业（农业）科研专项经费（201203033-2）和国家现代农业产业技术体系建设专项（CARS-03）资助。

This studywas supported by the Agricultural Science and Technology Innovation Program of CAAS，the China Special Fund for Agro-scientific Research in the Public Interest（201203033-2），and the China Agriculture Research System（CARS-03）.

通讯作者（Corresponding author）：王志敏，E-mail：Zhimin206@263. net

第一作者联系方式：E-mail：lifajily@163. com

jing 100081, *China*)

Abstract: The objective of this study was to understand the effects of different nitrogen (N) application amounts and split ratios on high efficiency of N uptake and utilization, as well as the response to different N treatments of Zhongmai 175 and Jingdong 17 planted in Wuqiao, Hebei, and Shunyi, Beijing in 2013—2014 and 2014—2015 cropping seasons. Nitrogen fertilizer was applied in different total and split (basal+jointing stage) amounts, namely 0, 60+0, 120+0, 120+60, 120+120, and 120+180kg·hm^{-2}. In the N range of 0～240kg·hm^{-2} in Wuqiao and 0－180kg·hm^{-2} in Shunyi, the canopy temperature depression (CTD), normalized difference vegetation index (NDVI), biomass of wheat population, and population spikelets increased with the increase of N application amount, as a result, higher yield at maturity was obtained; however, further more N application had a negative effect, showing decreased thousand-kernel weight (TKW), translocation amount (TA) and efficiency (TE) of dry matter accumulated before flow- ering to grain, contribution efficiency (CE), harvest index (HI), partial factor productivity from applied N (PFP_N), N uptake efficiency (NUpE) and N utilization efficiency (NUtE). Zhongmai 175 had higher yield and yield stability than Jingdong 17 in different N application treatments, showing higher levels of spike number (SN), stability of kernel number per spike (KNS), ker- nel weight, population vitality, biomass, HI, dry matter accumulation before flowering, TA, and NUpE. These characters might be the physiological basis of high yield and high efficiency in Zhongmai 175. Considering the return from yield and economic bene-fits, we suggest that the recommended N application amounts for Zhongmai 175 and Jingdong 17 should be 180～240kg·hm^{-2} in the northern part of Huang-Huai Rivers Valley Wheat Zone and 120～180 kg·hm^{-2} in the Northern Winter Wheat Zone. NDVI and CTD at middle to late grain filling stage can be used for rapid evaluation of varietal sensitivity to nitrogen because they are highly correlated with SN, yield, and biomass of wheat.

Keywords: Winter wheat; Nitrogen utilization efficiency; Dry matter; High yield; Stability

氮素是作物生长发育和增产的重要营养元素之一[1-2]，但大量施氮不仅增加生产成本，而且容易导致土壤养分失衡和环境污染[3]。近年来，在华北小麦主产区，氮肥过量施用导致增产效益和氮素利用效率下降的现象较为明显[4]，提高氮素利用效率是小麦生产可持续发展的必然要求[5]。因此，培育肥料利用效率高、高产稳产的新品种，并采取合理的施肥策略在作物生产中具有重要意义，并且节水、节肥、节药已成为未来我国粮食作物持续发展的战略选择。

为解决我国作物生产中肥料利用效率低、损失严重的问题，前人对如何通过合理的栽培措施以提高生产效率进行了大量研究，并提出了氮肥后移、以水控肥和以肥调水等高产优质栽培技术[6-7]。但是，由于对品种间肥料利用效率的遗传差异重视不够，导致其遗传改良进展缓慢[8]。已有的研究虽然在多方面探索了作物肥料吸收利用机制[9-10]，但尚不明确作物氮肥敏感性的品种评价指标，并缺乏大田条件下快速检测的方法。当前大面积推广的品种多在高肥条件下选育而成，对土壤肥力和施肥量要求较高[11]。生产中迫切需要能够适应不同肥力水平，特别是低肥条件下肥料利用效率高、不减产或减产少，高肥条件下耐肥性好、茎秆坚实不倒伏、高产高效的小麦新品种。中麦 175 通过了北部冬麦区水地和黄淮旱肥地 2 次国家审定以及北京、山西、河北、青海和甘肃 5 省（市）品种审定，并已在这些地区大面积推广，具有广泛的适应性[12]。肖永贵等[13]对京 411 及其 14 个衍生品种（系）在正常施肥和常年不施肥条件下比较表明，中麦 175 的产量皆最高。另外，对北部和黄淮麦区 64 份主要品种的氮和磷利用效率研究表明，中麦 175 在不施氮、不施磷及氮和磷皆不施的 3 种处理中，产量分别居参试品种的第 1、第 9 和第 1 位，说明中麦 175 为肥料高效型品种[14]。

为进一步探明中麦 175 对氮肥的反应特性及其机

制，本研究以中麦175和京冬17为材料，通过限量灌溉下不同氮肥处理，比较研究群体特性及氮素吸收和利用、干物质积累和转运特征，解析品种间和处理间产量变异原因，探讨两品种对不同氮肥环境的响应机制和需肥规律，以期为培育高产、稳产和广适性小麦新品种提供理论支撑和评价指标，并为氮高效品种大面积推广提供技术指导。

1 材料与方法

1.1 试验材料

中麦175和京冬17的生育期相近（相差1～2d）。京冬17于2007年通过北部冬麦区国家审定。

1.2 试验设计

2013—2014和2014—2015年分别在河北省沧州市吴桥县中国农业大学吴桥试验站（37.41°N，116.57°E，黄淮麦区北片）和北京市顺义区中国农业科学院作物科学研究所试验基地（40.16°N，116.35°E，北部冬麦区）进行试验。吴桥试验地为壤质底黏潮土，地下水埋深7m以上，播前0～20cm土壤含有机质10.4g·kg^{-1}、全氮0.83g·kg^{-1}、碱解氮41.9mg·kg^{-1}、速效磷23.9mg·kg^{-1}、速效钾93.23mg·kg^{-1}，pH为7.7；2013—2014和2014—2015年度小麦季降水量分别为121.9mm和128.0mm。顺义试验地为壤土，播前0～20cm土壤含有机质12.2g·kg^{-1}、全氮0.95g·kg^{-1}、碱解氮106.7mg·kg^{-1}、速效磷33.8mg·kg^{-1}、速效钾183.4mg·kg^{-1}，pH为7.9；2013—2014和2014—2015年小麦季降水量分别为125.6mm和127.0mm。

在吴桥试验点设6个处理，施氮量（基肥+拔节追肥）分别为纯氮0+0、60+0、120+0、120+60、120+120和120+180kg·hm^{-2}，随机区组设计，3次重复，小区面积30m^2。分别于2013年10月11日和2014年10月14日播种，行距16cm，基本苗均为345万·hm^{-2}。在顺义试验点设5个处理，施氮量（基肥+拔节追肥）分别为0+0、60+0、120+0、120+60和120+120kg·hm^{-2}，随机区组设计，3次重复，小区面积29.7m^2。分别于2013年9月27日和2014年9月28日播种，行距24cm，基本苗300万·hm^{-2}。以尿素作为氮肥，基肥除氮肥外，另施P_2O_5 105kg·hm^{-2}和K_2O 75kg·hm^{-2}；追肥在拔节期施用。在拔节期和开花期浇水，每次75mm。各小区除草、病虫害防治等田间管理措施均保持一致。

1.3 性状调查及其统计分析

于越冬期、拔节期、孕穗期、开花期和成熟期在各小区取2个0.5m样段，取样后剪掉根部，将茎鞘、嫩叶、老叶、穗部分开，105℃杀青30min，75℃烘干后称重，粉碎样品后用凯氏定氮法测定各器官中全氮含量。

开花期、灌浆前期和灌浆中后期，用GreenSeeker（Trimble，美国）测定归一化植被指数（normalized difference vegetation index，NDVI），令传感器于小麦冠层60cm上方保持平衡，在小区非边行区域顺播种方向往返测定2次。用REYTEK ST20XB型手持式红外测温仪（Reytek Corporation，美国）测定空气温度和冠层温度，用两者差计算气冠温差（canopy temperature depression，CTD），光谱通带为8～14μm，其灰度值为0.95，分别在开花期、灌浆前期和灌浆中后期选择晴朗无风少云的天气测定一次，测量时间为13：30—15：00[15]。

在成熟期调查穗数、穗粒数和生物量，在收获后测定产量、千粒重，计算干物质转运量和氮素利用效率。

花前干物质转运量=开花期干物质积累量－（成熟期干物质积累量－籽粒产量）[16]，花前干物质转运率=转运量/开花期干物质积累量[16]，花前干物质贡献率=干物质转运量/籽粒产量[16]；收获指数=籽粒产量/生物学产量[16]；氮肥偏生产力（PFP_N）=籽粒产量/施氮量[10]，氮素吸收效率（NUpE）=植株氮素积累量/施氮量[10]，氮素利用效率（NUtE）=籽粒产量/植株氮素积累量[10]。

利用SAS 9.3进行方差分析及处理间多重比较（Duncan's法）、品种间t测验，采用Microsoft Excel绘制线性关系图。

2 结果与分析

2.1 产量相关性状的方差分析

除开花期CTD外的其他17个性状基因型差异皆达极显著（$P<0.01$）水平；所有性状的环境和氮肥效应皆达极显著水平（$P<0.01$）；多数性状存在显著的基因型×环境、基因型×处理或基因型×

环境×处理互作效应（表 1）。在 2 个试验点，除个别性状外，绝大部分性状的基因型均方皆大于 1.5 倍基因型×年份互作均方，即基因型效应显著大于基因型×年份互作效应，因此可用均值表示产量相关性状。

2.2 不同施氮量下的产量性状

在 0～240kg・hm^{-2}（吴桥）或 0～180kg・hm^{-2}（顺义）施氮范围内，随施氮量增加两品种的产量皆呈递增趋势，且分别在 240kg・hm^{-2} 和 180kg・hm^{-2}时产量最高，过量施氮反而使产量降低。从产量三因素看，两品种的穗数在吴桥和顺义均随施氮量增加而增加，并分别在施氮 240kg・hm^{-2}和 180kg・hm^{-2}时最高；穗粒数随施氮量增加呈递增趋势，但施氮量超过 180kg・hm^{-2}后处理间差异不显著；千粒重均在低氮条件下较高，随施氮量的增加呈降低趋势。两品种在吴桥的穗数、千粒重和产量均高于顺义，而穗粒数则低于顺义，说明环境可通过影响穗数、穗粒数和千粒重进而影响产量。在吴桥，中麦 175 的产量显著高于京冬 17，主要是其穗数显著较高，其穗粒数和千粒重则低于京冬 17；在顺义，中麦 175 的穗数、千粒重和产量均高于京冬 17，穗粒数无明显差异（表 2）。从两地产量变异系数看，中麦 175 为 2.0%和 4.2%，京冬 17 为 2.5%和 5.1%，说明中麦 175 的产量稳定性优于京冬 17。

2.3 不同施氮量下的群体特性

在开花期和灌浆前期，随施氮量的增加，两品种的 NDVI 和 CTD 在 2 个环境下均呈增加趋势，而在灌浆中后期均呈先增加后降低趋势，且分别在施氮量 240kg・hm^{-2}（吴桥）和 180kg・hm^{-2}（顺义）时达最大值。随灌浆进程，NDVI 先保持稳定后逐渐降低，而 CTD 表现为先升高后降低，说明群体灌浆速率和活性随着灌浆的进行先升高后降低（表 3）。中麦 175 的 NDVI 和 CTD 在多数观测时间点高于京冬 17，说明其群体绿色覆盖度和群体活性更高且持久。

2.4 不同施氮量下的干物质积累

两品种开花前后的干物质积累量和生物量均随施氮量增加呈增加趋势，开花前和开花后干物质积累比例在不同施氮处理间差异较小；收获指数以不施氮处理最高，随施氮量增加呈降低趋势。两品种开花前和开花后的干物质积累量、生物量和收获指数均表现吴桥点高于顺义点，而顺义点的开花前干物质积累比例较高（表 4）。两品种相比，花前、花后和全生育期干物质积累量及收获指数均以中麦 175 高于京冬 17，说明中麦 175 在后期可能具有较强的抵抗环境胁迫能力和较高的物质转运能力。

2.5 不同施氮量下的物质转运

表 5 表明，两品种的开花前干物质积累量向籽粒转运量、转运率和对籽粒的贡献率在低氮下较高、随施氮量增加均呈降低趋势，说明两品种均具有一定适应低氮胁迫、通过调节干物质转运获得一定产量水平的能力。开花前干物质积累量向籽粒转运量和转运率在吴桥较高，而对籽粒的贡献率则是顺义较高，说明环境可以影响物质分配。中麦 175 在 2 个环境不同施氮条件下的开花前干物质积累量向籽粒转运量、转运率和对籽粒的贡献率均高于京冬 17，说明中麦 175 具有更强的花后干物质转运能力。

2.6 不同施氮量下的氮素吸收和利用效率

表 6 表明，氮肥偏生产力随施氮量增加而降低，中麦 175 的氮肥生产效率在吴桥和顺义皆高于京冬 17。从氮素吸收和利用效率来看，两品种在吴桥和顺义均随施氮量增加呈降低趋势，且在不同氮素处理间差异显著，说明在氮肥不足时冬小麦会增加对土壤氮素的吸收和利用能力。两品种在吴桥的氮素利用效率高于顺义，氮素吸收效率则低于顺义，说明环境可影响品种的氮素吸收和利用效率。中麦 175 在吴桥和顺义的氮素吸收和利用效率皆高于京冬 17，这与其氮肥偏生产力较高是一致的。

2.7 群体特征与产量性状的相关性分析

灌浆中后期的 NDVI 和 CTD 与穗数、产量和生物量均呈显著或极显著正相关，决定系数均在 0.6 以上（图 1）。因此，灌浆中后期的 NDVI 和 CTD 可作为快速评价冬小麦在不同施氮水平下产量表现即氮肥敏感性的指标。中麦 175 在不同处理与环境下灌浆中后期的 NDVI 和 CTD 更稳定，即在氮肥胁迫条件下能够保持较好的活力，因此产量也更稳定。

表 1　小麦冠层温差、干物质积累、产量及氮素利用相关性状的均方值

Table 1　Mean squares of wheat traits related to canopy temperature depression, dry matter accumulation, yield and nitrogen utilization

	性状 Trait	基因型 Genotype (G) (df: 1)	环境 Environment (E) (df: 3)	重复 Repeat (df: 2)	处理 Treatment (T) (df: 5)	基因型×环境 G×E (df: 3)	基因型×处理 G×T (df: 5)	基因型×环境×处理 G×E×T (df: 26)	误差 Error (df: 86)
产量组分 Yield component	产量 Yield	5 042 158.30**	33 876 552.70**	16 239.20	1 661 250.60**	150 321.60*	21 486.50	138 755.00**	19 209.90
	穗数 Spike number	54 940.06**	131 230.47**	226.00	6 756.91**	6 104.05**	411.18*	343.58**	166.56
	穗粒数 Kernel number spike	15.22**	75.23**	1.51*	11.35**	4.00**	0.54	0.92**	0.39
	千粒重 Thousand-kernel weight	10.24**	762.25**	0.03	17.76**	13.11**	0.47	1.90**	0.61
归一化植被指数 NDVI	开花期 Flowering stage	0.002 3**	0.044 6**	0.000 3	0.005 0**	0.002 1**	0.000 1	0.000 8**	0.000 1
	灌浆前期 Early stage of filling	0.003 9**	0.046 1**	0.000 2	0.010 4**	0.003 0**	0.000 3	0.000 7**	0.000 1
	灌浆中后期 Middle-late stage of filling	0.014 2**	0.024 7**	0.000 5	0.011 6**	0.001 1*	0.000 5	0.000 5	0.000 3
气冠温差 CTD	开花期 Flowering stage	0.12	13.72**	0.09	0.46**	9.23**	0.04	0.05	0.08
	灌浆前期 Early stage of filling	6.44**	9.50**	0.10	1.57**	1.22**	0.14	0.14	0.10
	灌浆中后期 Middle-late stage of filling	11.30**	10.49**	0.01	2.56**	2.49**	0.02	0.15**	0.05
干物质积累 Dry matter accumulation	生物量 Biomass	96 369.68**	351 575.57**	347.35	54 611.54**	87.62	655.66*	777.59**	230.83
	收获指数 Harvest index	14.27**	75.15**	0.15	7.48**	0.40	0.07	0.48*	0.28
	转运量 Translocation amount	10 179.05**	10 754.18**	28.66	1 611.92**	472.34**	34.57	111.36	97.35
	转运率 Translocation efficiency	41.56**	14.74**	0.05	36.32**	2.12*	0.09	1.27*	0.74
	贡献率 Contribution efficiency	46.20**	26.40**	0.11	56.61**	3.83	0.08	2.42	1.73
氮素利用效率 NUtE		5.73**	220.33**	0.23	86.94**	2.26*	2.21**	3.46**	0.60
自由度 df		1	3	2	4	3	4	20	70
氮肥偏生产力 PFP_N		287.69**	2 011.58**	0.19	36 014.80**	20.52**	43.99**	83.09**	1.62
氮素吸收效率 NUpE		0.115 8**	0.343 7**	0.001 9*	23.113 3**	0.012 0**	0.004 9**	0.042 2**	0.000 5

* $P<0.05$, ** $P<0.01$. NUtE: N utilization efficiency; PFP_N: partial factor productivity form applied N; NUpE: N uptake efficiency.

表 2　不同施氮处理对中麦 175 和京冬 17 产量相关性状的影响

Table 2　Effects of different N treatments on yield related traits in Zhongmai 175 and Jingdong 17

处理 Treatment	中麦 175 Zhongmai 175				京冬 17 Jingdong 17			
	穗数 SN (m^{-2})	穗粒数 KNS	千粒重 TKW (g)	产量 Yield ($kg \cdot hm^{-2}$)	穗数 SN (m^{-2})	穗粒数 KNS	千粒重 TKW (g)	产量 Yield ($kg \cdot hm^{-2}$)
河北吴桥 Wuqiao，Hebei								
N0	659.7c	30.4d	45.8a	8 126d	608.0c	30.9d	47.0a	7 608d
N60	687.6b	30.9cd	44.9ab	8 629c	625.7bc	31.9c	46.5ab	8 014c
N120	698.9b	31.3bc	44.3bc	8 810bc	645.0ab	32.6bc	45.9bc	8 149c
N120+60	723.8a	31.7ab	43.7c	8 966b	650.3a	33.4ab	45.5c	8 552ab
N120+120	729.5a	31.9a	43.4c	9 223a	649.9a	33.7a	45.2c	8 757a
N120+180	723.3a	32.2a	43.4c	8 888b	647.1ab	33.8a	45.0c	8 502b
平均值 Mean	703.8A	31.4B	44.2B	8 774A	637.7B	32.7A	45.8A	8 264B
北京顺义 Shunyi，Beijing								
N0	546.0d	32.6b	40.3a	6 848d	533.9b	32.6c	39.9a	6 479c
N60	552.6cd	33.7a	40.3a	6 997c	545.5ab	33.4ab	39.2a	6 716b
N120	562.9bc	33.0ab	39.2ab	7 050bc	549.6a	33.1bc	38.2b	6 788ab
N120+60	583.3a	33.7a	38.6b	7 221a	560.5a	34.1a	37.8b	6 915a
N120+120	575.6ab	33.6a	38.0b	7 120ab	557.2a	33.5ab	38.1b	6 838ab
平均值 Mean	564.1A	33.3A	39.3A	7 047A	549.3A	33.3A	38.6A	6 747B

同一试点，数据后不同小写字母表示处理间有显著差异，不同大写字母表示品种间差异显著（$P<0.05$）。

In each location, different lowercase and uppercase letters after data indicate significant difference among treatments and between cul- tivars at $P<0.05$, respectively. SN: spike number; KNS: kernel number per spike; TKW: thousand-kernel weight.

表 3　不同施氮处理对中麦 175 和京冬 17 NDVI 和 CTD 的影响

Table 3　Effects of different N treatments on NDVI and CTD in Zhongmai 175 and Jingdong 17

处理 Treatment	中麦 175 Zhongmai 175						京冬 17 Jingdong 17					
	NDVI1	NDVI2	NDVI3	CTD1	CTD2	CTD3	NDVI1	NDVI2	NDVI3	CTD1	CTD2	CTD3
河北吴桥 Wuqiao，Hebei												
N0	0.750c	0.737c	0.552c	5.37c	7.20c	4.34d	0.735b	0.728c	0.528c	4.93b	6.47b	3.55c
N60	0.767b	0.772b	0.591b	5.55bc	7.58b	4.95c	0.770a	0.760b	0.543bc	5.07ab	6.87a	3.97b
N120	0.781b	0.791a	0.608ab	5.67ab	7.65ab	5.06bc	0.773a	0.774ab	0.563ab	5.25a	6.87a	4.17ab
N120+60	0.796a	0.796a	0.616ab	5.68ab	7.85a	5.37ab	0.781a	0.769ab	0.577a	5.25a	6.92a	4.35a
N120+120	0.801a	0.808a	0.632a	5.82ab	7.73ab	5.41a	0.786a	0.782a	0.589a	5.33a	6.90a	4.45a
N120+180	0.799a	0.804a	0.624a	5.92a	7.63ab	5.31ab	0.789a	0.789a	0.579a	5.35a	7.17a	4.30a
平均值 Mean	0.782A	0.784A	0.604A	5.67A	7.61A	5.08A	0.772B	0.767B	0.563B	5.20B	6.86B	4.13B
北京顺义 Shunyi，Beijing												
N0	0.835b	0.822c	0.603c	6.05b	6.75b	3.43c	0.822b	0.810c	0.584d	6.39b	6.53b	3.15c
N60	0.840ab	0.827c	0.622b	6.08ab	6.84b	3.57c	0.830a	0.833b	0.605c	6.52ab	6.69b	3.35c
N120	0.843ab	0.837b	0.628b	6.15ab	6.83b	3.83b	0.832a	0.830b	0.620bc	6.74ab	6.93ab	3.67b
N120+60	0.842ab	0.850a	0.655a	6.37ab	7.60a	4.12a	0.837a	0.837ab	0.642a	6.78ab	7.27a	4.00a
N120+120	0.848a	0.852a	0.647a	6.39a	7.60a	4.02a	0.837a	0.847a	0.633ab	6.85a	7.30a	3.80ab
平均值 Mean	0.842A	0.837A	0.631A	6.20A	7.12A	3.79A	0.831B	0.831B	0.617B	6.65A	6.94A	3.59B

NDVI1、NDVI2 和 NDVI3 分别表示开花期、灌浆前期和灌浆中后期的归一化植被指数，CTD1、CTD2 和 CTD3 分别表示开花期、灌浆前期和灌浆中后期的气冠温差。同一试点，数据后不同小写字母表示处理间有显著差异，不同大写字母表示品种间差异显著（$P<0.05$）。

NDVI1, NDVI2, and NDVI3 indicate normalized difference vegetation index in flowering stage, early stage of grain filling and middle to late stage of grain filling, respectively; CTD1, CTD2, and CTD3 indicate canopy temperature depression in flowering stage, early stage of grain filling and middle to late stage of grain filling, respectively. In each location, different lowercase and uppercase letters after data indicate significant difference among treatments and between cultivars at $P<0.05$, respectively.

表 4　不同施氮处理对中麦 175 和京冬 17 干物质积累的影响

Table 4　Effects of different N treatments on dry matter accumulation in Zhongmai175 and Jingdong 17

处理	中麦 175 Zhongmai175						京冬 17 Jingdong17					
	播种一开花 Sowing-flowering		开花一成熟 Flowering-maturity		生物量 Biomass (kg・m^{-2})	收获指数 HI (%)	播种-开花 Sowing-flowering		开花-成熟 Flowering-maturity		生物量 Biomass (kg・m^{-2})	收获指数 HI (%)
	干重 DW (kg・m^{-2})	比例 Ratio (%)	干重 DW (kg・m^{-2})	比例 Ratio (%)			干重 DW (kg・m^{-2})	比例 Ratio (%)	干重 DW (kg・m^{-2})	比例 Ratio (%)		
河北吴桥 Wuqiao，Hebei												
N0	1. 11 d	65. 4 a	0. 59 c	34. 6 a	1. 70 d	50. 2 a	1. 06 d	65. 3 ab	0. 56 d	34. 7 bc	1. 63 e	49. 5 a
N60	1. 15 c	65. 3 a	0. 61 b	34. 7 a	1. 75 c	49. 6 b	1. 10c	65. 5 a	0. 58 c	34. 5 c	1. 68 d	48. 6 b
N120	1. 17b	65. 6 a	0. 61 b	34. 4 a	1. 78 b	49. 0 be	1. 12b	64. 8 be	0. 61 b	35. 2 ab	1. 73 c	48. 6 b
N120＋60	1. 18 ab	65. 0 a	0. 63 a	35. 0 a	1. 81 a	48. 6 cd	1. 14 a	65. 0 abc	0. 61 b	35. 0 abc	1. 76 b	48. 3 b
N120＋120	1. 18 ab	64. 8 a	0. 64 a	35. 2 a	1. 83 a	48. 7 cd	1. 14a	64. 5 c	0. 63 a	35. 5 a	1. 77 ab	48. 4 b
N120＋180	1. 19a	64. 8 a	0. 64 a	35. 2 a	1. 82 a	48. 2 d	1. 15 a	64. 7 c	0. 63 a	35. 3 a	1. 78 a	47. 8 b
平均值 Mean	1. 16 A	65. 1 A	0. 62 A	34. 9 A	1. 78 A	49. 1 A	1. 12B	65. 0 A	0. 60 A	35. 0 A	1. 73 B	48. 5 B
北京顺义 Shunyi，Beijing												
N0	1. 04 c	68. 1 a	0. 49 d	32. 0 a	1. 53 d	47. 3 a	1. 00 c	68. 3 ab	0. 47 b	32. 1 a	1. 47 d	46. 7 a
N60	1. 07 b	67. 9 a	0. 51 cd	32. 3 a	1. 58 c	46. 5 be	1. 03 b	68. 0 abc	0. 49 b	32. 3 a	1. 51 c	46. 0 b
N120	1. 08 b	67. 5 ab	0. 52 be	32. 4 a	1. 60 b	46. 2 c	1. 06 a	68. 0 a	0. 49 b	31. 6a	1. 55 b	45. 3 b
N120＋60	1. 09 a	66. 8 b	0. 55 a	33. 5 a	1. 64 a	46. 8 b	1. 07 a	67. 0 be	0. 52 a	32. 5 a	1. 59 a	45. 4 b
N120＋120	1. 10 a	67. 1 b	0. 54 ab	33. 0 a	1. 64 a	46. 1 c	1. 06 a	67. 2	0. 52 a	32. 7 a	1. 59 a	45. 2 b
平均值 Mean	1. 08 A	67. 4 A	0. 52 A	32. 6 A	1. 60 A	46. 5 A	1. 04 B	67. 8 A	0. 50 A	32. 2 A	1. 54 B	45. 7 B

同一试点，数据后不同小写字母表示处理间有显著差异，不同大写字母表示品种间差异显著（$P<0.05$）。

In each location, different lowercase and uppercase letters after data indicate signnificant difference among treatmengts and between culitivars at $P<0.05$, respectively. DW: dry weight; HI: harvest index.

表5 不同施氮处理对中麦175和京冬17物质转运的影响

Table 5 Effects of different N treatments on dry matter translocation in Zhongmai 175 and Jingdong 17

处理 Treatment	中麦175 Zhongmai 175			京冬17 Jingdong 17		
	转运量 TA（kg·m^{-2}）	转运率 TE（%）	贡献率 CE（%）	转运量 TA（kg·m^{-2}）	转运率 TE（%）	贡献率 CE（%）
河北吴桥 Wuqiao，Hebei						
N0	0.266a	23.9a	31.1a	0.242a	22.7a	30.0a
N60	0.261ab	22.8ab	30.1a	0.238ab	21.5b	29.0a
N120	0.259abc	22.2b	29.7a	0.233ab	20.8bc	27.7b
N120+60	0.247bcd	21.0c	28.1b	0.233ab	20.4c	27.5b
N120+120	0.245cd	20.7c	27.6b	0.228bc	20.0cd	26.6bc
N120+180	0.237d	20.1c	27.0b	0.222c	19.3d	26.1c
平均值 Mean	0.253A	21.8A	28.9A	0.233B	20.8B	27.8B
北京顺义 Shunyi，Beijing						
N0	0.237a	22.6a	32.6a	0.220a	21.9a	32.1a
N60	0.227b	21.2b	30.9b	0.212ab	20.7ab	30.4a
N120	0.220bc	20.3bc	29.7bc	0.211ab	19.9bc	30.1ab
N120+60	0.224bc	20.4bc	29.2c	0.200b	18.7c	27.6b
N120+120	0.215c	19.6c	28.6c	0.197b	18.6c	27.5b
平均值 Mean	0.224A	20.9A	30.2A	0.208B	19.9B	29.6B

同一试点，数据后不同小写字母表示处理间有显著差异，不同大写字母表示品种间差异显著（$P<0.05$）。

Ineach location, different lowercase and uppercase letters after data indicate significant difference among treatments and between cultivars at $P<0.05$, respectively. TA: translocation amount; TE: translocation efficiency; CE: contribution efficiency.

3 讨论

3.1 不同施氮量对冬小麦产量相关性状的影响

本研究表明，适量增加施氮量可提高产量，主要得益于单位面积穗数的提高。随施氮量的增加，开花前干物质积累量向籽粒转运量、转运率、对籽粒的贡献率、收获指数、氮素吸收和利用效率均呈降低趋势，这与张宏等[17]和Rampino等[18]的研究结果相符，可能与作物自身调节能力有关。在土壤养分亏缺时，作物通过提高肥料吸收和利用效率以满足自身需求，并提高开花前干物质积累量向籽粒转运量和转运率以获得一定产量；而在土壤养分充足时，作物体内硝酸还原酶、谷氨酰胺合成酶和谷氨酸合成酶的活性增加[19]，从而促进地上部生物量积累，为籽粒灌浆提供足够的养分储备，最终增加产量。

NDVI反映了地表绿色植被的覆盖比例，可用于评价作物生物量积累速率及叶片功能期长短。肖永贵等[13]发现冬小麦NDVI在不施氮和正常施氮处理间存在极显著差异，李升东等[20]认为不同基因型冬小麦各生育期的NDVI存在显著差异，抽穗期的NDVI值与其干旱产量指数呈正相关。本研究表明，限水条件下在施氮量0～240kg·hm^{-2}（吴桥）或0～180kg·hm^{-2}（顺义）范围内，增加施氮量可提高各时期干物质积累量并延缓衰老，即增加NDVI值。CTD作为反映环境变化对作物生长发育作用的重要指标，已越来越多地用于抗旱、耐热和养分亏缺的研究，一般认为增加施氮量可降低冠层温度即提高CTD，但不同时期结果有一定差异[21-22]。本研究表明，限水条件下在施用氮肥0～240kg·hm^{-2}（吴桥）或0～180kg·hm^{-2}（顺义）范围内，增加施氮量可提高冬小麦开花期、灌浆前期和灌浆中后期的CTD，并且随着时间的推移作用更为明显，这可能是适宜的施氮量提高了群体叶片活性，增加了单位面积生物量和叶片光合能力。NDVI和CTD在一定程度上反映了作物群体活性，不同氮肥处理间二者变化趋势与产量趋势基本一致，NDVI和CTD的稳定性对品种高产稳产具有重要作用。

表 6 不同施氮处理对中麦 175 和京冬 17 氮肥偏生产力、氮素吸收和利用效率的影响

Table 6 Effects of different N treatments on PFP_N, NUpE and NUtE in Zhongmai 175 and Jingdong 17 ($kg \cdot kg^{-1}$)

处理 Treatment	氮肥偏生产力 PFP_N		氮素吸收效率 NUpE		氮素利用效率 NUtE	
	中麦 175 Zhongmai 175	京冬 17 Jingdong 17	中麦 175 Zhongmai 175	京冬 17 Jingdong 17	中麦 175 Zhongmai 175	京冬 17 Jingdong 17
河北吴桥 Wuqiao，Hebei						
N0					41.21a	40.78a
N60	144.65a	133.56a	3.51a	3.40a	41.07a	39.31ab
N120	73.69b	67.91b	1.88b	1.83b	39.03b	37.10c
N120+60	49.81c	47.51c	1.33c	1.26c	37.40c	37.74bc
N120+120	38.43d	36.49d	1.03d	0.98d	37.11c	37.04c
N120+180	31.30e	29.33e	0.85e	0.82e	34.91d	34.63d
平均值 Mean	67.58A	62.96B	1.72A	1.66B	38.46A	37.77A
北京顺义 Shunyi，Beijing						
N0					36.37a	36.44a
N60	116.62a	111.94a	3.30a	3.17a	35.93a	34.95b
N120	58.75b	56.57b	1.73b	1.66b	34.35b	34.28b
N120+60	40.17c	38.42c	1.23c	1.16c	33.18c	33.04c
N120+120	29.67d	28.49d	0.95d	0.89d	32.17d	32.29c
平均值 Mean	61.29A	58.85B	1.80A	1.72B	34.40A	34.20A

同一试点，数据后不同小写字母表示处理间有显著差异，不同大写字母表示品种间差异显著（$P<0.05$）。

In each location, different lowercase and uppercase letters after data indicate significant difference among treatments and between cultivars at $P<0.05$, respectively. NUtE: N utilization efficiency; PFP_N: partial factor productivity from applied N; NUpE: N uptake efficiency.

3.2 不同品种间产量与氮素利用效率的差异

不同群体冠层结构对冠层氮素分布有明显影响，叶片平展型群体中叶片氮素含量随叶层下移而下降的速度要快于紧凑型[23]。中麦 175 具有株型紧凑、叶片较小且直立、分蘖多、冠层结构合理等优良特性，对促进氮素吸收、提高物质转运和增加产量形成具有重要作用。

中麦 175 穗数多，穗粒数和千粒重在不同环境下稳定性好，因而具有较高的丰产性和稳产性，相对于京冬 17，在多数环境下的 NDVI 和 CTD 较高，且在灌浆中后期不同氮素处理间稳定性较好，即在灌浆前期群体活力高，灌浆中后期对氮肥胁迫耐受性更好、活力更持久。小麦生物量积累可分为开花前、后 2 个时期，开花前贮藏碳水化合物再转运及开花后光合产物积累是籽粒产量的主要来源[24]。有研究表明，开花前的光合产物对籽粒产量的贡献占 3%～30%，对抵御后期高温、干旱胁迫、维持产量稳定具有重要作用[21]；开花后光合产物对籽粒的贡献率占 60%～80%，是高产的主要物质来源[25]。中麦 175 的花前和花后干物质积累量及收获指数均高于京冬 17，且开花前干物质积累量向籽粒转运量、转运率、对籽粒的贡献率均更高，说明中麦 175 具有更高的花前干物质积累量和更强的花后干物质转运能力，对保证高产稳产具有重要作用。另外，与京冬 17 相比，中麦 175 在不同氮肥环境下氮素吸收效率皆较高，氮素利用效率在施氮量 0～120$kg \cdot hm^{-2}$的低氮范围内更高。氮素吸收效率高是其总体肥料利用效率高的关键，这与肖永贵等[13]研究结果一致，与我们之前认为中麦 175 苗期具有较强耐低肥能力的结果相佐证[26]。

3.3 氮高效型品种筛选指标及施肥策略

对后代品系进行多点试验及抗旱、耐热、养分胁迫、抗病性鉴定已经成为国际玉米小麦改良中心（CIMMYT）小麦新品种选育的重要方式[22]。我国大多数育种单位还不能做到育种材料的多点鉴定和广泛筛选，导致缺乏抗逆性强、适应性广的高产稳产品种。群体特性在一定程度上反映了作物对不同环境的适应性，已有研究把幼苗早期活性、冠层温度、灌浆期叶片衰老等作为品种适应性和高世代育种选择标

准[27-30]。本研究表明，灌浆中后期（开花后 20 d 左右）的 NDVI 和 CTD 与穗数、产量和生物量在所有环境下均呈显著或极显著正相关，可作为快速评价品种肥料敏感性的指标。中麦 175 的育成及其在北部冬麦区水地、黄淮旱肥地及甘肃和青海春麦区的大面积推广表明，培育高产且广适的品种是可行的，其优良特性可为育种家培育新品种提供借鉴。

受地力、气候环境和种植模式等因素的影响，不同地区小麦对氮肥需求量有所不同[31]。北方小麦氮肥施用量一般为 180～250kg·hm^{-2}，淮北地区为 195～225kg·hm^{-2}，晋南旱作地区为 180kg·hm^{-2} 左右，四川丘陵旱地为 135～180kg·hm^{-2}[31-32]。本研究表明，在磷肥、钾肥一定的条件下，中麦 175 和京冬 17 在基施氮肥 120kg·hm^{-2}＋拔节期追施氮肥 60～120kg·hm^{-2}(吴桥)或基施氮肥 120kg·hm^{-2}＋拔节期追施氮肥 60kg·hm^{-2}(顺义)可获得较高产量。考虑到施氮量超过 240kg·hm^{-2}(吴桥)或 180kg·hm^{-2}(顺义)不仅减产，而且还显著减效的情况，我们推荐中麦 175 和京冬 17 在黄淮麦区(北片)的施氮量为 180～240kg·hm^{-2}，在北部冬麦区施氮量为 120～180kg·hm^{-2}。在此施氮量和采用春季浇 2 次水的节水栽培模式下，适宜的施肥模式是基肥＋拔节肥，其施肥比例应根据土壤肥力适当调整，如土壤肥力较差，应适当增加基肥比例，如土壤肥力较好，则应减少基肥比例。

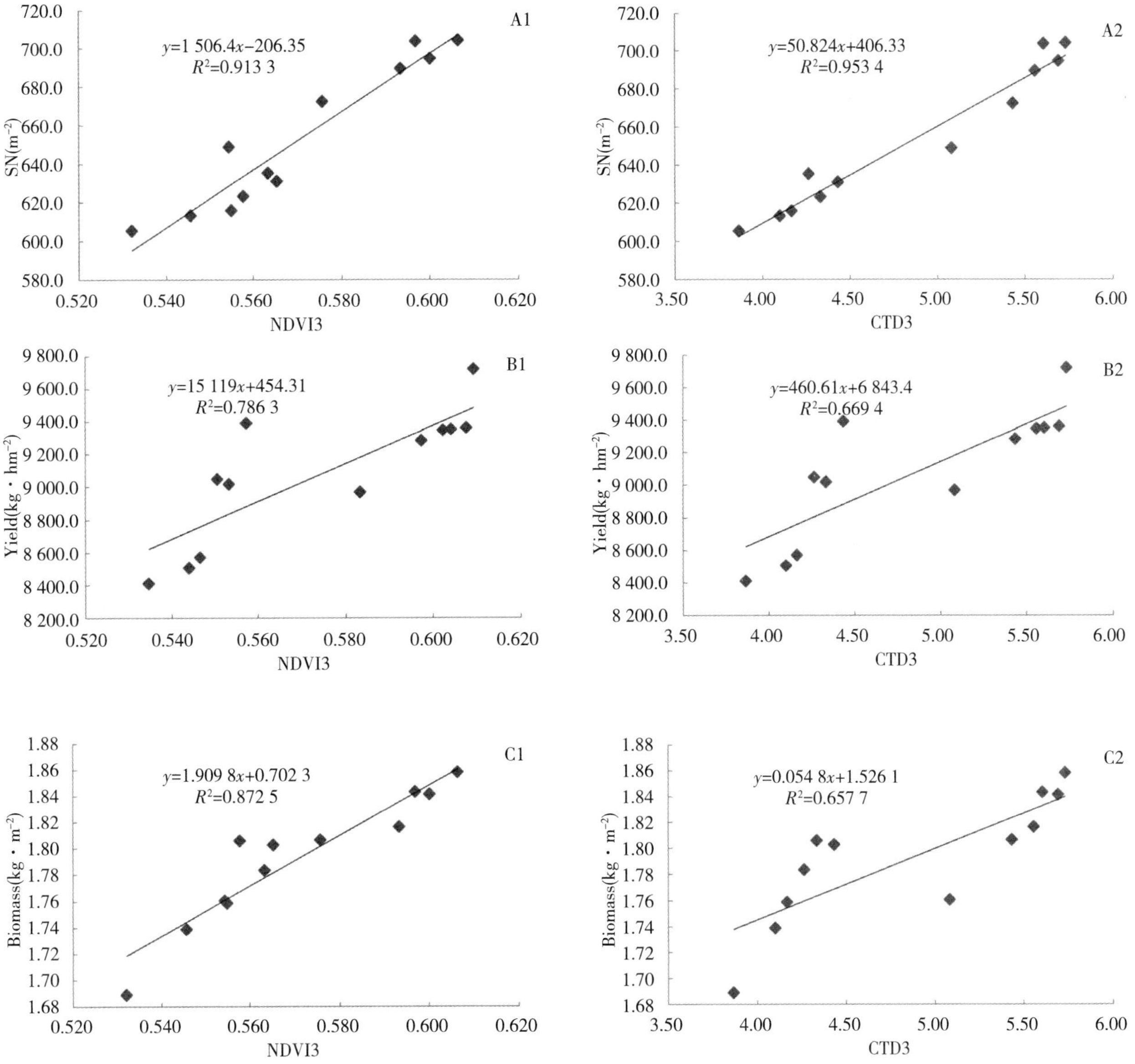

图 1 中麦 175 和京东 17 灌浆中后期 NDVI 和 CTD 与穗数、产量和生物量的线性关系

Fig. 1 Correlation of CTD, NDVI in middle to late stage of grain filling with SN, yield and biomass in Zhongmai 175 and Jingdong 17

4 结论

在不同地点和不同氮肥处理下，中麦 175 产量和氮肥生产率皆高于京冬 17。穗数多、穗粒重稳定性好、群体活力持久、生物量高、花前干物质积累和花后干物质转运能力强、氮素吸收效率高是中麦 175 在不同氮水平下高产高效的重要原因。灌浆中后期的 NDVI 和 CTD 与穗数、产量和生物量呈显著或极显著正相关，二者可协同作为评价品种肥料敏感性的指标。

参考文献

References

[1] Erisman J W, Galloway J A, Sutton M S, Klimont Z, Winiwater W. How a century of ammonia synthesis changed the world. *Nat Geosci*, 2008, 1: 636-639.

[2] 霍中洋，葛鑫，张洪程，戴其根，许轲，龚振凯．施氮方式对不同专用小麦氮素吸收及氮肥利用率的影响．作物学报，2004，30：449-454.

Huo Z Y, Ge X, Zhang H C, Dai Q G, Xu K, Gong Z K. Effect of different nitrogen application types on N-absorption and N-utilization rate of specific use cultivars of wheat. *Acta Agron Sin*, 2004, 30: 449-454. (in Chinese with English abstract)

[3] Goulding K W T. Minimising losses of nitrogen from UK agri- culture. *J R Agric Soc Eng*, 2004, 165: 1-11.

[4] 张福锁，崔振岭，王激清，李春俭，陈新平．中国土壤和植物养分管理现状与改进策略．植物学通报，2007，24：687-694.

Zhang F S, Cui Z L, Wang J Q, Li C J, Chen X P. Current status of soil and plant nutrient management in China and improvement strategies. *Chin Bull Bot*, 2007, 24: 687-694. (in Chinese with English abstract)

[5] 王志敏，王璞，李绪厚，李建民，鲁来清．冬小麦节水省肥高产简化栽培理论与技术．中国农业科技导报，2006，8（5）：38-44.

Wang Z M, Wang P, Li X H, Li J M, Lu L Q. Principle and tech-nology of water-saving ferilizer-saving high-yielding and simple cultivation in winter wheat. *Rev China Agric Sci Technol*, 2006, 8 (5): 38-44. (in Chinese with English abstract)

[6] 潘庆民，于振文，王月福，田奇卓．公顷产 9 000kg 小麦氮素吸收分配的研究．作物学报，1999，25：541-547.

Pan Q M, Yu Z W, Wang Y F, Tian Q Z. Studies on uptake and distribution of nitrogen in wheat at the level of 9000 kg per hec- tare. *Acta Agron Sin*, 1999, 25: 541-547. (in Chinese with English abstract)

[7] 王晨阳，朱云集，夏国军，朱家永，李久星，王永华，罗毅．氮肥后移对超高产小麦产量及生理特性的影响．作物学报，1998，24：978-983.

Wang C Y, Zhu Y J, Xia G J, Zhu J Y, Li J X, Wang Y H, Luo Y. Effects of application nitrogen at the later stage on grain yield and plant physiological characteristics of super-high-yielding winter wheat. *Acta Agron Sin*, 1998, 24: 978-983. (in Chinese with English abstract)

[8] Cormier F, Faure S, Dubreuil P, Heumez E, Beauchene K, La- farge S, Praud S, Gouis J L. A multi-environ-mental study of re- cent breeding progress on nitrogen use efficiency in wheat (*Triti cum aestivum* L.). *Theor Appl Genet*, 2013, 126: 3035-3048.

[9] Celine M D, Francoise D V, Julie D, Fabien C, Laure G, Akira S. Nitrogen uptake, assimilation and remobilization in plants: challenges for sustainable and productive agriculture. *Ann Bot*, 2010, 105: 1141-1157.

[10] Xu G H, Fan X R, Miller A J. Plant nitrogen assimilation and use efficiency. *Plant Biol*, 2012, 63: 153-182.

[11] Masclaux-Daubresse C, Reisdorf-Cren M, Orsel M. Leaf nitrogen remobilisation for plant development and grain filling. *Plant Biol*, 2008, 10: 23-36.

[12] 李兴茂，倪胜利．不同水分条件下广适性小麦品种中麦 175 的农艺和生理特性解析．中国农业科学，2015，48：4374-4380.

Li X M, Ni S L. Agronomic and physiological characterization of the wide adaptable wheat cultivar Zhongmai 175 under two different irrigation conditions. *Sci Agric Sin*, 2015, 48: 4374-4380. (in Chinese with English abstract)

[13] 肖永贵，李思敏，李法计，张宏燕，陈新民，王德森，夏先春，何中虎．两种施肥环境下冬小麦京 411 及其衍生系产量和生理性状的遗传分析．作物学报，2015，41：1333-1342.

Xiao Y G, Li S M, Li F J, Zhang H Y, Chen X M, Wang D S, Xia X C, He Z H. Genetic analysis of yield and physiological traits in elite parent Jing 411 and its derivatives under two fertilization environments. *Acta Agron Sin*, 2015, 41: 1333-1342. (in Chinese with English abstract)

[14] 何中虎，陈新民，王德森，张艳，肖永贵，李法计，张勇，李思敏，夏先春，张运宏，庄巧生．中麦 175 高产高效广适特性解析与育种方法思考．中国农业科学，2015，48：3394-3403.
He Z H，Chen X M，Wang D S，Zhang Y，Xiao Y G，Li F J，Zhang Y，Li S M，Xia X C，Zhang Y H，Zhuang Q S. Characterization of wheat cultivar Zhongmai 175 with high yielding potential，high water and fertilizer use efficiency，and broad adaptability. *Sci Agric Sin*，2015，48：3394-3403.（in Chinese with English abstract）

[15] Reynolds M P，Nagarajan S，Razzaque M A，Ageeb O A A. Heat tolerance. In：Reynolds M P，Ortiz-Monasterio J I，McNab A，eds. Application of Physiology in Wheat Breeding. Mexico：CIMMYT Press，2001. pp 124-135.

[16] Dordas C A，Sioulas C. Dry matter and nitrogen accumulation，partitioning，and retranslocation in safflower（*Carthamus tinto- rius* L.）as affected by nitrogen fertilization. *Field Crops Res*，2009，110：35-43.

[17] 张宏，周建斌，刘瑞，张鹏，郑险峰，李生秀．不同栽培模式及施氮对半旱地冬小麦/夏玉米氮素累积、分配及氮肥利用率的影响．植物营养与肥料学报，2011，17：1-8.
Zhang H，Zhou J B，Liu R，Zhang P，Zheng X F，Li S X. Effects of different cultivation patterns and nitrogen fertilizer on accu- mulation，distribution and use efficiency of nitrogen in winter wheat/summer maize rotation system on semi-dry land farming. Plant *Nutr Fert Sci*，2011，17：1-8.（in Chinese with English abstract）

[18] RampinoP，Stefano Pataleo，Carmela Gerardi，Ginvanni Mita，Carla Perrotta. Drought stress response in wheat：physiological and molecular analysis of resistant and sensitive genotypes. *Plant*，*Cell & Environ*，2006，29：2143-2152.

[19] 赵鹏，何建国，熊淑萍，马新明，张娟娟，王志强．氮素形态对专用小麦旗叶酶活性及自理蛋白质和产量的影响．中国农业大学学报，2010，15（3）：29-34.
Zhao P，He J G，Xiong S P，Ma X M，Zhang J J，Wang Z Q. Studies on the effects of different nitrogen forms on enzyme activity in flag leaves in wheat and protein and yield of grain for specialized end-uses. *J China Agric Univ*，2010，15（3）：29-34.（in Chinese with English abstract）

[20] 李升东，王法宏，司纪升，孔令安，冯波，张宾．不同基因型小麦 NDVI 值与产量的关系．干旱地区农业研究，2008，26（6）：47-50.
Li S D，Wang F H，Si J S，Kong L A，Feng B，Zhang B. The rela- tionship between normalized difference vegetation index and yield of different genotype wheat varieties. *Agric Res Arid Areas*，2008，26（6）：47-50.（in Chinese with English abstract）

[21] Blad B L，Bauer A，Hatfield J L，Hubbard K G，Kanemasu E T，Reginato R. J. Influence of water and nitrogen levels on canopy temperatures of winter wheat grown in the North American Great Plains. *Agric For Meteorol*，1988，44：159-173.

[22] Hegde D M. Effect of irrigation and N fertilization on water rela- tion，canopy temperature，yield，N uptake and water use of onion. Indian *J Agric Sei*，1986，56：858-867.

[23] Anten N P R，Schieving F，Werger M J A. Patterns of light and nitrogen distribution in relation to whole canopy carbon gain in C_3 and C_4 mono and dicotyledonous species. *Oecologia*，1995，101：504-513.

[24] Ehdaie B，Alloush G A，Madore M A，Waines J G. Genotype variation for stem reserves and mobilization in wheat：II. Postan- thesis changes in internode water-soluble carbohydrates. *Crop Sci*，2006，46：2094-2104.

[25] Mullen R W，Freeman K W，Raun W L，Raun W R，Johnson G V，Stone M L，Solie J B. Identifying an in-season response index and the potential to increase wheat yield with nitrogen. *Agron J*，2003，95：347-351.

[26] 李法计，肖永贵，金松灿，夏先春，陈新民，汪洪，何中虎．京 411 及其衍生系苗期氮和磷利用效率相关性状的遗传分析．麦类作物学报，2015，35：737-746.
Li F J，Xiao Y G，Jin S C，Xia X C，Chen X M，Wang H，He Z H. Genetic analysis of nitrogen and phosphorus utilization efficiency related traits at seeding stage of Jing 411 and its derivatives. *J Triticeae Crops*，2015，35：737-746.（in Chinese with English abstract）

[27] Babar M A，van Ginkel M，Klatt A R，Prasad B，Reynolds M P. The potential of using spectral reflectance indices to estimate yield in wheat grown under reduced irrigation. *Euphytica*，2006，150：155-172.

[28] Chen J，LiangY，Hu X，Wang X，Tan F，Zhang H，Ren Z，Luo P. Physiological characterization of stay green wheat cultivars during the grain filling stage under field growing conditions. *Acta Physiol Plant*，2010，32：875-882.

[29] Christopher J T，Manschadi A M，Hammer G，Bor-

rell A K. De- velopmental and physiological traits associated with high yield and stay-green phenotype in wheat. *Aust J Agric Res*, 2008, 59: 354-364.

[30] Fan T, Balta M, Rudd J, Payne W A. Canopy temperature depres- sion as a potential selection criterion for drought resistance in wheat. *Agric Sci China*, 2005, 4: 793-800.

[31] 吴中伟，樊高琼，王秀芳，郑亭，陈溢，李金刚，郭翔. 不同氮肥用量及其生育期分配比例对四川丘陵区带状种植小麦氮素利用的影响，植物营养与肥料学报，2014，20：1338-1348.

Wu Z W, Fan G Q, Wang X F, Zheng T, Chen Y, Li J G, Guo X. Effects of nitrogen ferilizer levels and application stages on ni-trogen utilization of strip-relay-intercropping wheat in Sichuan Hilly Areas. *J Plant Nutr Fert Sci*, 2014. 20: 1338-1348. (in Chinese with English abstract)

[32] 李廷亮，谢英荷，洪坚平，冯倩，孙巫鸿，王志伟. 施氮量对晋南旱地冬小麦光合特性、产量及氮素利用的影响. 作物学报，2013，39：704-711.

Li T L, Xie Y H, Hong J P, Feng Q, Sun C H, Wang Z W. Effects of nitrogen application rate on photosynthetic characteristics, yield and nitrogen utilization in rainfed winter wheat in southern Shanxi. *Acta Agron Sin*, 2013, 39: 704-711. (in Chinese with English abstract)

QTL mapping for leaf senescence-related traits in common wheat under limited and full irrigation

Xing-Mao Li • Zhong-Hu He • Yong-Gui Xiao • Xian-Chun Xia • Richard Trethowan • Hua-Jun Wang • Xin-Min Chen

X. -M. Li Z. -H. He Y. -G. Xiao X. -C. Xia X. -M. Chen

Institute of Crop Science/National Wheat Improvement Center, Chinese Academy of Agricultural Sciences (CAAS), Beijing 100081, China

e-mail: chenxinmin@caas.cn

X. -M. Li

Key Laboratory of High Efficiency Water Utilization in Dry Farming Region, Gansu Academy of Agricultural Sciences, Lanzhou 730070, China

Z. -H. He

CIMMYT China Office, c/o CAAS, Beijing 100081, China

R. Trethowan

Plant Breeding Institute, University of Sydney, Private Bag 4011, Narellan, NSW 2567, Australia

H. -J. Wang

Gansu Provincial Key Laboratory of Aridland Crop Science, Lanzhou 730070, China

Abstract: Leaf senescence is an important trait for yield improvement under stress. In the present study, 207 $F_{2:4}$ random inbred lines (RILs) derived from the Jingdong 8/Aikang 58 cross were investigated under limited and full irrigation environments at two locations during the 2011-2012 and 2012-2013 cropping seasons. The RILs were genotyped with 149 SSR markers and QTLs for leaf senescence-related traits and heading dates (HD) were analyzed by inclusive composite interval mapping. The broad sense heritabilities of normalized difference vegetation index at Zadoks47 (NDVIv) and at Zadoks75 (NDVIg), leaf senescence rate (LSR), leaf senescence scored visually (LSS), leaf area index (LAI) and HD were 0.37-0.54, 0.39-0.48, 0.4-0.45, 0.56-0.58, 0.64-0.79 and 0.82-0.86, respectively. There were significant correlations between NDVIg and LSR ($r=-0.55$ to -0.70), NDVIg and LSS ($r=-0.61$ to -0.61), and LSS and LSR ($r=0.48$-0.68). NDVIv and LAIv explained 18.5% and 19.4% of the variation in grain yield under limited irrigation, respectively. Forty five QTLs were distributed on 15 chromosomes. The respective numbers of QTLs for NDVIv, NDVIg, LSR, LSS and HD were 10, 10, 9, 9 and 7 across all eight environments. Previously unreported QTLs were found on chromosomes 1A, 2D, 5B, 7A and 7D for NDVI, on 6D for LSR and 5B for LSS. *QNDVIv.caas-4A* explained 23.7%-56.6% of the phenotypic variation (PV) for NDVIv and was stably expressed in four environments. In contrast, *QNDVIg.caas-4B.2* explained 13.2%-16.0% of PV but was only ex-

Published in Euphytica, 2015, 203: 569-582.

pressed in full irrigation. Pleiotropic QTLs were detected; *QNDVIv. caas-5B*, explaining 22.9%-35.9% of PV, had an effect on LSS, and *QNDVIg. caas-4D*, accounting for 11.5%-28.5% of PV, also influenced NDVIv. QTLs controlling LSR, *QLSR. caas-4D* and *QLSR. caas-3B*, also increased thousand kernel weight and grain yield, indicating that rapid senescence increased grain filling rate. Some QTLs such as *QNDVIg. caas-1A. 1*, *QLSR. caas-2A*, *QLSS. caas-2B*, *QLSS. caas-3B*, *QLSS. caas-*5B and *QLSS. caas-2D. 1* were detected only under limited irrigation. These drought stress induced QTLs could be valuable for improving drought resistance in wheat.

Keywords: *Triticum aestivum* • Quantitative trait locus • Leaf senescence • Reduced irrigation

Abbreviations:

GY	Grain yield
HD	Heading date
LAIg	Leaf area index at Zadoks75
LAIv	Leaf area index at Zadoks47
LSR	Leaf senescence rate
LSS	Leaf senescence score
NDVIg	Normalized difference vegetation index at Zadoks75
NDVIv	Normalized difference vegetation index at Zadoks47
TKW	Thousand kernel weight

Introduction

Crop production faces huge challenges due to climate change and population growth. Yields are predicted to either stagnate or decline across 24%-39% of major crop areas globally (Ray et al. 2012), especially in regions where abiotic stress is a major constraint. Wu et al. (2012) suggested that manipulation of leaf senescence was essential for crop improvement to either achieve yield potential or to stabilize yield under stress conditions.

Leaf senescence is a highly regulated developmental process that culminates with the programmed death of cells (Thomas et al. 2003). It is affected by internal factors including rate of development and age, but is also influenced by environmental factors such as darkness/shade, heat/light, drought, low nitrogen availability, and disease stress (Gregersen et al. 2013). Delayed senescence is physiologically characterized as stay-green which likely results from inhibition of chlorophyll breakdown, and extended photosynthetic capacity (Thomas and Howarth 2000). Jiang et al. (2004) classified stay green as either functional stay green, where leaves photosynthesize longer, or non-functional stay green where plants remain green, but lack photosynthetic competence. Hereafter, functional stay-green is referred to as delayed onset of senescence or a slower kinetics of senescence (Thomas and Smart 1993). Delayed onset of senescence in wheat (Spano et al. 2003; Bogard et al. 2011) and slow kinetics of senescence were reported previously (Hafsi et al. 2000; Zhang et al. 2006; Christopher et al. 2008; Chen et al. 2010).

Flag leaves with longer maintenance of chlorophyll and photosynthesis extend the assimilatory capacity of the canopy, thus contributing more assimilates to grain filling, delaying leaf senescence and ultimately increasing grain yield (Spano et al. 2003; Luo et al. 2006; Zhang et al. 2006; Blake et al. 2007; Chen et al. 2010). The effect is particularly strong under irrigated conditions (Christopher et al. 2008; Chen et al. 2011; Guttieri et al. 2013). Crop varieties with delayed leaf senescence show multiple beneficial effects, such as increased photosynthetic competence, higher water- use efficiency (Gorny and Garczynski 2002), and improved resistance to stem rot and spot blotch (Joshi et al. 2007). Delayed senescence may have cumulative effects on dry matter, thus improving adaptation to stress (Lopes and Reynolds 2012). Genotypes with delayed onset of leaf senescence increase the total net photo-assimilates available for grain growth and are therefore valuable for wheat breeding.

Under drought, the rate of CO_2 fixation in the chloroplast is often insufficient to consume all absorbed light energy, which leads to formation of reactive oxygen species (ROS) (Foyer and Noctor 2005). Many antioxidant enzymes show reduced activity during senescence, which also contributes to increased ROS accumulation (Tian et al. 2013). Therefore, drought induced premature senescence results in loss of yield. A slower rate of senescence associated with genetic yield increases under drought has been reported in sorghum (Borrell et al. 2000a, b), maize (Baenziger et al. 1999) and durum wheat (Benbella and Paulsen 1998; Hafsi et al. 2000). Spano et al. (2003) reported-that stay-green wheat varieties had higher yield than their parental genotypes under drought. In wheat, longer leaf area duration was also positively correlated with yield under water stress (Verma et al. 2004; Snape et al. 2007). Therefore, delayed senescence was considered an important target trait for improving yield, particularly under water stress.

Higher rates of leaf photosynthesis have notbeen linked to improved yield in specific environments, largely because the source is less limited than the sink (Abbad et al. 2004). Naruoka et al. (2012) reported that stay-green tends to be negatively associated with yield under wet and cool conditions. Zhang et al. (2012) found that genotypes with early flag leaf senescence had more roots, stems, leaves, faster grain filling rates and higher grain yields. Sykorova et al. (2008) even reported the expression of a senescence-induced gene that delayed senescence, but with no positive effect on yield or related components. Early anthesis is often correlated with early senescence. Therefore, the relationship between leaf senescence and yield can be confounded (Bogard et al. 2011). Clearly, the relationship between delayed senescence and yield in different genetic backgrounds and environments needs to be further explored.

The genetic basis of stay green has been investigated in crops such as sorghum and maize (Tao et al. 2000; Gentinetta et al. 1986). Green leaf duration is determined by the genetic background (Wingler et al. 1998) and was verified as a quantitative trait in various crop species (Xu et al. 2000; Cha et al. (2002); Jiang et al. 2004). Genetic control of stay-green has been reported in wheat (Simon 1999) and additive gene effects were observed under optimal conditions (Joshi et al. 2007). Identification of QTLs affecting senescence would allow this trait to be manipulated through breeding. QTL associated with flag leaf senescence were detected on the long arms of chromosomes 2D under drought stress and 2B under irrigated conditions (Verma et al. 2004). QTL influencing senescence were also reported on chromosomes 1AS, 2A, 6A, 6B, 3A, 3BS, 7A and 7DS under high temperatures (Vijayalakshmi et al. 2010; Kumar et al. 2010), and 2A, 1D, 2D and 3B under water stress (Barakat et al. 2013; Saleh et al. 2014). It was not clear whether those QTL can be detected under full irrigation. There are few studies on senescence QTL under full irrigation and no studies in contrasting environments have been reported so far.

In the current study, senescence traits were measured by visually and physiologically using spectral characteristics to evaluate senescence in a RIL population derived from two Chinese winter wheat varieties, Jingdong 8 and Aikang 58. Aikang 58 shows delayed senescence in comparison to Jingdong 8. The objectives of this study were to (1) evaluate different senescence traits using the RIL population, (2) identify QTL for senescence-related traits and heading date (HD), and (3) distinguish between QTL for drought- induced senescence and senescence resulting from normal maturity.

Materials and methods

Two hundred and ten wheat genotypes, including 207 $F_{2:4}$ lines derived from a cross between Jingdong 8 and Aikang 58, the two parents, and a check variety, Zhongmai 175, were evaluated in field experiments. $F_{2:4}$ lines were used in the 2011-2012 cropping season and the harvested seeds from these $F_{2:4}$ lines in Beijing were mixed and used in 2012-2013 cropping season. Aikang 58, carries dwarfing genes *Rht-D1b* and *Rht8* and the 1B.1R translocation, and is characterized by high yield potential and excellent lodging resistance. It currently occupies around 2 million ha in the southern part of the Yellow and Huai Valleys Facultative Wheat Region under full irrigation. Jingdong8, carries *Rht*8 and the1B.1R translocation, and was a leading variety in the Northern Winter Wheat Region under both irrigated and rainfed conditions during the past 15 years.

Field experiments were conducted inthe 2011-2012 and 2012-2013 cropping seasons at the CAAS, Beijing (B) Experimental Station, and at Gaoyi (G) in Hebei province as detailed in Li et al. (2014). A randomized complete block design with two replications was employed for each experiment. Each plot consisted of 4×2m rows with 30cm between rows (Li et al. 2014). The eight field trials are denoted as BD1, BI1, BD2, BI2, GD1, GI1, GD2, GI2, representing Beijing limited and full irrigation (FI) 2011-2012, Beijing limited and FI 2012-2013, Gaoyi limited and FI 2011-2012, and Gaoyi limited and FI 2012-2013, respectively. Sufficient rainfall was recorded in September, thus excellent germination was achieved at both locations during both cropping seasons. Limited irrigation (LI), considered as a drought environment, as applied at the pre-overwintering stage to ensure winter survival. Full irrigation (FI) was given four times at pre-overwintering, stem elongation in spring, post-stem longation, and at grain-filling. Water supplied was around 70 mm per irrigation. Rainfall and irrigations from sowing to harvesting and traits such as plant height of eight field trials in two cropping seasons are presented in Table 1.

The leaf senescence score (LSS) was captured at adoks73, according to the CIMMYT Physiological Breeding Field Guide (Pask et al. 2012), and was measured visuallyon a scale of 0 to 10, according to the percentage of dead leaf area to total leaf area: 0 is equivalent to no observed dead leaf tissue and values increase incrementally by 10% up to a score of 10 or 100% dead leaf tissue. NDVI was measured at Zadoks47 (NDVIv) and at Zadoks75 (NDVIg) using a portable spectroradiometer (GreenSeeker, Ntech Industries, Inc., Ukiah, CA, USA). Leaf senescence rate (LSR) was evaluated as follows: LSR = (NDVIv-NDVIg) /NDVIv. Leaf area index (LAI) was calculated at Zadoks47 (LAIv) and at Zadoks75 (LAIg) using photosynthetically active radiation (AccuPAR LP-80, Decagon, Inc. USA). Thousand kernel weight (TKW) was measured by weighing two samples of 500 kernels from each plot. Grain yield (GY) was determined as the weight of grain harvested per unit area (kg/ha). Heading date was determined as the number of days from sowing to heading.

Table 1 Water supply, average grain yield, plant height and heading date of the trials in the 2011—2012 and 2012—2013 crop seasons

Year	Location	Code	Rainfall from sowing to harvest (mm)	Irrigation (mm)	Days to heading	Plant height (cm)	Grain yield (kg/ha)
2011—2012	Beijing	BD1, LI	150	70	207	76.7	4 613.4
		BI1, FI	150	280	209	83.6	6 100.1
	Gaoyi	GD1, LI	106	70	194	75.3	5 086.2
		GI1, FI	106	210	197	81.9	5 497.4

(continued)

Year	Location	Code	Rainfall from sowing to harvest (mm)	Irrigation (mm)	Days to heading	Plant height (cm)	Grain yield (kg/ha)
2012—2013	Beijing	BD2, LI	174	70	213	64.0	3 186.1
		BI2, FI	174	280	214	83.9	5 621.6
	Gaoyi	GD2, LI	127	70	205	71.7	4 328.6
		GI2, FI	127	210	207	83.5	5 748.5

BD Beijing drought/limited irrigation (LI), *BI* Beijing irrigated/full irrigation (FI), *GD* Gaoyi drought/limited irrigation (LI), *GI* Gaoyi irrigated/full irrigation (FI)

QTL analysis

More than 50 leaves from each genotype were mixed for DNA extraction, in order to reduce residual heterozygosity in the population. Linkage maps were constructed using QTL IciMapping ver. 3.2 software (http://www.isbreeding.net). Based on the consensus map, 149 SSR markers were located on 20 wheat chromosomes. QTL analysis was performed using inclusive composite interval mapping (ICIM). The detailed information is provided in Li et al. (2014).

Statistical analysis

Analysis of variance (ANOVA) of phenotypic data and correlation coefficients among environments were performed with the Statistical Analysis System (SAS Institute 2000). The PROC GLM was used in ANOVA for all traits. The PROC MIXED procedure was used in ANOVA to estimate the effects of markers on traits. correlation analysis between parameters was performed using the " PROC CORR " procedure. Broadsense heritabilities were estimated in all environments as

$$h^2=\sigma_g^2/\left(\sigma_g^2+\frac{\sigma_{ge}^2}{e}+\frac{\sigma_\varepsilon^2}{re}\right)$$

where the genetic variance $\sigma_g^2=(MS_f-MS_{fe})/re$ genotype×environment interaction variance

$$\sigma_{ge}^2=(MS_{fe}-MS_f)/r$$

error variance $\sigma_\varepsilon^2=MS_e$, MS_f = genotype mean square, S_{fe} = genotype × environment interaction square, MS_e = error mean square and r and e were the numbers of replicates and environments, respectively.

Results

Phenotypic evaluation and correlation

Significant differences were observed among the $F_{2:4}$ lines for NDVI, LAI, HD, LSR and LSS by analyses of variance. Transgressive segregation occurred for all traits, indicating quantitative inheritance (Table 2). The average NDVIv, LAI, LSR and LSS of Aikang 58 were smaller than those of Jingdong 8 with both limited and FI; however, NDVIg and HD for Aikang 58 were higher than Jingdong 8. The values of NDVIv, NDVIg, LAI, LSS and HD were lower under limited irrigation whereas LSR was greater under limited irrigation (Table 2). The broad-sense heritability of HD was 0.86 and 0.82 under limited and FI, respectively. Heritabilities of 0.58, 0.54 and 0.48 were calculated for LSS, NDVIv and NDVIg, respectively, under limited irrigation. These were higher than the corresponding values of 0.56, 0.37 and 0.39 observed under FI. LAI showed higher heritability (0.64-0.79) than LSR (0.40-0.45).

Grain yield was significantly and positively correlated with NDVIv and LAIv under limited irrigation, explaining 18.5 and 19.4% ofthe variation, respectively (Table 3). The correlation between NDVIv and LAIv was 0.47 under limited irrigation, and 0.48 and 0.68 between LSS and LSR under full and limited irrigation, respectively. NDVIg was significantly correlated with LSS and LSR under both limited and FI (ranging from −0.55 to −0.70). HD was significantly correlated with LSS, LSR and

NDVIg with respective correlations of −0.65, −0.46 and 0.43 under limited irrigation, and −0.49, −0.4 and 0.5 under FI (Table 3). No significant correlations were observed between yield and LSR, and LSS, although under FI, the correlation coefficient between LSR and GY was 0.40 (a = 0.01) in Beijing 2012. The results indicate that NDVIv can be used to assist yield selection in LI, while NDVIg and LSR are useful for evaluating leaf senescence.

Table 2 Summary statistics (mean, maximum, minimum, and standard deviation) and heritabilities for all traits measured on the Jingdong 8/Aikang 58 population

Trait	Environment	Jingdong8	Aikang58	Average	Min—max	h^2
NDVIv	LI	0.72	0.71	0.68±0.09	0.42—0.83	0.54
	FI	0.79	0.75	0.77±0.07	0.52—0.86	0.37
NDVIg	LI	0.48	0.55	0.48±0.11	0.24—0.68	0.48
	FI	0.66	0.71	0.67±0.06	0.50—0.79	0.39
LAIv	LI	3.82	3.18	3.26±0.39	2.33—4.08	0.67
	FI	5.04	4.05	5.02±0.53	3.82—6.61	0.64
LAIg	LI	3.58	2.85	3.07±0.45	1.99—4.78	0.69
	FI	3.06	3.83	3.99±0.59	2.67—5.85	0.79
LSS	LI	7.30	1.30	4.56±2.27	1.00—10.00	0.58
	FI	10.0	6.00	8.66±3.20	5.00—10.00	0.56
LSR	LI	0.32	0.20	0.27±0.13	0.02—0.68	0.45
	FI	0.17	0.09	0.13±0.04	0.01—0.25	0.40
HD (d)	LI	204.8	205.4	204.7±6.9	190.0—215.0	0.86
	FI	206.9	208.3	206.8±6.2	192.0—215.0	0.82

LI, Limited irrigation at two locations; FI full irrigation at two locations; NDVIv, normalized difference vegetation index at Zadoks47; NDVIg, normalized difference vegetation index at Zadoks75; LAIv, leaf area index at Zadoks47; LAIg, leaf area index at Zadoks 75; HD, heading date; LSR, leaf senescence rate; LSS, leaf senescence score

QTLs for NDVIv

Ten QTLs for NDVIv identified in the Jingdong8/ Aikang58 population were located on chromosomes 1A (2), 2D, 3B (2), 4A, 4B, 4D, 5B and 7A (Table 4). The alleles increasing NDVIv at the 2D, 4B and 4D loci came from Aikang 58, and those increasing NDVIv at 4A, 5B and 7A loci were from Jingdong 8.

The QTL of largest effect located on chromosome 4A, designated *QNDVIv. caas-4A*, was identified in four environments and explained 23.7 to 56.6% of PVE in FI and 31.6 to 32.8% of PVE in LI, respectively. Other QTLs included *QNDVIv. caas-4D*, which accounted for 18.4% of PVE in GD1 and17.5% in BI1; *QNDVIv. caas-5B*, which was detected across three environments and explained 22.9, 35.9 and 27.9% of PVE in BI2, GI2 and GD1, respectively; and *QNDVIv. caas-3B. 1* which explained 33.1% of PVE in BD1 and 23.3% in BI2. In addition, six QTLs were each identified in one fully irrigated environment. No NDVIv QTL was detected in BD2, GD2 and GI1 (Table 4).

QTLs for NDVIg

Ten QTLs for NDVIg were identified on chromosomes 1A (2), 2D (2), 3B (2), 4B (2), 4D and 7D (Table 4). The alleles increasing NDVIg at eight loci on chromosomes 1A (2), 2D (2), 3B, 4B, 4D and 7D were contributed by Aikang 58 whereas an allele at the *NDVIg. caas-4B. 1* locus, also increasing NDVIg, was contributed by Jingdong 8. Both parents contributed alleles that increased NDVIg at the *QNDVIg. caas-3B. 1* locus in different environments.

Aikang 58 made a larger contribution to NDVIg than Jingdong 8.

QTLs included *QNDVIg. caas-1A. 1* which accounted for 15. 1% of PVE in BD1 and 7. 8% in GD2 in limited irrigation; *QNDVIg. caas-4B. 2* which explained 16% of PVE in BI2 and 13. 2% in GI2 under FI; *QNDVIg. caas-2D. 2* found in GD2 (10. 2% of PVE) and GI2 (8. 7%); *QNDVIg. caas-4D* found in BD1 (11. 5%) and BI1 (28. 5%); *QNDVIg. caas-3B. 1* found in GDI (32. 8%), GI1 (14. 9%) and BI2 (22. 1%); *QNDVIg. caas-4B. 1* found in GD2 (5. 3%), and *QNDVIg. caas-7D* found in BD1 (8. 7%). *QND- VIg. caas-2D. 1* and *QNDVIg. caas-3B. 2* were identified in only one fully irrigated environment (Table 4).

Table 3 Correlation coefficients among physiological traits and grain yield

Trait	NDVIv	NDVIg	LAIv	LAIg	HD	LSR	LSS	GY
NDVIv		0. 46 **	0. 47 **	0. 23 **	−0. 02	0. 23 **	0. 07	0. 43 **
NDVIg	0. 39 **		0. 39 **	0. 29 **	0. 43 **	−0. 70 **	−0. 61 **	0. 39 **
LAIV	0. 27 **	0. 22 **		0. 23 **	−0. 09	−0. 06	−0. 02	0. 44 **
LAIg	0. 25 **	0. 21 **	0. 63 **		−0. 02	−0. 17 *	−0. 13	0. 29 **
HD	0. 09	0. 50 **	0. 22 **	0. 22 **		−0. 46 **	−0. 65 **	−0. 10
LSR	0. 43 **	−0. 55 **	0. 05	0. 01	−0. 40 **		0. 68 **	−0. 09
LSS	−0. 07	−0. 61 **	−0. 11	−0. 13 *	−0. 49 **	0. 48 **		−0. 08
GY	0. 35 **	0. 19 **	0. 15 *	0. 11	0. 13	0. 14	0. 03	

Average data from four environments was used to calculate correlation coefficients. Upper values represent limited irrigation, and lower values represent full irrigation

NDVIv Normalized difference vegetation index at Zadoks47, *NDVIg* normalized difference vegetation index at Zadoks75, *LAIv* leaf area index at Zadoks47, *LAIg* leaf area index at Zadoks75, *HD* heading date, *LSR* leaf senescence rate, *LSS* leaf senescence score

**, * Significant at $a=0.001$ and $a=0.01$, respectively

QTLs for LSS

Nine QTLs for LSS were identified on chromosomes 1A, 2B, 2D (2), 3A, 3B, 4D, 5B and 7D (Table4). Alleles increasing LSS at *QLSS. caas-2D. 1*, *QLSS. caas-2B*, and *QLSS. caas-3A*, originated from Aikang 58 whereas alleles increasing LSS at other loci were contributed by Jingdong 8.

QLSS. caas-1A was identified in three environments and explained 10. 9, 18. 0 and 5. 7% of PVE in BD1, BD2 and BI1, respectively. Other QTL included: *QLSS. caas-4D*, which accounted for 25. 5% of the PVE in BD1, 7. 3% in BD2 and 23. 6% in BI1; *QLSS. -caas-2D. 1*, *QLSS. caas-3B* and *QLSS. caas-5B*, which explained 27. 5, 30. 3 and 16. 4% of PVE, respectively, in BD1; *QLSS. caas-2D. 2* and *QLSS. caas-7D* only found in BD2; and *QLSS. caas-2B* and *QLSS. caas-3A* found only in GD2. The latter four QTLs accounted for 6. 2-11. 1% of the PVE (Table 4).

QTLs for LSR

Nine QTLs for LSR were identified on chromosomes1A (2), 2A, 2D (2), 3B, 4D, 6A and 6D (Table 4). Alleles that increased LSR were found at loci on chromosomes 1A, 3B, 6A and 6D and were contributed by Jingdong 8, whereas that increasing LSR at a locus on chromosome 2A was contributed by Aikang 58. Alleles that increased LSR at *QLSR. caas-2D. 1* and *QLSR. caas-2D. 2* came from Aikang 58 and Jingdong 8, respectively. Both parents contributed alleles that increased LRS at the *QLRS. caas-4D* locus in different environments.

Five QTLs were detected in only one environment. These included *QLSR. caas-2D. 1* and *QLSR. caas-2D. 2* identified in GD1 and GD2, respectively; and *QLSR. caas-1A. 2*, *QLSR. caas-6A* and *QLSR. caas-6D* found in BI1, GI2 and GI1, respectively. Other QTL detected included *QLSR. caas-2A* found in the LI environments

BD1 (PVE 10.4%) and GD2 (15.4%); *QLSR. caas-3B* found only in the fully irrigated environments BI2 (6.2%), GI1 (6.1%) and GI2 (5.7%); *QLSR. caas-1A. 1* which explained 14.1% of PVE in BI2, was stably detected in LI and accounted for 15.2% of PV in BD1, 8.1% in BD2 and 10.7% in GD2. The Jingdong 8 allele at *QLSR. caas-4D* showed positive effects in BD1 (6.7% of PVE) and BI1 (15.2%), however, the effect was negative in GD1 (31.6%) and GI1 (31.5%).

QTLs for HD

Seven QTLs for HD were identified on chromosomes 1A, 2B, 3A, 4B (2), 4D and 5B (Table 4). Alleles that delayed HD at loci on chromosomes 4B and 4D were contributed by Aikang 58 whereas alleles that delayed HD on chromosomes 1A, 2B, 3A and 5B originated from Jingdong 8.

A stable QTL, *QHD. caas-5B*, explained 9.8-27.3% of PV across six environments (BD1, BD2, GD2, BI2, GI1 and GI2). Other QTL detected included *QHD. caas-3A* which explained 7.5-12.7% of PV in four environments (BD1, BI2, GD2 and GI2); *QHD. caas-4B. 1* and *QHD. caas-4B. 2* which accounted for 8-11.8% and 5.5-9.1% of PV, respectively, in three environments; *QHD. caas-2B* (7.3-8.3%) and *QHD. caas-4D* (9.8-11.2%) found in two environments; and *QHD. caas-1A* which was detected only in GD1 (Table 4).

Table 4 QTLs detected for senescence-related traits

Trait	Environment	QTL	Nearest marker	Distance	LOD[a]	PVE (%)[b]	Add[c]	Dom[d]
NDVIv	BD1	*QNDVIv. caas-3B. 1*	*Xgwm547*	4	4.4	33.1	0.12	−6.34
		QNDVIv. caas-4A	*Xgwm350*	4	4.5	31.6	0.29	−5.84
	BI1	*QNDVIv. caas-4B*	*Xgwm495*	7	4.8	10.4	−0.57	0.03
		QNDVIv. caas-4D	*Xwmc89*	1.5	9.0	17.5	−0.67	−0.22
	BI2	*QNDVIv. caas-1A. 1*	*Xbarc269*	2.8	3.4	23.0	0.31	−6.84
		QNDVIv. caas-2D	*Xwmc170*	0.7	6.3	23.1	−0.4	−12.06
		QNDVIv. caas-3B. 1	*Xgwm547*	2	3.7	23.3	−0.05	−8.35
		QNDVIv. caas-3B. 2	*Xwmc418*	0.5	3.5	6.4	0.72	−1.52
		QNDVIv. caas-4A	*Xgwm350*	2	4.6	23.7	0.05	−7.14
		QNDVIv. caas-5B	*Xbarc1120*	0.8	4.2	22.9	0.69	−9.72
		QNDVIv. caas-7A	*Xbarc219*	1.2	5.9	13.0	1.7	1.08
	GD1	*QNDVIv. caas-4A*	*Xgwm350*	5	3.5	32.8	0.06	−4.84
		QNDVIv. caas-4D	*Xwmc89*	2.5	7.3	18.4	−1.51	0.09
		QNDVIv. caas-5B	*Xbarc1120*	1.8	4.3	27.9	0.2	−6.16
	GI2	*QNDVIv. caas-4D*	*Xwmc89*	4.5	6.8	14.8	−0.57	−0.38
		QNDVIv. caas-1A. 2	*Xwmc84*	6.6	7.4	55.1	−0.08	−11.14
		QNDVIv. caas-4A	*Xgwm350*	6	6.7	56.6	0.75	−10.81
		QNDVIv. caas-5B	*Xbarc110*	1.9	4.6	35.9	−0.11	−11.14
NDVIg	BD1	*QNDVIg. caas-1A. 1*	*Xwmc120*	3.3	7.6	15.1	−3.62	0.2
		QNDVIg. caas-4D	*Xbarc105*	3.4	6.3	11.5	−3.04	−1.23
		QNDVIg. caas-7D	*Xwmc264*	7.5	3.3	8.7	−0.19	−4.22
	BI1	*QNDVIg. caas-1A. 2*	*Xwmc84*	1.4	4.3	8.2	−0.93	1.31
		QNDVIg. caas-4D	*Xbarc105*	10.5	9.7	28.5	−1.77	−1.43
	BI2	*QNDVIg. caas-2D. 1*	*Xwmc170*	2.7	3.7	24.3	−0.16	−4.01
		QNDVIg. caas-3B. 1	*Xgwm547*	2	4.2	22.1	−0.15	−3.87
		QNDVIg. caas-4B. 2	*Xgwm495*	7.8	5.1	16.0	−0.99	−0.1
	GD1	*QNDVIg. caas-3B. 1*	*Xgwm547*	1	15.5	32.8	0.21	−26.1

(continued)

Trait	Environment	QTL	Nearest marker	Distance	LOD[a]	PVE (%)[b]	Add[c]	Dom[d]
	GD2	*QNDVIg. caas-1A. 1*	*Xwmc120*	0.3	4.3	7.8	−2.03	0.49
		QNDVIg. caas-2D. 2	*Xwmc144*	2.3	5.0	10.2	−1.99	−1
		QNDVIg. caas-4B. 1	*Xwmc141*	0	3.0	5.3	1.61	−0.91
	GI1	*QNDVIg. caas-3B. 1*	*Xgwm547*	1	5.2	14.9	0.13	−7.11
		QNDVIg. caas-3B. 2	*Xwmc418*	5.5	3.8	8.8	−0.48	0.36
	GI2	*QNDVIg. caas-2D. 2*	*Xwmc144*	5.7	3.1	8.7	−1.01	1.92
		QNDVIg. caas-4B. 2	*Xgwm495*	8.8	3.8	13.2	−1.67	1.01
LSS	BD1	*QLSS. caas-1A*	*Xwmc120*	1.3	5.7	10.9	0.46	0.12
		QLSS. caas-2D. 1	*Xwmc170*	7	4.0	27.5	−0.05	−1.68
		QLSS. caas-3B	*Xgwm547*	5	4.7	30.3	0.17	−1.71
		QLSS. caas-4D	*Xwmc89*	2.5	10.8	25.5	0.61	0.41
		QLSS. caas-5B	*Xbarc110*	0.9	3.6	16.4	0.09	−2.31
	BD2	*QLSS. caas-1A*	*Xwmc120*	2.3	10.2	18.0	1.11	−0.85
		QLSS. caas-2D. 2	*Xwmc144*	0.3	5.1	7.5	0.75	0.04
		QLSS. caas-4D	*Xwmc89*	1.5	4.8	7.3	0.46	1.00
		QLSS. caas-7D	*Xbarc260*	13	3.6	11.1	0.10	1.49
	BI1	*QLSS. caas-1A*	*Xwmc120*	0.7	3.4	5.7	0.55	0.30
		QLSS. caas-4D	*Xwmc89*	4.5	12.4	23.6	1.11	0.32
	GD2	*QLSS. caas-2B*	*Xgwm429*	4	3.5	7.4	−0.52	−0.12
		QLSS. caas-3A	*Xbarc67*	1	3.6	6.2	−0.41	0.22
LSR	BD1	*QLSR. caas-1A. 1*	*Xwmc120*	2.3	7.4	15.2	0.049	−0.016
		QLSR. caas-2A	*Xwmc177*	3	4.4	10.4	−0.003	0.065
		QLSR. caas-4D	*Xwmc89*	0.5	3.7	6.7	0.027	0.028
	BD2	*QLSR. caas-1A. 1*	*Xwmc120*	0.6	4.0	8.1	0.018	−0.004
	BI1	*QLSR. caas-1A. 2*	*Xwmc84*	0.1	5.4	10.8	0.011	−0.009
		QLSR. caas-4D	*Xbarc105*	10	5.4	15.2	0.014	0.009
	BI2	*QLSR. caas-1A. 1*	*Xwmc120*	8	5.3	14.1	0.015	−0.003
		QLSR. caas-3B	*Xbarc251*	0	3.3	6.2	0.010	−0.005
	GD1	*QLSR. caas-2D. 1*	*Xwmc181*	0	3.4	6.4	−0.008	0.009
		QLSR. caas-4D	*Xbarc105*	13	7.7	31.6	−0.018	−0.017
	GD2	*QLSR. caas-1A. 1*	*Xgwm135*	5	5.5	10.7	0.035	-0.025
		QLSR. caas-2A	*Xwmc177*	11	4.0	15.4	−0.015	0.073
		QLSR. caas-2D. 2	*Xwmc144*	12	7.0	20.0	0.050	0.015
	GI1	*QLSR. caas-3B*	*Xbarc251*	2	3.8	6.1	0.006	−0.003
		QLSR. caas-4D	*Xbarc105*	10	13.2	31.5	−0.012	−0.007
		QLSR. caas-6D	*Xbarc365*	0.2	3.8	5.8	0.006	0.003
	GI2	*QLSR. caas-3B*	*Xbarc251*	0	3.0	5.7	0.016	−0.004
		QLSR. caas-6A	*Xwmc256*	0.1	3.2	6.0	0.017	−0.004
HD	BD1	*QHD. caas-3A*	*Xbarc1044*	0.5	4.6	8.0	0.50	−0.18
		QHD. caas-4B. 2	*Xwmc47*	6.5	3.1	5.8	−0.39	−0.33
		QHD. caas-4D	*Xwmc89*	1.5	5.8	11.2	−0.57	−0.09
		QHD. caas-5B	*Xgwm499*	4.9	4.9	16.3	0.60	0.42
	BD2	*QHD. caas-5B*	*Xgwm499*	3.9	5.7	15.3	0.38	0.04
	BI1	*QHD. caas-4B. 1*	*Xgwm251*	7.3	4.8	11.8	−0.49	0.10

(continued)

Trait	Environment	QTL	Nearest marker	Distance	LOD[a]	PVE (%)[b]	Add[c]	Dom[d]
		QHD. caas-4D	*Xwmc89*	1. 5	4. 5	9. 8	−0. 41	−0. 17
	BI2	*QHD. caas-3A*	*Xbarc1044*	5. 5	3. 5	8. 5	0. 35	−0. 08
		QHD. caas-4B. 2	*Xwmc47*	1. 5	4. 7	9. 1	−0. 35	−0. 16
		QHD. caas-5B	*Xgwm499*	5. 9	6. 0	27. 3	0. 38	0. 60
	GD1	*QHD. caas-1A*	*Xwmc84*	2. 4	3. 2	18. 6	0. 02	4. 21
	GD2	*QHD. caas-2B*	*Xgwm429*	5. 9	3. 6	8. 3	0. 54	0. 24
		QHD. caas-3A	*Xbarc1044*	1. 5	4. 3	7. 5	0. 47	−0. 17
		QHD. caas-4B. 2	*Xwmc47*	1. 5	3. 3	5. 5	−0. 41	−0. 18
		QHD. caas-5B	*Xgwm499*	3. 9	6. 0	20. 0	0. 63	0. 54
	GI1	*QHD. caas-4B. 1*	*Xgwm251*	16. 7	3. 4	8. 0	−0. 88	−0. 16
		QHD. caas-5B	*Xgwm499*	5. 1	4. 8	9. 8	1. 08	−0. 11
	GI2	*QHD. caas-2B*	*Xgwm429*	10	3. 2	7. 3	0. 32	0. 22
		QHD. caas-3A	*Xbarc1044*	8. 5	5. 1	12. 7	0. 41	−0. 11
		QHD. caas-4B. 1	*Xgwm251*	8. 3	4. 9	11. 6	−0. 39	−0. 04
		QHD. caas-5B	*Xgwm499*	0. 9	7. 1	14. 1	0. 45	0. 16

BD1, BI1, BD2 and BI2 represent Beijing 2011-2012 for limited irrigation, 2011-2012 for full irrigation, 2012-2013 for limited irrigation and 2012-2013 for full irrigation, respectively; GD1, GI1, GD2 and GI2 represent Gaoyi 2011-2012 for limited irrigation, 2011-2012 for full irrigation, 2012-2013 for limited irrigation and 2012-2013 for full irrigation, respectively

NDVIv Normalized difference vegetation index at Zadoks 47, *NDVIg* normalized difference vegetation index at Zadoks 75, *HD* heading date, *LSR* leaf senescence rate, *LSS* leaf senescence score

[a] LOD thresholds of 3. 0 were used for declaration of QTL

[b] Phenotypic variation explained by QTL

[c] QTLs with positive additive effects indicate increasing effects from Jingdong 8, whereas those with negative additive effects indicate increasing effects from Aikang 58

[d] Dominance effect of QTL

Discussion

Phenotypic evaluation

Various parameters such as chlorophyll content, photochemical efficiency, senescence-associated enzyme activity, protein content and gene expression could be used to evaluate senescence at the single leaf level. However, plant breeders require a more simple and rapid leaf senescence evaluation method for field selection. Flag leaf area duration can be evaluated visually and expressed as percent green leaf area remaining at a given time; however, this measure is subjective and influenced by the observer. Computer image analysis software allows precise and quantitative determination of stay-green and is amenable to high throughput screening (Hafsi et al. 2000); how- ever, this in vitro method damages the plant. In this study, we used several parameters (NDVI in mid-grain filling; LSR and LSS assessed visually) to measure senescence.

NDVI measurement is influenced by develop- mental stage. Measuring NDVI at the appropriate growth stage can improve the accuracy of senescence evaluation. LSR was calculated as theratio of NDVI reduction during senescence to NDVI at the onset of senescence. However, leaf senescence in wheat generally occurs during grain filling and sometimes immediately after flowering in drought affected environments. It was difficult to measure NDVI at the onset of senescence, largely due to its short duration. When flag leaves are fully expanded before heading, NDVIv is at a peak. This peak time was longer and could be used as a measure for estimating LSR. LSR was significantly correlated with LSS under both limited and FI (r =0. 48-0. 68, Table 3), indicating that LSR provided a reliable estimate of leaf senescence in the present study.

Measuring NDVI was more accurate and efficient in the field than LSS. NDVI at physiological maturity was used to calculate LSR independently from phenology (Lopes and Reynolds 2012). NDVIg was significantly correlated with LSS in both limited and FI in this study ($r = -0.61$ to -0.61). Therefore, NDVI assessed at mid-grain fill could be used to replace LSS for evaluating senescence.

Indirect selection for GY using physiological traits has been considered for wheat breeding (Reynolds et al. 2001). Grain yield was significantly and positively correlated with NDVI and LAI, particularly under limited irrigation (Table 3), but the correlation between NDVIg and yield in FI was lower. Measuring NDVI was less labor-intensive than assessing LAI in the field, thus NDVI captured at Zadoks47 was identified as the optimal trait/development combination for assessing leaf senescence in wheat breeding.

Pleiotropic effects of QTLs

QTLs distributed on chromosomes 1A (2), 2B, 2D (2), 3A, 3B (2), 4B, 4D, 5B (2), 6A, and 6D, showed pleiotropic effects (Table 5), and both parents con- tributed alleles that delayed senescence (Table 4). Pleiotropic QTLs could be attributed to the confounding relationship between LSS and LSR and GY. Clearly, grain might not be fully filled due to early leaf senescence, and the Jingdong 8 alleles that increased LSR at *QLSR. caas-1A. 1*, *QLSR. caas-1A. 2* and *QLSR. caas-6D* also reduced TKW, whereas *QLSS. caas-5B* not only reduced TKW but GY as well. However, the positive correlations between yield and delayed senescence were not consistent across all environments. Alleles increasing LSS at *QLSS. caas-3B* and *QLSS. caas-3A* also increased GY, and alleles increasing LSR at *QLSR. caas-3B* and *QLSR. caas-6A* increased either TKW or GY. Alleles decreasing NDVIg at *QNDVIg. caas-3B* and *QNDVIg. caas-4B* were observed to increase TKW. These pleiotropic QTLs for senescence, TKW and/or GY may be associated with the rate of grain filling as confirmed by Wang et al. (2009) and Barakat et al. (2011). For example, based on the consensus map (Somers et al. 2004) *Xwmc256*, closely linked to *QLSR. caas-6A*, was 5 cM from QTL identified for grain filling rate at *Xgwm132* on chromosome 6A (Barakat et al. 2011). However, *QLSS. caas-5B* and *QLSR. caas-6D*, which are independent of HD, are newly reported.

HD was significantly correlated with LSS and LSR, indicating that development stage had a significant effect on leaf senescence. *QLSR. caas-1A. 2*, *QLSS. caas-2B*, *QLSS. caas-3A* and *QLSR. caas-4D* co-located with QTLs for HD. *QLSR. caas-2A*, *QLSR. caas-2D. 2*, *QLSR. caas-3B. 1*, *QLSR. caas-3B. 2*, *QLSR. caas-6A* and *QLSR. caas-6D* were independent on HD; clearly, alleles at these loci with large PVE may contribute to senescence rate. Alleles at *QLSS. caas-1A*, which is independent of HD and colocated with *QLSR. caas-1A. 1*, affected both the onset of senescence and the senescence rate.

Alleles decreasing LSS at *QLSS. caas-2B*, *QLSS. caas-3A* and *QLSS. caas-4D* delayed HD in accordance with the significant negative correlations between LSS and HD observed at the phenotypic level (Table 3). An allele delaying HD at *QHD. caas-4D* also decreased LSR in Beijing, but increased LSR at Gaoyi. Coincidentally, grain filling duration at Gaoyi was shorter than that at Beijing, thus it is likely that this HD-delaying gene that increased LSR, may be associated with grain filling. One example was an allele at *Xbarc288* associated with maximum grain filling rate, average grain filling rate and yield components (Wang et al. 2009). This locusis close to *Xbarc98* and *Xwmc89* on chromosome 4D based on the Synthetic/Opata mapping population (Sourdille et al. 2004). Therefore, senescence-related QTLs, such as *QLSS. caas-4D*, could be important for improving the rate of grain filling.

QTLs for NDVI on chromosomes 1B, 2B, 3B, 4B, 4A and 7B were also reported previously (Pinto et al. 2010; Bennett et al. 2012), whereas those on 1A,

2D, 4D, 5B, 7A and 7D were newly identified in the present study. Of the new QTLs, *QND- VIg. caas-1A. 1* and *QNDVIv. caas-5B* with large PVE were independent on HD and GY, but co-located with QTL for leaf senescence and TKW. All QTLs that delayed HD also decreased grain weight. *QHD. caas-5B* with a large PVE was stably detected in six of the eight environments. The Jingdong 8 allele at this locus delayed HD and increased kernel number per spike under limited irrigation. QTL previously reported for frost resistance, flowering time, grain filling rate and GY are located close to this locus (Tóth et al. 2003; Quarrie et al. 2005; Wang et al. 2009; Galaeva et al. 2013). These pleiotropic QTLs can be a target for MAS in wheat breeding programs.

Table 5 Summary of pleiotropic QTLs in the Jingdong 8/Aikang 58 $F_{2:4}$ population

Chromosome	Marker	LSR	LSS	HD	NDVIv	NDVIg	TKW	GY
1A	*Xgwm135/Xwmc120*	+C	+C			−C	−C	
1A	*Xwmc84*	+F		+L	−C	−F	−F	
2B	*Xgwm429/Xcfe67*		−L	+C				
2D	*Xwmc144*	+L	+L			−C		
2D	*Xwmc170*		−L		−F	−F		
3A	*Xbarc1044/Xbarc67*		−L	+C				−F
3B	*Xgwm547*		+L		+L/ −F(B)	+C(G)/ −F(B)		+L
3B	*Xwmc418/Xbarc251*	+F			+F	−F	+F	+L
4B	*Xgwm495−Xgwm251*			−F	−F	−F	+L	
4D	*Xbarc105/Xwmc89*	+C (B) / −C (G)	+C	−C	−C	−C	+F	+F/−F
5B	*Xbarc74/Xgwm499*			+C			−C	
5B	*Xbarc110*		+L		−F		−F	−L
6A	*Xwmc256−Xbarc103*	+F					+C	
6D	*Xbarc365−Xbarc54*	+F					−C	

−, Increasing effect came from Aikang58; +, increasing effect came from Jingdong 8; L, limited irrigation; F, full irrigation; C, limited irrigation and full irrigation; B, Beijing; G, Gaoyi

NDVIv Normalized difference vegetation index at Zadoks47, *NDVIg* normalized difference vegetation index at Zadoks75, *HD* heading date, *LSR* leaf senescence rate *LSS* leaf senescence score, *TKW* thousand kernel weight, *GY* grain yield

Drought-induced senescence

The maintenance of healthy green leaves with high chlorophyll concentration and a similar leaf canopy area under stress can be interpreted as stress tolerance (Olivares-Villegas et al. 2007). For example, high chlorophyll concentration was reported to reduce post-anthesis drought induced senescence in sorghum (Karen et al. 2007). In the present study, the significant correlation between chlorophyll concentration and LSS in LI ($r = -0.39$, data not shown) indicated that high chlorophyll concentration also reduced LSR under stress.

A functional stay-green wheat mutant was reported to improve drought tolerance by accumulating more soluble sugars and proteins under water stress (Tian et al. 2012). A longer duration of grain filling in drought resistant varieties increases mobilization of stem reserves, thus limiting yield loss under stress conditions, as compared to sensitive genotypes (Gupta et al. 2011). Moreover, the contribution of remobilized assimilates to GY tended to increase as water stress intensified (Ercoli et al. 2008). Therefore, under limited irrigation, carbohydrate accumulation and remobilization plays an important role during wheat senescence. Although carbohydrate assessment was not included in the current study, co-located QTL can provide information for analyzing trait association. The leaf senescence QTL *QLSS. caas-2D. 1* was also attributed to drought-induced flag leaf senescence by Barakat et al. (2013). *QLSS. caas-1A* was colocated with *QNDVIg. caas-1A* and *QLRS. caas-1A*, where QTL for stem water-soluble carbohydrates and its remobilization efficiency, and QTL for grain filling rate, were also identified based on *Xgwm357*, which was 2 cM from *Xgwm135* (Yang et al. 2007; Wang et al. 2009; Somers et al. 2004). *QLSS. caas-7D* and *QLSS. caas-5B* were only detected under limited irrigation in the current study. However, *QLSS. caas-7D* was detected in the severe drought environment BD2, and this chromosomal region was associated with stem water-soluble carbohydrate accumulation efficiency based on *Xgwm44*, which is only 2 cM from *Xbarc126* (Yang et al. 2007; Somers et al. 2004). The Jingdong 8 allele that increased LSS at *QLSS. caas-5B* also reduced TKW and GY in this study. A QTL for stem water-soluble carbohydrate remobilization was reported near *QLSS. caas-3B* based on the common marker *Xgwm547* (Yang et al. 2007). *QLSR. caas-3B*, which was associated with TKW and GY under limited irrigation, is located in a similar region to QTL previously reported for TKW and GY (Cuthbert et al. 2008; Bennett et al. 2012). These genes are potential selection targets for improving drought tolerance of wheat.

❖ Acknowledgments

This work was supported by the CGIAR Generation Challenge Program (GCP, G7010. 02. 01), National Natural Science Foundation of China (31161140346), International Collaboration in Science and Technology (2014DFG31690), and China Agriculture Research System (CARS-3-1-3).

❖ References

Abbad H, Jaafari SE, Bort J, Araus JL (2004) Comparison of flag leaf and ear photosynthesis with grain yield of durum wheat under various water conditions and genotypes. Agronomie 24: 19-28.

Baenziger M, Edmeades GO, Lafitte HR (1999) Selection for drought tolerance increases maize yields across a range of nitrogen levels. Crop Sci 39: 1035-1040.

Barakat MN, Al-Doss AA, Elshafei AA, Moustafa KA (2011) Identification of new microsatellite marker linked to thegrain filling rate as indicator for heat tolerance genes in F_2 wheat population. Aust J Crop Sci 5: 104-110.

Barakat MN, Wahba LE, Milad SI (2013) Molecular mapping of QTLs for flag leaf senescence under water stressed conditions in wheat (*Triticum aestivum* L.). Biol Plant 57: 79-84.

Benbella M, Paulsen GM (1998) Efficacy of treatments for delaying senescence of wheat leaves: II. Senescence and grain yield under field conditions. Agron J 90: 332-338.

Bennett D, Reynolds M, Mullan D, Izanloo A, KuchelH, Langridge P, Schnurbusch T (2012) Detection of two major grain yield QTL in bread wheat (*Triticum aestivum* L.) under heat, drought and high yield potential environments. Theor Appl Genet 125: 1473-1485.

Blake NK, Lanning SP, Martin JM, Sherman JD, Talbert LE (2007) Relationship of flag leaf characteristics to economically important traits in two spring wheat crosses. Crop Sci 47: 491-496.

Bogard M, Jourdan M, Allard V, Martre P, Perretant MR, Ravel C, Heumez E, Orford S, Snape J, Griffiths S, Gaju O, Foulkes J, LeGouis J (2011) Anthesis date mainly explained correlations between post-anthesis leaf senescence, grain yield, and grain protein concentration in a winter wheat population segregating for flowering time QTLs. J Exp Bot 62: 3621-3636.

Borrell AK, Hammer GL, Douglas ACL (2000a) Doesmaintaining green leaf area in sorghum improve yield under drought? I Leaf growth and senescence. Crop Sci 40: 1026-1037.

Borrell AK, Hammer GL, Henzell RG (2000b) Does maintaining green leaf area in sorghum improve yield under drought? II. Dry matter production and yield. Crop Sci 40: 1037-1048.

Cha KW, Lee YJ, Koh HJ, Lee BM, Nam YW, Paek NC (2002) Isolation, characterization, and mapping of the stay green mutant in rice. Theor Appl Genet 104: 526-532.

Chen J, Liang Y, Hu X, Wang X, Tan F, Zhang H, Ren Z, Luo P (2010) Physiological characterization of wheat cultivars during the grain filling stage under field growing conditions. Acta Physiol Plant 32：875-882.

Chen CC，Han GQ，He HQ，Westcott M（2011）Yield，protein，and remobilization of water-soluble carbohydrate and nitrogen of three spring wheat cultivars as influenced by nitrogen input. Agron J 103：786-795.

Christopher JT，Manschadi AM，Hammer G，Borrell AK（2008）Developmental and physiological traits associated with high yield and stay-green phenotype in wheat. Aust J Agr Res 59：354-364.

Cuthbert JL，Somers DJ，Brûlé-Babel AL，Brown PD，Crow GH（2008）Molecular mapping of quantitative trait loci for yield and yield components in spring wheat（*Triticum aestivum* L.）. Theor Appl Genet 117：595-608.

Ercoli L，Lulli L，Mariotti M，Mosani A，Arduini I（2008）Post anthesis dry matter and nitrogen dynamics in durum wheat as affected by nitrogen supply and soil water availability. Eur J Agron 28：138-147.

Foyer CH，Noctor G（2005）Redox homeostasis and antioxidant signaling：a metabolic interface between stress perception and physiological responses. Plant Cell 17：1866-1875.

Galaeva MV，Fayt VI，Chebotar SV，Galaev AV，Sivolap YM（2013）Association of microsatellite loci alleles of the group-5 of chromosomes and the frost resistance of winter wheat. Cytol Genet 47：261-267.

Gentinetta E，Ceppi D，Lepori C，Perico G，Motto M，Salamini F（1986）A major gene for delayed senescence in maize. Pattern of photosynthate accumulation and inheritance. Plant Breed 97：193-203.

Gorny AG，Garczynski S（2002）Genotypic and nutritional dependent variation in water use efficiency and photosynthetic activity of leaves in winter wheat. J Appl Genet 43：145-160.

Gregersen PL，Culetic A，Boschian L，Krupinska K（2013）Plant senescence and crop productivity. Plant Mol Biol 82：603-622.

Gupta AK，Kaur K，Kaur N（2011）Stem reserve mobilization and sink activity in wheat under drought conditions. Amer J Plant Sci 2：70-77.

Guttieri MJ，Stein RJ，Waters BM（2013）Nutrient partitioning and grain yield of *TaNAM-RNAi* wheat under abiotic stress. Plant Soil 371：573-591.

Hafsi M，Mechmeche W，Bouamama L，Djekoune A，Zaharieva M，Monneveux P（2000）Flag leaf senescence，as evaluated by numerical image analysis，and its relationship with yield under drought in durum wheat. J Agr Crop Sci 185：275-280.

Institute SAS（2000）SAS user's guide：statistics. SAS Institute Inc，Cary

Jiang GH，He YQ，Xu CG，Li XH，Zhang Q（2004）The geneticbasis of stay-green in rice analyzed in population of dihybrid lines derived from *indica* by *japonica* cross. Theor Appl Genet 108：688-698.

Joshi AK，Kumari M，Singh VP，Reddy CM，Kumar S，Rane J，Chand R（2007）Stay green trait：variation，inheritance and its association with spot blotch resistance in spring wheat（*Triticum aestivum* L.）. Euphytica 153：59-71.

Karen H，Subudhi PK，Borrell A，Jordan D，Rosenow D，Nguyen H，Klein P，Klein R，Mullet J（2007）Sorghum stay green QTL individually reduce post-flowering drought-induced leaf senescence. J Exp Bot 58：327-338.

Kumar U，Joshi AK，Kumari M，Paliwal R，Kumar S，Röder MS（2010）Identification of QTLs for stay green trait in wheat（*Triticum aestivum* L.）in the 'Chirya 3' & # x00D7； 'Sonalika' population. Euphytica 174：437-445.

Li XM，Xiao YG，Chen XM，Xia XC，Wang DS，He ZH，Wang HJ（2014）Identification of QTLs for seedling vigor in winter wheat. Euphytica 198：199-209.

Lopes MS，Reynolds MP（2012）Stay-green in spring wheat can be determined by spectral reflectance measurements（normalized difference vegetation index）independently from phenology. J Exp Bot 63：3789-3798.

Luo P，Ren Z，Wu X，Zhang H，Zhang H，Feng J（2006）Structural and biochemical mechanism responsible for the stay-green phenotype in common wheat. Chin Sci Bull 51：2595-2603.

Naruoka Y，Sherman JD，Lanning SP，Blake NK，Martin JM，Talbert LE（2012）Genetic analysis of green leaf duration in spring wheat. Crop Sci 52：99-109.

Olivares-Villegas JJ，Reynolds MP，McDonald GK（2007）Drought adaptive attributes in the Seri/Babax hexaploid wheat population. Funct Plant Biol 34：189-203.

Pask AJD，Pietragalla J，Mullan DM，Reynolds MP（2012）Physiological breeding II：a field guide to wheat phenotyping. CIMMYT，Mexico.

Pinto RS，Reynolds MP，Mathews KL，McIntyre CL，Olivares- Villegas JJ，Chapman SC（2010）Heat and drought adaptive QTL in a wheat population designed to minimize confounding agronomic effects. Theor Appl

Genet 121: 1001-1021.

Quarrie SA, Steed A, Calestani C, Semikhodskii A, Lebreton C, Chinoy C, Steele N, Pljevljakusic' D, Waterman E, Weyen J, Schondelmaier J, Habash DZ, Farmer P, Saker L, Clarkson DT, Abugalieva A, Yessimbekova M, Turuspekov Y, Abugalieva S, Tuberosa R, Sanguineti MC, Hollington PA, Aragués R, Royo A, Dodig D (2005) A highdensity genetic map of hexaploid wheat (*Triticum aestivum* L.) from the cross Chinese Spring 9 SQ1 and its use to compare QTLs for grain yield across a range of environ- ments. Theor Appl Genet 110: 865-880.

Ray DK, Ramankutty N, Mueller ND, West PC, Foley JA (2012) Recent patterns of crop yield growth and stagnation. Nature Comm 3: 1293.

Reynolds MP, Ortiz-Monasterio JI, McNab A (2001) Application of physiology in wheat breeding. CIMMYT, Mexico Saleh MS, Al-Doss AA, Elshafei AA, Moustafa KA, Al-Qurainy FH, Barakat MN (2014) Identification of new TRAP markers linked to chlorophyll content, leaf senescence, and cell membrane stability in water-stressed wheat. Biol Plant 58: 64-70.

Simon MR (1999) Inheritance of flag-leaf angle, flag-leaf area and flag leaf area duration in four wheat crosses. Theor Appl Genet 98: 310-314.

Snape JW, Foulkes MJ, Simmonds J, Leverington M, Fish LJ, Wang Y, Ciavarrella M (2007) Dissecting gene×environmental effects on wheat yields via QTL and physio- logical analysis. Euphytica 154: 401-408.

Somers DJ, Isaac P, Edwards K (2004) A high-density micro- satellite consensus map for bread wheat (*Triticum aestivum* L.). Theor Appl Genet 109: 1105-1114.

Sourdille P, Singh S, Cadalen T, Brown-Guedira GL, Gay G, Qi L, Gill BS, Dufour P, Murigneux A, Bernard M (2004) Microsatellite-based deletion bin system for the establishment of genetic-physical map relationship in wheat (*Triticum aestivum* L.). Funct Integr Genomics 4: 12-25.

Spano G, Di Fonzo N, Perrotta C, Platani C, Ronga G, Lawlar DW Napier JA, Shewry PR (2003) Physiological characterization of 'stay green' mutant in durum wheat. J ExpBiol 54: 1415-1420.

Sykorova B, Kuresova G, Daskalova S, Trckova M, Hoyerova K, Raimanova I, Motyka V, Travnickova A, ElliottMC, Kaminek M (2008) Senescence-induced ectopic expre sion of the *A. tumefaciens* ipt gene in wheat delays leaf senescence, increases cytokinin content, nitrate influx, and nitrate reductase activity, but does not affect grain yield. J Exp Bot 59: 377-387.

Tao YZ, Hanzell RG, Jordan DR, Butler DG, Kelly AM, McIntyre CL (2000) Identification of genomic regions associated with stay green in sorghum by testing RILs in multiple environments. Theor Appl Genet 100: 1225-1232.

Thomas H, Howarth CJ (2000) Five ways to stay green. J Exp Bot 51: 329-337.

Thomas H, Smart CM (1993) Crops that stay green. Ann Appl Biol 123: 193-219.

Thomas H, OughamHJ WagstaffC, Stead AD (2003) Defining senescence and death. J Exp Bot 54: 1127-1132.

Tian F, Gong J, Zhang J, Zhang M, Wang G, Li A, Wang W (2013) Enhanced stability of thylakoid membrane proteins and antioxidant competence contribute to drought stress resistance in the *tasg1* wheat stay-green mutant. J Exp Bot 64: 1509-1520.

Tian F, Gong J, Wang G, Wang G, Fan Z, Wang W (2012) Improved drought resistance in a wheat stay-green mutant *tasg1* under field conditions. Biol Plant 56: 509-515.

Tóth B, Galiba G, Feher E, Sutka Y, Snape JW (2003) Mapping genes affecting flowering time and frost resistance on chromosome 5B ofwheat. TheorApplGenet 107: 509- 514.

Verma V, Foulkes MJ, Worland AJ, Sylvester-Bradley R, Caligari PDS, Snape JW (2004) Mapping quantitative trait loci for flag leaf senescence as a yield determinant in winter wheat under optimal and drought-stressed environments. Euphytica 135: 255-263.

Vijayalakshmi K, Fritz AK, Paulsen GM, Bai G, Pandravada S, Gill BS (2010) Modeling and mapping QTL for senescence-related traits in winter wheat under high temperature. Mol Breeding 26: 163-175.

Wang RX, Hai L, Zhang XY, You GX, Yan CS, Xiao SH (2009) QTL mapping for grain filling rate and yield-related traits in RILs of the Chinese winter wheat population Heshangmai 3/Yu8679. Theor Appl Genet 118: 313-325.

Wingler A, von Schaewen A, Leegood RC, Lea PJ, Quick WP (1998) Regulation of leaf senescence by cytokinin, sugars, and light. Plant Physiol 116: 329-335.

Wu XY, Kuai BK, Jia JZ, Jing HC (2012) Regulation of leaf senescence and crop genetic improvement. J Integr Plant Biol 54: 936-952.

Xu W, Subudhi PK, Crasta OR, Rosenow DT, Mullet JE, Nguyen HT (2000) Molecular mapping for QTLs conferring stay-green in grain sorghum (*Sorghum bicolor*

L. Moench). Genome 43：461-469.

Yang DL，Jing RL，Chang XP，Li W（2007）Quantitative traitloci mapping for chlorophyll fluorescence and associatedtraits in wheat（*Triticum aestivum* L.）. J Integr Plant Biol 49：646-654.

Zhang CJ，Chen GX，Gao XX，Chu CJ（2006）Photosynthetic decline in flag leaves of two field-grown spring wheat cultivars with different senescence properties. S Afr J Bot 72：15-23.

Zhang SW，Wang CF，Miao F，Zhou CJ，Yao YH，Li GX（2012）Photosynthetic characteristics and its significance of topmost three leaves at fruiting stage in wheat with presenile flag leaf. Acta Agron Sin 38：2258-2266.

Fine mapping of a stripe rust resistance gene *YrZM175* in bread wheat

Jingchun Wu[1,2] • Dengan Xu[3] • Luping Fu[4,5] • Ling Wu[6] • Weihao Hao[1] • Jihu Li[7] • Yan Dong[2] • Fengju Wang[2] • Yuying Wu[8] • Zhonghu He[2,9] • Hongqi Si[1] • Chuanxi Ma[1] • Xianchun Xia[1,2]

Communicated by Urmil Bansal.

Chuanxi Ma machuanxi@ahnu.edu.cn

Xianchun Xiaxiaxianchun@caas.cn

[1] *College of Agronomy, Anhui Agricultural University*, 130 *Changjiang West Road, Hefei* 230036, *Anhui province, China*

[2] *Institute of Crop Sciences, National Wheat Improvement Centre, Chinese Academy of Agricultural Sciences* (*CAAS*), 12 *Zhongguancun South Street, Beijing* 100081, *China*

[3] *College of Agronomy, Qingdao Agricultural University*, 700 *Changcheng Road, Qingdao* 266109, *Shandong province, China*

[4] *Jiangsu Key Laboratory of Crop Genetics and Physiology/Jiangsu Key Laboratory of Crop Cultivation and Physiology, Agricultural College of Yangzhou University, Yangzhou* 225009, *Jiangsu province, China*

[5] *Jiangsu Co-Innovation Centre for Modern Production Technology of Grain Crops, Yangzhou University, Jiangsu province, Yangzhou* 225009, *China*

[6] *Crop Research Institute, Sichuan Academy of Agricultural Sciences*, 4 *Shizishan Road, Chengdu* 610011,

Sichuan province, China

[7] *Crop Research Institute, Shandong Academy of Agricultural Sciences*, 202 *Gongye North Road, Jinan* 250100,

Shandong province, China

[8] *College of Agronomy, Henan Agricultural University*, 63 *Agricultural Road, Zhengzhou* 450002, *Henan province, China*

[9] *International Maize and Wheat Improvement Centre* (*CIMMYT*) *China Office c/o*, *CAAS*, *Beijing* 100081, *China*

Received: 12 April 2022 / Accepted: 3 August 2022

Key message: A stripe rust resistance gene *YrZM175* in Chinese wheat cultivar Zhongmai 175 was mapped to a genomic interval of 636.4 kb on chromosome arm 2AL, and a candidate gene was predicted.

Abstract: Stripe rust, caused by *Puccinia striiformis* f. sp. *tritici* (PST), is a worldwide wheat disease that causes large losses in production. Fine mapping and cloning of resistance genes are important for accurate marker-assisted breeding. Here, we report the fine mapping and candidate gene analysis of stripe rust resistance gene *YrZM175* in a Chinese wheat cultivar Zhongmai 175. Fifteen F_1, 7 325 F_2

Published in Theoretical and Applied Genetics, 2022, 135, 3485-3496.

plants and 117 $F_{2:3}$ lines derived from cross Avocet S/Zhongmai 175 were inoculated with PST race CYR32 at the seedling stage in a greenhouse, and $F_{2:3}$ lines were also evaluated for stripe rust reaction in the field using mixed PST races. Bulked segregant RNA-seq (BSR-seq) analyses revealed 13 SNPs in the region 762. 50-768. 52 Mb on chromosome arm 2AL. By genome mining, we identified SNPs and InDels between the parents and contrasting bulks and mapped *YrZM175* to a 0. 72-cM, 636. 4-kb interval spanned by *YrZM175-InD1* and *YrZM175-InD2* (763, 452, 916-764, 089, 317 bp) including two putative disease resistance genes based on IWGSC RefSeq v1. 0. Collinearity analysis indicated similar target genomic intervals in Chinese Spring, *Aegilops tauschii* (2D: 647. 7-650. 5 Mb), *Triticum urartu* (2A: 750. 7-752. 3 Mb), *Triticum dicoccoides* (2A: 771. 0-774. 5 Mb), *Triticum turgidum* (2B: 784. 7-788. 2 Mb), and *Triticum aestivum* cv. Aikang 58 (2A: 776. 3-778. 9 Mb) and Jagger (2A: 789. 3-791. 7 Mb). Through collinearity analysis, sequence alignments of resistant and susceptible parents and gene expression level analysis, we predicted *TRITD2Bv1G264480* from *Triticum turgidum* to be a candidate gene for map-based cloning of *YrZM175*. A gene-specific marker for *TRITD2Bv1G264480* co-segregated with the resistance gene. Molecular marker analysis and stripe rust response data revealed that *YrZM175* was different from genes *Yr1*, *Yr17*, *Yr32*, and *YrJ22* located on chromosome 2A. Fine mapping of *YrZM175* lays a solid foundation for functional gene analysis and marker-assisted selection for improved stripe rust resistance in wheat.

Introduction

Wheat (*Triticum aestivum* L.) is a staple food crop worldwide, providing substantial amounts of nutrients for humans (Shewry 2009; Shewry and Hey 2015). The estimated global wheat production in 2021 was 770 million tonnes (FAO, http://www.fao.org/worldfoodsituation/en/). Wheat stripe rust (or yellow rust), caused by *Puccinia striiformis* f. sp. *tritici* (PST), is a widely distributed disease that causes millions of tonnes of yield losses every year (Chen and Kang 2017). Stripe rust occurred in 1. 78 million ha of wheat in China during 2011-2016 and has since increased to about 5 million ha annually (Huang et al. 2020). Resistant cultivars are the most environment-friendly approach to control stripe rust (Xia et al. 2007). Although many resistance genes have been identified, relatively few of them, such as *Yr5*, *Yr15*, and *Yr18*, continue to confer resistance in China. Therefore, it is necessary to mine new stripe rust resistance genes and to deploy them in wheat cultivars, preferably in multiple gene combinations.

More than 80 stripe rust resistance genes (*Yr1*-*Yr83*) have been formally catalogued (McIntosh et al. 2017; Feng et al. 2018; Nsabiyera et al. 2018; Gessese et al. 2019; Pakeerathan et al. 2019; Li et al. 2020). Among formally catalogued and other named genes nine have been cloned (*Yr5*/*YrSP*, *Yr7*, *Yr10*, *Yr15*, *Yr18*, *Yr36*, *Yr46*, *YrAS2388*, and *YrU1*). *Yr18*, *Yr36*, and *Yr46* confer adult-plant resistance and encode a putative ATP-binding cassette (ABC) transporter (Krattinger et al. 2009), a protein with a kinase domain and a lipid binding domain (Fu et al. 2009), and a hexose trans- porter (Moore et al. 2015), respectively. *Yr5*/*YrSP* (alleles), *Yr7*, *Yr10*, *Yr15*, *YrAS2388* and *YrU1* are all-stage resistance genes. *Yr15* encodes a protein with kinase-pseudokinase domains, and the others encode various nucleotide-binding site and leucine-rich repeat (NBS-LRR) proteins. *Yr5*/*YrSP* and *Yr7* belong to a gene cluster originated from *T. spelta* (Marchal et al. 2018); *YrAS2388* came from *Ae. tauschii* (Liu et al. 2014; Zhang et al. 2019); *Yr15* was derived from *T. dicoccoides* (Klymiuk et al. 2018) and *YrU1* from *T. urartu* (Wang et al. 2020).

Wheat cultivar Zhongmai 175 is widely grown in the North Winter Wheat Zone of China; it has high yield and good level of resistance to stripe rust. It also shows moderate to high resistance in Sichuan and Gansu provinces, the hotspots for stripe rust epidemics and year-round survival in China. One of the parents of Zhongmai 175 was the French line VPM1 which had high resistance to powdery mildew, stem rust, leaf rust and stripe rust, and was widely used as a disease resistant germplasm by breeders. Zhongmai 175 carries *Yr17* from VPM1 and a putatively new stripe rust resistance gene *YrZM175* that was located in chromosome 2A (Lu et al. 2016). The aims of this study were to fine-map *YrZM175*, predict a candidate gene for further analysis, and to develop gene-specific markers for research and breeding.

Materials and methods

Plant materials

A total of 7,325 F_2 plants derived from the cross of Avocet S/Zhongmai 175 were used for fine mapping of *YrZM175*. One hundred and seventeen random $F_{2:3}$ lines were evaluated for stripe rust reaction in the greenhouse and field to confirm the results from the F_2 population.

Evaluation of stripe rust reaction

Fifteen F_1, 7,325 F_2 plants and 117 $F_{2:3}$ lines of Avocet S/ Zhongmai 175 and the parents were inoculated at the seedling stage in controlled greenhouse conditions with PST race CYR32, kindly provided by Dr. Gangming Zhan (North-west A & F University, Yangling, Shaanxi). CYR32 is avirulent to Zhongmai 175 and virulent to Avocet S. Ten to 15 seeds were planted in a 9×9×9 cm plastic pot, along with three seeds of susceptible cultivar Mingxian 169 as control. Seedlings were inoculated with CYR32 at 1.5-leaf stage (7-8 days old) by brushing fresh urediniospores from fully infected Mingxian 169 leaves. Inoculated plants were kept in plastic boxes with 100% humidity at 10℃ in darkness for 24 h, then transferred to 15±2℃ conditions with 16 h of light (20,000 lx) daily. Infection types (ITs) were scored 14-18 days after inoculation based on a 0-4 scale (0-2 as resistant, 3-4 as susceptible) (Bariana and McIntosh 1993).

Fifteen F_1 plants and 117 $F_{2:3}$ lines along with both parents with 30-40 seedlings each were also evaluated for stripe rust reaction at Pidu Experimental Station, Sichuan Academy of Agricultural Sciences in Chengdu, in the 2017-2018 and 2018-2019 cropping seasons. $F_{2:3}$ lines were sown in two replications. Avocet S was used as a susceptible control and spreader and inoculated with mixed PST races (CYR32, CYR33, and CYR34). ITs and disease severity (DS) were recorded twice at 4- to 5-day intervals when the DS of Avocet S reached 90—100%. DS was scored as the percentage of infected leaf area. ITs were scored similarly to those at seedling stage mentioned above.

Table 1 Seedling reactions of Zhongmai 175, Avocet S, F_1, F_2 plants and $F_{2:3}$ lines from the cross of Avocet S/ Zhongmai 175 to PST race CYR32

Material	Total no.	Infection type						Expected ratio	χ^2	P
		0	0;	1	2	3	4			
Zhongmai 175	15	11	4							
Avocet S	12					3	9			
F_1	15	10	5							
F_2	7,325	1,277	3,817	283	155	774	1,019	3∶1	1.06	0.30
		RR[a]		Rr[b]		rr[c]				
$F_{2:3}$	117	33		56		28		1∶2∶1	0.64	0.73

[a] Homozygous resistant

[b] Segregating

[c] Homozygous susceptible

DNA and RNA extraction

Leaf samples of plants were used to extract genomic DNA using the CTAB method (Saghai-Maroof et al. 1984). For $F_{2:3}$ lines, equal amounts of leaf tissue from 8-10 plants of each line were mixed for DNA extraction. Because of large genome differences between Zhongmai 175 and Avocet S and the absence of some genes in the target region in Avocet S, we later chose Mingxian 169 as a susceptible control for RNA isolation. Fresh leaf tissues of Zhongmai 175 and Mingxian 169 inoculated with CYR32 were collected before inoculation (0 h) and at every two hours in 12 h post-inoculation (hpi), and at 10 a. m daily in 1 to 14 days post-inoculation (dpi) with three biological replications for RNA extraction using the Plant RNA Kit (Omega Biotek, Inc. Norcross, Georgia USA).

BSR-seq assay

Based on phenotypic evaluations at 14 dpi, 50 resistant (IT=0) and 50 susceptible plants (IT=4) were separately mixed to form resistant (R) and susceptible (S) pools for RNA extraction and sequencing. The pooled leaf samples and parents were sent to Novogene (Tianjin, China) for RNA-seq. After quality testing, single RNA libraries were constructed for each sample and sequenced in an Illumina HiSeq4000 platform, generating 150 bp paired-end reads. BWA (Burrows-Wheeler Aligner) (Li et al. 2009) was used to align the clean reads of each sample against the reference genome of Chinese Spring, *Ae. tauschii* ssp. *strangulata* accession AL8/78 (Luo et al. 2017), *T. urartu* accession G1812(Ling et al. 2018), *T. turgidum* ssp. *dicoccoides* accession Zavitan (Avni et al. 2017), *T. turgidum* ssp. *durum* cv. Svevo (Maccaferri et al. 2019), and *T. aestivum* cv. Aikang 58 and Jagger (Walkowiak et al. 2020) (https://www.ncbi.nlm.nih.gov/assembly/?term = triticeae), respectively, after a series of quality control (QC) procedures. SNPs/InDels calling between the two pools was performed using the Unified Genotyper function in GATK (McKenna et al. 2010). The homozygous SNPs/InDels were extracted to calculate the SNP/InDel index. The sliding window method was used to present the SNP/InDel index of polymorphic sites in whole genome. The differences in SNP/InDel indices between the two pools were calculated as the delta SNP/InDel index to determine the target region of the resistance gene. SNPs/InDels with delta SNP/InDel index of 1 or −1 are considered highly reliable.

Marker development through BSR-seq and genome mining

Highly reliable SNPs from BSR-seq analyses were converted into cleaved amplified polymorphic sequence (CAPS) markers following Thiel et al. (2004) based on their 700-1,000 bp flanking sequences identified in the Chinese Spring reference genome RefSeq v1.0 (IWGSC 2018) and the six other genome sequences mentioned above. Chromosome-specific primers for each SNP were designed with DNAMAN 8.0 software. Due to the limited number of SNPs from BSR-seq, more variations between the parents were identified from gene sequences in the target region in reference to the RefSeq v1.0 (IWGSC 2018). According to gene annotations from RefSeq v1.0 (IWGSC 2018) and the six other genome sequences, gene-specific primers were designed from putative disease resistance genes in the target scaffolds to isolate gene sequences in the parents. PCR sequencing results for the parents were aligned to identify the polymorphic SNPs/ InDels using Geneious Prime 2019.0.3 software. InDels between the parents were converted to sequence-tagged site (STS) markers, whereas CAPS markers were developed from SNPs.

Genetic map construction

Chi-squared (χ^2) tests were conducted to determine the goodness-of-fit of the observed and expected ratios of segregation in the F_2 and $F_{2:3}$ populations using SAS 9.2 (SAS Institute, Cary, NC, USA). All susceptible F_2 plants were tested by molecular markers in the target region for linkage analysis with the resistance gene. Genetic distances between polymorphic SNP, STS and SSR markers linked to the stripe rust resistance gene were calculated using the Kosambi mapping function (Kosambi 1943) computed by Mapmaker 3.0 with an LOD threshold of 3.0 (Lincoln et al. 1993). The linkage map was graphically drawn with MapDraw V2.1 (Liu and Meng 2003).

Collinearity analysis of homologous sequences

Collinear alignment of genes and homoeologous intervals was performed using TGT website (http://wheat.cau.edu.cn/TGT/) by blasting markers flanking *YrZM175* to obtain physical positions of homologous sequences in related species in *Triticum* multi-omics center databases on WheatOmics 1.0 (http://wheatomics.sdau.edu.cn/) and to confirm corresponding homoeologous intervals. Gene annotations within homoeologous intervals were searched and downloaded in EnsemblPlants (http://plants.ensembl.org/index.html), and related disease resistance genes were checked.

Genome walking to get sequences of candidate genes

Genome walking is an effective technique to obtain unknown sequences adjacent to known sequences. To find the physical position of the start point with different sequences in the target region between Zhongmai 175 and Chinese Spring to perform genome walking, we designed chromosome-specific primers at 763.83 Mb upstream of *TraesCS2A02G562800* in Chinese Spring RefSeq v1.0 to perform PCR with Zhongmai 175 and Chinese Spring as templates. If sequences of PCR products were common between Zhongmai 175 and Chinese Spring we continued to design primers to amplify downstream sequences from 763.83 Mb until sequences of PCR products showed polymorphisms between Zhongmai 175 and Chinese Spring. Then the position of the polymorphic sequence was set as the start point for genome walking using the Genome Walking Kit (TaKaRa, Dalian). Walking primer designing and PCR amplifications were performed following the manufacturer's protocol. Purified PCR products were ligated with pEASY-T5 Blunt-Zero cloning vectors at 25℃ for 2 h and transformed into Trans1-T1 competent cells by the heat shock (TransGen Biotech Co., Ltd., Beijing). After coating the plate and incubating at 37℃ for 24 h, 6-10 positive clones from each PCR product were randomly selected and sequenced by Shanghai Sangon Biotech Co., Ltd (http://www.sangon.com/).

Quantitative real-time PCR

First-strand cDNA was synthesized using the PrimeScript RT Reagent Kit with gDNA Eraser (TaKaRa, Kyoto, Japan). The cDNA products of Zhongmai 175 and Mingxian 169 were used for quantitative real-time PCR (qPCR) assays in a BioRad CFX system with the iTaq Universal SYBR Green Supermix (BioRad). Gene-specific primers for candidate genes and the *ACTIN* gene as internal control were designed using DNAMAN 8.0 software. Each sample had three biological replicates. Relative gene expression was calculated using the $2^{-\Delta\Delta Ct}$ equation (Schmittgen and Livak 2008).

Results

Stripe rust resistance of Zhongmai 175 at seedling stage

Zhongmai 175 confers resistance to a broad spectrum of PST races (Lu et al. 2016) and displays high resistance at all growth stages (Fig. S1). Among 7,325 F_2 plants derived from Avocet S/Zhongmai 175, 5,532 and 1,793 plants were resistant and susceptible, respectively, conforming with a 3∶1 ratio ($\chi^2=1.06$, $P=0.30$). The $F_{2:3}$ lines segregated 33 homozygous resistant, 56 segregating and 28 homozygous susceptible, conforming with a 1∶2∶1 ratio ($\chi^2=0.35$, $P=0.84$) (Table 1). Thus, the stripe rust resistance in Zhongmai 175 was conferred by a single dominant gene.

Fine mapping of *YrZM175*

BSR-seq analysis revealed 13 highly reliable SNPs in the genomic region 762.50-768.52 Mb on chromosome arm 2AL based on the Chinese Spring RefSeq v1.0 (IWGSC 2018) (Fig. 1A and S2). We initially screened 35 simple sequence repeat (SSR) markers on chromosome 2A and identified nine markers polymorphic between the parents; among them, *Xwmc658*, *Xg-*

wm382, and *Xgwm311* on 2AL were linked to *YrZM175* with genetic distances of 6.51 cM, 5.12 cM, and 5.12 cM, respectively (Fig. 1B and S3). The physical positions of *Xwmc658* and *Xgwm382* (*Xgwm311*) are at 771.16 Mb and 772.96 Mb, respectively, in agreement with the results of BSR-seq. The 13 polymorphic SNPs were mainly concentrated in two physical positions, i.e., four SNPs at 762.50-762.64 Mb and six around 768.52 Mb on 2AL (Table S1). We developed CAPS markers *YrZM175-CAPS1* and *YrZM175-CAPS2* (Table S2), based on a SNP at 762.50 Mb and another at 768.52 Mb, with genetic distances of 0.81 cM and 2.57 cM from *YrZM175*, respectively.

Many polymorphic SNPs/InDels between parents were identified by sequencing genes in the target region and were converted to CAPS or STS markers (Table S2). *YrZM175* was then mapped in a genetic interval of 1.41 cM between *YrZM175-CAPS1* and *YrZM175-InD1*, spanning a 1.58 Mb (762.50-764.08 Mb) genomic region containing 35 highconfidence genes, including six putative disease resistance genes (Table S3) based on IWGSC RefSeq v1.0. After designing gene-specific PCR primers (Table S4) and sequencing six putative resistance genes in Zhongmai 175, Avocet S and Mingxian 169, *TraesCS2A02G560600* showed a SNP between Zhongmai 175 and Mingxian 169 in the CDS domain, but the gene was absent in Avocet S. During genome walking and mining, we identified two polymorphic sites and converted them to markers *YrZM175-CAPS5* and *YrZM175-InD2* (Fig. S3) with genetic distances of 0.27 cM and 0.12 cM from the resistance gene, respectively. Finally, we mapped *YrZM175* in a genetic interval of 0.72 cM between *YrZM175-InD1* and *YrZM175-InD2*, spanning a 636.4 kb (763,452,916-764,089,317bp) on chromosome arm 2AL (Figs. 1B and 2A). There were nine high-confidence genes in this genomic region based on IWGSC RefSeq v1.0, including putative disease resistance genes *TraesCS2A02G562800* and *TraesCS2A 02G563200* (Table S5).

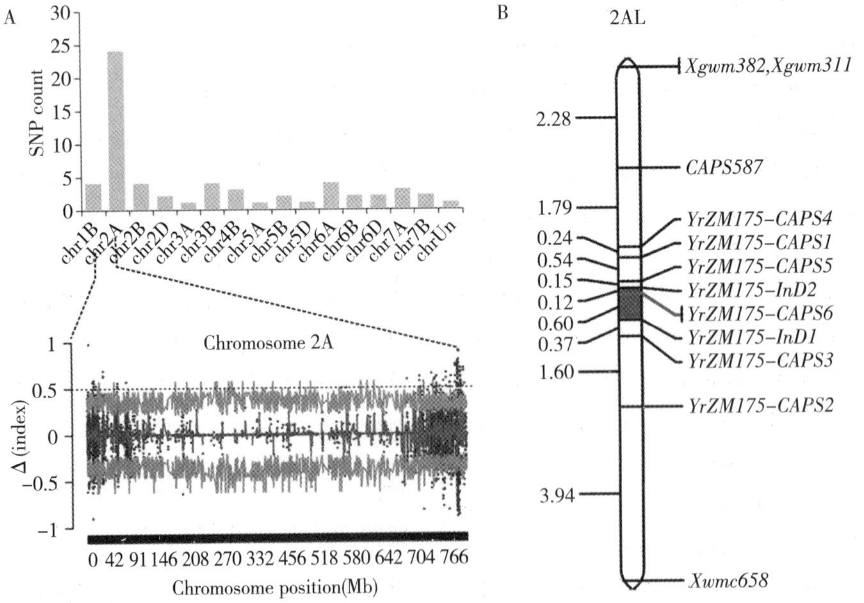

Fig. 1 Distribution of SNPs and linkage map of *YrZM175*

A. Distribution of SNPs in each chromosome based on BSR-seq analysis and physical positions of SNPs on chromosome 2A. **B.** Linkage map of *YrZM175* determined from data of the Avocet S/Zhongmai 175 F_2 population. Marker names are displayed at the right. Genetic distance intervals (cM) are shown at the left. *CAPS6* co-segregated with *YrZM175*.

Analysis of genes in the 636.4 kb interval of Chinese Spring

The NBS-LRR gene *TraesCS2A02G563200* showed no difference between Zhongmai 175, Avocet S and Mingxian 169, indicating that it was not a candidate. No PCR product was obtained following amplification of kinase gene *TraesCS2A02G562800* in all three genotypes, exhibiting that this gene also was not a candidate. The genome sequence was quite different or absent in the three cultivars compared with Chinese Spring (Fig. 2B). After two cycles of genome walking from 763.83 Mb upstream of *TraesCS2A02G562800* we obtained a 3,860-bp genome sequence from Zhongmai 175. Blasting of this sequence against the *Triticum* genome databases in WheatOmics 1.0 we found that this sequence was identical with a receptor-like kinase gene *TRITD2Bv1G265970* on chromosome arm 2BL in *T. turgidum* ssp. *durum* cv. Svevo (Maccaferri et al. 2019). Isolation of the entire 7, 739-bp open-reading frame (ORF) of *TRITD2Bv1G265970* in Zhongmai 175, Mingxian 169 and Avocet S with gene-specific primers (Table S6) according to the reference genome of Svevo found no sequence difference between Zhongmai 175 and Avocet S in the ORF domain, whereas the sequence of the ORF in Mingxian 169 was different from that in Zhongmai 175. Sequencing following amplification of a 3,000-bp promoter sequence upstream of the ATG start codon by genomic-specific primers (Table S7) referring to Svevo revealed three SNPs, single 1-bp and 11-bp insertions, and single 9-bp, 11-bp and 23-bp deletions in Avocet S compared with Zhongmai 175 (Fig. S4). Annotation of the promoter sequence of Zhongmai 175 on database PLACE (https://www.dna.affrc.go.jp/PLACE/?action=newplace) indicated that the SNPs/InDels could change several important cis-acting regulators such as ABA responsive element (ABRE), carbohydrate metabolite signal responsive element 1 (CMSRE-1) and salicylic acid regulatory element GT-1 motif. However, tests of all susceptible F_2 plants with a gene-specific marker developed from a polymorphic site between Zhongmai 175 and Avocet S in the promoter, and the marker located 0.21 cM from the resistance gene failed to show an association. We concluded that *TRITD2Bv1G265970* was not the candidate of *YrZM175*.

Collinearity analysis of the target genomic interval

Based on *Triticum* genome databases (*Ae. tauschii*, *T. urartu*, *T. dicoccoides*, *T. turgidum* and 10+ Genomes) at TGT website, we performed collinearity analysis of the closest marker *YrZM175-InD2* and the targeted genomic region according to the genome sequence of Chinese Spring RefSeq v1.0 (IWGSC 2018). The 636.4-kb mapping interval showed a good collinearity with 10+ Genome varieties, whereas most of the wild relatives had inversions, duplications, and deletions of genes compared with Chinese Spring (Fig. S5). The genomic interval flanked by *YrZM175-CAPS5* and *YrZM175*-InD1 in Chinese Spring RefSeq v1.0 was similar to corresponding intervals in *Ae. tauschii* ssp. *strangulata* accession AL8/78 (2D: 647.7-650.5Mb), *T. urartu* accession G1812 (2A: 750.7-752.3 Mb), *T. turgidum* ssp. *dicoccoides* accession Zavitan (2A: 771.0-774.5 Mb), *T. turgidum* ssp. *durum* cv. Svevo (2A: 773.2-775.1 Mb), *T. turgidum* ssp. *durum* cv. Svevo (2B: 784.7-788.2 Mb), and *T. aestivum* cv. Aikang 58 (2A: 776.3-778.9 Mb) and Jagger (2A: 789.3-791.7 Mb), including eight, five, 12, eight, 11, three and four putative disease resistance genes in the target intervals, respectively (Fig. S6).

Analysis of genes in the collinear interval of other wheat varieties and related species

Based on BSR-seq analysis using genome sequence information from five species (*Ae. tauschii*, *T. urartu*, *T. dicoccoides*, *T. turgidum*, and *T. aestivum* cv. Aikang 58 and Jagger), there was an obvious single peak on homoeologous group 2 chromosomes in all species (Fig. S7). Based on two polymorphic SNPs between R and S pools from BSR-seq analysis in the collinear interval on chromosome 2AL referring *T. dicoccoides* and *T. turgidum* genome sequences, respectively, molecular markers were developed and tested on susceptible F_2 plants from cross Avocet S/Zhongmai 175, and the linkage distances of two SNPs with *YrZM*175 were 0.12 cM. The results confirmed sequence similarity between Chinese Spring, *T. dicoccoides* and *T. turgidum*. Based on collinearity and homology analysis of genes in the collinear interval at website TGT, there were ten putative disease resistance genes in the region upstream of marker *YrZM*175-*InD*2 on chromosomes 2A or 2D based on four *Triticum* reference genomes (*T. turgidum*, *T. dicoccoides*, *T. urartu*, and *Ae. tauschii*), whereas these genes were absent in Chinese Spring (Fig. 3). Amplifying and sequencing all ten genes (*TRITD2Av1G295090*, *TRITD2Av1G295160*, *TRITD2Av1G295250*, *TRIDC2AG082340*, *TRIDC2AG082360*, *TRIDC2AG082430*, *TuG1812G0200006348*, *TuG1812G0200006350*, *AET2Gv21292200*, *AET2Gv21292500*) in Zhongmai 175 and Avocet S indicated that these genes were also not present in Zhongmai 175 and Avocet S. Therefore, we inferred that Zhongmai 175 does not contain these gene sequences on chromosome arm 2AL.

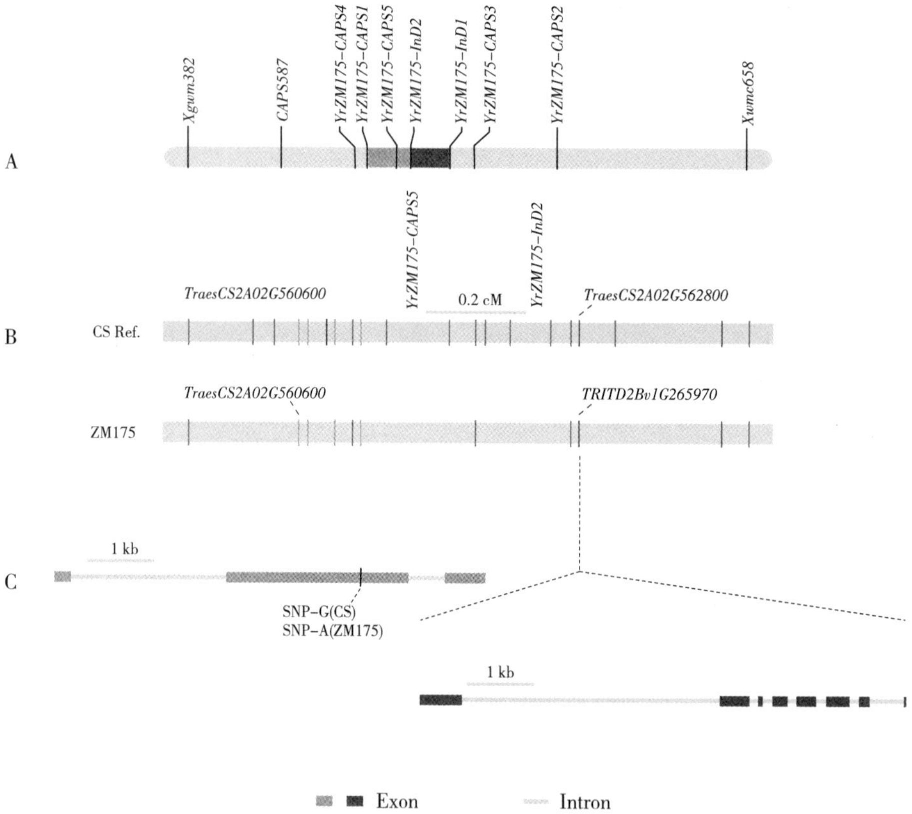

Fig. 2 Physical map of chromosome arm 2AL segments harboring *YrZM175* in Zhongmai 175 and Chinese Spring

A. Linkage map of *YrZM175*. The 0.72 cM genetic interval containing *YrZM175* is highlighted in red and the adjacent 0.69 cM genetic interval is highlighted in orange. **B.** Schematic of genes within the genomic region of ZM175 and CS. Orange rectangles represent four putative disease resistance genes *TraesCS2A02G560600*, *TraesCS2A02G560700*, *TraesCS2A02G560900* and *TraesCS2A02G561100* in the adjacent 0.69 cM genetic interval. Only *TraesCS2A02G560600* showed difference; a SNP in the CDS domain between Zhongmai 175 and Mingxian 169. The red rectangles represent two putative disease resistance genes *TraesCS2A02G562800* and *TraesCS2A02G563200* in the 0.72 cM (636.4 kb region, 763.45-764.08 Mb) genetic interval. The gene in red corresponds to *TRITD2Bv1G265970* in *T. turgidum*. Hollow rectangles indicate genes that are absent in Zhongmai 175 relative to CS. C. Gene structures of *TraesCS2A02G560600* and *TRITD2Bv1G265970*. *TraesCS2A02G560600* has three exons and two introns. *TRITD2Bv1G265970* has eight exons and seven introns. The black bar in *TraesCS2A02G560600* represents the SNP between Zhongmai 175 (SNP-A) and Mingxian 169 or Chinese Spring (SNP-G). ZM175: Zhongmai 175; CS: Chinese Spring.

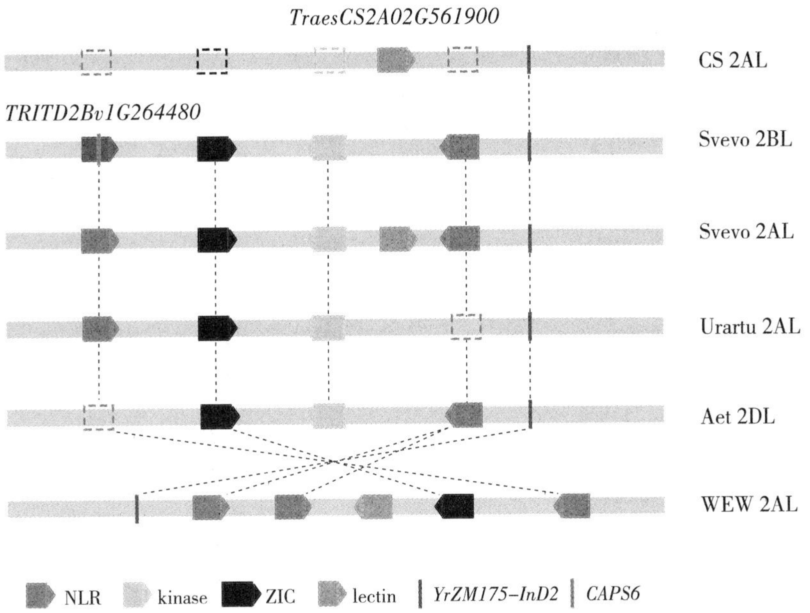

Fig. 3 Collinearity of genes in the target genomic interval

Genes upstream of marker *YrZM175-InD2* (red bar) in genomic regions collinear with chromosome arm 2AL of Chinese Spring, 2AL of Svevo, 2BL of Svevo, 2AL of *T. urartu*, 2DL of *Ae. tauschii* and 2AL of *T. dicoccoides*. Dotted boxes represent genes that are absent in Chinese Spring, *T. urartu* and *Ae. tauschii*. *YrZM175-InD2* is close to gene *TraesCS2A02G561900* in Chinese Spring. The red box indicates candidate gene *TRITD2Bv1G264480* on chromosome 2BL of *T. turgidum*, and the green bar within red box indicates the cosegregating marker *CAPS6*. CS: Chinese Spring; Aet: *Ae. tauschii*; WEW: *T. dicoccoides*

A putative disease-resistance-like gene isolated in the target interval of chromosome arm 2AL in Zhongmai 175 by genome walking was almost the same (with two SNPs in the intron) as a kinase gene *TRITD2Bv1G265970* on chromosome arm 2BL in *T. turgidum*. This indicated that chromosome arm 2AL in Zhongmai 175 might carry genes originating from chromosome arm 2BL in*T. turgidum*. Analysis of 11 putative disease resistance genes (Table S8) in the collinear interval of 2BL in *T. turgidum* in Zhongmai 175 and Mingxian 169 detected differential expression of *TRITD2Bv1G264470* and *TRITD2Bv1G264480*. The expression levels of *TRITD2Bv1G264470* were significantly higher at 5-14 dpi than at 0 dpi in both cultivars, with higher expression in Mingxian 169 (Fig. 4A). The expression of *TRITD2Bv1G264480* was first observed in Zhongmai 175 at 2 hpi, then significantly increased at 1-14 dpi, with constantly higher expression level than in Mingxian 169 (Fig. 4B). *TRITD2Bv1G264470* and *TRITD2Bv1G264480* are NBS-LRR genes. Amplifying parental lines with gene-specific primers (Table S9) indicated no sequence difference at CDS domain of *TRITD2Bv1G264470* between Zhongmai 175 and Mingxian 169 or Avocet S. The ORF sequence of *TRITD2Bv1G264480* was 23, 559 bp with a 5,661-bp CDS domain in *T. turgidum*. The homoeologous CDS sequences of *TRITD2Bv1G264480* were amplified in the target interval on chromosome arm 2AL in Zhongmai 175, Mingxian 169 and Avocet S (Fig. S8). There were 30 SNPs at CDS domain between Zhongmai 175 and Mingxian 169, resulting in 19 amino acid variations. The deletion of 3, 890 bp in Mingxian 169 led to a frameshift mutation (Fig. S9). Based on one SNP at CDS domain between Zhongmai 175 and Avocet S, we developed a molecular marker *CAPS6* (Table S2) and tested all susceptible F_2 plants. *CAPS6* co-segregated with the resistance gene. Therefore, we considered that *TRITD2Bv1G264480* was a candidate of *YrZM175*.

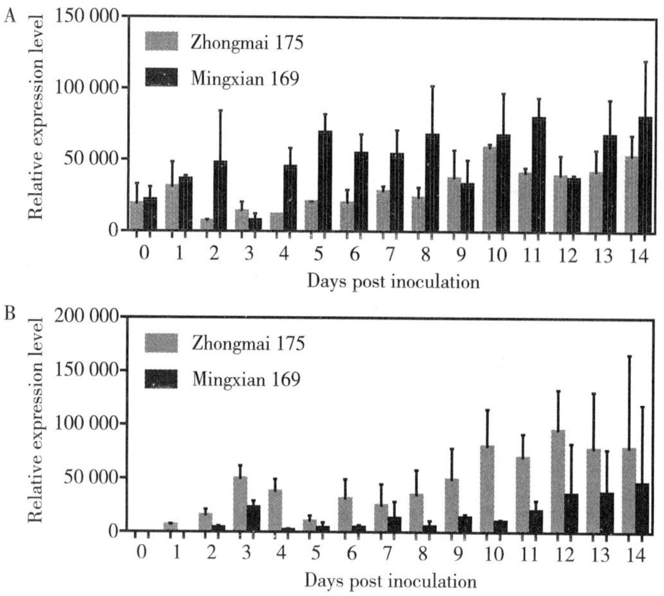

Fig. 4 Relative expression levels of *TRITD2Bv1G264470* and *TRITD2Bv1G264480* in Zhongmai 175 and Mingxian 169 at 0-14 days post inoculation

A. Relative expression level of *TRITD2Bv1G264470* in Zhongmai 175 and Mingxian 169 at 0-14 dpi. **B.** Relative expression level of *TRITD2Bv1G264480* in Zhongmai 175 and Mingxian 169 at 0-14 dpi. dpi: days post inoculation.

Discussion

Identification of candidate genes for *YrZM175*

The targeted region of *YrZM175* according to the Chinese Spring genome contained a cluster of putative resistance genes. After gene cloning and sequence analysis of six putative disease resistance genes in this interval, we performed transgenic assays of *TraesCS2A02G560600* using two transgenic vectors; one was the CDS domain driven by Ubi promoter and another was the 6,251-bp ORF (including coding and intron region) with a 2,770-bp presumed native promoter and 1,500 bp terminator (3'-UTR). Unfortunately, all the positive T_1 plants were susceptible to race CYR32 in seedling stage tests, indicating that *TraesCS2A02G560600* was not a candidate of *YrZM175*. We amplified five and two genes upstream and downstream, respectively, of *TraesCS2A02G562800* according to the Chinese Spring genome and again failed to obtain PCR products, indicating great sequence differences of this genomic region between Zhongmai 175 and the Chinese Spring reference genome. With the release of reference genomes such as *T. turgidum* ssp. *durum* cv. Svevo (Maccaferri et al. 2019), *T. aestivum* cv. Zang1817 (Guo et al. 2020), 10 + Genome (Walkowiak et al. 2020), and *T. aestivum* cv. Fielder (Sato et al. 2021) we were able to extend our investigation. We performed genome walking to obtain the unknown gene sequences in the 636.4-kb interval on chromosome arm 2AL of Zhongmai 175. Firstly, we cloned gene sequences at 763.83 Mb of chromosome arm 2AL in Zhongmai 175 referring to Chinese Spring genome to confirm that it was definitely from the genome sequence of chromosome arm 2AL. Then, we cloned the downstream gene sequences from 763.83 Mb step by step with genome-specific primers (Table S10) and determined the positions of amplified gene sequences based on Chinese Spring reference genome. Finally, after two steps of genome walking, we obtained a gene sequence that was almost the same as a genome sequence (*TRITD2Bv1G265970*) on chromosome arm 2BL of durum wheat Svevo. However, the linkage distance between *TRITD2Bv1G265970* and the resistance gene was 0.21 cM, indicating that *TRITD2Bv1G265970* was not the candidate of *YrZM175*.

We amplified and sequenced ten putative disease resistance genes in the collinear chromosome arm 2AL interval in five *Triticum* genomes in Zhongmai 175 and obtained no candidate gene sequences. Finally, we assessed *TRITD2Bv1G264480* on chromosome arm 2BL in *T. turgidum* as a candidate gene based on a co-segregating marker, sequence alignment of resistant and susceptible parents and gene expression patterns. The marker *CAPS6* for *TRITD2Bv1G264480* was co-segregating with *YrZM175* in the F_2 population with 7,325 plants (Fig. S10). *CAPS6* was a dominant marker that could not distinguish homozygous susceptible plants and heterozygous plants. We sequenced the *CAPS6* PCR products of 10 recombinant susceptible F_2 plants that were identified by the marker *YrZM175-InD2* with a genetic distance of 0.12 cM to *YrZM175*, and the result indicated that 10 plants contained the homozygous susceptibility allele at *TRITD2Bv1G264480* locus (Fig. S11). Our research indicated that Zhongmai 175 has complicated genetic origins that caused difficulties in mapping the target genes according to single reference genome. In addition to the Chinese Spring genome sequence, it is also very necessary to refer other available genome sequences of wheat and related species for fine mapping and cloning of disease resistance genes in bread wheat.

Candidate gene *TRITD2Bv1G264480*

The putative resistance gene *TRITD2Bv1G264480* encodes an NBS-LRR protein that belongs to two overlapping homologous superfamilies: P-loop containing nucleoside triphosphate hydrolase, and apoptotic protease-activating factors (helical domain) (http://www.ebi.ac.uk/inter pro/ entry/InterPro/IPR002182/). The predicted protein of *TRITD2Bv1G264480* is closest to *TRIDC2BG0 90350* on chromo- some 2B from *T. dicoccoides* (Fig. S12). The paralogous genes on chromosome 2A in genomes of seven bread wheat varieties clustered together phylogenetically (Fig. S12), with 89% and 66% sequence similarities in the CDS domain to *TRITD2Bv1G264480* on chromosome 2B and the paralogous gene *TRITD2Av1G294930* on chromosome 2A in *T. turgidum*, respectively, indicating much closer phylogenetic relationship with *TRITD2Bv1G264480* from chromosome 2B in *T. turgidum*, whereas the orthologous genes on chromosome 2B in *T. aestivum* cv. CDC Landmark, cv. CDC Stanley, cv. Fielder and cv. Julius showed 67.4% similarity with *TRITD2Bv1G264480*. In addition, paralogous genes on chromosomes 2A in *T. Urartu* and 2D in *Ae. tauschii* and orthologous genes on chromosome 2B in *T. spelta* also had quite distant phylogenetic *relationships with* *TRITD2Bv1G264480* (Fig. S12). Therefore, we inferred *TRITD2Bv1G264480* was transferred from chromosome 2B to 2A during evolution from an old crossover event of bread wheat. The CDS domain of candidate gene in Zhongmai 175 showed 97.8% and 96.8% similarities with *TRITD2Bv1G264480* (*T. turgidum*) and *TRIDC2BG090350* (*T. dicoccoides*), respectively, whereas the sequence similarities with homologues genes on chromosome 2A in other bread wheat varieties (Zang1817, Jagger, Fielder, Mace, CDC Landmark and Norin 61) were relatively low (all below 88.3%), indicating multiple allelic variations or haplotypes of the candidate gene present in bread wheat during evolution.

The relationship between *YrZM175* and other *Yr* genes on chromosome 2A

Zhongmai 175 has been widely grown in the North Winter Wheat Zone of China since 2008. It was derived from the cross between Jing 411 and BPM27 (Fig. S13). BPM27 is a disease-resistant line developed by Prof. Zuomin Yang at China Agricultural University in the 1980s; one of its parents was VPM1, a French line with 2NS translocation derived from crosses of *Ae. ventricosa*, *T. turgidum* L. var. *carthlicum* (*T. persicum*) and common wheat cv. Marne (Fig. S13). Bariana and McIntosh (1994) reported that the stripe rust resistance of VPM1 was conferred by *Yr17* derived from *Ae. Ventricosa*.

Lu et al. (2016) mapped *YrZM175* on chromosome arm 2AS using 344 F_2 plants and 147 $F_{2:3}$ lines derived from cross Lunxuan 987/Zhongmai 175 tested with CYR29. The present study used many more F_2 plants from a cross of Avocet S and Zhongmai 175 but employed race CYR32. The discrepancy of gene locations can be attributed to different mapping populations, molecular markers and PST races used in the two studies. Unfortunately, the PST race CYR29 used previously (Lu et al. 2016) was no longer available to test the present population, thus we are not very sure whether the resistance genes are the same or not. The race CYR32 is virulent to VPM1 (*Yr17*) (Table S11). Screening of the 1,793 susceptible F_2 plants in the Avocet S/Zhongmai 175 population with the marker URIC/LN2 (Helguera et al. 2003) as proxy for *Yr17* indicated linkage of 55.2 cM. *YrZM175* is located at the distal end (763,452,916-764,089,317bp) of chromosome arm 2AL, whereas *Yr17* is located at the distal (25.0-38.0 cM linkage interval) of chromosome arm 2AS (Cruz et al. 2016) (Fig. S14). Lu et al. (2016) performed multi-pathotype test on VPM1 (*Yr17*) and Zhongmai 175 (*Yr17* and *YrZM175*), the two lines showed different reactions to seven PST races at the seedling stage (Table S11). A test with the marker URIC/LN2 in 92 susceptible F_2 plants indicated that the linkage distance between *Yr17* and *YrZM175* was 39.0 cM (Lu et al. 2016), being similar to the present study.

To date, three *Yr* genes have been mapped on chromosome arm 2AL, i.e., *Yr1*, *Yr32* and *YrJ22* (Lupton and Macer 1962; Eriksen et al. 2004; Chen et al. 2016). *Yr1* is linked to *Xgwm382* (Fig. 5B), with a genetic distance of 5.6 cM (Zheng et al. 2017; Bansal et al. 2009). *Xgw m382* (at 772.96 Mb) was tested on all 1, 793 susceptible F_2 plants from Avocet S/Zhongmai 175; it was linked to *YrZM175* (763, 452, 916-764, 089, 317 bp) with a genetic distance of 5.1 cM (Fig. 5C). The EST-SSR marker *BU099658* (763.69 Mb) may accurately show the presence of *Yr1* (Hasancebi et al. 2014), which locates at almost the same physical position as *YrZM175*. Because Zhongmai 175 and Chinese 166 have different pedigrees and different responses to PST races (Lu et al. 2016; Chen et al. 2016) (Table S11), we believe *YrZM175* and *Yr1* are different. Based on the linkage maps and physical positions of linked markers, the physical position of *YrZM175* is close to *Yr1*, possibly indicating allelism. To clarify their relationship, tests of allelism and gene cloning should be carried out. *Xgwm382* is also closely linked to *YrJ22* with a genetic distance of 2.5 cM (Fig. 5D) (Chen et al. 2016). The physical position of *YrJ22* is at 768.0-769.0 Mb on 2AL (Personal communication with Dr. Can Chen), a 5 Mb physical distance from *YrZM175*. As Jimai 22 and Zhongmai 175 have different pedigrees and different arrays of reactions to PST races (Table S11), we assume that *YrZM175* and *YrJ22* are different genes or alleles.

Eriksen et al. (2004) found that the marker *Xcdo678* linked with *Yr32* in a genetic distance of 35 cM, while *Xcdo678* is co-segregating with *Pm4* (Ma et al. 1994). *Pm4* and *Yr1* are closely linked to each other with a genetic distance of 2.0 cM (McIntosh and Arts 1996). As a result, the genetic distance between *Yr32* and *Yr1* is about 33 cM (Fig. 5A) (Yang et al. 2019), indicating that *Yr32* is far away from *YrZM175* and *Yr1*. Based on linkage maps and molecular marker analyses, we conclude that *YrZM175* is different from *Yr1*, *Yr17*, *Yr32*, and *YrJ22*.

Complexity of the distal genome region of chromosome 2AL

Structural genomic variation such as presence-absence variations (PAVs) and copy number variations (CNVs) has been recognized to have the potential to generate phenotypic variation in maize (Springer et al. 2013) and barley (Muñoz-Amatriaín et al. 2013). Megabase-scale PAVs of *Tripsacum* origin were confirmed to be under selection during maize domestication and adaptation (Huang et al. 2021). Rimbert et al. (2018) also found high frequencies of PAVs in the distal regions of wheat chromosomes. Akhunov et al. (2003a, b) found that duplicated loci were most frequently located in the distal regions of chromosomes, and their distribution was positively correlated with recombination rate. The higher recombination rate indicates that these regions are fast-evolving in adapting to biotic and abiotic stresses. Besides, there are higher polymorphisms in the distal regions of chromosomes with the decreased levels of synteny between homoeologous chromosomes as distance from centromeres increase (Akhunov et al. 2003a, b). Our experimental results on chromosome arm 2AL in Zhongmai 175 were in agreement with these related reports mentioned above; this is reflected by the result that the candidate gene *TRITD2Bv1G264480* for *YrZM175* likely originated from chromosome arm 2BL of *T. turgidum*. The distal region of chromosome has great structural genomic variations and therefore is easier to evolve new alleles that trigger phenotypic variation.

In conclusion, *YrZM175* is a stripe rust resistance gene that confers moderate to high resistance to stripe rust in the field. We fine mapped *YrZM175* and identified a candidate gene based on a high-resolution linkage map, collinearity analysis and gene expression analysis. Transgenic assays are ongoing to validate the functions of candidate of *YrZM175*. The identification and fine mapping of *YrZM175* provide options for deployment in combination with other effective resistance genes.

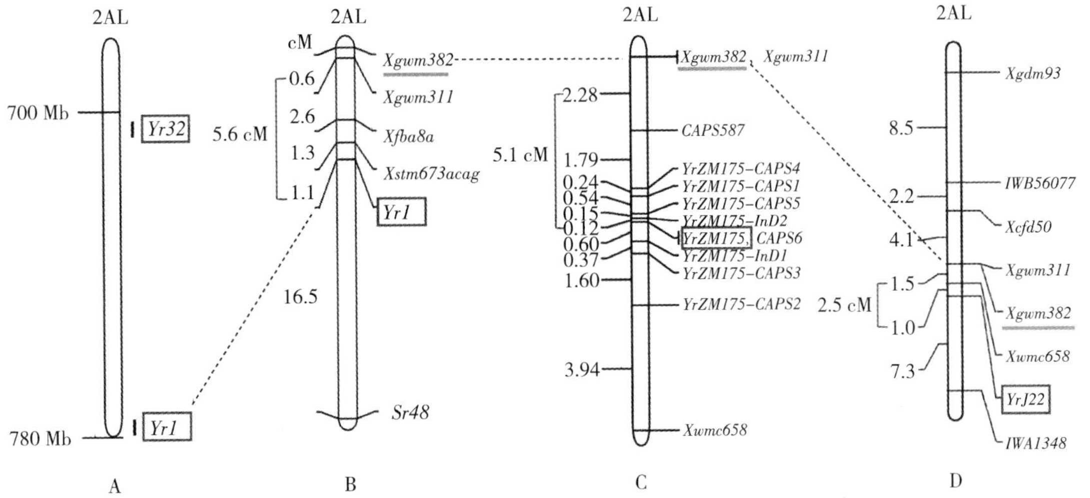

Fig. 5 Relationship between *YrZM175* and *Yr1*, *YrJ22* or *Yr32*

A. Physical positions of *Yr32* and *Yr1* (Yang et al. 2019). **B.** Linkage map of *Yr1* (Bansal et al. 2009). C. Linkage map of *YrZM175* in the present study. D. Linkage map of *YrJ22* (Chen et al. 2016). The SSR marker *Xgwm382* linked to three genes *Yr1*, *YrZM175* and *YrJ22* is underlined in orange. The red lines and numbers indicate genetic distances between *Xgwm382* and *Yr1*, *YrZM175*, or *YrJ22*.

Acknowledgements

We thank Prof. R. A. McIntosh, Plant Breeding Institute, University of Sydney, for critical review of this manuscript. This work was funded by the National Natural Science Foundation of China (31971929, 31961143007), the National Key Research and Development Program of China (2016YFD0101802, 2016YFE0108600), and CAAS Science and Technology Innovation Program.

References

Akhunov ED, Akhunova AR, Linkiewicz AM, Dubcovsky J, Hummel D, Lazo G, Chao SM, Anderson OD, David J, Qi LL, Echalier B, Gill BS, Miftahudin GJP, Rota ML, Sorrells ME, Zhang DS, Nguyen HT, Kalavacharla V, Hossain K, Kianian SF, Peng JH, Lapitan NLV, Wennerlind EJ, Nduati V, Anderson JA, Sidhu D, Gill KS, McGuire PE, Qualset CO, Dvorak J (2003a) Synteny perturbations between wheat homoeologous chromosomes caused by locus duplications and deletions correlate with recombination rates. Proc Natl Acad Sci U S A 100: 10836-10841.

Akhunov ED, Goodyear AW, Geng S, Qi LL, Echalier B, Gill BS, Miftahudin GJP, Lazo G, Chao SM, Anderson OD, Linkiewicz AM, Dubcovsky J, Rota ML, Sorrells ME, Zhang DS, Nguyen HT, Kalavacharla V, Hossain K, Kianian SF, Peng JH, Lapitan NLV, Gonzalez-Hernandez JL, Anderson JA, Choi DW, Close

TJ, Dilbirligi M, Gill KS, Walker-Simmons MK, Steber C, McGuire PE, Qualset CO, Dvorak J (2003b) The organization and rate of evolution of wheat genomes are correlated with recombination rates along chromosome arms. Genome Res 13: 753-763.

Avni R, Nave M, Barad O, Baruch K, Twardziok SO, Gundlach H, Hale I, Mascher M, Spannagl M, Wiebe K, Jordan KW, Golan G, Deek J, Ben-Zvi B, Ben-Zvi G, Himmelbach A, MacLachlan RP, Sharpe AG, Fritz A, Ben-David R, Budak H, Fahima T, Korol A, Faris JD, Hernandez A, Mikel MA, Levy AA, Steffenson B, Maccaferri M, Tuberosa R, Cattivelli L, Faccioli P, Ceriotti A, Kashkush K, Pourkheirandish M, Komatsuda T, Eilam T, Sela H, Sharon A, Ohad N, Chamovitz DA, Mayer KFX, Stein N, Ronen G, Peleg Z, Pozniak CJ, Akhunov ED, Distelfeld A (2017) Wild emmer genome architecture and diversity elucidate wheat evolution and domestication. Science 357: 93-97.

Bansal UK, Hayden MJ, Keller B, Wellings CR, Park RF, Bariana HS (2009) Relationship between wheat rust resistance genes *Yr1* and *Sr48* and a microsatellite marker. Plant Pathol 58: 1039-1043.

Bariana HS, McIntosh RA (1993) Cytogenetic studies in wheat XV Location of rust resistance in VPM1 and their genetic linkage with other disease resistance genes in chromosome 2A. Genome 36: 476-482.

Bariana HS, McIntosh RA (1994) Characterisation and origin of rust and powdery mildew resistance genes in VPM1 wheat. Euphytica 76: 53-61.

Chen XM, Kang ZS (2017) Chapter 7 - Stripe rust research and control: conclusions and perspectives. In: Chen XM, Kang ZS (eds) Stripe rust. Springer, Dordrecht, pp 601-630.

Chen C, He ZH, Lu JL, Li J, Ren Y, Ma CX, Xia XC (2016) Molecular mapping of stripe rust resistance gene *YrJ22* in Chinese wheat cultivar Jimai 22. Mol Breeding 36: 118.

Cruz CD, Peterson GL, Bockus WW, Kankanala P, Dubcovsky J, Jordan KW, Akhunov E, Chumley F, Baldelomar FD, Valent B (2016) The 2NS Translocation from *Aegilops ventricosa* Confers Resistance to the *Triticum* Pathotype of *Magnaporthe oryzae*. Crop Sci 56: 990-1000.

Eriksen L, Afshari F, Christainsen MJ, McIntosh RA, Jahoor A, Wellings CR (2004) *Yr32* for resistance to stripe (yellow) rust present in the wheat cultivar Carstens V. Theor Appl Genet 108: 567-575.

Feng JY, Wang MN, Deven RS, Chao SM, Zheng YL, Chen XM (2018) Characterization of Novel Gene *Yr79* and four additional quantitative trait loci for all-stage and high-temperature adult-plant resistance to stripe rust in spring wheat PI 182103. Phytopathology 108: 737-747.

Fu DL, Uauy C, Distelfeld A, Blechl A, Epstein L, Chen XM, Sela H, Fahima T, Dubcovsky J (2009) A Kinase-START gene confers temperature-dependent resistance to wheat stripe rust. Science 323: 1357-1360.

Gessese M, Bariana H, Wong D, Hayden M, Bansal U (2019) Molecular mapping of stripe rust resistance gene *Yr81* in a common wheat landrace Aus27430. Plant Dis 103: 1166-1171.

Guo WL, Xin MM, Wang ZH, Yao YY, Hu ZR, Song WJ, Yu KH, Chen YM, Wang XB, Guan PF, Appels R, Peng HR, Ni ZF, Sun QX (2020) Origin and adaptation to high altitude of Tibetan semi wild wheat. Nat Commun 11: 5085.

Hasancebi S, Mert Z, Ertugrul F, Akan K, Aydin Y, Senturk Akfirat F, Altinkut Uncuoglu A (2014) An EST-SSR marker, bu099658, and its potential use in breeding for yellow rust resistance in wheat. Czech J Genet Plant Breed 50: 11-18.

Helguera M, Khan IA, Kolmer J, Lijavetzky D, Li ZQ, Dubcovsky J (2003) PCR assays for cluster of rust resistance genes and their use to develop isogenic hard red spring wheat lines. Crop Sci 43: 1839-1847.

Huang C, Jiang YY, Li CG (2020) Analysis on the occurrence, damage and evolution of main wheat diseases and insect pests in my country from 1987 to 2018. Plant Prot 46: 186-193 (in Chinese with English abstract).

Huang YM, Huang W, Meng Z, Braz GT, Li YF, Wang K, Wang H, Lai JS, Jiang JM, Dong ZB, Jin WW (2021) Megabase-scale presence- absence variation with *Tripsacum* origin was under selection during maize domestication and adaptation. Genome Biol 22: 237.

International Wheat Genome Sequencing Consortium (IWGSC) (2018) Shifting the limits in wheat research and breeding using a fully annotated reference genome. Science 361: eaar7191.

Klymiuk V, Yaniv E, Huang L, Raats D, Fatiukha A, Chen SS, Feng LH, Frenkel Z, Krugman T, Lidzbarsky G, Chang W, Jääskeläinen M, Schudoma C, Paulin L, Laine P, Bariana H, Sela H, Saleem K, Sørensen CK, Hovmøller MS, Distelfeld A, Chalhoub B, Dubcovsky J, Korol AB, Schulman AH, Fahima T (2018) Cloning of the wheat *Yr15* resistance gene sheds light on the plant tandem kinase-pseudokinase family. Nat Commun 9: 3735.

Kosambi DD (1943) The estimation of map distances from recombination values. Ann Eugen 12: 172-175.

Krattinger SG, Lagudah ES, SpielmeyerW, Singh RP, Huerta-Espino J, McFadden H, Bossolini E, Selter LL, Keller B (2009) A putative ABC transporter confers durable resistance to multiple fungal pathogens in wheat. Science 323: 1360-1363.

Li H, Handsaker B, Wysoker A, Fennell T, Ruan J, Homer N, Marth G, Abecasis G, Durbin R, 1000 Genome Project Data Processing Subgroup (2009) The sequence alignment/map (SAM) format and SAMtools. Bioinformatics 25 (16): 2078-2079.

Li JB, Dundas L, Dong CM, Li GR, Trethowan R, Yang ZJ, Hoxha S, Zhang P (2020) Identification and characterization of a new stripe rust resistance gene *Yr83* on rye chromosome 6R in wheat. Theor Appl Genet 133: 1095-1107.

Lincoln SE, Daly MJ, Lander ES (1993) Constructing genetic linkage maps with Mapmaker/eXP3. 0. Whitehead Institute Techn rep, 3rd edn. Whitehead Institute, Cambridge.

Ling HQ, Ma B, Shi X, Liu H, Dong L, Sun H, Cao Y, Gao Q, Zheng S, Li Y, Yu Y, Du H, Qi M, Li Y, Lu H, Yu H, Cui Y, Wang N, Chen C, Wu H, Zhao Y, Zhang J, Li Y, Zhou W, Zhang B, Hu W, van Eijk MJT, Tang J, Witsenboer HMA, Zhao S, Li Z, Zhang A, Wang D, Liang C (2018) Genome sequence of the progenitor of wheat A subgenome *Triticum urartu*. Nature 557: 424-428.

Liu RH, Meng JL (2003) MapDraw: a microsoft excel macro for drawing genetic linkage maps based on given genetic linkage data. Hereditas 25: 317-321.

Liu W, Frick M, Huel R, Nykiforuk CL, Wang XM, Gaudet DA, Eudes F, Conner RL, Kuzyk A, Chen Q, Kang ZS, Laroche A (2014) The stripe rust resistance gene *Yr10* encodes an evolutionary- conserved and unique CC-NBS-LRR sequence in wheat. Mol Plant 7: 1740-1755.

Lu JL, Chen C, Liu P, He ZH, Xia XC (2016) Identification of a new stripe rust resistance gene in Chinese winter wheat Zhongmai 175. J Integr Agr 15: 2461-2468.

Luo MC, Gu YQ, Puiu D, Wang H, Twardziok SO, Deal KR, Huo N, Zhu T, Wang L, Wang Y, McGuire PE, Liu S, Long H, Ramasamy RK, Rodriguez JC, Van SL, Yuan L, Wang Z, Xia Z, Xiao L, Anderson OD, Ouyang S, Liang Y, Zimin AV, Pertea G, Qi P, Bennetzen JL, Dai X, Dawson MW, Müller HG, Kugler K, RivarolaDuarte L, Spannagl M, Mayer KFX, Lu FH, Bevan MW, Leroy P, Li P, You FM, Sun Q, Liu Z, Lyons E, Wicker T, Salzberg SL, Devos KM, Dvořák J (2017) Genome sequence of the progenitor of the wheat D genome *Aegilops tauschii*. Nature 551: 498-502.

Lupton FGH, Macer RCF (1962) Inheritance of resistance to yellow rust (*Puccinia glumarum* Erikss. & Henn.) in seven varieties of wheat. Trans Br Mycol Soc 45: 21-45.

Ma ZQ, Sorrells ME, Tanksley SD (1994) RFLP markers linked to powdery mildew resistance genes *Pm1*, *Pm2*, *Pm3*, and *Pm4* in wheat. Genome 37: 871-875.

Maccaferri M, Harris N, Twardziok SO, Pasam RK, Gundlach H, Spannagl M, Ormanbekova D, Lux T, Prade VM, Milner SG, Himmelbach A, Mascher M, Bagnaresi P, Faccioli P, Cozzi P, Lauria M, Lazzari B, Stella A, Manconi A, Gnocchi M, Moscatelli M, Avni R, Deek J, Biyiklioglu S, Frascaroli E, Corneti S, Salvi S, Sonnante G, Desiderio F, Marè C, Crosatti C, Mica E, Özkan H, Kilian B, Vita PD, Marone D, Joukhadar R, Mazzucotelli E, Nigro D, Gadaleta A, Chao S, Faris JD, Melo AT, Pumphrey M, Pecchioni N, Milanesi L, Wiebe K, Ens J, MacLachlan RP, Clarke JM, Sharpe AG, Koh CS, Liang KY, Taylor GJ, Knox R, Budak H, Mastrangelo AM, Xu SS, Stein N, Hale I, Distelfeld A, Hayden MJ, Tuberosa R, Walkowiak S, Mayer KF, Ceriotti A, Pozniak CJ, Cattivelli L (2019) Durum wheat genome highlights past domestication signatures and future improvement targets. Nat Genet 51: 885-895.

Marchal C, Zhang JP, Zhang P, Fenwick P, Steuernagel B, Adamski NM, Boyd L, Mclntosh R, Wulff BB, Berry S, Lagudah E, Uauy C (2018) BED-domain-containing immune receptors confer diverse resistance spectra to yellow rust. Nat Plants 4: 662-668.

McIntosh RA, Arts CJ (1996) Genetic linkage of the *Yr1* and *Pm4* genes for stripe rust and powdery mildew resistance in wheat. Euphytica 89: 401-403.

McIntosh RA, Dubcovsky J, Rogers WJ, Morris C, Xia XC (2017) Catalogue of gene symbols for wheat: 2017 Supplement. http: www. shigen. nig. ac. jp/wheat/komugi/genes/macgene/supplement2017. pdf.

McKenna A, Hanna M, Banks E, Sivachenko A, Cibulskis K, Kernytsky A, Garimella K, Altshuler D,

Gabriel S, Daly M, DePristo MA (2010) The genome analysis toolkit: a MapReduce framework for analyzing next-generation DNA sequencing data. Genome Res 20: 1297-1303.

Moore JW, Herrera-Foessel S, Lan CX, Schnippenkoetter W, Ayliffe M, Huerta-Espino J, Lillemo M, Viccars L, Milne R, Periyannan S, Kong XY, Spielmeyer W, Talbot M, Bariana H, Patrick JW, Dodds P, Singh R, Lagudah E (2015) A recently evolved hexose transporter variant confers resistance to multiple pathogens in wheat. Nat Genet 47: 1494-1498.

Muñoz-Amatriaín M, Eichten SR, Wicker T, RichmondTA, Mascher M, Steuernagel B, Scholz U, Ariyadasa R, Spannagl M, Nuss- baumer T, Mayer KFX, Taudien S, Platzer M, Jeddeloh JA, Springer NM, Muehlbauer GJ, Stein N (2013) Distribution, functional impact, and origin mechanisms of copy number variation in the barley genome. Genome Biol 14: R58.

Nsabiyera V, Bariana HS, Qureshi N, Wong D, Hayden MJ, Bansal UK (2018) Characterisation and mapping of adult plant stripe rust resistance in wheat accession Aus27284. Theor Appl Genet 131: 1459-1467.

Pakeerathan K, Bariana H, Qureshi N, Wong D, Hayden M, Bansal U (2019) Identification of a new source of stripe rust resistance *Yr82* in wheat. Theor Appl Genet 132: 3169-3176.

Rimbert H, Darrier B, Navarro J, Kitt J, Choulet F, Leveugle M, Duarte J, Rivière N, Eversole K, International Wheat Genome Sequencing Consortium (2018) High throughput SNP discovery and genotyping in hexaploid wheat. PloS One 13: e0186329.

Saghai-Maroof MA, Soliman KM, Jorgensen RA, Allard RW (1984) Ribosomal DNA spacer-length polymorphisms in barley: Mendelian inheritance, chromosomal location, and population dynamics. Proc Natl Acad Sci U S A 24: 8014-8018.

Sato K, Abe F, Mascher M, Haberer G, Gundlach H, Spannagl M, Shirasawa K, Isobe S (2021) Chromosome-scale genome assembly of the transformation-amenable common wheat cultivar 'Fielder.' DNA Res 28: dsab008.

Schmittgen TD, Livak KJ (2008) Analyzing real-time PCR data by the comparative CT method. Nat Protoc 3: 1101-1108.

Shewry PR (2009) Wheat. J Exp Bot 60 (6): 1537-1553.

Shewry PR, Hey SJ (2015) The contribution of wheat to human diet and health. Food Energy Secur 4 (3): 178-202.

Springer NM, Ying K, Fu Y, Ji TM, Yeh CT, Jia Y, Wu W, Richmond T, Kitzman J, Rosenbaum H, Iniguez AL, Barbazuk WB, Jeddeloh JA, Nettleton D, Schnable PS (2013) Maize inbreds exhibit high levels of copy number variation (CNV) and presence/absence variation (PAV) in genome content. PLoS Genet 5: e1000734.

ThielT, Kota R, Grosse I, Stein N, Graner A (2004) SNP2CAPS: a SNP and indel analysis tool for CAPS marker development. Nucleic Acids Res 32: e5.

Walkowiak S, Gao L, Monat C, Haberer G, Kassa MT, Brinton J, Ramirez-Gonzalez RH, Kolodziej MC, Delorean E, Thambugala D, Klymiuk V, Byrns B, Gundlach H, Bandi V, Siri JN, Nilsen K, Aquino C, Himmelbach A, Copetti D, Ban T, Venturini L, Bevan M, Clavijo B, Koo D-H, Ens J, Wiebe K, N'Diaye A, Fritz AK, Gutwin C, Fiebig A, Fosker C, Fu BX, Accinelli GG, Gardner KA, Fradgley N, Gutierrez-Gonzalez J, Halstead-Nussloch G, Hatakeyama M, Koh CS, Deek J, Costamagna AC, Fobert P, Heavens D, Kanamori H, Kawaura K, Kobayashi F, Krasileva K, Kuo T, McKenzie N, Murata K, Nabeka Y, Paape T, Padmarasu S, Percival-Alwyn L, Kagale S, Scholz U, Sese J, Juliana P, Singh R, Shimizu-Inatsugi R, Swarbreck D, Cockram J, Budak H, Tameshige T, Tanaka T, Tsuji H, Wright J, Wu J, Steuernagel B, Small I, Cloutier S, Keeble-Gagnère G, Muehlbauer G, Tibbets J, Nasuda S, Melonek J, Hucl PJ, Sharpe AG, Clark M, Legg E, Bharti A, Langridge P, Hall A, Uauy C, Mascher M, Krattinger SG, Handa H, Shimizu KK, Distelfeld A, Chalmers K, Keller B, Mayer KFX, Poland J, Stein N, McCartney CA, Spannagl M, Wicker T, Pozniak CJ (2020) Multiple wheat genomes reveal global variation in modern breeding. Nature 588: 277-283.

Wang H, Zou SH, Li YW, Lin FY, Tang DZ (2020) An ankyrin-repeat and WRKY-domain-containing immune receptor confers stripe rust resistance in wheat. Nat Commun 11: 1353.

Xia XC, Li ZF, Li GQ, He ZH, Singh RP (2007) Stripe rust resistance in Chinese bread wheat cultivars and lines. Wheat Production in Stressed Environments, pp. 77-82.

Yang MY, Li GR, Wan HS, Li LP, Li J, Yang WY, Pu ZJ, Yang ZJ, Yang EN (2019) Identification of QTLs for stripe rust resistance in a recombinant inbred line population. Int J Mol Sci 20: 3410.

Zhang CZ, Huang L, Zhang HF, Hao QQ, Lyu B, Wang MNL, Epstein L, Liu M, Kou CL, Qi J, Chen FJ, Li MK, Gao G, Ni F, Zhang LQ, Hao M, Wang JR, Chen XM, Luo MC, Zheng YL, Wu JJ, Liu DC, Fu DL (2019) An ancestral NB-LRR with duplicated 3' UTRs confers stripe rust resistance in wheat and barley. Nat Commun 10: 4023.

Zheng SG, Li YF, Lu L, Liu ZH, Zhang CH, Ao DH, Li LR, Zhang CY, Liu R, Luo CP, Wu Y, Zhang L (2017) Evaluating the contribution of *Yr* genes to stripe rust resistance breeding through markerassisted detection in wheat. Euphytica 213: 50.

麦瘟病研究进展与展望

何心尧[1]　郝元峰[2]　周益林[3]　何中虎[2,4]

([1] CIMMYT，Apdo. Postal 6-641，06600，Mexico，D. F.，Mexico；
[2] 中国农业科学院作物科学研究所，北京 100081；
[3] 中国农业科学院植物保护研究所，北京 100193；
[4] 国际玉米小麦改良中心（CIMMYT）中国办事处，北京 10008)

摘要：由 *Magnaporthe oryzae* 引起的麦瘟病是一种毁灭性小麦真菌病害，过去仅在南美流行，可造成10%～100%减产。2016年该病害首次在亚洲出现，给世界小麦生产带来重大潜在威胁。本文对麦瘟病病原生物学与病害流行学、小麦抗性材料筛选、麦瘟病的抗病性机制和综合治理等进行评述，并介绍了该领域国际合作研究的成功经验，以期为国内开展类似工作提供借鉴。尽管我国尚无麦瘟病报道，但南方部分地区为潜在适生区，异常气候可能会导致其大范围流行，因此需高度警惕。建议与国际麦瘟病协作网合作，尽快开展麦瘟病相关研究，建立对此病害的监测预警和防治关键技术储备体系，以保障我国小麦生产安全。

关键词：麦瘟病；*Magnaporthe oryzae*；病害综合治理

Progress and Perspective in Wheat Blast Research

HE Xin-Yao[1]，HAO Yuan-Feng[2]，ZHOU Yi-Lin[3]，and HE Zhong-Hu[2,4]

(*[1] CIMMYT，Apdo. Postal 6-641，06600，Mexico，D. F.，Mexico；[2] Institute of Crop Science，Chinese Academy of Agricultural Sciences (CAAS)，Beijing 100081，China；[3] Institute of Plant Protection，CAAS，Beijing 100193，China；[4] CIMMYT-China Office，c/o CAAS，Beijing 100081，China*)

Abstract: Wheat blast is an emerging deadly disease which can cause 10%-100% of yield losses. The disease primarily devastating only in South America，was first identified in Asia in 2016，posing a great potential threat to world wheat production. Here，we reviewed the features of wheat blast pathology and epidemiology，the resistance sources，the resistance mechanisms，and the integrated disease management strategies. The success of international collaboration on fighting the disease is also addressed，which provides proof of concept model that we can follow in our studies. Wheat blast has not been found

---

原文发表在《作物学报》，2017，43 (8)：1105-1114.

本研究由国家重点研发计划专项（2016YFE0108600）和中央级科研院所基本科研业务费专项（Y2017XM09）资助。

This research was supported by the National Key Research and Development Program of China (2016YFE0108600) and Core Funds for Public Scientific Institution (Y2017XM09).

通讯作者（Corresponding author）：何中虎，E-mail：zhhecaas@163.com

第一作者联系方式：E-mail：x.he@cgiar.org

in China, but we need to be very cautious and highly alert since there is a risk that the disease can outbreak in certain areas of southern China. The unusual weather pattern will put the situation even worse. To keep national wheat production safe, we need collaborate with International Wheat Blast Consortium to carry out necessary researches and build early warning and disaster preparedness systems to facilitate mitigation of the disease of high consequences and importance.

Keywords: Wheat blast; *Magnaporthe oryzae*; Integrated disease management

麦瘟病是由子囊菌 *Magnaporthe oryzae* 的 *Triticum* 致病型（*MoT*）引起的真菌病害，主要流行于南美热带地区[1]。最早于 1985 年在巴西 Parana 州发现[2]，随后在玻利维亚（1996 年）、巴拉圭（2002 年）及阿根廷东北部（2007 年）被报道，流行面积达 300 万 hm^2，病害减产率达 10%～100%，产量损失依年份、品种和播期不同而异[3]。1992 年以前，巴西主要种植高感品种，以 Anahuac 为例，产量损失为 11%～55%；随后种植了具有一定耐病性的品种，但是对控制病害损失没有取得显著成效，如 2005 年，在 2 次喷施杀菌剂的情况下，2 个主推品种的产量损失仍达到 14%～32%[4]。2009 年麦瘟病在巴西大流行导致很多地块绝产，引起国际社会广泛重视。同年，美国农业部资助的麦瘟病项目启动，由堪萨斯州立大学牵头，来自美国、巴西、玻利维亚、巴拉圭等国家的 10 余家单位参与。项目分两个阶段，第一阶段主要进行麦瘟病的基因组学诊断和抗性种质鉴定，第二阶段建立麦瘟病综合治理新策略，至 2016 年资助经费共计 650 万美元。2010 年 5 月 3 日至 5 日，国际玉米小麦改良中心（CIMMYT）和巴西农业研究院（EMBRAPA）在巴西 Passo Fundo 市召开“首届国际麦瘟病研讨会”，主题为麦瘟病——全球小麦生产的潜在威胁，来自 11 个国家的 50 余名代表参会。2011 年 5 月 18 日，美国肯塔基大学土壤学家 Lloyd Murdock 博士在该校普林斯顿试验基地偶然发现一例麦瘟病病穗，基因组序列结果显示其病原菌并非 *MoT*，而是 *M. oryzae* 的 *Lolium* 致病型（*MoL*），是由异常气候变化造成 *MoL* 从寄主黑麦草跳跃到小麦上引起的，这是该病害首次在南美以外的地区被发现[5]。

通过与南美流行区气候相似性比较，Duveiller 等[6]指出印度中部、孟加拉国、埃塞俄比亚等地区为该病潜在暴发区。曹学仁等[7]也进行过类似预测，认为我国云南、广东、海南等省的少数地区为该病害适宜发病区；并且彭居俐等[8]指出，随气候变暖，我国病害适宜区域可能会进一步扩大。Cruz 等[9]预测美国有 40% 的冬小麦种植区域适合麦瘟菌生存，25%的区域有麦瘟病暴发可能，其中路易斯安那、密西西比、佛罗里达 3 个州可能性最大。

2016 年 2 月，麦瘟病首次在孟加拉国被发现，随后迅速扩展，至 3 月发病面积已达 1.5 万 hm^2，验证了之前的预测。国际上有 2 个研究团队对此事件率先做出反应，第一个是由 CIMMYT、美国农业部外来病害杂草研究中心（马里兰）、肯塔基大学、堪萨斯州立大学、孟加拉国农业研究所小麦研究中心等组成的团队，通过对来自 3 个不同发病地区菌株（BdBar16-1、BdJes16-1 和 BdMeh16-1）的基因组测序分析，发现其与南美麦瘟病菌株有极高遗传相似性，证实孟加拉国发生病害为麦瘟病，为 *MoT* 致病型[10]；第二个团队由英国 The Sainsbury Laboratory 及来自孟加拉国、瑞士、法国、澳大利亚、巴西和德国的 14 个研究机构组成，他们利用被称为 Field Pathogenomics 的快速检测技术，对孟加拉国采集的病原菌和寄主混合样品进行转录组测序，并在 Open Wheat Blast 网站（http://www.wheatblast.net/）公布序列信息，最先确定病害为麦瘟病，菌株序列与巴西菌株相似，推测病原菌可能来自南美[11]。

在孟加拉国，小麦是仅次于水稻的重要主粮，但产量不及需求的三分之一，麦瘟病的流行进一步恶化了该国的小麦供给。麦瘟病在孟加拉国的暴发表明该病害可在南美以外气候适宜区域发生，这对周边的印度、巴基斯坦、尼泊尔及我国都是严重威胁，一旦大面积暴发，会严重影响这些地区的粮食安全，这也是国际社会做出快速反应的主要原因。

1 病原生物学与病害流行学

1.1 症状

麦瘟病症状可以表现在叶部和茎部，但以穗部为主，故归为穗部病害[1]。病原菌在穗部的侵染点为穗轴，并迅速造成侵染点维管组织坏死，使营养物质和水分无法运送至侵染点上方小穗，导致这些小穗在短时间内白化枯死[5]。被侵染的穗轴节片表现为褐色至黑色，侵染后期可见亮黑色孢子[5]。在开花期和籽粒发育早期的侵染可造成侵染点上方籽粒完全败育，在灌浆期轻度侵染籽粒表现为皱缩和低容重，高度侵染籽粒则完全干瘪，颜色暗淡，而发育后期的侵染对籽粒影响较小[11-12]。叶部感染一般只见于高感材料，初期表现为灰绿色水浸状斑点，随着病情发展，病斑扩大为淡黄褐色坏死斑[11,13]。

麦瘟病的穗部症状与赤霉病相似，尤其在较为干燥的条件下赤霉病亦可导致白化穗，但赤霉病的侵染点在小花而非穗轴，故可造成不连续小穗感染[5]。此外，赤霉病的典型特征是在颖壳上产生粉红色霉层，从而与麦瘟病较易区分[5]。这两种病害主要在巴西和巴拉圭同时发生，但麦瘟病偏好低纬度地区，而赤霉病多在较高纬度发生，所以两者共发区域并不大[12]。

1.2 病原菌

麦瘟病和稻瘟病的病原菌均为 *Magnaporthe oryzae*（无性世代为 *Pyricularia oryzae*），最初有人认为麦瘟病的病原菌来自稻瘟病菌，但随后研究表明两者是由不同致病型引起的[14]。根据不同寄主，*M. oryzae* 至少可分为4种致病型，分别是侵染水稻的 *Oryza* 致病型（*MoO*），侵染小麦、大麦和小黑麦的 *MoT*，侵染燕麦的 *Avena* 致病型（*MoA*）及侵染黑麦草的 *MoL*[1,15]。侵染马唐草的病原菌为 *M. grisea*，与 *M. oryzae* 亲缘关系较远[16]。最近，Castroagudín 等[15]根据分子分类学证据提出，麦瘟菌实际上由 *MoT* 和另一个被命名为 *Magnaporthe graminis-tritici* 的新种组成，但本文仍沿用传统命名法，将二者统称为 *MoT*。*MoT* 与其他致病型均为梨形三胞分生孢子，没有形态上的差异，只能通过分子标记区分[17]。

一般来说，某一致病型只侵染相应寄主，但也有交叉侵染，尤其是在温室试验中[18]。如温室接种条件下 *MoO* 可侵染小麦[15,18]，但没有 *MoT* 侵染水稻的报道，一般自然条件下不会发生这样的交叉侵染[1,15]。某些 *MoT* 菌株与 *MoO* 相似性较高，可能源自后者的寄主跳跃[20]；而大部分 *MoT* 菌株与 *MoO* 遗传相似度较低，可能为独立起源。*MoT* 表现很大的遗传多样性和进化速度，可能缘于其有性生殖或拟有性重组[4,19]，这给防治麦瘟病带来极大困难。另外，美国的 *MoL* 与巴西部分 *MoT* 具有较近亲缘关系，在温室试验中可以发生寄主交叉侵染[5,21-22]。在美国，*MoL* 和小麦存在季节和空间的隔离，其导致麦瘟病的可能性不大，但肯塔基州的麦瘟病病穗证明是由 *MoL* 在反常气候条件下导致的，这为气候变化背景下麦瘟病向非流行区蔓延提出了警示[5]。

1.3 病害流行条件

麦瘟病在田间的最适发病条件尚不清楚，迄今为止的大流行多与厄尔尼诺现象共发，其共性为开花期连阴雨，日均温度18～25℃，随后湿热和高光照[3]。控温和控湿试验发现，温、湿度对 *M. oryzae* 的产孢都有显著影响，但其互作不显著，最适产孢条件为相对湿度90%以上、温度28℃左右[23]。Cardoso 等[24]认为，麦瘟病的最适发病条件为25～30℃和长时间的湿润环境；如果保湿时间少于10h，则无论温度如何都不会发病。Ha 等[25]的研究结果表明，最适发病温度为26～32℃，最佳保湿时间为24h。为明确影响麦瘟病的各种环境因子及菌株间最佳发病条件差异，尚需做更多工作，从而准确预测病害的发生和采取必要的防治措施。

1.4 病害循环

由于自然条件下麦瘟病偶发，所以迄今为止对麦瘟病的病害循环所知甚少，最受关注的是病原菌初侵染源问题。在2009年巴西 Parana 州的大流行中，大片地块中播期一致的小麦同时均匀发病，表明有大量的风传病原菌在小麦发育的易感期集中侵染，但病菌来源尚不清楚[1]。自然条件下尚未在寄主上发现 *M. oryzae* 的有性世代，但在实验条件下一些菌株可进行有性杂交[26]。有一种假说认为病原菌在麦田周边的一种或多种杂草上越夏，巴西麦田周边的杂草种类很多，如 *Cenchrus echinatus* 、*Eleusine indic*、*Digitaria sanguinali*、*Brachiaria plantagine*、*Echinocloa crusgalli* 、*Pennisetum setosu* 、*Setaria geniculata*、*Hyparrhenia rufa* 以及 *Rhynchelytrum roseum* 等，但它们在麦瘟病流行中的作用尚未明确[3]。此外，与小麦轮作或者种植在相邻地块上的禾谷类作物也可能成为麦瘟病的寄主。Urashima 等[20]报道了一系列可被 *MoT* 侵染的禾谷类作物，包括大麦、黑麦、小黑麦、谷子、燕麦、玉米和高粱。Cruz 等[27]发现小麦基部衰老叶片上的 *MoT* 孢子数目远高于上部未衰老叶片，并且其产孢时期与抽穗期重叠，提出基部老叶片上孢子是穗部侵染的重要菌源。

麦瘟病也可通过种子传播，经商品粮进出口及种质交换等途径使远距离传播成为可能[19]。但一般认为种子传播在麦瘟病流行中的作用较小，侵染穗部的菌源主要来自其他寄主上的气传无性孢子。另外，考虑到子囊孢子在实验室条件下很容易产生，其在麦瘟病流行中的作用也许被低估[22]。Maciel 等[19]认为子囊孢子在提高 *MoT* 遗传多样性及其在病害扩散中起重要作用。

2 小麦抗性材料筛选

2.1 鉴定方法

麦瘟病鉴定分苗期叶片和成株期穗部鉴定两种。由于苗期温室鉴定成本低、周期短以及叶部症状更清晰易读和可重复性强，很多抗病鉴定研究以苗期叶片为主。但苗期叶片抗性与成株穗部抗性的相关性还存在争议，一些研究表明二者显著相关[28-29]，但也有结果显示二者无相关性[18-19,25]。Maciel 等[19]认为叶部和穗部的抗性没有必然联系，但由于大多数材料在这两个部位均表现为感病，导致两者在统计上达到显著相关的假象。因此，在以后的研究中应加强对成株穗部抗性的鉴定。

麦瘟病大田评价指标以严重度为主，即感病小穗占总小穗数的百分比；也有研究者以病穗率为指标，即单位面积内感病的穗数占总穗数的百分比。Maciel 等[30]建立了一套基于 ImageJ 软件的麦瘟病严重度评价体系，与人工评价的结果高度吻合（$0.82<r^2<0.90$），有助于高通量的表型鉴定。Arruda 等[28]在试验中发现病穗率和严重度高度相关（$r>0.9$），据此提出在麦瘟病评价体系中可用前者取代后者，因为病穗率的评价相对容易且更适合大面积评估。

麦瘟病鉴定的另一问题是小麦生育期对病害的影响。Duveiller 等[1]认为正是生育期的不同导致了一些材料在不同试验中表现出很大抗性差异，接种时间不合适会错过易感期，进而导致对其抗性水平的高估。因此在病害筛选试验中，建议针对生育期不同的材料进行分组接种，保证每一组内的材料都具有相似的生育期以避免其对抗性鉴定的干扰。

2.2 普通小麦中的抗源

抗性材料的鉴定和利用对麦瘟病流行区的小麦生产至关重要，对非流行麦区的预防作用也不容忽视[3]。早期研究报道过一些抗病或耐病推广品种，但这些材料在后续试验中却为感病，可能是由于小种变异引起的[12,31]。Urashima 等[32]和 Cruz 等[33]分别鉴定了 20 份和 70 份小麦材料对 72 个和 18 个菌株的苗期抗性，结果没有鉴定到对所有菌株都表现抗性的材料。Prestes 等[34]在温室对 85 份巴西小麦材料进行成株期鉴定，其中 18 份抗性较好，但没有发现免疫材料。Kohli 等[3]在其综述中推荐了一些稳定的中等抗性材料，包括 BH1146、BR18、IPR85、CD113、CNT8 和 OCEPAR 选系，可作为抗源在育种中应用。

鉴于该病害有可能在美国流行，美国农业部外来病害杂草研究中心（马里兰）和堪萨斯州立大学先后设立了生物安全第三等级实验室，引入巴西、玻利维亚、巴拉圭等南美麦瘟病流行地区的菌株到美国，以鉴定美国及引进小麦材料对麦瘟病的抗性。利用 1988 年采自巴西的 *MoT* 菌株 T-25 对 85 份美国材料进行鉴定，Cruz 等[29]发现了一些高抗材料（病害严重度低于 3%），如 Postrock、JackPot、Jagger、Santa Fe、Overley 和 Jagalene，其中前 4 份材料也具有很好的苗期抗性。据小麦遗传资源信息系统（GRIS，http：//wheatpedigree.net/）的信息，除 JackPot 外，其余 5 份材料系谱均含有 Jagger，说明 Jagger 为麦瘟病有效抗源。随后 Cruz 等[35]利用分子标记证明以上材料均为小麦-偏凸山羊草 2NS（2AS）易位系，且该易位系能显著降低麦瘟病发病程度，对 T-25 及当前毒力更强菌株 B-71（2012 年采自玻利维亚）均有效。2NS 易位系来自法国材料 VPM1，由偏凸山羊草与波斯小麦杂交后代与普通小麦 Marne 回交自交选育而成，在法国作为育种亲本被广泛使用。由于其白粉病抗性突出，VPM1 及其衍生材料通过国际白粉病鉴定圃发放到世界各地并得到广泛应用。Jagger 为美国硬红冬小麦，是近 20 年来美国中部大平原种植面积最大的品种，系谱为 KS82W418/Stephens，没有 VPM1，其 2NS 易位系来源尚待确认。在美国东部软红冬麦区，2NS 易位系在佐治亚大学的小麦育种中得到广泛应用，分子标记检测表明高达 86%的品种和高代品系含 2NS，源自佐治亚较早选育的 GA84202 和 GA881130 品系，系谱中均有 VPM1，可作为抗麦瘟病的种质材料（Jerry Johnson，个人交流，2016）。

在 CIMMYT 材料中，Milan 及其衍生品种在南美表现较好的麦瘟病抗性，已在玻利维亚、巴西和巴拉圭等国育种中应用[3]。分子标记检测显示 Milan 及其衍生品种 Kachu、Mutus 等均为 2NS 易位系，其中

Milan面包加工品质突出，Kachu为印度国家小麦区域试验的对照品种，在生产上均有重要利用价值。澳大利亚大面积推广品种Mace也含2NS易位片段，但对麦瘟病的抗性有待鉴定（Ravi Singh，个人交流，2016）。2NS易位系在我国应用成功的实例较少，分子标记检测只有中国农业科学院作物科学研究所育成的中麦175含有该易位系，推测其来自法国材料Fr81-4（未发表资料），中国科学院成都生物研究所育成的川育18、四川省农业科学院作物研究所育成的川麦36和川麦39等有Milan背景，是否含2NS需进一步确认。

全球气候变化可能会使麦瘟病的流行区域进一步扩大[1,7-8]，因此对于非流行区小麦材料进行标记检测和麦瘟病田间抗性鉴定很有必要。截至目前，田间鉴定过的材料仍以南美材料为主，也有部分美国材料，而对于其他地区小麦材料的抗性还知之甚少。CIMMYT与麦瘟病流行区的玻利维亚、巴西、巴拉圭等国以及美国堪萨斯州立大学合作，正在对其种质库收藏的材料进行鉴定，并配置了相关群体用于抗性遗传研究[1]。

2.3 小麦近缘种中的抗源

除已成功利用的2NS易位系外，对小麦近缘物种的抗源发掘工作尚未广泛开展。Cruz等[36]鉴定了20份人工合成小麦（由硬粒小麦或二粒小麦与粗山羊草杂交并经染色体加倍获得），发现其中5份材料具有较好抗性。Bockus等[37]对10份粗山羊草进行了麦瘟病抗性试验，发现其中3份的抗性接近抗病对照。

3 麦瘟病的抗病性机制

3.1 抗病性的遗传机制

3.1.1 非寄主抗性 小麦对非*MoT*类*M. oryzae*的抗性为非寄主抗性，遵循基因对基因假说，已发现的此类抗性基因包括*Rmg1*、*Rmg4*、*Rmg5*、*Rmg6*和*RmgTd*（*t*）。*Rmg1*亦称*Rwt4*，是小麦中发现的第1个*M. oryzae*抗性基因，在日本品种农林4号中发现并定位于1D染色体，特异性识别无毒基因*PWT4*。*Rmg1*主要表现对*MoA*的抗性，而对广泛存在的*MoO*和*MoT*不起作用，因为只在*MoA*中发现其对应的无毒基因*PWT4*[38]。Nga等[39]在4A和6D染色体上分别定位了*Rmg4*和*Rmg5*，特异性识别*M. oryzae*的*Digitaria*致病型（*MoD*）。由于能够感染*MoD*的小麦品种很罕见[39]，因此推测*Rmg4*和*Rmg5*在普通小麦中分布很普遍。Vy等[40]在1D染色体上发现了小麦对*MoL*的主效抗性基因*Rmg6*，特异识别*MoL*中的A1位点。Cumagun等[41]在野生二粒小麦的7B染色体上发现抗性基因*RmgTd*（*t*），特异识别*MoA*的无毒基因*PWT3*，产生过敏性坏死斑。

3.1.2 寄主抗性 小麦对*MoT*的抗性为寄主抗性，在某些情况下也遵循基因对基因假说，尤其在苗期，但不如在非寄主抗性中表现明显，可归因于部分抗性基因的存在[1,19]。迄今共发现4个寄主抗性基因，包括*Rmg2*、*Rmg3*、*Rmg7*和*Rmg8*。*Rmg2*和*Rmg3*从小麦品种Thatcher中发现，分别位于7A和6B染色体[42]。*Rmg7*源自于二粒小麦品系St17、St24和St25[43]，定位在2A染色体，其对应的无毒基因为*AVR-Rmg7*[44]。*Rmg8*来源于普通小麦品系S-615，位于2B染色体，对应于无毒基因*AVR-Rmg8*[44]。目前只在苗期验证了*Rmg2*和*Rmg3*的抗性，其在穗部的效应还有待研究[42]；而*Rmg7*和*Rmg8*则在苗期和成株期均表达抗性[44]。此外，已在生产上得到应用的2NS易位系具有较好的成株期抗性，可使穗部严重度平均降低64%～81%[37]。该易位片段携带*Lr37*、*Sr38*、*Yr17*、*Rkn3*、*Cre5*等多个抗性基因，其相关分子标记可用来初步筛选麦瘟病抗源[37]。

3.2 抗病性的细胞和生化机制

Tufan等[45]应用2个毒性小种和一个无毒小种分别接种小麦品种Renan，比较其在细胞和转录组水平上的变化。接种无毒小种和毒性小种均可导致以细胞壁增厚为代表的抗性反应。接种72h后，无毒小种可在接种点的表皮细胞内生成菌丝，但绝大多数无法侵染至邻近表皮和叶肉细胞；而毒性小种的菌丝在相同时间内可在接种点及其附近的多个细胞内生成大量菌丝。在转录组水平，无毒小种和毒性小种均可导致大规模的转录重编程，但两者诱导产生的转录本有很大不同，包括与细胞防御、新陈代谢、细胞运输和调控相关的基因及大量未知功能基因。

Debona等[46]比较了抗、感两种基因型在接种后的叶片中产生的不同化学物质，抗性材料中的超氧化物歧化酶（SOD）、过氧化物酶（POD）、抗坏血酸过氧化物酶（APX）、谷胱甘肽-S-转移酶（GST）、谷胱甘肽还原酶（GR）以及过氧化氢酶（CAT）浓度的增幅远大于感病材料，因而认为这些酶可及时降解由病菌侵染产生的活性氧进而避免或降低细胞损伤，从而表现出抗病反应。

4 麦瘟病综合治理

鉴于目前大面积推广品种多为感病，少数抗病品种抗源单一，对新出现菌株有丧失抗性的风险，仅靠品种本身抗性不足以将损失控制在可接受范围，生产上应实施综合治理策略，包括杀菌剂、化学调控、耕作措施和生物防治在内的多种措施，有望取得较好防治效果。

4.1 化学防治

自1985年麦瘟病在巴西暴发以来，研究人员一直积极寻找高效的防治药剂，多种杀菌剂被用于防治麦瘟病的试验，不同成分和配比防治效果差异很大[3-4,12,31,47-48]。这些试验多集中在开花期前后，杀菌剂喷施2～3次，间隔10～12d，防控效果在0～70%，尚无任何药剂或配方能达到100%的防效。综合不同试验结果，防病效果较好的杀菌剂有三唑类（戊唑醇或羟菌唑）、甲氧基丙烯酸酯类（QoI）、代森锰、三环唑等[47-48]。研究发现杀菌剂对具有一定抗性水平的品种防控效果较好，而对感病品种效果较差[3-4]。需要指出的是，长期依赖单一成分杀菌剂容易诱导病原菌抗药性，近期报道麦瘟病菌已对QoI类杀菌剂产生了抗药性[49-50]。多种杀菌剂轮换使用可防止或减缓病原菌抗性的产生。

除了杀菌剂成分、剂量和施用方法外，合适的喷施设备也至关重要。与叶部喷施杀菌剂不同，针对穗部病害需要特制喷头，并控制喷施角度才能取得较为理想的防治效果。在这方面可借鉴赤霉病的相关研究成果[51-52]。

尽管麦瘟病主要是一种穗部病害，针对叶部的杀菌剂喷施也可能有助于控制病害。基部衰老叶片上的*MoT*孢子是穗部麦瘟病的重要侵染源，应用5种配方的杀菌剂对叶片进行喷雾，均可显著降低叶片的菌量，降幅达62%～77%[27]。据此建议，在抽穗前对叶部增施一次杀菌剂[27]，以达到控制穗部侵染菌源的目的。

如前所述，带病籽粒是麦瘟病初侵染和远距离传播的重要因子，因此应用杀菌剂处理种子是一种有效的防控方法。与大田杀菌剂防控效果相比，种子处理通常可以达到很好的效果。Goulart和Paiva[53]报道了14种杀菌剂对种子处理的研究，均可以达到有效清除麦瘟菌的目的。

4.2 诱导抗病性

硅（Si）元素在植物抗病性中有重要作用。Xavier-Filha等[13]用硅灰石处理小麦，发现处理组较对照组的麦瘟病叶部症状减轻31%，且处理组的潜伏期延长28%。这种Si诱导的抗病性在随后的研究中得到验证[54-56]，其抗病机制可能是植物体内Si的增加可以激活谷胱甘肽还原酶（GR），有效清除细胞内活性氧而减缓病害症状[54]；同时，麦瘟菌菌丝在施用Si的叶片细胞内的生长受到显著抑制[57-59]，并且酚类和黄酮类物质在这些细胞内积累[58-59]。Cruz等[58]还发现，在施Si处理的小麦材料中，抗病相关基因的表达强度是对照的2～3倍。上述报道的研究对象都是麦瘟病在叶部的症状，Pagani等[47]首次报道了Si对于提高穗部抗性的有效性，但其效果受品种和环境效应影响显著。Cruz等[60]通过光学显微镜和扫描电镜观察，发现施用Si后穗轴上菌丝体的生长较对照受到显著抑制，为Si提高穗部抗性提供了进一步证据。

植物抗性诱导剂是一类化学物质的总称，其本身并无杀菌活性，但可诱导植物产生系统获得抗性。与化学杀菌剂相比，其主要优点是不会导致病原菌的抗药性。Cruz等[61]测试了茉莉酸、脱乙酸壳聚糖、硅酸钾、acibenzolar-S-methyl、乙烯利和磷酸钾对叶部麦瘟病的效果，发现只有磷酸钾和乙烯利表现抗病激活效应。Rios等[62]做了类似试验，发现喷施茉莉酸和乙烯利可有效诱导抗性，而acibenzolar-S-methyl的效果不显著。Pagani等[47]亦报道了磷酸钾对穗部麦瘟病的有效控制。

施肥对小麦抗麦瘟病也有影响。氮素诱导感病现象在小麦和水稻中普遍存在，过量施氮与麦瘟病和稻瘟病的发生紧密相关[63]，因此合适的施氮量有助于对麦瘟病的控制。Debona等[64]发现Mg施用量与叶部麦瘟病抗性呈负相关关系，指出这可能与Mg引起的Ca含量下调有关。

4.3 栽培防治

利用栽培措施防治麦瘟病需考虑品种布局、耕作模式、气候条件、农事操作习惯等多种因素，根据当地生产的具体情况，灵活采用关键技术，其中控制播期是一项最重要的措施之一。在巴西，早播材料通常感病更重，因其开花期前后的气候特别适宜麦瘟病发生，故应尽量避免在4月10日前播种[4,65]。在孟加拉国，根据2016年的发病情况，晚播材料比早播材料感病更重（Bahadur Meah，私人交流，2016），这一规律尚待证实。

轮作是控制病害的常用方法，但这种方法在南美对麦瘟病不大适用，因为与小麦轮作的作物如小黑麦、大麦、玉米、燕麦和谷子都是 *MoT* 的寄主，无法起到降低田间菌量的作用[3,20]。在孟加拉国，利用作物轮作来防治麦瘟病是否可行尚待研究。理论上，小麦播种前实施深耕和去除田间其他寄主也可减少田间菌量，进而降低小麦感染麦瘟病的风险[12]。然而，当前生产中广泛倡导保护性耕作，提倡免耕和少耕，与深耕防病策略相悖，并且去除田间其他寄主费工费力，很难大范围推广。

4.4 生物防治

迄今尚无生物防治措施应用于麦瘟病。考虑到 *MoT* 与 *MoO* 的相似性，可借鉴在稻瘟病上相关研究结果和成功经验，尝试利用对稻瘟病有效的生物制剂，例如无毒的 *MoO* 菌株，真菌类的 *Exserohilus monoceras*，拮抗细菌 *Pseudomonas fluorescens*、*Bacillus polymyxa*、*Bacillus lincheniformis*，以及链霉素类等[66-67]。首先应检测这些菌株或抗生素对麦瘟病菌的防控效果，然后进行田间试验，因生物防治效果受环境影响很大，只有在大面积田间试验中表现出显著效果后才可应用于生产。另外，*Bipolaris sorokiniana* 被报道是 *MoO* 的拮抗菌[66]，但也是小麦蠕孢叶枯病（spot blotch）的病原菌，这种病害同样在南美和南亚流行，因此该真菌不适合作为麦瘟病的生防菌。

5 展望

在过去三十余年，人们积累和总结了大量麦瘟病防控技术和经验，在一定程度上掌握了麦瘟病发病规律、寄主和病原菌互作和抗病性遗传机制，但仍未达到有效控制病害的目标。比如，现有抗源仍很有限，且抗性不高，在不同环境条件下表现也不稳定；可用于辅助选择的分子标记很少，在育种实践中利用十分有限；化学药剂防效尚不理想。与此同时，麦瘟菌仍在持续进化，其流行范围不断扩大，全球变暖增加了该病害在南美和南亚以外区域发生的可能，防治麦瘟病形势十分严峻。

麦瘟病在孟加拉国暴发后，国际社会给予高度关注，并采取了积极有效行动，包括4个方面。第一，反应迅速，措施得力。该病害2016年2月中旬发生，3月初在媒体报道，3月23日病样送达英国 The Sainsbury Laboratory，4月14日科学家证实病害为麦瘟病，4月28日国际著名期刊 *Nature* 发文报道该病害首次在亚洲出现[68]，5月 CIMMYT 在玻利维亚建立国际麦瘟病鉴定圃，并扩繁抗病品种供孟加拉国使用。病害发生后，孟加拉国政府对发病较轻地块喷施农药，严重地块焚烧秸秆以控制病原菌传播，并禁止发病区小麦留种。第二，多国跨学科协作。面对突发病情，有来自 CIMMYT、英国、美国、孟加拉国等国家或国际组织的科学家参加，涉及植物病理、育种、分子遗传、栽培及市场推广等多个学科，还有种子企业及政府人员参与。为便于种质、菌种等资源交换和信息交流，在 CIMMYT 推动下，国际麦瘟病协作网（International Wheat Blast Consortium）续签了为期5年（2016年7月4日至2021年7月3日）的合作协议，为联合防控麦瘟病提供框架指南。该协作网成立于2011年，共13个单位参加。第三，技术储备发挥重要作用。麦瘟病在南美出现以后，美国即引入巴西菌株用于研究，在美国农业部持续资助下，堪萨斯州立大学2009年开始建立生物安全实验室，专门对麦瘟病进行研究，发现2NS易位系抗麦瘟病，且已在育种中使用；黑麦草致病型 *MoL* 可侵染小麦；确定肯塔基州发现麦瘟病病穗是 *MoL* 寄主跳跃引起；发现粗山羊草等麦瘟病新抗源；模型预测美国东南部为麦瘟病易发区等。堪萨斯州立大学及其合作者8年的持续研究为全球抵抗麦瘟病做出巨大贡献。如果没有这些技术储备，孟加拉国麦瘟病大暴发极有可能引起世界恐慌。第四，宣传和培训。为普及麦瘟病知识，CIMMYT（http://www.cimmyt.org/wheat-blast/）和堪萨斯州立大学（https://www.k-state.edu/wheatblast/）分别建立麦瘟病信息网。2016年4月9日在巴西召开"第二届国际麦瘟病研讨会"，为控制麦瘟病在孟加拉国甚至更大范围蔓延开展广泛交流。7月26日至27日，CIMMYT 在尼泊尔组织召开了区域协商研讨会，提出防治麦瘟病的中长期行动计划。2017年2月4日至16日，由孟加拉国农业研究所及 CIMMYT 等共同组织"麦瘟病监测预警及防控"专题培训班，有来自孟加拉国、印度和尼泊尔的 40 余名学员参加。经国际同行共同努力，麦瘟病在孟加拉国得到了较好控制，虽然 2017 年也有发生，但规模较 2016 年已有明显下降。上述应对机制和措施非常值得我国学习和借鉴，对维护我国粮食作物安全生产和可持续发展有重要意义。

孟加拉国暴发麦瘟病也给我国小麦育种敲响了警钟。我们利用 2NS 特异分子标记对国内主要麦区的1 000多份品种和苗头品系进行检测，发现 2NS 频率极低（未发表资料），预测国内抗麦瘟病资源非常有限。鉴于中国还没有发现麦瘟病，国内科研单位独立研究非常困难，因此应尽早加入国际麦瘟病协作网，积极与 CIMMYT 等单位合作，尽快开展病害生物学及流行学、抗病品种筛选和培育及杀菌剂研制等方面研究，建立病害预测预警和防治关键技术储备体系。借鉴国际成功经验，建议开展以下六方面工作：（1）建立病原菌快速监测及预警系统，利用基因组学等手段及时侦察麦瘟菌可能通过风媒等传入中国。（2）成立生物安全实验室，主动引入麦瘟菌用于研究。（3）与麦瘟病流行地区合作，鉴定我国小麦种质资源，发掘新抗源和培育抗病品种，携带 2NS 的小麦品种对现在南美菌株的抗性比对 30 年前菌株的抗性已有所下降，未来有失去抗性的风险。（4）与抗赤霉病研究结合，开发防治穗部病害杀菌剂及喷施设备。（5）收集气象数据，预测厄尔尼诺发生年份及麦瘟病在国内发生的可能区域。（6）研究 *MoO* 及 *MoL* 在我国的分布及对小麦的交叉侵染。Islam 等[11]和 Farman 等[5]均发现黑麦草 *MoL* 菌株与部分南美麦瘟病流行菌株在遗传上有极大相似性，*MoL* 的广泛存在势必给小麦生产带来潜在威胁，一旦满足发病条件，即可发生寄主跳跃，侵染小麦，对此我们必须有所准备。培育抗病品种是防治麦瘟病最经济有效的方法，在抗源品种匮乏的现阶段，建议育种中加大对 Milan、Jagger、中麦 175 等品种的利用，从目前 2NS 广泛应用看，该易位片段应无不良连锁累赘；同时，需挖掘其他新抗源，并尽快导入到大面积推广品种中。

参考文献

References

[1] Duveiller E, He X, Singh P K. Wheat Blast: An emerging disease in South America potentially threatening wheat production. In: Bonjean A, van Ginkel M, eds. The World Wheat Book. Vol 3. Paris: Lavoisier, 2016. pp 1107-1122.

[2] Igarashi S, Utiamada C M, Igarashí L, Kazuma A H, Lopes R. *Pyricularia* em trigo: 1. Ocorrencia de *Pyricularla* sp. no estado do Parana. *Fitopatol Bras*, 1986, 11: 351-352.

[3] Kohli M M, Mehta Y R, Guzman E, De Viedma L, Cubilla L E. Pyricularia blast: a threat to wheat cultivation. *Czech J Genet Plant*, 2011, 47: S130-S134.

[4] Urashima A S, Grosso C RF, Stabili A, Freitas E G, Silva C P, Netto D C S, Franco I, Bottan J H M. Effect of Magnaporthe grisea on seed germination, yield and quality of wheat. Advances in Genetics, Genomics and Control of Rice Blast Disease: Springer, 2009. pp 267-277.

[5] Farman M, Peterson G L, Chen L, Starnes J H, Valent B, Bachi P, Murdock L, Hershman D E, Pedley K F, Fernandes J M C, Ba- varesco J. The *Lolium* pathotype of *Magnaporthe oryzae* recovered from a single blasted wheat plant in the United States. *Plant Dis*, 2017, 101: 684-692.

[6] Duveiller E, Hodson D, Sonder K, Tiedermann AV. An international perspective on wheat blast. *Phytopathology*, 2011, 101: S220.

[7] 曹学仁，陈林，周益林，段霞瑜．基于 MaxEnt 的麦瘟病在全球及中国的潜在分布区预测．植物保护，2011，37（3）：80-83.
Cao X R, Chen L, Zhou Y L, Duan X Y. Potential distribution of *Magnaporthe grisea* in China and the world, predicted by MaxEnt. *Plant Prot* (Beijing), 2011, 37 (3): 80-83. (in Chinese with English abstract)

[8] 彭居俐，周益林，何中虎．警惕麦瘟病全球扩散．麦类作物学报，2011，31：989-993.
Peng J L, Zhou Y L, He Z H. Global warning on the spread of wheat blast. *J Triticeae Crops*, 2011, 31: 989-993. (in Chinese with English abstract)

[9] Cruz C D, Magarey R D, Christie D N, Fowler G A, Fernandes J M, Bockus WW, Valent B, Stack J P. Climate suitability for *Magnaporthe oryzae-Triticum* pathotype in the United States. *Plant Dis*, 2016, 100: 1979-1987.

[10] Malaker P K, Barma N C, Tewari TP, Collis W J, Duveiller E, Singh P K, Joshi A K, Singh R P, Braun H J, Peterson G L. First report of wheat blast caused by *Magnaporthe oryzae* pathotype *Triticum* in Bangladesh. *Plant Dis*, 2016, 100: 2330.

[11] Islam M T, Croll D, Gladieux P, Soanes D M, Persoons A, Bhat- tacharjee P, Hossain M S, Gupta D R, Rahman M M, Mahboob M G, Cook N, Salam M U, Surovy M Z, Sancho V B, Maciel J L N, Júnior A N, Castroagudín V L, de Assis Reges J T, Ceresini P C, Ravel S, Kellner R, Fournier E, Tharreau D, Lebrun M H, McDonald B A, Stitt T, Swan D, Talbot N J, Saunders D G O, Win J, Kamoun S. Emergence of wheat blast in Bangladesh was caused by a South American lineage of *Magnaporthe oryzae*. *BMC Biol*, 2016, 14: 84.

[12] Igarashi S. Update on wheat blast (*Pyricularia oryzae*) in Brazil. In: Saunders D A, ed. Wheat for the Nontraditional Warm Areas: A Proceedings of the International Conference. Foz do Iguaçu, Brazil: CIMMYT, 1990. pp 480-485.

[13] Xavier-Filha M S, Rodrigues F A, Domiciano GP, Oliveira H V, Silveira P R, Moreira W R. Wheat resistance to leaf blast mediated by silicon. *Aust Plant Pathol*, 2011, 40: 28-38.

[14] Prabhu A, Filippi M, Castro N. Pathogenic variation among isolates of *Pyricularia oryzae* affecting rice, wheat, and grasses in Brazil. *Int J Pest Manag*, 1992, 38: 367-371.

[15] Castroagudín V L, Moreira S I, Pereira D A S, Moreira S S, Brunner P C, Maciel J L N, Crous PW, McDonald B A, Alves E, Ceresini P C. *Pyricularia graminis-tritici*, *a new Pyricularia* species causing wheat blast. Persoonia, 2016, 37: 199-216.

[16] Couch B C, Kohn L M. A multilocus gene genealogy concordant with host preference indicates segregation of a new species, *Magnaporthe oryzae*, from *M. grisea*. *Mycologia*, 2002, 94: 683-693.

[17] Pieck M L, Ruck A, Farman M, Peterson G L, Stack J P, Valent B, Pedley K F. Genomics-based marker discovery and diagnostic assay development for wheat blast. *Plant Dis*, 2017, 101: 103-109.

[18] Urashima A, Kato H. Pathogenic relationship between isolates of *Pyricularia grisea* of wheat and other hosts at different host development stages. *Fitopatol Bras*, 1998, 23: 30-35.

[19] Maciel J L N, Ceresini P C, Castroagudin V L, Zala M, Kema G H J, McDonald B A. Population structure and pathotype diversity of the wheat blast pathogen *Magnaporthe oryzae* 25 years after its emergence in Brazil. *Phytopathology*, 2014, 104: 95-107.

[20] Urashima A S, Stabili A, Galbieri R. DNA fingerprinting and sexual characterization revealed two distinct populations of *Magnaporthe grisea* in wheat blast from Brazil. *Czech J Genet Plant*, 2005, 41: 238-245.

[21] Farman M L. *Pyricularia grisea* isolates causing gray leaf spot on perennial ryegrass (*Lolium perenne*) in the United States: relationship to *P. grisea* isolates from other host plants. *Phytopathology*, 2002, 92: 245-254.

[22] Urashima A S, Igarashi S, Kato H. Host range, mating type, and fertility of *Pyricularia grisea* from wheat in Brazil. *Plant Dis*, 1993, 77: 1211-1216.

[23] Alves K JP, Fernandes J M C. Influence of temperature and relative air humidity on the sporulation of *Magnaporthe grisea* on wheat. *Fitopatol Bras*, 2006, 31: 579-584.

[24] Cardoso C A de A, Reis E M, Moreira E N. Development of a warning system for wheat blast caused by *Pyricularia grisea*. *Summa Phytopathol*, 2008, 34: 216-221.

[25] Ha X, Wei T, Koopmann B, von Tiedemann A. Microclimatic requirements for wheat blast (*Magnaporthe grisea*) and characterisation of resistance in wheat (meeting abstract). In: Tielkes E, ed. Resilience of Agricultural Systems against Crises. Tropentag, September 19-21, 2012, Göttingen-Kassel/Witzenhausen. p155.

[26] Tosa Y, Hirata K, Tamba H, Nakagawa S, Chuma I, Isobe C, Osue J, Urashima A, Don L, Kusaba M. Genetic constitution and pathogenicity of *Lolium* isolates of *Magnaporthe oryzae* in comparison with host species-specific pathotypes of the blast fungus. *Phytopathology*, 2004, 94: 454-462.

[27] Cruz C D, Kiyuna J, Bockus WW, Todd T C, Stack J P, Valent B. *Magnaporthe oryzae* conidia on basal wheat leaves as a potential source of wheat blast inoculum. *Plant Pathol*, 2015, 64: 1491-1498.

[28] Arruda M A, Bueno C R, Zamprogno K C, Lavorenti N A, Urashima A S. Reaction of wheat to *Magnaporthe grisea* at different stages of host development. *Fitopatol Bras*, 2005, 30: 121-126.

[29] Cruz C D, Bockus W W, Stack J P, Tang X, Valent B, Pedley K F, Peterson G L. Preliminary assessment of resistance among US wheat cultivars to the *Triticum* pathotype of *Magnaporthe oryzae*. *Plant Dis*, 2012, 96: 1501-1505.

[30] Maciel J L N, Danelli A L D, Boaretto C, Forcelini C A. Diagrammatic scale for the assessment of blast on wheat spikes. *Summa Phytopathol*, 2013, 39: 162-166.

[31] Urashima A S, Kato H. Varietal resistance and chemical control of wheat blast fungus. *Summa Phytopathol*, 1994, 20: 107-112.

[32] Urashima A S, Lavorent N A, Goulart A C, Mehta Y R. Resistance spectra of wheat cultivars and virulence diversity of *Magnaporthe grisea* isolates in Brazil. *Fitopatol Bras*, 2004, 29: 511-518.

[33] Cruz M F A, Maciel J L, Prestes A M, Bombonatto E A, Pereira J F, Consoli L. Molecular pattern and virulence of *Pyricularia grisea* isolates from wheat. *Trop Plant Pathol*, 2009, 34: 393-401.

[34] Prestes A, Arendt P, Fernandes J, Scheeren P. Resistance to *Magnaporthe grisea* among Brazilian wheat genotypes. In: Buck H T, Nisi J E, Salomon N, eds. Wheat Production in Stressed Environments. Springer, 2007. pp 119-123.

[35] Cruz C D, Peterson G L, Bockus WW, Kankanala P, Dubcovsky J, Jordan K W, Akhunov E, Chumley F, Baldelomar F D, Valent B. The 2NS translocation from *Aegilops ventricosa* confers resistance to the *Triticum* pathotype of *Magnaporthe oryzae*. *Crop Sci*, 2016, 56: 990-1000.

[36] Cruz M F A, Prestes A M, Maciel J L, Scheeren P L. Partial resistance to blast on common and synthetic wheat genotypes in seedling and in adult plant growth stages. *Trop Plant Pathol*, 2010, 35: 24-31.

[37] Bockus W, Cruz C, Kalia B, Gill B, Stack J, Pedley K, Peterson G, Valent B. Reaction of selected accessions of *Aegilops tauschii* to wheat blast, 2011. Plant Disease Management Reports No. 6, The American Phytopathological Society, St. Paul, MN, 2012, p CF005.

[38] Hau V T B, Hirata K, Murakami J, Nakayashiki H, Mayama S, Tosa Y. *Rwt4*, a wheat gene for resistance to *Avena* isolates of *Magnaporthe oryzae*, functions as a gene for resistance to *Panicum* isolates in Japan. *J Gen Plant Pathol*, 2007, 73: 22-28.

[39] Nga N T T, Hau V T B, Tosa Y. Identification of genes for resistance to a *Digitaria* isolate of *Magnaporthe grisea* in common wheat cultivars. *Genome*, 2009, 52: 801-809.

[40] Vy T T P, Hyon G S, Nga N T T, Inoue Y, Chuma I, Tosa Y. Genetic analysis of host-pathogen incompatibility between *Lolium* isolates of *Pyricularia oryzae* and wheat. *J Gen Plant Pathol*, 2014, 80: 59-65.

[41] Cumagun C J R, Anh V L, Vy T T P, Inoue Y, Asano H, Hyon G S, Chuma I, Tosa Y. Identification of a hidden resistance gene in tetraploid wheat using laboratory strains of *Pyricularia oryzae* produced by backcrossing. *Phytopathology*, 2014, 104: 634-640.

[42] Zhan S W, Mayama S, Tosa Y. Identification of two genes for resistance to *Triticum* isolates of *Magnaporthe oryzae* in wheat. *Genome*, 2008,

51: 216-221.

[43] Tagle A G, Chuma I, Tosa Y. *Rmg7*, a new gene for resistance to *Triticum* isolates of *Pyricularia oryzae* identified in tetraploid wheat. *Phytopathology*, 2015, 105: 495-499.

[44] Anh V L, Anh N T, Tagle A G, Vy T T P, Inoue Y, Takumi S, Chuma I, Tosa Y. *Rmg8*, a new gene for resistance to *Triticum* isolates of *Pyricularia oryzae* in hexaploid wheat. *Phytopathology*, 2015, 105: 1568-1572.

[45] Tufan H A, McGrann G R, Magusin A, Morel J B, Miché L, Boyd L A. Wheat blast: histopathology and transcriptome reprogramming in response to adapted and nonadapted *Magnaporthe isolates*. *New Phytol*, 2009, 184: 473-484.

[46] Debona D, Rodrigues F Á, Rios J A, Nascimento K J T. Bio-chemical changes in the leaves of wheat plants infected by *Pyricularia oryzae*. *Phytopathology*, 2012, 102: 1121-1129.

[47] Pagani A P S, Dianese A C, Café-Filho A C. Management of wheat blast with synthetic fungicides, partial resistance and silicate and phosphite minerals. *Phytoparasitica*, 2014, 42: 609-617.

[48] Rocha J R A S C, Pimentel A J B, Ribeiro G, de Souza M A. Efficiency of fungicides in wheat blast control. *Summa Phytopathol*, 2014, 40: 347-352.

[49] Castroagudín V L, Ceresini P C, de Oliveira S C, Reges J T A, Maciel J L N, Bonato A LV, Dorigan A F, McDonald B A. Resistance to QoI fungicides is widespread in Brazilian populations of the wheat blast pathogen *Magnaporthe oryzae*. *Phytopathology*, 2015, 105: 284-294.

[50] de Oliveira S C, Castroagudín V L, Nunes Maciel J L, Santos Pereira D A, Ceresini P C. Cross-resistance to QoI fungicides azoxystrobin and pyraclostrobin in the wheat blast pathogen *Pyricularia oryzae* in Brazil. *Summa Phytopathol*, 2015, 41: 298-304.

[51] Mesterhazy A, Toth B, Varga M, Bartok T, Szabo-Hever A, Farady L, Lehoczki-Krsjak S. Role of fungicides, application of nozzle types, and the resistance level of wheat varieties in the control of *Fusarium* head blight and deoxynivalenol. *Toxins*, 2011, 3: 1453-1483.

[52] Lehoczki-Krsjak S, Varga M, Mesterházy A. Distribution of prothioconazole and tebuconazole between wheat ears and flag leaves following fungicide spraying with different nozzle types at flowering. *Pest Manag Sci*, 2014, 71: 105-113.

[53] Goulart A C P, Paiva F A. Controle de *Pyricularia oryzae* e *Helminthosporium sativum* pelo tratamento de sementes de trigo com fungicidas. *Pesquisa Agropecuária Brasileira*, 1991, 26: 1983-1988.

[54] Debona D, Rodrigues F, Rios J A, Nascimento K, Silva L. The effect of silicon on antioxidant metabolism of wheat leaves infected by *Pyricularia oryzae*. *Plant Pathol*, 2014, 63: 581-589.

[55] Rios J A, Rodrigues F A, Debona D, Silva L C. Photosynthetic gas exchange in leaves of wheat plants supplied with silicon and infected with *Pyricularia oryzae*. *Acta Physiol Plant*, 2014, 36: 371-379.

[56] Perez C E A, Rodrigues F Á, Moreira W R, DaMatta F M. Leaf gas exchange and chlorophyll a fluorescence in wheat plants sup- plied with silicon and infected with *Pyricularia oryzae*. *Phytopathology*, 2014, 104: 143-149.

[57] Sousa R S, Rodrigues F A, Schurt D A, Souza N F A, Cruz M F A. Cytological aspects of the infection process of *Pyricularia oryzae* on leaves of wheat plants supplied with silicon. *Trop Plant Pathol*, 2013, 38: 472-477.

[58] Cruz M F A, Debona D, Rios J A, Barros E G, Rodrigues F A. Potentiation of defense-related gene expression by silicon in- creases wheat resistance to leaf blast. *Trop Plant Pathol*, 2015, 40: 394-400.

[59] da Silva W L, Cruz M F A, Fortunato A A, Rodrigues F Á. Histochemical aspects of wheat resistance to leaf blast mediated by silicon. *Sci Agric*, 2015, 72: 322-327.

[60] Cruz M F A, Silva L A F, Rios J A, Debona D, Rodrigues F Á. Microscopic aspects of the colonization of *Pyricularia oryzae* on the rachis

of wheat plants supplied with silicon. *Bragantia*, 2015, 74: 207-214.

[61] Cruz M F A, Diniz A P C, Rodrigues F A, de Barros E G. Foliar application of products on the reduction of blast severity on wheat. *Trop Plant Pathol*, 2011, 36: 424-428.

[62] Rios J A, Rodrigues F Á, Debona D, Resende R S, Moreira W R, Andrade C C L. Induction of resistance to *Pyricularia oryzae* in wheat by acibenzolar-S-methyl, ethylene and jasmonic acid. *Trop Plant Pathol*, 2014, 39: 224-233.

[63] Ballini E, Nguyen T T, Morel J B. Diversity and genetics of nitrogen-induced susceptibility to the blast fungus in rice and wheat. *Rice*, 2013, 6: 1-13.

[64] Debona D, Rios J A, Nascimento K J T, Silva L C, Rodrigues F A. Influence of *Magnesium* on physiological responses of wheat infected by *Pyricularia oryzae*. *Plant Pathol*, 2015, 65: 114-123.

[65] MehtaY,Riede C, Campos L, Kohli M. Integrated management of major wheat diseases in Brazil: an example for the Southern Cone region of Latin America. *Crop Prot*, 1992, 11: 517-524.

[66] Manandhar H K, Jørgensen H J L, Mathur S B, Smedegaard- Petersen V. Suppression of rice blast by preinoculation with avirulent *Pyricularia oryzae* and the nonrice pathogen *Bipolaris sorokiniana*. *Phytopathology*, 1998, 88: 735-739.

[67] Skamnioti P, Gurr S J. Against the grain: safeguarding rice from rice blast disease. *Trends Biotechnol*, 2009, 27: 141-150.

[68] Callaway E. Devastating wheat fungus appears in Asia for first time. *Nature*, 2016, 532: 421-422.

加工品质与营养特性

中国鲜面条耐煮特性及评价指标

张　艳[1]　阎　俊[2]　肖永贵[1]　王德森[1]　何中虎[1,3]

([1]中国农业科学院作物科学研究所/国家小麦改良中心，北京 100081；
[2]中国农业科学院棉花研究所，河南安阳 455112；
[3]国际玉米小麦改良中心（CIMMYT）中国办事处，北京 100081)

摘要：以我国北部和黄淮冬麦区的46份主栽小麦品种和育成品系为材料，分析了品质性状与煮熟面条冲洗水中总有机物含量（TOM）、干物质蒸煮损失率、面条吸水性和黏性等面条耐煮性指标的关系。结果表明，小麦品种的磨粉品质、面团流变学特性、淀粉品质及TOM值、蒸煮损失率和黏性等面条耐煮性指标存在较大变异。拉伸面积和最大抗延阻力与TOM值呈显著负相关，相关系数分别为−0.66（$P<0.01$）和−0.56（$P<0.01$）；稳定时间、拉伸面积和最大抗延阻力与面条煮6min和10min后鲜重的相关系数为−0.55～−0.63（$P<0.01$），耐揉指数与二者的相关系数分别为0.67（$P<0.01$）和0.69（$P<0.01$）；糊化温度与面条煮10min后鲜重呈极显著正相关（$r=0.60$，$P<0.01$），说明提高小麦面粉的蛋白质含量、面筋强度可以显著改善面条耐煮特性，蛋白质特性是影响面条耐煮性的主要品质因子，淀粉糊化参数对面条耐煮性也有一定影响。TOM值与面条煮6min和10min后鲜重呈显著正相关，相关系数分别为0.66（$P<0.01$）和0.69（$P<0.01$）；面条煮6min与煮10min后鲜重也呈高度正相关（$r=0.86$，$P<0.01$）。建议将10g鲜面条煮10min后的鲜重≤21.0g作为优质鲜面条耐煮性的主要评价指标。

关键词：普通小麦；面筋质量；面条耐煮特性；面条煮后鲜重

Characteristics and Evaluation Parameters Associated with Cooking Quality of Chinese Fresh Noodle

ZHANG Yan[1], YAN Jun[2], XIAO Yong-Gui[1], WANG De-Sen[1], and HE Zhong-Hu[1,3]

([1] *Institute of Crop Sciences / National Wheat Improvement Center, Chinese Academy of Agricultural Sciences, Beijing* 100081, *China*; [2] *Cotton Research Institute, Chinese Academy of Agricultural Sciences, Anyang* 455000, *China*; [3] *CIMMYT China Office, Beijing* 100081, *China*)

Abstract: Noodle cooking quality plays an important role in assessing processing quality of Chinese fresh noodle. Forty-six Chinese wheat cultivars and advanced lines from the Northern Plain and the Huang-Huai River Valleys Winter Wheat Regions were used to determine the relationship between wheat

---

原文发表在《作物学报》，2012，38（11）：2078—2085.
本研究由国家自然科学基金项目（31171547）和引进国际先进农业科学技术计划（948计划）重大国际合作项目（2011-G3）资助。
通讯作者（Corresponding author）：何中虎，E-mail：zhhecaas@163.com
第一作者联系方式：E-mail：zhycaas@yahoo.com.cn

quality characters and evaluation parameters of Chinese fresh noodle cooking quality including total organic matter (TOM), cooking losses, water sorption, and noodle stickiness. The results indicated that large variations were observed in milling quality, dough rheology characteristics, starch properties, and noodle cooking quality parameters including TOM value, cooking losses, and noodle stickiness. Extensogram energy and maximum resistance contributed negatively to TOM value, with correlation coefficients of −0.66 ($P<0.01$) and −0.56 ($P<0.01$), respectively. Correlation coefficients between Farinogram stability, Extensogram energy and maximum resistance and cooked noodle weights with optimal cooking (6 min) and overcooking (10 min) ranged from −0.55 to −0.63 ($P<0.01$). Farinogram mixing tolerance index was significantly and positively correlated with cooked noodle weights for 6 min and 10 min, with correlation coefficients of 0.67 ($P<0.01$) and 0.69 ($P<0.01$), respectively. Starch pasting temperature was significantly and positively correlated with cooked noodle weights for 10 min ($r=0.60$, $P<0.01$). This suggested that increased flour protein content and dough gluten strength contributed positively to noodle cooking quality, flour protein property was the major factor in determining noodle cooking quality, and noodle cooking quality was also affected slightly by starch pasting parameters. TOM value was significantly and positively correlated with cooked noodle weights for 6 min and 10 min, with correlation coefficients of 0.66 ($P<0.01$) and 0.69 ($P<0.01$), respectively. Correlation coefficient between cooked noodle weight with 6 min and 10 min cooking time was 0.86 ($P<0.01$). Therefore, it was recommended that cooked noodle weight for 10 min could be an important parameter for evaluation of noodle cooking quality. The cooked noodle weight for 10 min (10 g fresh noodle) should be no more than 21.0 g for good noodle cooking quality in Chinese wheat samples.

Keywords: Bread wheat; Gluten quality; Noodle cooking quality; Cooked noodle weight

面条是我国的传统食品，在东南亚、日本、朝鲜等地也广为消费，在欧美等国的消费量正迅速增加。面条的种类很多，主要分为干面条（又称挂面）、鲜面条和方便面。鲜面条口感好，是主要消费类型之一。近年来，国内对小麦品种的面粉特性和面条品质的关系等进行了深入研究，基本明确了面条遗传改良的选种指标[1-2]，开发并验证了主要选种指标的分子标记[3]，面条感官评价方法研究也有一定进展[4-6]。虽然已培育和推广了济麦 19、济麦 20 和豫麦 34 等优质面条品种，但面粉加工企业和消费者普遍反映用中国小麦品种制作的面条耐煮性差，主要表现在煮面时间稍长，韧性等口感品质下降、黏性增加，面汤变得黏稠；尤其是很多中国消费者喜欢食热汤面条，煮熟的面条在热汤中很快变软、丧失咬劲和弹性，严重影响面条的食用品质。因此进行耐煮性研究对全面改良中国面条品质具有重要意义。

在国内外有关面条耐煮性研究中，所用的评价指标有煮熟面条冲洗水中总有机物含量（TOM）值[7]、干物质蒸煮损失率、面条吸水性、面条的硬度和黏性等[8]。TOM 值和干物质蒸煮损失率越低，表示煮面过程落入面汤中和附着在面条表面的物质越少，面条耐煮性越好；煮后面条重量越低，表示面条的吸水性不强，过度煮面后面条仍然能保持一定的硬度和较低的黏性，耐煮性也较好。国外多数针对意大利通心面（Spaghetti）的研究表明，蛋白质含量和质量与 TOM 值、干物质蒸煮损失率及面条黏性呈显著负相关[8-10]，提高直链淀粉含量可以降低通心面的吸水性[11]。张玲等[12]用 TOM 值评价面条的煮面品质，认为蛋白质含量是影响 TOM 值的主要因子，赵振东等[13]也认为可以用 TOM 值鉴定中国面条的煮面品质。张国权等[14]认为小麦品种的籽粒硬度、出粉率、湿面筋含量和粉质仪吸水率与面条吸水性呈显著负相关，干物质蒸煮损失率与品质性状相关不显著。张剑等[15]发现小麦品种的沉降值、形成时间和稳定时间与干物质蒸煮损失率呈显著负相关，总淀粉含量对其也有一定影响，蛋白质品质是影响面条耐煮性的主要因子，但没有提出优质耐煮面条的选择范围。还有一些学者以商业面粉为材料，研究了破损淀粉和淀粉组分对干物质蒸煮损失率和面条吸水性的影响，认为破损淀粉率、直链和支链淀粉含量对干物质蒸煮损失率有显著影响[16-17]。

需要指出的是，上述研究存在两个突出问题，一是实验材料局限性大，样品数量偏少，代表性不强；二是没有明确上述耐煮性指标之间的关系，也没有给出具体的选择指标，导致育种家在品质改良中对耐煮性考虑不够，这是造成我国小麦品种面条耐煮性差的重要原因。由于TOM值、干物质蒸煮损失率和面条吸水性测定操作步骤和测试效率差异较大，因此迫切需要明确不同方法间的关系，向育种单位推荐快速准确的面条耐煮性评价方法。本文分析我国北部和黄淮冬麦区46份主栽小麦品种（系）的品质性状与TOM值、干物质损失率、面条吸水性和黏性等面条耐煮性指标的关系，进一步明确影响面条耐煮特性的品质因子及优质耐煮面条的选择指标，旨在为我国小麦面条品质的遗传改良提供理论依据。

1 材料与方法

1.1 试验材料与制粉

选用2009—2010年度种植于北京、河北、山东、河南和陕西的小麦样品46份（表1），这些样品多数为目前的主栽品种和一些育成的苗头品系，基本反映了我国主产区小麦生产和育种的现状。样品未受穗发芽和霉变影响，清理后测定籽粒含水量和硬度，调节水分含量，硬质小麦为16.5%，混合小麦为15.5%，软质小麦为14.5%，润麦20h，用Buhler MLU 202实验磨（Buhler Bros，Ltd，Uzwil，瑞士）按AACC 26-21A方法制粉，出粉率均为60%。制取60%出粉率的方法是首先计算60%出粉率的面粉总重量，再按皮1-心1-皮2-心2-皮3-心3的粉路顺序依次称取面粉，前一道粉路的面粉全部取完再取下一道粉路的面粉，直至取到60%出粉率所需的面粉重量。

1.2 小麦品质性状测定方法

用单籽粒谷物硬度仪（SKCS 4100，Perten Instruments AB，瑞典）测定籽粒硬度，该值越大表示硬度越大。用近红外分析仪（Foss 1241，瑞典）测定籽粒和面粉蛋白质含量。分别按AACC方法08-01、54-21和54-10测定面粉灰分、粉质仪和拉伸仪参数。用快速黏度测试仪（RVA，Super 3，Newport Scientific，澳大利亚），参照Batey等[18]方法测定峰值黏度和稀澥值等参数。

1.3 面条制作与感官评价

面条制作时实验室温度为（22±2)℃，相对湿度50%～60%。按照叶一力等[19]的方法进行面条制作和感官评价。

1.4 面条耐煮性指标测定

1.4.1 TOM值

将100g压好的面条放在盛有1 000mL沸腾蒸馏水的锅中煮面6min，之后按张玲等[12]介绍的方法继续操作。

1.4.2 干物质蒸煮损失率 参照张剑等[15]方法，并略作修改。将测定TOM值（100g面条）煮面所剩的面汤回收，测量其体积（mL）。用玻璃棒搅拌面汤使干物质均匀分布在汤中，然后立即取5mL放入已称重的玻璃培养皿中，在105℃烘箱中烘至完全干燥。称重，计算蒸煮损失率。

1.4.3 面条吸水性 参照Matsuo等[20]方法，并略作修改。称取7份10g面条，分别放入小型不锈钢漏篮中，将7个漏篮放在盛有2 000mL沸水的大锅中煮面4、5、6、7、8、9和10min。捞出面条后静置5min（此时面条表面的水分完全沥干），称重。用面条煮后的鲜重表示其吸水性。

1.4.4 面条黏性 在面条感观评价的同时，用TA-XT2i型质构仪（Stable Micro System，英国）按TPA程序测定面条黏性。

1.5 统计分析

用SAS 8e（Statistical Analysis System）进行基本统计量计算、相关和多重比较分析。采用*t*测验方法检验显著性。

2 结果与分析

2.1 品质性状

供试小麦样品的磨粉品质、面团流变学特性和淀粉品质的多数性状都存在较大变异（表2），其中籽粒蛋白质含量和粉质仪吸水率变幅较小，变异系数小于10%；形成时间、稳定时间、耐揉指数、拉伸面积和最大抗延阻力变幅最大，其变异系数在60.7%～85.8%之间；籽粒硬度及淀粉糊化参数低谷黏度、稀澥值和最终黏度也存在较大变幅，变异系数均大于20%，说明供试样品具有很好的代表性。

供试小麦样品的面条加工品质性状也存在一定变异，面条耐煮性指标TOM值、蒸煮损失率、黏性和面条感官色泽的变幅较大，变异系数均大于15%，其中TOM值的变异系数最大，为54.2%。面条吸水性参数煮面6min和10min面条重量及感官评价指标表观状况、软硬度、黏弹性、光滑性、食味和总分的变幅较小，变异系数小于10%（表3）。总体来看，供试小麦品种面条加工品质的变异小于磨粉、蛋白质和淀粉等品质性状的变异。

2.2 面条吸水性分析

煮面时间从4、5、6、7、8、9到10min，所有样品的面条煮后鲜重都依次增加（数据未列出），说明随着煮面时间延长，面条的吸水性增加，耐煮性变差。但不同样品在煮面时间不足（4min）、适宜（6min）和过度（10min）阶段所表现的吸水性存在较大差异，选择吸水性最弱（西农979）和最强（临麦4号）及变化趋势不同的2个样品（太空6号和周麦23），分析其面条煮不同时间后的鲜重和吸水性的变化（表4）。临麦4号在煮4、6和10min后的面条鲜重最高，分别为18.2、19.9和23.5g，显著高于其他3个样品，说明临麦4号的面条吸水性高，耐煮特性较差。太空6号和周麦23在煮4min后面条鲜重没有显著差异，分别为17.1g和17.0g，在煮6min后面条鲜重也相近，分别为18.6g和18.5g，但在煮10min后面条鲜重（22.2g和20.2g）达5%显著水平，说明在最适宜煮面时间内太空6号和周麦23的面条吸水性相近、耐煮特性基本一致，但随着煮面时间延长，太空6号的耐煮性明显劣于周麦23。西农979在煮面的3个时间段面条鲜重都最低，分别为16.2、17.7和19.8g，表明其面条耐煮性最好。由此可见，西农979和周麦23的面条耐煮性优于太空6号和临麦4号。

以面条煮10min后鲜重<21.0g、21.0～22.0g和>22.0g为标准，将46份样品分为3类。第1类为面条耐煮性优质类，包括西农979、周麦23、石B05-7338、中优335、中麦349、汶农14、冀师02-1、郑麦9023、舜麦1718、中优206、10CA11、新麦18、西农9871、北京0045、10CA006、中麦895、山农055843、中优629和济麦20共19个样品，其中只有汶农14属于弱筋小麦，中麦349、北京0045和中麦895三个样品属于中筋小麦，其他样品均为强筋小麦；第2类为面条耐煮性中等类，含有17个样品，其中8个属于弱筋小麦，6个属于中筋小麦，3个属于强筋小麦；第3类为面条耐煮性较差类，共有10个样品，其中7个属于弱筋小麦，2个属于中筋小麦，1个属于强筋小麦。面条耐煮性优质、中等和较差类样品面条煮6min后鲜重的平均值分别为17.9、18.5和19.5g，面条煮10min后鲜重的平均值分别为20.6、21.4和22.6g，差异均达到1%的显著水平。总体来看，面筋质量好的小麦制作的面条耐煮性好。

2.3 面条耐煮性指标比较

面条耐煮性指标TOM值与面条煮6min和10min后鲜重呈显著正相关（图1-A），相关系数分别为0.66（$P<0.01$）和0.69（$P<0.01$）；面条煮6min后鲜重与煮10min后鲜重也呈高度正相关（图1-B），$r=0.86$（$P<0.01$）；但干物质蒸煮损失率和面条黏性与TOM值和面条煮后鲜重的相关不显著。其原因可能是TOM值和面条煮后鲜重主要受面筋强度及部分淀粉糊化特性的影响，而影响干物质蒸煮损失率和面条黏性的因子是蛋白质含量和面筋的延展性。TOM值的测定需要经过煮面、煮后面条冲洗、冲洗液干燥、用试剂溶解和发生化学反应及中和滴定等步骤，操作复杂、费时费力；面条煮后鲜重只需要将煮熟面条表面的水分沥干称重即可，操作简单、快速，用面条煮10min后鲜重评价面条耐煮特性比面条煮6min后鲜重更有效；干物质蒸煮损失率则需要把面汤中的水分烘干或通过冷冻干燥去除水分，耗费时间较长；面条黏性一般用质构仪测定，但此设备价格昂贵，目前许多单位无法购置。综上所述，建议以面条煮10min后鲜重为面条耐煮特性评价主要指标。以面条耐煮性优质类最低值为标准，推荐10g优质耐煮样品面条煮10min后鲜重≤21.0g。

表1 供试小麦品种

Table 1 Name and origin of tested wheat cultivars

编号 Code	品种名称 Cultivar	来源地 Origin	编号 Code	品种名称 Cultivar	来源地 Origin
1	北京0045 Beijing 0045	北京 Beijing	3	中麦415 Zhongmai 415	北京 Beijing
2	中麦175 Zhongmai 175	北京 Beijing	4	中麦548 Zhongmai 548	北京 Beijing

（续）

编号 Code	品种名称 Cultivar	来源地 Origin	编号 Code	品种名称 Cultivar	来源地 Origin
5	中优 206 Zhongyou 206	北京 Beijing	26	烟农 23 Yannong 23	山东 Shandong
6	中优 335 Zhongyou 335	北京 Beijing	27	临麦 4 号 Linmai 4	山东 Shandong
7	中优 629 Zhongyou 629	北京 Beijing	28	缁麦 12 Zimai 12	山东 Shandong
8	轮选 987 Lunxuan 987	北京 Beijing	29	潍麦 8 号 Weimai 8	山东 Shandong
9	石 05-7338 Shi 05-7338	河北 Hebei	30	汶农 14 Wennong 14	山东 Shandong
10	冀师 02-1 Jishi 02-1	河北 Hebei	31	良星 66 Liangxing 66	山东 Shandong
11	邯 6172 Han 6172	河北 Hebei	32	中麦 155 Zhongmai 155	山东 Shandong
12	舜 1718 Shun 1718	山西 Shanxi	33	中麦 349 Zhongmai 349	河南 Henan
13	泰山 21 Taishan 21	山东 Shandong	34	中麦 895 Zhongmai 895	河南 Henan
14	泰山 23 Taishan 23	山东 Shandong	35	08CA101	河南 Henan
15	泰山 24 Taishan 24	山东 Shandong	36	10CA006	河南 Henan
16	泰山 223 Taishan 223	山东 Shandong	37	10CA11	河南 Henan
17	泰农 2413 Tainong 2413	山东 Shandong	38	郑麦 366 Zhengmai 366	河南 Henan
18	济南 17 Jinan 17	山东 Shandong	39	郑 9023 Zheng 9023	河南 Henan
19	济麦 19 Jimai19	山东 Shandong	40	太空 6 号 Taikong 6	河南 Henan
20	济麦 20 Jimai 20	山东 Shandong	41	矮抗 58 Aikang 58	河南 Henan
21	济麦 22 Jimai 22	山东 Shandong	42	周麦 18 Zhoumai 18	河南 Henan
22	济宁 16 Jining 16	山东 Shandong	43	周麦 23 Zhoumai 23	河南 Henan
23	鲁麦 23 Lumai 23	山东 Shandong	44	新麦 18 Xinmai 18	河南 Henan
24	山农 055843 Shannong 055843	山东 Shandong	45	西农 979 Xinong 979	陕西 Shaanxi
25	烟农 19 Yannong 19	山东 Shandong	46	西农 9871 Xinong 9871	陕西 Shaanxi

表 2　小麦品质性状参数的平均值、变幅和变异系数

Table 2　Mean, range, and coefficients of variance for quality parameters among tested samples

类型 Type	变量 Parameter	均值 Mean	变幅 Range	变异系数 *CV*（%）
磨粉品质	籽粒硬度 Grain hardness	54.3	9.4～76.0	33.9
Milling quality	籽粒蛋白质含量 Grain protein content（14% MB，%）	13.1	11.4～15.4	7.4
	面粉灰分 Flour ash（%）	0.46	0.38～0.60	11.5
面团流变学特性	吸水率 Water absorption（%）	64.5	53.9～72.5	6.4
Dough rheology	形成时间 Development time（min）	4.4	1.3～19.0	70.8
	稳定时间 Stability（min）	6.6	1.1～27.0	85.8
	耐揉指数 Mixing tolerance index（BU）	63.9	4.0～195.0	66.1
	拉伸面积 Energy（cm^2）	57.3	2.7～158.2	60.7
	延展性 Extensibility（mm）	165.3	64.1～234.0	16.3
	最大抗延阻力 Maximum resistance（BU）	254.3	23.9～722.0	64.9

（续）

类型 Type	变量 Parameter	均值 Mean	变幅 Range	变异系数 *CV*（%）
淀粉糊化特性 Starch pasting property	峰值黏度 Peak viscosity（RVU）	166.9	82.3～230.1	17.9
	低估黏度 Trough（RVU）	127.5	35.3～189.2	23.2
	稀懈值 Breakdown（RVU）	39.5	20.8～61.8	22.0
	最终黏度 Final viscosity（RVU）	194.7	74.3～270.8	20.9
	糊化温度 Pasting temperature（℃）	73.3	66.8～89.8	10.9

表3 面条加工品质性状参数的平均值、变幅和变异系数

Table 3 Mean, range, and coefficients of variance for processing parameters of noodles

类型 Type	变量 Parameter	均值 Mean	变幅 Range	变异系数 *CV*（%）
耐煮性指标 Cooking quality parameter	TOM值 TOM value	1.00	0.09～2.12	54.2
	蒸煮损失率 Cooking loss（%）	5.58	3.73～7.65	16.3
	面条黏性 Noodle stickiness	−1.96	−4.30～−1.05	35.4
	面条煮6 min后鲜重 Cooked weight at 6 min（g）	18.5	17.2～20.6	4.2
	面条煮10 min后鲜重 Cooked weight at 10 min（g）	21.3	19.8～23.5	4.0
感官评价指标 Sensory evaluation parameter	色泽 Color（15 score）	9.2	6.0～11.8	15.2
	表观状况 Appearance（10 score）	7.1	5.5～8.3	7.7
	软硬度 Firmness（20 score）	13.2	11.5～16.0	9.8
	黏弹性 Viscoelasticity（30 score）	19.4	15.8～22.9	8.2
	光滑性 Smoothness（15 score）	10.1	7.8～12.3	9.5
	食味 Flavor（10 score）	7.0	5.7～7.9	8.2
	总分 Total score（100 score）	66.0	54.4～75.5	7.5

表4 不同煮面时间的样品面条鲜重

Table 4 Cooked noodle weight of samples at different cooking time（g）

样品 Sample	煮面时间 Noodle cooking time						
	4 min	5 min	6 min	7 min	8 min	9 min	10 min
临麦4号 Linmai 4	18.2 a	18.8 a	19.9 a	20.7 a	21.4 a	22.1 a	23.5 a
太空6号 Taikong 6	17.1 b	17.6 c	18.6 b	19.8 b	20.2 b	21.0 b	22.2 b
周麦23 Zhoumai 23	17.0 b	17.9 b	18.5 b	18.9 c	19.5 c	20.1 c	20.2 c
西农979 Xinong 979	16.2 c	16.7 d	17.7 c	18.2 d	18.5 d	19.2 d	19.8 d

数据后不同字母表示经 t 测验样品间有显著差异（$P<0.05$）

Values followed by different letters are significantly different among samples（$P<0.05$）according to t-test

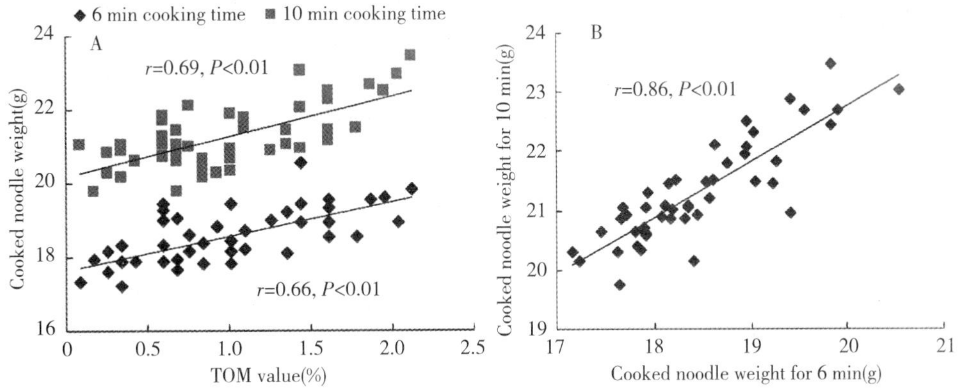

图 1 面条耐煮性指标的相关性

Fig. 1 Relationship among evaluation parameters of noodle cooking quality

A：TOM 值和面条煮后鲜重的关系；B：面条煮 6min 和 10min 后鲜重之间的关系

A：relationship between TOM value and cooked noodle weight；B：relationship of noodle weights between optimum cooked time and overcooked

表 5 小麦品质性状与面条耐煮性指标间的相关系数

Table 5 Correlation coefficients between wheat quality traits and cooking quality parameters of noodle

性状 Trait	TOM 值 TOM value	蒸煮损失率 Cooking loss	煮后面条鲜重 Cooked noodle weight		面条黏性 Noodle stickiness
			煮 6min For 6min	煮 10min For 10min	
籽粒硬度 Grain hardness	ns	ns	−0.37*	−0.47**	ns
蛋白质含量 Protein content	ns	−0.51**	ns	ns	−0.41**
形成时间 Development time	−0.33*	ns	−0.34*	−0.36*	ns
稳定时间 Stability	−0.42**	ns	−0.59**	−0.58**	ns
耐揉指数 MTI	0.49**	ns	0.67**	0.69**	ns
拉伸面积 Energy	−0.66**	ns	−0.55**	−0.61**	ns
延展性 Extensibility	−0.31*	−0.51**	ns	ns	ns
最大抗阻 Max. resistance	−0.56**	ns	−0.60**	−0.63**	ns
峰值黏度 Peak viscosity	−0.44**	ns	−0.36*	ns	ns
低谷黏度 Trough viscosity	−0.36*	ns	−0.50**	−0.30*	ns
最终黏度 Final viscosity	−0.43**	ns	−0.52**	−0.45**	ns
糊化温度 Pasting temperature	ns	ns	0.48**	0.60**	ns

* $P<0.05$，** $P<0.01$；MTI：mixing tolerance index of Farinograph；ns：no significant

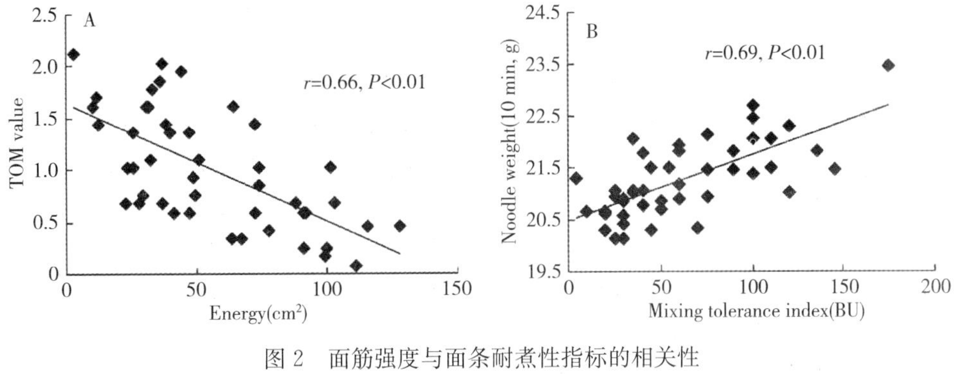

图 2 面筋强度与面条耐煮性指标的相关性

Fig. 2 Relationship between gluten strength and parameters of noodle cooking quality

A：拉伸面积与面条 TOM 值的关系；B：耐揉指数与面条煮后鲜重的关系

A：relationship between energy and TOM value of cooked noodle；B：relationship between mixing tolerance index and overcooked noodle weight

2.4 小麦品质性状对面条耐煮性的影响

粉质仪、拉伸仪和淀粉糊化等主要参数与 TOM 值呈显著相关（表 5 和图 2），其中拉伸面积和最大抗延阻力与 TOM 值的相关系数分别为－0.66（$P<0.01$）和－0.56（$P<0.01$），说明面筋强度越大，TOM 值越低。蛋白质含低和面团的延展性与蒸煮损失率呈显著负相关，相关系数分别为－0.51（$P<0.01$）和－0.51（$P<0.1$）。籽粒硬度、而筋强度和淀粉糊化参数与面条煮后鲜重呈显著相关，其中籽粒硬度与面条煮 6min 后鲜重的相关系数为－0.47（$P<0.01$）稳定时间、拉伸面积和最大抗延阻力与面条煮 6 min 和 10 min 后鲜重的相关系数为－0.55～－0.63（$P<0.01$），耐揉指数与二者的相关系数分别为 0.67（$P<0.01$）和 0.69（$P<0.01$），说明籽粒越硬、面筋强度越强，面条吸水性越低、耐煮性越好。淀粉糊化参数低谷黏度和最终黏度与面条煮 6min 后鲜重的相关系数分别为－0.50（$P<0.01$）和－0.52（$P<0.01$），糊化温度与面条煮 10min 后鲜重呈极显著正相关（$r=0.60.P<0.01$），说明面粉具有较高的糊化黏度和较低的糊化温度，其面条的吸水性较低。综上所述，提高小麦面粉的蛋白质含量、面筋强度和延展性可以改善面条的耐煮特性，蛋白质特性是影响面条耐煮性的主要品质因子，籽粒硬度和淀粉糊化参数对面条耐煮性也有一定影响。

3 讨论

中国面条品质评价指标包括面条颜色、表面状况、软硬度、黏弹性和光滑性等感官评价参数，面条耐煮特性也是决定面条成品品质的关键指标[6,12,21]。本研究表明，蛋白质含量和面筋强度是决定面条耐煮性的主要因素，随着蛋白质含量和面筋强度的增加，面条 TOM 值、干物质蒸煮损失率和吸水性明显降低，耐煮性增强，这与前人的研究结果[8-10,12-15]基本一致。尹寿伟等[16]和王晓曦等[17]以一个商业面粉为试验材料，认为淀粉破损率与干物质蒸煮损失率呈显著正相关，淀粉总量和直、支链淀粉对面条吸水性和干物质蒸煮损失率也有显著影响。Huang 等[22]研究表明，添加谷元粉可以显著提高熟面条在热水中的抗软化能力；添加淀粉在适宜煮面时间内，其中一个样品的面条硬度显著增大，另一个样品面条硬度无明显变化，但在热水中浸泡 20min 后添加淀粉对 2 个样品面条硬度的影响均不显著，可见淀粉特性对面条耐煮性的影响随着样品和煮面时间（或在热水中浸泡时间）不同而不同。本试验表明淀粉糊化参数与 TOM 值和面条吸水性存在一定相关性，与干物质蒸煮损失率和面条黏性无显著关系，因此淀粉品质对面条耐煮性也有一定影响。这与 Huang 等[22]的结果有所不同，可能的原因一是所用材料不同，二是煮面时间不同，本研究是采用适宜煮面时间 6min 和过度煮面 10min，而 Huang 等[22]是采用适宜煮面时间 5min 和在热水中浸泡 20min。

前人对面条耐煮特性评价所用的指标包括 TOM 值、面条吸水性（或面条煮后鲜重）、干物质蒸煮损失率和面条黏性等[7-17]，但这些指标间的关系如何、哪个指标评价面条耐煮特性最简单有效并没有明确结果。本研究表明 TOM 值与面条煮 6min 和 10min 后鲜重呈显著正相关，面条煮 6min 与煮 10min 后鲜重也呈高度正相关（$r=0.86$，$P<0.01$），由此说明以其中之一为指标评价面条耐煮性是完全可行的；干物质蒸煮损失率和面条黏性与 TOM 值和煮后面条重量的相关不显著，可能原因是 TOM 值、煮面 6min 和 10min 面条重量主要受面筋强度和部分淀粉糊化特性的影响，而干物质蒸煮损失率和面条黏性则受蛋白质含量和面团延展性的影响。另外，一些样品煮面 6min 时其吸水性无明显差异，但煮面时间延长至 10min 时其吸水性的变化差异显著，因此用面条煮 10min 后鲜重评价面条吸水性的效果优于面条煮 6min 后鲜重。TOM 值测定操作步骤复杂、所需时间长，每天最多测定 10 份样品；面条吸水性测定操作简单、快速，每天可测定 20 份样品；干物质蒸煮损失率每天测定的样品数介于前两者之间，虽然每天可以得到 20 份样品的面汤，但烘干或冷冻干燥去掉面汤中的水分至少延长一天时间；面条黏性需用价格昂贵的质构仪测定，所以基于本研究结果，建议以面条煮 10min 后鲜重为耐煮特性评价主要指标。以面条耐煮性优质类最低值为标准，推荐 10g 优质耐煮样品面条煮 10min 后鲜重≤21.0g。建议育种单位在面条品质感官评价时，辅助测定面条煮 10min 后鲜重，从而可以全面评价面条的加工品质。

4 结论

小麦面粉的蛋白质含量和面筋强度对面条耐煮特性有较大的正向作用，蛋白质特性是影响面条耐煮性的主要品质因子，淀粉糊化参数对面条耐煮性也有一定影响。耐煮性评价指标以面条煮后鲜重测定操作简单快速，用面条过度煮10min后鲜重评价面条吸水性的效果优于适宜煮6min后鲜重，建议将10g鲜面条煮10min后的鲜重≤21.0g作为优质鲜面条耐煮性的主要评价指标。

参考文献

References

[1] Wei Y-M（魏益民），Zhang G-Q（张国权），Ou-Yang S-H（欧阳韶晖），Xi M-L（席美丽），Hu X-Z（胡新中），Sietz W. Influence of processing parameters on quality properties of spaghetti. *J Chin Cereals Oils Assoc*（中国粮油学报），1998，13（5）：42-45.（in Chinese with English abstract）

[2] Liu J-J（刘建军），He Z-H（何中虎），Zhao Z-D（赵振东），Liu A-F（刘爱峰），Song JM（宋建民），Peña R J. Investigation on relationship between wheat quality traits and quality parameters of dry white Chinese noodles. *Acta Agron Sin*（作物学报），2002，28（6）：738-742.（in Chinese with English abstract）

[3] He Z-H（何中虎），Xia X-C（夏先春），Chen X-M（陈新民），Zhang Y（张艳），Zhang Y（张勇），Wang D-S（王德森），Xia L-Q（夏兰琴），Zhuang Q-S（庄巧生）. Wheat quality improvement：history，progress，and prospects. *Sci Agric Sin*（中国农业科学），2007，40（suppl-1）：91-98.（in Chinese with English abstract）

[4] Li S-B（李硕碧），Shan M-Z（单明珠），Wang Y（王怡），Li B-Y（李必运），Zhang S-G（张蜀光）. Evaluation of wheat quality for wet noodle making. *Acta Agron Sin*（作物学报），2001，279（3）：334-339.（in Chinese with English abstract）

[5] Lei J（雷激），Zhang Y（张艳），Wang D-S（王德森），Yan J（阎俊），He Z-H（何中虎）. Methods for evaluation of quality characteristics of dry white Chinese noodles. *Sci Agric Sin*（中国农业科学），2004，37（12）：2000—2005.（in Chinese with English abstract）

[6] Zhang Y（张艳），Yan J（阎俊），Yoshida H，Wang D-S（王德森），Chen D-S（陈东升），Nagamine T，Liu J-J（刘建军），He Z-H（何中虎）. Standardization of laboratory processing of Chinese white salted noodle and its sensory evaluation system. *J Triticeae Crops*（麦类作物学报），2007，27（10）：158-165.（in Chinese with English abstract）

[7] Dexter J E，Matsuo R R，MacGregor A W. Relationship of instrumental assessment of spaghetti cooking quality to the type and the amount of material rinsed from cooked spaghetti. *J Cereal Sci*，1985，3：39-53.

[8] D'Egidio M G，Mariani B M，Nardi S，Novaro P，Cubadda R. Chemical and technological variables and their relationships：a predictive equation for pasta cooking quality. *Cereal Chem*，1990，67：275-281.

[9] Malcolmson L J，Matsuo R R，Balshaw R. Textural optimization of spaghetti using response surface methodology：effects of drying temperature and durum protein level. *Cereal Chem*，1993，70：417-423.

[10] Del Nobile M A，Baiano A，Conte A，Mocci G. Influence of protein content on spaghetti cooking quality. *J Cereal Sci*，2005，41：347-356.

[11] Soh H N，Sissons M J，Turner M A. Effect of starch size distribution and elevated amylose content on durum dough rheology and spaghetti cooking quality. *Cereal Chem*，2006，83：513-519.

[12] Zhang L（张玲），Wang X-Z（王宪泽），Yue Y-S（岳永生）. TOM being a new assessment method for Chinese noodle cooking quality and effects of wheat quality characteristics on it. *J Chin Cereals Oils Assoc*（中国粮油学报），1998，13（1）：49-53.（in Chinese with English abstract）

[13] Zhao Z-D（赵振东），Liu J-J（刘建军），Dong J-Y（董进英），Zhang L（张玲），Li Q（李群）. The relation between TOM and cooking quality of Chinese noodle. *Acta Agron Sin*（作物学报），1998，24（6）：738-741.（in

Chinese with English abstract)
[14] Zhang G-Q（张国权），Wei Y-M（魏益民），Ou-Yang S-H（欧阳韶晖），Xi M-L（席美丽），Hu X-Z（胡新中），Sietz W. Noodle quality characters of wheat varieties. *J Chin Cereals Oils Assoc*（中国粮油学报），2000，15（3）：5-8.（in Chinese with English abstract)
[15] Zhang J（张剑），Li M-Q（李梦琴），Gong X-Z（龚向哲），Jiang L-Y（姜琳瑶）. Relationship between wheat flour traits and quality indexes of fresh-wet noodles. *J Chin Cereals Oils Assoc*（中国粮油学报），2008，23（2）：20-24.（in Chinese with English abstract)
[16] Yin S-W（尹寿伟），Lu Q-Y（陆启玉），Yang X-G（杨秀改）. Study on effect of damaged starch on cooking quality of noodles. *Food Sci Technol*（食品科技），2005，(10)：68-70.（in Chinese with English abstract)
[17] Wang X-X（王晓曦），Lei H（雷宏），Qu Y（曲艺），Liu X（刘鑫），Shi J-F（史建芳）. Effect of starch composition in flour on noodle cooking quality. *J Henan Univ Technol*（Nat Sci Edn）（河南工业大学学报·自然科学版），2010，31（2）：24-27.（in Chinese with English abstract)
[18] Batey I L，Curtin B M，Moore S A. Optimization of Rapid Visco Analyser test conditions for predicting Asian noodle quality. *Cereal Chem*，1997，74：497-501.
[19] Ye Y-L（叶一力），He Z-H（何中虎），Zhang Y（张艳）. Effect of different water addition levels on Chinese white noodle quality. *Sci Agric Sin*（中国农业科学），2010，43（4）：795-804.（in Chinese with English abstract)
[20] Matsuo R R，Dexter J E，Boudreau A，Daun J K. The role of lipids in determining spaghetti cooking quality. *Cereal Chem*，1986，63：484-489.
[21] Du W（杜巍），Wei Y-M（魏益民），Zhang G-Q（张国权）. Study on the effect of wheat character to noodle quality. *Food Sci Technol*（食品科技），2001，(2)：54-56.（in Chinese with English abstract)
[22] Huang S D，Liu J J，Quail K. Effect of starch quality and gluten and starch fortification on noodles' resistance to softening. In：Proceeding of the 61st Australian Cereal Chemistry Conference，Royal Australian Chemistry Institute，Gold Coast，Australia，2012.（in press)

Characterization of A-and B-type starch granules in Chinese wheat cultivars

ZHANG Yan[1], GUO Qi[1], FENG Nan[1], WANG Jinrong[2], WANG Shujun[2], HE Zhonghu[1,3]

[1] *Institute of Crop Science/National Wheat Improvement Center, Chinese Academy of Agricultural Sciences, Beijing 100081, P. R. China*

[2] *Key Laboratory of Food Nutrition and Safety, Ministry of Education/Tianjin University of Science & Technology, Tianjin 300457, P. R. China*

[3] *International Maize and Wheat Improvement Center (CIMMYT) China Office, Beijing 100081, P. R. China*

Abstract: Starch is the major component of wheat flour and serves as a multifunctional ingredient in food industry. The objective of the present study was to investigate starch granule size distribution of Chinese wheat cultivars, and to compare structure and functionality of starches in four leading cultivars Zhongmai 175, CA12092, Lunxuan 987, and Zhongyou 206. A wide variation in volume percentages of A- and B-type starch granules among genotypes was observed. Volume percentages of A- and B-type granules had ranges of 68.4—88.9% and 9.7—27.9% in the first cropping seasons, 74.1—90.1% and 7.2—25.3% in the second. Wheat cultivars with higher volume percentages of A-and B-type granules could serve as parents in breeding program for selecting high and low amylose wheat cultivars, respectively. In comparison with the B-type starch granules, the A-type granules starch showed difference in three aspects: (1) higher amount of ordered short-range structure and a lower relative crystallinity, (2) higher gelatinization onset (T_o) temperatures and enthalpies (ΔH), and lower gelatinization conclusion temperatures (T_c), (3) greater peak, though, and final viscosity, and lower breakdown viscosity and pasting temperature. It provides important information for breeders to develop potentially useful cultivars with particular functional properties of their starches suited to specific applications.

Keywords: bread wheat, A- and B-type starch granules, short-range molecular order, relative crystallinity, gelatinization and pasting properties

1 Introduction

Starch is the major component of wheat flour and serves as a multifunctional ingredient in food industry. Starch occurs as granules in the endosperm of wheat grain. The granule shape, size and its hierarchical structure are important determinants of starch functionality (Lindeboom *et al.* 2004; Park *et al.* 2004). Wheat starch consists of two distinct forms of granules: A-and B-type granules. The A-type starch granules are disk-like or lenticular in shape with a diameter of >10 μm, while the B-type starch granules are less than 10 μm in diameter and spherical or polygonal in shape (Vermeylen *et al.* 2005; Ao and Jane 2007; Kim and Huber 2008; Wang *et al.* 2014). In

Published in Journal of Integrative Agriculture, 2016, 15 (10): 2203-2214.

wheat, A-type granules contribute to more than 70% total weight of the starch (Bechtel *et al.* 1990; Peng *et al.* 1999; Shinde *et al.* 2003), whereas B-type granules comprise up to 90% of granules in number (Raeker *et al.* 1998).

The proportions of B-and A-starch granules, by weight, volume and number, differ among genotypes (Raeker *et al.* 1998; Li *et al.* 2001). A wide range of variation (17-50%) for B-starch granule volume was observed in bread wheat, suggesting possibilities of genetic manipulation of granule size distribution (Stoddard 1999). The B-granules occupied volumes in a range of 28.5-56.2% for hard red winter and hard red spring wheat (Park *et al.* 2009). The volume percentages of A- and B-type starch granules were 52.7%-65.5% and 34.5%-47.3% in seven Chinese wheat cultivars (Dai *et al.* 2009). In several studies, environmental stress such as temperature (Liu *et al.* 2011), water deficit (Dai *et al.* 2009; Zhang *et al.* 2010), nutrient supplementation (Ni *et al.* 2012; Li *et al.* 2013), and light intensity (Li *et al.* 2010) significantly changed starch granule size distribution and amylose content in wheat.

Wheat starch A- and B-type granules differ in composition, chain length distribution of amylopectin, relative crystallinity, microstructure (e.g., surface pores, channels, cavities), and they have been summarized in details (Soulaka and Morrison 1985; Fortuna *et al.* 2000; Chiotelli and Le Meste 2002; Bertolini *et al.* 2003; Shinde *et al.* 2003; Van Hung and Morita 2005; Geera *et al.* 2006; Kim and Huber 2008; Kim 2009; Salman *et al.* 2009). The differences in these structural characteristics lead to variations in swelling, gelatinization, retrogradation and pasting properties of the two types of starch granules (Eliasson and Kaelsson 1983; Fortuna *et al.* 2000; Chiotelli and Le Meste 2002; Shinde *et al.* 2003; Geera *et al.* 2006; Soh *et al.* 2006; Kim 2009). A-type granules have higher gelatinization enthalpy, amylose content, pasting parameters such as peak, trough, breakdown, final and setback viscosities, and lower gelatinization onset and peak temperatures, whereas B-type granules have higher lipid-complexed amylose content and swelling power, broader gelatinization ranges, and lower gelatinization enthalpy (Sahlström *et al.* 2003; Geera *et al.* 2006; Soh *et al.* 2006; Kim and Huber 2010a; Yin *et al.* 2012). Shinde *et al.* (2003) and Soh *et al.* (2006) observed that peak and final viscosities of wheat starch reduced with increasing proportion of B-type granules. Thus, the proportion of A- and B-type granules impacts wheat starch structural characteristics and functional properties. However, there are yet inconsistent reports on amylopectin chain-length distribution, relative crystallinity, and microstructure of A- and B-type wheat starch granules (Vermeylen *et al.* 2005; Liu *et al.* 2007; Salman *et al.* 2009). Much of this inconsistency is likely attributable to the different genotypes used in previous reports. In addition, there are very little information available on amylopectin chain-length distribution, relative crystallinity and microstructure of A- and B-type starch granules in Chinese cultivars.

The objective of the present study was to investigate starch granule size distribution in Chinese wheat cultivars, and to assess A- and B-type starch granule characteristics of four leading cultivars in morphology, amylose content, chain length distribution of amylopectin, short-range molecular order, relative crystallinity, and gelatinization and pasting properties. The results of this study will help breeders to develop potentially useful cultivars with particular functional properties of their starches suited to specific applications.

2 Materials and methods

2.1 Experimental materials

A total of 345 Chinese leading cultivars and advanced lines, including 66 from the Northern China Plain zone, 251 from the Yellow and Huai Valley zone, 12 from the middle and low Yangtze Valley zone, 3 from the southwestern China zone, and 13 introductions from other countries were grown in Anyang, Henan

Province in two seasons, including 245 and 208 genotypes in 2010—2011 and 2011-2012 cropping seasons, respectively. Only 108 genotypes were grown in two consecutive cropping seasons. They were used to determine the distribution of A- and B- type starch granules. Among them, four leading cultivars Zhongmai 175, CA12092, Lunxuan 987, and Zhongyou 206 collected from the same field in 2012—2013 season in Beijing were used to analyze morphology, amylose content, chain length distribution of amylopectin, short-range molecular order, relative crystallinity, and gelatinization and pasting properties of the A- and B-type starch granules. Zhongmai 175, Lunxuan 987 and Zhongyou 206 have been released in the Northern China Plain zone, and CA12092 is an advanced line currently including in the regional yield trials. Zhongmai 175 is characterized by high yielding potential and broad adaptation, soft kernel, excellent noodle, and steamed bread qualities, and is currently a leading cultivar and also serves as a check cultivar in the regional yield trials. Zhongyou 206 possessed hard kernel, strong gluten and excellent bread-making quality. CA12092 and Lunxuan 987 showed high yielding potential and broad adaptation, hard kernel, but poor gluten quality, and averaged qualities for noodles and steamed bread.

2.2 Flour milling

Grain hardness was measured on 300-kernel sample with a Perten Single Kernel Characterization System (SKCS) 4100 (Perten Instruments, Springfield, IL, USA). The tested samples were tempered overnight to 14.5 and 16.5% moisture for soft and hard wheat, respectively. 200g grain samples from each genotype were milled using a Brabender Quadrumat Junior mill (Brabender Inc., Duisberg, Germany). Grain samples of Zhongmai 175 and CA12092 were tempered overnight to 14.5% moisture, and those of Lunxuan 987 and Zhongyou 206 were also tempered overnight to 16.5% moisture and milled using the same mill as mentioned above.

2.3 Starch isolation

Starch was extracted according to the methods of Liu *et al.* (2007) and Park *et al.* (2005) with minor modifications, the tailings were centrifuged twice and all the starch portions were combined. To separate gluten from starch, dough was made by mixing 6 g of flour with 4 g of distilled water, stood for 10 min, and then washed with 60 mL of water. The gluten was washed twice with 20 mL of water to ensure the complete separation of all the starch. The combined starch suspensions were filtered using a nylon cloth (75 μm openings). The resulting starch filtrate was centrifuged at 2 500×g for 15 min, and the supernatant was discarded. The precipitate was divided into two portions and the upper gray-colored tailings were transferred to another tube. Water was added into the lower light-colored portions and slurries were centrifuged again. These steps were repeated until there were no gray-colored tailings on top of the starch. The collected tailings from each repeat were re-suspended and centrifuged twice. Then, the top layer was discarded as described above. The resulting starch from the above steps was combined and freeze-dried. The dried starch granules were ground lightly with a mortar and pestle and passed a 100-mesh sieve.

2.4 Fractionation of A- and B- type starch granules

Large A- and small B-type starch granules were separated from prime starch by repeated suspensions in six cycles (Park *et al.* 2005). The sediment comprised mainly A-type starch granules and the supernatant contained mainly B-type granules. The sediment and supernatant were centrifuged for 15 min (4 000×g), and the A- and B- starch granule factions were collected, respectively. The fractions were frozen, lyophilized, and ground with a household coffee mill to pass through a 149μm mesh sieve.

2.5 Granule size distribution

The proportion of A- and B-type granules in wheat starch was determined using a Sympatec Helos/Rodos laser diffraction particle size analyzer (Sympatec GmbH, Clausthal-Zeller-feld, Germany), and the data were calculated as the volume percentage (%)

occupied by starch granules. Granules with size of < 10.0 μm and between 10.1-35.0 μm in diameter were classified as B- and A-type starch granules, respectively (Peng *et al*. 1999). Granules with diameters > 35.0 μm were considered to be impurities or compound granules. Each sample was measured twice, and the differences between two measurements of B-type granule contents were less than 0.5%.

2.6 Granule morphology

Starch samples were fixed onto the surface of double-sided, carbon-coated adhesive tape attached to an aluminium stub. The mounted starch samples were coated with gold prior to imaging in a scanning electron microscope (SU1510, Hitachi High-technologies Corporation, Japan). The accelerating voltage was 10.0 kV.

2.7 Starch crystallinity

Starch crystallinity was measured using a Panalytical X' Pert Pro X-ray diffractometer (PANalytical, Holland) with a Co-K_α source (λ=0.1789 nm) operating at 45 kV and 35 mA. The detailed operating conditions and sample treatment before measurement were described elsewhere (Wang *et al*. 2009). The relative crystallinity was quantitatively estimated as a ratio of the crystalline area to the total area between 4-40° (2θ) using the Origin software (ver. 7.5, Microcal Inc., Northampton, MA, USA).

2.8 Chain length distribution of amylopectin

Chain length distribution of amylopectin was analyzed using a high-performance anion-exchange chromatography equipped with a pulsed amperometric detector (HPAEC-PAD) (Dionex Corporation, Sunnyvale, CA) according to the method of Liu *et al*. (2007). 18 mg starch was dispersed in 6 mL of sodium acetate buffer (pH 3.5) by stirring in a boiling water bath for 20 min. After cooling, isoamylase solution (10 μL) was added. The sample was incubated at 37℃ with slow stirring for 16 h. The enzyme was inactivated by boiling the samples for 15 min. The sample was filtered (0.45 μm nylon syringe filter) and injected into the HPAEC-PAD System.

The HPAEC-PAD System consisted of a Dionex DX2500 equipped with an ED50 electrochemical detector with a gold working electrode, GP50 gradient pump, LC30 chromatography oven, and AS40 automated sampler (Dionex Corporation, Sunnyvale, CA, USA). The standard quintuple potential waveform was employed, with the following periods and pulse potentials: $T1 = 0.20$ s, $E1 = 0.1$ V; $T2=0.40$ s, $E2=0.1$ V; $T3=0.41$ s, $E3=-2$ V; $T4=0.43$ s, $E4=0.6$ V; $T5=0.44$ s, $E5=-0.1$ V. Data were collected using Chromeleon software, ver. 8.00 (Dionex Corporation, Sunnyvale, CA, USA). The mobile phase was prepared in deionized water with helium sparging and contained eluent A (100mmol · L^{-1} NaOH) and eluent B (50mmol · L^{-1} sodium acetate in 100mmol · L^{-1} NaOH). Flow rate was 1.0mL · min^{-1}. Linear components were separated on a Dionex CarboPac™ PA100 column with gradient elution (0-5 min, 40% A; 15-50 min, 70% A; 15—50 min, 70% A+30% B) at a column temperature of 26℃ and a flow rate of 1 mL · min^{-1}. A CarboPac™ PA100 guard column was installed in front of the analytical column.

2.9 Fourier transform infrared spectroscopy (FT-IR)

The FT-IR spectra of wheat starch samples were obtained using a Tensor 27 FT-IR spectrometer (Bruker, Germany) equipped with a DLATGS detector. The sample preparation and operation conditions were described elsewhere (Wang *et al*. 2014). The ratios of absorbance at 1 045 cm^{-1}/ 1 022 cm^{-1} were used to characterize the short-range ordered structure of starch.

2.10 Amylose content

0.1 mg of starch sample was dissolved in 1 mL of 95% ethanol (v/v) and 9 mL of NaOH solution (1 N) and placed overnight in a refrigerator at 4℃. Distilled water was added to make up 100 mL in a volumetric flask containing the dissolved starch sample. The amount of amylose was determined using an automated chemistry analyzer (FS3100, OI Analytical, USA) according to the operator's manual. Rice amylose and amylopectin from the China National Rice Research Institute, Chinese Academy of Agricultural Sciences were used to establish a calibration curve using a set of starches with amylose concentrations of 1.5, 10.4, 16.2, 19.3, and 22.5%. An additional rice starch with 26.5% amylose was used as the control sample.

2.11 Thermal analysis

Thermal transition analysis of starch samples was made using a differential scanning calorimeter (DSC 200 F3, NETZSCH, Germany) equipped with a thermal analysis data station and data recording software, according to the method of Wang and Copeland (2012). About 3 mg of starch granules were weighed into 40 μL aluminum pans. Distilled water was added to the starch with a microsyringe to obtain a starch/water ratio of 1 : 3 in the DSC pans. Care was taken to ensure that the starch samples were completely immersed in the water by gentle shaking before the pans were sealed, reweighed and left overnight at room temperature before analysis. An empty pan was used as a reference. The pans were heated from 30 to 115℃ at a scanning rate of 10℃ min^{-1}. The instrument was calibrated using indium as a standard. The onset (T_o), peak (T_p) and conclusion (T_c) temperatures and the enthalpy change (ΔH) were determined through data recording software.

2.12 Pasting properties

The pasting profiles were analyzed using a newport scientific rapid visco analyser 3 (RVA-3) (Newport Scientific, Australia). Starch slurries containing 8% (w/w) starch (dry weight) in a total weight of 28 g were held at 50℃ for 1 min before heating at a rate of 6℃ min^{-1} to 95℃, holding at 95℃ for 5 min, and then cooling at a rate of 6℃ min^{-1} to 50℃ and held at 50℃ for 2 min. The speed of the mixing paddle was 960 r • min^{-1} for the first 10 s, then 160 r • min^{-1} for the remainder of the experiment. Peak viscosity (PV), viscosity at trough (also known as minimum viscosity, MV) and final viscosity (FV) were recorded, and breakdown (BD, which is PV minus MV) and setback (SB, which is FV minus MV) were calculated using the Thermocline software provided with the instrument.

2.13 Statistical analysis

All testings were conducted at least twice and the results were reported as the mean value and standard deviations excluding the X-ray diffraction, chain length distribution of amylopectin and FT-IR measurements. Mean, standard deviation, variable coefficient and Duncan's test ($P<0.05$) were conducted using the SPSS 17.0 Statistical Software Program (SPSS Inc. Chicago, IL, USA).

3 Results

3.1 Starch granule size distribution of Chinese cultivars

Significant differences in granule size distribution for A- and B-type granules were observed in the tested cultivars (Table 1). The volume percentages of A- and B-type granules had a range of 68.4-88.9% and 9.7-27.9% in 2010—2011 cropping season, 74.1-90.1% and 7.2-25.3% in 2011—2012 cropping season. This suggested that improvement for starch quality could be achieved through breeding given the wide variation for starch granule distribution present in Chinese wheat cultivars. The A-starch granules in most cultivars exhibited the volume percentage of 74.0-82.0% and 78.0-86.0%, while the B-type starch granules in most cultivars showed the volume percentage of 14.0-20.0% and 12.0-18.0% in 2010—2011 and 2011—2012 cropping seasons, respectively (Fig. 1). This indicated that cropping season had impacts on starch granule distribution in Chinese cultivars.

Twenty-two cultivars with higher volume percentage of A-type granules and lower volume percentage of B-type granules were selected. Other 22 cultivars also showed higher volume percentage of B-type granules and lower volume percentage of A-type granules. Name of these cultivars are listed in Table 2. Among them, Wheatear, Zhou 9811-1, Ruzhou 0319, and Chuanmai 42-white with higher volume percentage of A-type granules and lower volume percentage of B-type granules, and 07CA266 and Bainong 64 with higher volume percentage of B-type granules and lower volume percentage of A-type granules were found in two consecutive cropping seasons. They could serve as crossing parents for selecting high amylose cultivars in breeding programs due to A-type granules with more amylose relative to B-type granules (Liu *et al*. 2007), which increase the resistant starch in wheat suitable for improving public health (Rahman *et al*. 2007). Meanwhile 07CA266 and Bainong 64 could also become crossing parents for selecting low amylose wheat cultivars that can positively contribute to noodle quality improvement. For the above other cultivars only grown in single cropping season, further investigation will be needed to confirm starch granule distribution of them.

Table 1 Mean, range, standard deviation (SD), and coefficient of variance (CV%) of the volume percentages of A-and B-type starch granules in the tested cultivars

Season	No. of cultivars	Starch granule[1)]	Mean	Range	SD	CV (%)
2010—2011	245	A	79.0	68.4—88.9	3.46	4.4
		B	18.9	9.7—27.9	3.13	16.5
2011—2012	208	A	82.6	74.1—90.1	3.02	3.7
		B	16.4	7.2—25.3	3.21	19.6

[1)] A, the volume percentages of A-type starch granules (%); B, the volume percentages of B-type starch granules (%).

3.2 Characterization of A- and B-type starch granules in four cultivars

Granule morphology The granular morphologies of A- and B-type granules separated from starch of Zhongmai 175, CA12092, Lunxuan 987, and Zhongyou 206 were similar, as observed from the representative images of Zhongmai 175 in Fig. 2. A-type starch granules displayed a disk-like shape with diameter in the range of 10-30 μm. In contrast, B-type granules displayed a spherical shape with diameter less than 10 μm, and granules with diameter of about 5 μm were predominant. The morphology of the starch was in agreement with previous reports (Jane *et al*. 1994; Song and Jane 2000; Yoo and Jane 2002; Wang *et al*. 2014).

FT-IR spectroscopy Similar FT-IR patterns were also observed for unfractionated, A- and B-type starch granules from four cultivars, hence only FT-IR spectra of Zhongmai 175 is presented (Fig. 3-A). The deconvoluted FT-IR spectrum in the range of 800-1 200 cm^{-1} of the B-type starch granules differed greatly from that of unfractionated and A-type starch granules (Fig. 3-B). The 1047 cm^{-1}/1 022 cm^{-1} ratios of unfractionated starch, A- and B-type starch granules are presented in Table 2. Interestingly, A-type large granules showed the highest 1 047 cm^{-1}/1 022 cm^{-1} ratios, whereas B-type small granules presented the lowest values. This indicated that A-type large granules had a larger amount of ordered short-range structure than B-type small granules.

3.3 Branch chain-length distribution of amylopectin

The amylopectin chain length distribution of the starches was classified into four categories: short chains with degree of polymerization (DP) 6-12, medium length chains with DP 13-24, long chains with DP 25-36, and very long chains with DP>36 (Table 3). Amylopectin of unfractionated, A- and B-type starch granules contained a higher proportion of medium chains with DP 13-24 (47.1-65.8%) and short chains with DP 6-12 (16.3-36.5%), and a smaller proportion of long chains with DP 25-36 (11.1-17.9%) and very long chains with DP>36 (0-1.9%). Differences in chain length distribution of amylopectin of A- and B-type granules separated from four cultivars were observed. The A-type granules of Zhongmai 175 and Zhongyou 206 contained less branch chains of DP 6-12 and more branch chains of DP 13-24 than did the B-type granules, in general agreement with the results from Ao and Jane (2007) and Salman et al. (2009). However, the A-type granules of CA12092 and Lunxuan 987 contained more branch chains of DP 6-12 and less branch chains of DP 13-24 than did the B-type granules, consistent with the report by Vermeylen et al. (2005). In contrast to B-type granules, the A-type granules of four cultivars had lower branch chains of DP 25-36 and DP > 36. These data indicate that amylopectin molecules of the A- and B-type granules have distinct fine structures, and they are likely genetically controlled during their biosynthesis (Peng *et al*. 2000).

3.4 X-ray crystallinity

Unfractionated, A- and B-type starch granules of four cultivars presented similar XRD patterns, as observed from the representative patters of Zhongmai 175 (Fig. 4). Four characteristic diffraction peaks were noted at 2θ 15.5, 17.4, 18.7, and 23° (Fig. 4), indicating the presence of A-type crystalline polymorphs in wheat starches. The relative crystallinity of B-type starch granules was the highest, while it was the lowest for A-type starch granules (Table 3). The result indicated that B-type starch granules had a larger amount of crystallites than A-type starch granules, consistent with Ao and Jane (2007).

3.5 Thermal properties

DSC data showed that the gelatinization temperatures of unfractionated, A- and B-type starch granules of four wheat cultivars varied from 54.2 to 69.3℃, and the gelatinization enthalpy changes were in the range of 8.1-11.1 (Table 4). No significant difference was found for gelatinization parameters changes of the A- and B-type starch granules among four cultivars. A-type starch granules relative to B-type starch granules exhibited higher gelatinization onset (T_o), temperature and enthalpy changes (ΔH), but lower gelatinization conclusion temperatures (T_c).

3.6 Pasting properties

Amylose content and pasting parameters of unfractionated, A- and B-type starch granules from four wheat cultivars are shown in Table 5. A-type starch granules had a higher amylose content (31.4-33.3%) than B-type starch granules (29.9-32.5%), in agreement with reports by Soulaka and Morrison (1985), Peng *et al*. (1999) and Shinde *et al*. (2003). A-type starch granules exhibited greater peak, trough, and final viscosities, and lower breakdown viscosity and pasting temperature than B-type starch granules. This was consistent with previous reports (Franco *et al*. 2002; Shinde *et al*. 2003; Ao and Jane 2007), suggesting greater swelling power of A-type starch granules.

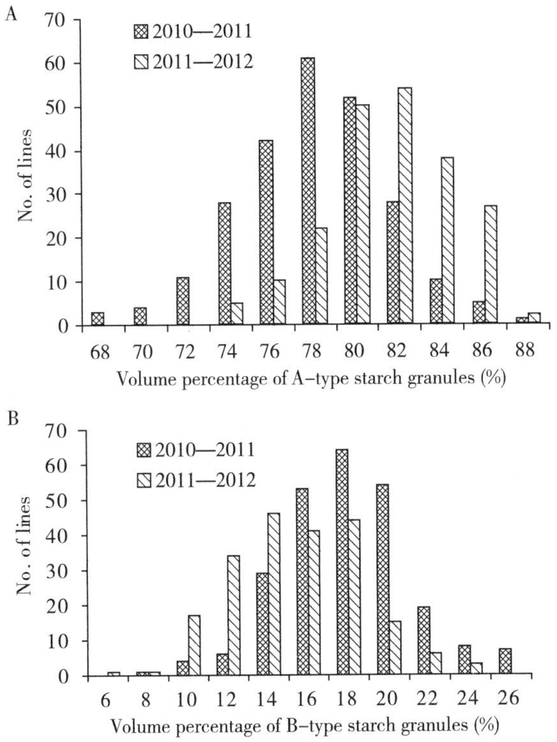

Fig. 1 The frequency distribution of percentage volumes of A- and B-type starch granules in the tested cultivars. The horizontal axes indicate the percentage volumes of A- or B-type starch granules (%)

Table 2 Cultivars with higher volume percentage of A-and B-type granules in two consecutive cropping seasons

Cropping season	Cultivars[1)]	A	B	Cultivars[2)]	A[3)]	B[4)]
2010—2011	Wheatear	88.9	9.7	Yumai 34	71.7	27.9
	Zhoumai 26	87.4	10.7	07CA266	72.4	27.2

(continued)

Cropping season	Cultivars1)	A	B	Cultivars2)	A3)	B4)
2011—2012	Fumai 8	87. 1	11. 8	Xin 9408	73. 3	26. 7
	Zhi4001	86. 8	10. 8	BAY24	72. 8	26. 7
	Zhou9811-1	86. 5	11. 8	Jagger	73. 1	26. 5
	U07-6308	86. 0	12. 9	AAY08	71. 8	26. 1
	04Zhong70	85. 4	13. 3	CA9507-dwarf	72. 9	26. 0
	Zhou 18	85. 2	13. 3	08CA137	68. 6	25. 3
	Yubao 2	85. 1	12. 5	10CS4801	74. 8	24. 8
	Ruzhou 0319	84. 8	13. 8	Xin 05-1241	74. 8	24. 7
	Chuanmai 42-white	84. 7	13. 5	Bainong 64	68. 4	24. 3
	Zhouheimai 1	90. 1	7. 2	Tainong 2413	74. 1	25. 3
	Wheatear	90. 0	8. 4	Luomai 05123	74. 6	24. 8
	Brula	87. 9	11. 3	Luyuan 502	75. 2	24. 3
	Ruzhou 0319	87. 9	10. 3	Shixin 733	76. 8	22. 6
	09CA86	87. 5	10. 6	Kenong 2011	76. 9	22. 6
	Luomai 6082	87. 2	11. 8	Pubing 3228	76. 1	22. 6
	Luomai 24	87. 2	10. 6	Zimai 12	76. 1	22. 5
	Chuanmai 42-white	87. 1	11. 0	Shixin 828	77. 3	22. 3
	Yan 4110	87. 1	11. 8	Tainong 9862	77. 2	22. 2
	Zhou 24	87. 1	11. 2	07CA266	77. 8	21. 6
	Zhou 9811-1	86. 7	11. 5	Bainong 64	77. 7	21. 2

1) Cultivars with higher volume percentage of A-type granules and lower volume percentage of B-type granules.

2) Cultivars with higher volume percentage of B-type granules and lower volume percentage of A-type granules.

3) A, the volume percentages of A-type starch granules (%).

4) B, the volume percentages of B-type starch granules (%).

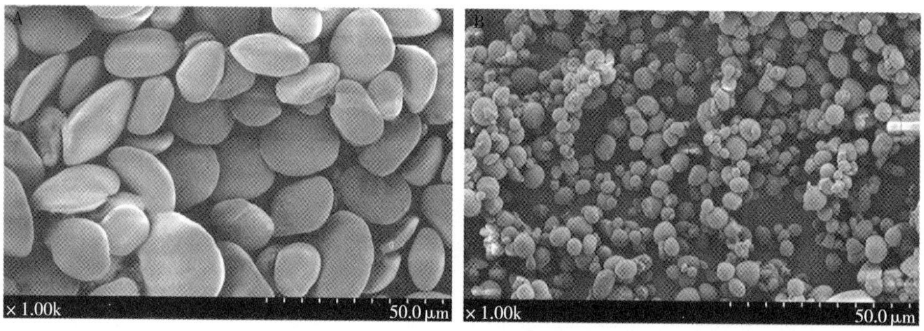

Fig. 2 Scanning electron micrographs of A-type (left) and B-type (right) granules in Zhongmai 175. Scale bar=5. 0 μm

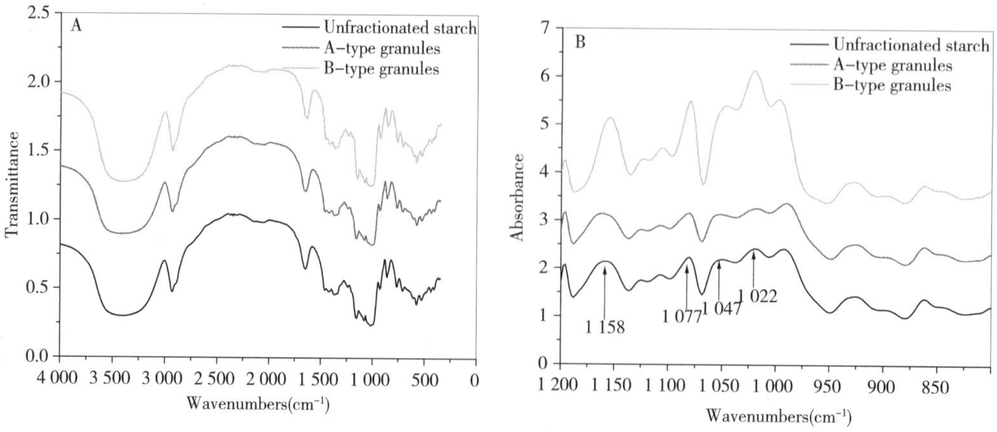

Fig. 3 Fourier transform infrared spectroscopy FT-IR average spectra (A) and the subtraction spectra (B) of unfractionated starch, A- and B- starch granules in Zhongmai 175

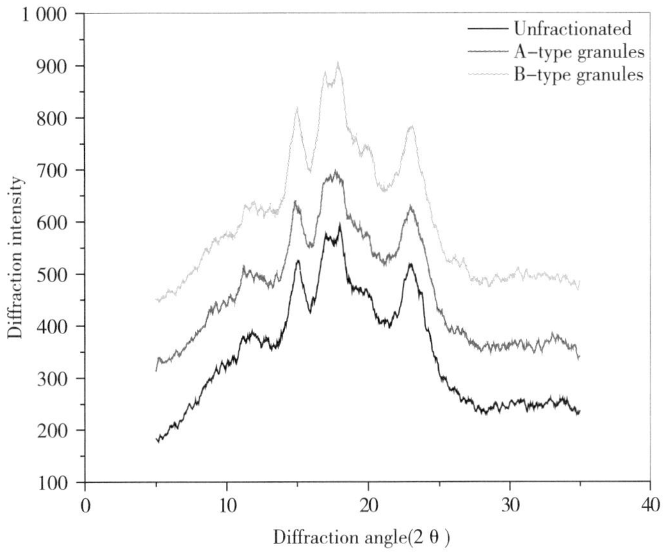

Fig. 4 X-ray diffraction patterns of unfractionated starch, A- and B-starch granules in Zhangmai 175

Table 3 Ratio between 1 047 and 1 022 cm^{-1} of fourier transform infrared spectroscopy (FT-IR) spectra, relative crystallinity and branch chain length distribution for unfractionated starch, A- and B-starch granules of four wheat cultivars

Cultivar	Starch source	1 047 cm^{-1}/ 1 022 cm^{-1}	Relative crystallinity (%)	Distribution (%)[1)]			
				DP 6-12	DP 13-24	DP 25-36	DP>36
Zhongmai 175	Unfractionated starch	0.86a	25.6b	25.0a	59.2b	14.4c	1.5c
	A-granules	0.90a	24.7c	23.3c	60.4a	14.6b	1.7b
	B-granules	0.74b	27.3a	24.7b	58.4c	15.1a	1.8a
CA12092	Unfractionated starch	0.88a	25.1b	28.3a	58.7c	12.0c	1.0c
	A-granules	0.89a	23.6c	26.7b	59.4b	12.8b	1.1b
	B-granules	0.79b	26.7a	16.3c	65.8a	16.1a	1.9a
Lunxuan 987	Unfractionated starch	0.88a	24.8b	34.7a	53.3c	11.1b	0.9b
	A-granules	0.90a	23.7c	31.8b	56.4b	10.9c	0.9b
	B-granules	0.79b	25.4a	20.2c	63.8a	14.6a	1.5a
Zhongyou 206	Unfractionated starch	0.89a	24.6b	20.8c	61.4a	17.9a	0b
	A-granules	0.90a	23.4c	26.8b	59.4b	13.1c	0b
	B-granules	0.80b	25.8a	36.5a	47.1c	16.4b	0.7a

1) DP, degree of polymerization

Values followed by the same letter in the same column are not significantly different ($P<0.05$). The same as below.

Table 4 Thermal properties of unfractionated starch, A- and B-starch granules in four wheat cultivars[1)]

Cultivar	Starch	T_o (℃)	T_p (℃)	T_c (℃)	ΔH (J g^{-1})
Zhongmai 175	Unfractionated starch	57.4 a	62.1 a	67.7 a	9.0 a
	A-granules	56.5 a	61.5 b	66.6 b	9.2 a
	B-granules	54.5 b	61.0 c	68.6 a	8.1 b
CA12092	Unfractionated starch	56.7 a	62.4 a	68.4 b	10.8 a
	A-granules	56.6 a	62.0 b	67.5 c	11.1 a

(continued)

Cultivar	Starch	T_o (℃)	T_p (℃)	T_c (℃)	ΔH (J g^{-1})
	B-granules	54.3 b	62.3 a	70.1 a	9.2 b
Lunxuan 987	Unfractionated starch	56.5 a	62.6 a	68.9 b	10.5 a
	A-granules	56.7 a	62.1 a	68.6 b	10.6 a
	B-granules	54.2 b	61.6 b	70.4 a	9.6 b
Zhongyou 206	Unfractionated starch	57.0 a	63.1 a	69.3 b	10.4 a
	A-granules	56.5 a	62.6 b	68.6 c	10.3 b
	B-granules	55.3 b	62.1 c	69.9 a	10.2 b

1) T_o, T_p and T_c, onset, peak and complete temperature, respectively; ΔH, enthalpy change.

Table 5 Amylose and rapid visco analyzer (RVA) parameters of unfractionated starch, A-and B-starch granules in four wheat cultivars

Cultivar	Sample	Amylose (%)	Peak viscosity (cP)	Though (cP)	Breakdown (cP)	Final viscosity (cP)	Pasting temperature (℃)
Zhongmai 175	Unfractionated starch	29.7 b	1 848 b	1 326 b	523 b	1 929 b	92.5 a
	A-granules	32.5 a	2 454 a	2 043 a	412 b	2 497 a	88.4 c
	B-granules	30.2 b	1 232 c	430 c	802 a	1 275 c	90.6 b
CA12092	Unfractionated starch	31.8 a	2 002 b	1 326 b	588 a	2 201 b	87.1 b
	A-granules	31.9 a	2 230 a	2 043 a	275 c	2 358 a	87.2 b
	B-granules	30.4 b	1 534 c	430 c	442 b	1 688 c	95.4 a
Lunxuan 987	Unfractionated starch	32.3 b	2 045 b	1 739 b	306 a	2 295 b	88.1 b
	A-granules	33.3 a	2 370 a	2 134 a	236 b	2 609 a	84.4 c
	B-granules	32.5 ab	860 c	560 c	300 a	988 c	95.2 a
Zhongyou 206	Unfractionated starch	31.2 a	2 306 b	1 813 b	493 b	2 567 b	89.7 a
	A-granules	31.4 a	2 643 a	2 333 a	310 c	2 857 a	84.9 c
	B-granules	29.9 b	1 398 c	754 c	645 a	1 548 c	85.6 b

4 Discussion

4.1 Starch granules size distribution

Previous studies indicated that starch granule size distribution was mainly affected by environmental factors such as growing season temperature, rainfall patterns and humidity, growth locations, and sustained or episodic environmental stresses (Raeker *et al*. 1998; Dai *et al*. 2009; Beckles *et al*. 2014), generally consistent with our observations that some differences were found in the volume percentage of A- and B-type starch granules of most cultivars grown at the same location among two cropping seasons. Moreover, a wide variation in volume percentages of A- and B-type starch granules among cultivars grown in the same environment was observed in the present study, indicating genotype can play a major role in determining the variation in granule size distribution of wheat starch, in agreement with previous reports (Dengate and Meredith 1984; Stoddard 2000, 2003).

4.2 FT-IR spectroscopy, X-ray crystallinity and branch chain-length distribution of amylopectin

The FT-IR spectrum of starch is sensitive to changes in structure of a short-range molecular level (double helices). The absorbance bands at 1 022 and 1 047 cm^{-1} are characteristics of amorphous and ordered structures in starch (Van Soest *et al*. 1995). Thus, the ratio of 1 047 cm^{-1}/1 022 cm^{-1} can be used to characterize the short-range molecular order of double helices in starches (Van Soest *et al*. 1995; Capron *et al*. 2007).

In the present study, A-type starch granules presented a higher ratio of 1 047 cm^{-1}/1 022 cm^{-1} than B-type starch granules, indicating the presence of a larger amount of ordered short-range double helices in A-type starch granules. This was in general agreement with the chain length distribution of amylopectin in Zhongmai 175 and Zhongyou 206, showing that A-type starch granules have a higher proportion of intermediate chains (DP 13-24). The length of amylopectin chains acceptable for the formation of double helices is in the range of 10-25 glucosyl residues, preferentially 12-18 (Genkina *et al.* 2007). For A-type starch granules, higher proportion of intermediate chains would result in the formation of more double helices, causing a higher IR ratio of 1 047 cm^{-1}/1 022 cm^{-1} compared with B-type starch granules. However, inconsistent results were observed for starches from CA 12092 and Lunxuan 987. These inconsistencies observed between molecular order and chain length distribution of starch granules suggested that the molecular order of starch granules is not determined solely by starch chains that may form double helices. B-type starch granules presented higher relative crystallinity compared with A-type starch granules, in agreement with reports from Ao and Jane (2007) and Sahal and Jackson (1996). Generally, the higher IR ratio of 1 047 cm^{-1}/1 022 cm^{-1}, the higher relative crystallinity starch granules have. In this study, A-type starch granules with higher IR ratio of 1 047 cm^{-1}/1 022 cm^{-1} yet presented lower relative crystallinity. The different result might be due to the impact of amylose molecules on the conformation of amylopectin double helices. Kozlov *et al.* (2007) found that an increase in amylose content is accompanied by accumulation of amylose tie-chains in amylopectin clusters forming defects in crystalline lamellae. Disordered ends of amylopectin double helices not participating in the formation of crystals are also proposed to be contributing factors for defects of the crystalline regions and for greater disorder in the packing of the lamellar structure (Koroteeva *et al.* 2007a, b). A-type starch granules have higher amylose content, which would result in the formation of more defects of the crystalline regions, thus leading to lower relative crystallinity. Our results also corroborated the fact that starch crystallites are formed by the long-range regular array of double helices, and that double helices content are higher than relative crystallinity.

4.3 Thermal and pasting properties

DSC measures the heat required for the melting of doublehelices or starch crystallites following granule swelling during gelatinization (Wang and Copeland 2012; Wang and Copeland 2013). DSC gelatinization parameters are influenced by the molecular structure of amylopectin, amylose/amylopectin ratio, crystalline/amorphous ratio, or a combination thereof (Noda *et al.* 1998). The enthalpy change primarily reflects the loss of molecular order (double helices) rather than the melting of starch crystallites (Cooke and Gidley 1992). A-type starch granules presented higher onset temperature and enthalpy change than did B-type starch granules. The higher enthalpy change of A-type starch granules could be attributed to the higher amount of ordered short-range molecular structure. Our results were in agreement with Ao and Jane (2007). However, Liu *et al.* (2007) reported that A-type starch granules had lower gelatinization temperature, and Ghiasi *et al.* (1982) found that A- and B-type starch granules had similar gelatinization temperature regimes. Thus, further investigation is needed to confirm the association between starch granules and gelatinization property.

A-type starch granules had greater peak, trough and final viscosities, and lower breakdown viscosity and pasting temperature than B-type starch granules, in agreement with reports by Sahlstrom *et al.* (2003), Ao and Jane (2007) and Kim and Huber (2010b). Pasting properties of starch are affected by starch granule size, amylose and lipid content, and amylopectin structure. Amylopectin is primarily responsible for granule swelling, whereas amylose and lipid restrict the swelling (Tester and Morrison 1990). The B-type starch granules had more lipids than the A-type starch granules, and lipids can form helical complexes with amylose, which restricted granule swelling (Ao and Jane 2007). Thus, the B-type starch granules developed lower peak viscosity at a higher pasting temperature.

5 Conclusion

A wide variation in volume percentages of A- and B-type starch granules was observed in Chinese wheat cultivars. Volume percentage of A- and B-type granules had ranges of 68.4-88.9% and 9.7-27.9% in 2010—2011, 74.1-90.1% and 7.2-25.3% in 2011—2012. Wheatear, Zhou 9811-1, Ruzhou 0319 and Chuanmai 42-white with higher volume percentage of A-type granules and lower volume percentage of B-type granules, and 07CA266 and Bainong 64 with higher volume percentage of B-type granules and lower volume percentage of A-type granules could serve as parents in breeding program for selecting high and low amylose wheat cultivars, respectively. The unfractionated, A- and B-type starch granules in four leading cultivars presented significant differences in granule morphology, IR ratio of 1 047 cm^{-1}/1 022 cm^{-1}, amylopectin chain length distribution, relative crystallinity, gelatinization, and pasting properties. Some differences were also observed for amylopectin chain length distribution of A- and B-type granules separated from starches of four tested cultivars. The A-type granules starch had a higher amount of ordered short-range structure, but a lower relative crystallinity as compared with the B-type starch granules. The A-type granules starch displayed higher gelatinization onset (T_o) temperatures and enthalpies (ΔH), and lower gelatinization conclusion temperatures (T_c) than the B-type starch granules. The A-type starch granules compared to the B-type granules starch had greater peak, though, and final viscosity, and lower breakdown viscosity and pasting temperature.

Acknowledgements

The authors gratefully acknowledge the financial support from the National Natural Science Foundation of China (31171547, 31401651).

References

Ao Z, Jane J L. 2007. Characterization and modeling of the A- and B-granule starches of wheat, triticale, and barley. *Carbohydrate Polymers*, 67, 46-55.

Bechtel D B, Zayas I, Kaleikau L, Pomeranz Y. 1990. Size- distribution of wheat starch granules during endosperm development. *Cereal Chemistry*, 67, 59-63.

BecklesD M, Thitisaksakul M. 2014. How environmental stress affects starch composition and functionality in cereal endosperm. *Starch/Stärke*, 66, 58-71.

Bertolini A C, Souza E, Nelson J E, Huber K C. 2003. Composition and reactivity of A- and B-type starch granules of normal, partial waxy, and waxy wheat. *Cereal Chemistry*, 80, 544-549.

Capron I, Robert P, Colonna P, Brogly M, Planchot V. 2007. Starch in rubbery and glassy states by FTIR spectroscopy. *Carbohydrate Polymers*, 68, 249-259.

Chiotelli E, Le Meste M. 2002. Effect of B- and A-wheat starch granules on thermomechanical behavior of starch. *Cereal Chemistry*, 79, 286-293.

Cooke D, Gidley M J. 1992. Loss of crystalline and molecular order during starch gelatinization: Origin of the enthalpic transition. *Carbohydrate Research*, 227, 103-112.

Dai Z M, Yin Y P, Wang Z L. 2009. Starch granule size distribution from seven wheat cultivars under different water regimes. *Cereal Chemistry*, 86, 82-87.

Dengate H, Meredith P. 1984. Variation in size distribution of starch granules from wheat grain. *Journal of Cereal Science*, 2, 83-90.

Eliasson A C, Karlsson R. 1983. Gelatinization properties of different size classes of wheat starch granules measured with differential scanning calorimetry. *Starch/Stärke*, 35, 130-133.

Fortuna T, Januszewska R, Juszczak L, Kielski A, Palasinski M. 2000. The influence of starch pore characteristics on pasting behaviour. *International Journal of Food Science and Technology*, 35, 285-291.

Franco C M L, Wong K S, Yoo S H, Jane J L. 2002. Structural and functional characteristics of selected soft wheat starches. *Cereal Chemistry*, 79, 243-248.

Geera B P, Nelson J E, Souza E, and Huber K C. 2006. Composition and properties of A- and B-type starch granules of wild-type, partial waxy, and waxy soft wheat. *Cereal Chemistry*, 83, 551-557.

Ghiasi K, Hoseney R C, Varriano-Marston E. 1982. Gelatinization of wheat starch III. Comparison by differential scanning calorimetry and light microscopy. *Cereal Chemistry*, 59, 258-262.

Genkina N K, Wikman J, Bertoft E, Yuryev V P. 2007. Effects of structural imperfection on gelatinization characteristics of amylopectin starches with A- and B-type crystallinity. *Biomacromolecules*, 8, 2329-2335.

Jane J, Kasemsuwan T, Leas S, Zobel H, Robyt J F. 1994. Anthlolgy of starch granule morphology by scanning electron microscopy. *Starch/Stärke*, 46, 121-129.

Kim H S. 2009. Wheat starch A- and B-type granule microstructure and reactivity. Ph D thesis, University of Idaho, USA.

Kim H S, Huber K C. 2008. Channels within soft wheat starch A- and B-type granules. *Journal of Cereal Science*, 48, 159-172.

Kim H S, Huber K C. 2010a. Impact of A/B-type granule ratio on reactivity, swelling, gelatinization, and pasting properties of modified wheat starch. Part I: Hydroxypropylation. *Carbohydrate Polymers*, 80, 94-104.

Kim H S, Huber K C. 2010b. Physicochemical properties and amylopectin fine structures of A- and B-type granules of waxy and normal soft wheat starch. *Journal of Cereal Science*, 51, 256-264.

Koroteeva D A, Kiseleva V I, Krivandin A V, Shatalova O V, Blaszczak W, Bertoft E. 2007a. Structural and thermodynamic properties of rice starches with different genetic background. Part 2. Defectiveness of different supramolecular structures in starch granules. *International Journal of Biological Macromolecules*, 41, 534-547.

Koroteeva D A, Kiseleva V I, Sriroth K, Piyachomkwan K, Bertoft E, Yuryev P V. 2007b. Structural and thermodynamic properties of rice starches with different genetic background. Part 1. Differentiation of amylopectin and amylose defects. *International Journal of Biological Macromolecules*, 41, 391-403.

Kozlov S S, Krivandin A V, Shatalova O V, Noda T, Bertoft E, Fornal J. 2007. Structure of starches extracted from near-isogenic wheat lines. Part II. Molecular organization of amylopectin clusters. *Journal of Thermal Analysis and Calorimetry*, 87, 575-584.

Li J H, Vasanthan T, Rossnagel B, Hoover R. 2001. Starch from hull-less barley: I. Granule morphology, composition and amylopectin structure. *Food Chemistry*, 74, 395-405.

Li W, Yan S, Yin Y, Wang Z. 2010. Starch granules size distribution in wheat grain in relation to shading after anthesis. *Journal of Agricultural Science*, 148, 183-189.

Li W H, Shan Y L, Xiao X L, Zheng J M. 2013. Effect of nitrogen and sulfur fertilization on accumulation characteristics and physicochemical properties of A- and B-wheat starch. *Journal of Agricultural and Food Chemistry*, 61, 2418-2425.

Lindeboom N, Chang P R, Tyler R T. 2004. Analytical, biochemical and physicochemical aspects of starch granule size with emphasis on B- granule starches: A review. *Starch/ Stärke*, 56, 89-99.

Liu P, Guo W, Jiang Z, Pu H. 2011. Effects of high temperature after anthesis on starch granules in grains of wheat (*Triticum aestivum* L.). *Journal of Agricultural Science*, 149, 159-169.

Liu Q, Gu Z, Donner E, Tetlow I, Emes M. 2007. Investigation of digestibility *in vitro* and physicochemical properties of A- and B-type starch from soft and hard wheat flour. *Cereal Chemistry*, 84, 15-21.

Ni Y, Wang Z, Yin Y, Li W. 2012. Starch granules size distribution in wheat grain in relation to phosphorus fertilization. *Journal of Agricultural Science*, 150, 45-52.

Noda T, Takahata Y, Sato T, Suda I, Morishita T, Ishiguro K, Yamakawa O. 1998. Relationships between chain length distribution of amylopectin and gelatinization properties within the same botanical origin for sweet potato and buckwheat. *Carbohydrate Polymers*, 37, 153-158.

Park S H, Chung O K, Seib P A. 2005. Effects of varying weight ratios of A- and B- wheat starch granules on experimental straight-dough bread. *Cereal Chemistry*, 82, 166-172.

Park S H, Wilson J D, Chung O K, Seib P A. 2004. Size distribution and properties of wheat starch granules in relation to crumb grain score of pup-loaf bread. *Cereal Chemistry*, 81, 699-704.

Park S H, Wilson J D, Seabourn B W. 2009. Starch granule size distribution of hard red winter and hard red spring wheat: Its effects on mixing and bread-making quality. *Journal of Cereal Science*, 49, 98-105.

Peng M, Gao M, Abdel-Aal E S M, Hucl P, Chibbar R N. 1999. Separation and characterization of A- and B-type starch granules in wheat endosperm. *Cereal Chemistry*, 76, 375-379.

Peng M, Gao M, Baga M, Hucl P, Chibbar R N. 2000. Starch- branching enzymes preferentially associated with A-type starch granules in wheat endosperm. *Plant Physiology*, 124, 265-272.

Raeker M O, Gaines C S, Finney P L, Donelson T. 1998. Granule size distributions and chemical composition of starches from 12 soft wheat cultivars. *Cereal Chemistry*, 75, 721-728.

Rahman S, Bird A, Regina A, Li Z Y, Ral J P, McMaugh S, Topping D, Morell M. 2007. Resistant starch in cereal: Exploiting genetic engineering and genetic variation. *Journal of Cereal Science*, 46, 251-260.

Sahal D, Jackson D S. 1996. Structural and chemical properties of native corn starch granules. *Starch/Stärke*, 48, 249-255.

Sahlström S, Bævre A B, Bråthen E. 2003. Impact of starch properties on health bread characteristics. II. Purified A- and B-granule fractions. *Journal of Cereal Science*, 37, 285-293.

Salman H, Blazek J, Lopez-Rubio A, Gilbert E P, Hanley T, Copeland L. 2009. Structure-function relationships in A and B granules from wheat starches of similar amylose content. *Carbohydrate Polymers*, 75, 420-427.

Shinde S V, Nelson J E, Huber K C. 2003. Soft wheat starch pasting behavior in relation to A- and B-type granule content and composition. *Cereal Chemistry*, 80, 91-98.

Van Soest J J G, Tournois H, De Wit D, Vliegenthart J F G. 1995. Shot-range structure in (partially) crystalline potato starch determined with attenuated total reflectance Fourier- transform IR spectroscopy. *Carbohydrate Research*, 279, 201-214.

Soh H N, Sissons M J, Turner M A. 2006. Effect of starch granule size distribution and elevated amylose content on durum dough rheology and spaghetti cooking quality. *Cereal Chemistry*, 83, 513-519.

Song Y, Jane J. 2000. Characterization of barley starches of waxy, normal and high amyloase varieties. *Carbohydrate Polymers*, 41, 365-377.

Soulaka A B, Morrison W R. 1985. The amylose and lipid contents, dimensions, and gelatinization characteristics of some wheat starches and their A- and B-granule fractions. *Journal of the Science of Food and Agriculture*, 36, 709-718.

Stoddard F L. 1999. Survey of starch particle-size distribution in wheat and related species. *Cereal Chemistry*, 76, 145-149.

Stoddard F L. 2000. Genetics of wheat starch B-granule content. *Euphytica*, 112, 23-31.

Stoddard F L. 2003. Genetics of starch granule size distribution in tetraploid and hexaploid wheats. *Australian Journal of Agricultural Research*, 54, 637-648.

Tester R, Morrison W R. 1990. Swelling and gelatinization of cereal starches. I. Effects of amylopectin, amylose and lipids. *Cereal Chemistry*, 67, 551-557.

Van H P, Morita N. 2005. Physicochemical properties

of hydroxypropylated and cross-linked starches from A-type and B-type wheat starch granules. *Carbohydrate Polymers*, 59, 239-264.

Vermeylen R, Goderis B, Reynaers H, Delcour J A. 2005. Gelatinization related structural aspects of B- and A- wheat starch granules. *Carbohydrate Polymers*, 62, 170-181.

Wang S, Copeland L. 2012. Effect of alkali treatment on structure and function of pea starch gramles. *Food Chemistry*, 135, 1635-1642.

Wang S, Copeland L. 2013. Molecular disassembly of starch granules during gelatinization and its effect on starch digestibility: A review. *Food & Function*, 4, 1564-1580.

Wang S, Yu J, Zhu Q, Yu J, Jin F. 2009. Granular structure and allomorph position in C-type Chinese yam starch granule revealed by SEM, 13C CP/MAS NMR and XRD. *Food Hydrocolloids*, 23, 426-433.

Wang S J, Luo H, Zhang J, Zhang Y, He Z H, Wang S. 2014. Alkali-induced changes in functional properties and *in vitro* digestibility of wheat starch: The role of surface proteins and lipids. *Journal of Agricultural and Food Chemistry*, 62, 3636-3643.

Yin Y A, Qi J C, Li W H, Cao L P, Wang Z B. 2012. Formation and developmental characteristics of A- and B-type starch granules in wheat endosperm. *Journal of Integrative Agriculture*, 11, 73-81.

Yoo S H, Jane J. 2002. Structural and physical characteristics of waxy and other wheat starches. *Carbohydrate Polymers*, 49, 297-305.

Zhang T, Wang Z, Yin Y, Cai R. 2010. Starch content and granule size distribution in grains of wheat in relation to post-anthesis water deficits. *Journal of Agronomy and Crop Science*, 196, 1-8.

Effects of Wheat Starch Granule Size Distribution on Qualities of Chinese Steamed Bread and Raw White Noodles

Qi Guo,[1,2] Zhonghu He,[1,3] Xianchun Xia,[1] Yanying Qu,[2] and Yan Zhang[1,4]

[1] *Institute of Crop Science/National Wheat Improvement Center, Chinese Academy of Agricultural Sciences (CAAS), 12 Zhongguancun South Street, Beijing 100081, China.*

[2] *College of Agronomy, Xinjiang Agricultural University, 42 Nanchang Road, Urumqi 830052, Xinjiang, China.*

[3] *CIMMYT China Office, C/O CAAS, 12 Zhongguancun South Street, Beijing 100081, China.*

[4] *Corresponding author. Phone: +86-10-82108741. E-mail: zhangyan07@caas.cn*

Abstract: Starch is a crucial component determining the processing quality of wheat-based products such as Chinese steamed bread (CSB) and raw white noodles (RWN). Flour from wheat cultivar Zhongmai 175 was used for fractionation into starch, gluten, and water solubles by hand washing. The starch fraction was successfully separated into large (>10μm diameter) and small starch granules (<10μm diameter) by repeated sedimentation. Flour fractions were reconstituted to original levels in the flour by using constant gluten and water solubles and varying the weight ratio of large and small starch granules. As the proportion of small granules increased in the reconstituted flours, farinograph water absorption increased, and amylose content, pasting peak viscosity, trough, and final viscosity decreased. Starch granule size distribution significantly affected processing quality of CSB and RWN. Superior crumb structure score (12.0) was observed in CSB made from reconstituted flour with 35% small starch granules. CSB made from reconstituted flours with 30 and 35% small starch granules exhibited the highest total scores, with values of 85.4 and 83.3, respectively. Significant improvements in color, viscoelasticity, and smoothness of RWN were obtained with an increase in small starch granule content, and reconstituted flours with 30-40% small starch granules produced RWN with moderate firmness.

China is the largest wheat (*Triticum aestivum* L.) producer and consumer in the world (FAO 2014). Noodles and steamed bread are the dominant products consumed in China, each sharing about 40% of wheat consumption. As living standards increase, genetic improvement of noodle and steamed bread qualities becomes a major breeding objective (He et al 2010a, 2010b). Chinese steamed bread (CSB) is mostly consumed in northern China, although southern-style steamed bread is also consumed (He et al 2003). Previous studies indicated that protein quality and starch pasting properties were associated with northern-style CSB quality (Huang et al 1996; Crosbie 1998; Liu et al 2000). Many varieties of noodles are consumed across China; however, raw white noodles (RWN) are the most popular type. A number of studies showed that protein content and quality, starch properties, and color-associated traits such as polyphenol oxidase activity and yellow pigment were responsible for RWN quality (Ge et al 2003; Liu et al 2003b; He et al

Published in Cereal Chemistry, 2014.9 (16): 623-630.

2005, 2010b). Crosbie (1991) indicated that starch swelling power was significantly correlated with total texture score of the cooked noodles. Liu et al (2003a) reported that flour with high pasting peak viscosity, breakdown, and flour swelling volume contributed to good noodle quality. Therefore, starch properties are an important factor in determining both steamed bread and noodle qualities.

Starch is the main component of endosperm in wheat grain. Mature wheat endosperm contains at least two populations of starch granules: large A-type starch granules with a diameter of 10-35μm and small B-type starch granules with an average diameter of <10 μm, thus showing a bimodal granule size distribution (Evers 1973; Bechtel et al 1990; Raeker et al 1998; Peng et al 1999). Differences in granule morphology, structure, and composition of the two populations of starch granules during synthesis lead to significant differences in their functional properties (Parker 1985; Bechtel et al 1990; Bechtel and Wilson 2003; Kim and Huber 2010; Wei et al 2010). A-type granules have higher gelatinization enthalpy, amylose content, and pasting parameters such as peak viscosity, trough, breakdown, final viscosity, and setback and lower gelatinization onset and peak temperature, whereas B-type granules have higher lipid-complexed amylose and swelling power, broader gelatinization ranges, and lower gelatinization enthalpy (Sahlström et al 2003a, 2003b; Geera et al 2006; Soh et al 2006; Kim and Huber 2010; Yin et al 2012).

Differences in starch granule size distributions had significant effects on textural and processing properties of end products in wheat. Edwards et al (2002) reported that a higher proportion of small starch granules increased dough elasticity. Binding more water, B-type granules were likely to increase dough stiffness and reduce the flow properties of the dough during fermentation and baking. Small A-granules (size about 12μm) had a large impact on weight and form ratio of bread (Sahlström et al 1998). The optimum proportion of B-granules was 25-30% by weight of total starch; beyond that, loaf volume decreased (Soulaka and Morrison 1985). Park et al (2005) found that bread made from reconstituted flour with 30% small granule and 70% large granule starch by weight of total starch had the best crumb grain and the highest peak fineness value. As the proportion of small granules increased in the reconstituted flour, bread made from these flours became softer in texture but had an extended storage life. These inconsistent results were probably because of differences between the experimental approaches used to prepare granule fractions and in the baking methods used by researchers. Other reports showed that significant increases in pasta firmness and slight reductions in stickiness were observed when B-type granule content increased. The optimum proportion of B-type granules for pasta ranged from 32 to 44% by volume (Soh et al 2006). A higher proportion of A-granules within the range 60-73% contributed to the positive quality of white salted noodles (Black et al 2000).

These resultsindicated that starch granule size and size distribution were important factors in determining CSB and RWN quality. However, there was insufficient information to demonstrate the effect of wheat starch granule distribution on processing quality of CSB and RWN. Thus, any association of starch granule size with CSB and RWN qualities remains unknown. The objective of this study was to understand the effects of varying the ratio of large and small starch granules on flour characteristics and processing quality of CSB and RWN by using the reconstitution method.

MATERIALS AND METHODS

Wheat Grain. Zhongmai 175, characterized by high yielding potential, good noodle and steamed bread qualities, and broad adaptation, was released in the Northern China Plain Winter Wheat Region and rainfed area in the Yellow and Huai Valley's Facultative Wheat Region by the Ministry of Agriculture of China. It is currently a leading cultivar in the Northern

China Plain and also serves as a check cultivar in the Regional Cultivar Testing Trial. Zhongmai 175 was grown under local management at Shunyi Station, Crop Science Institute, Chinese Academy of Agricultural Sciences, during the 2011—2012 cropping season and harvested for the present study. The grain was tempered to 14% moisture content for 16 h and milled with a Buhler MLU 202 laboratory experimental mill (Buhler, Uzwil, Switzerland) following AACC International Approved Method 26－21.02 to give flour with the straight-run extraction rate.

Fraction Preparation. To meet the needs of sensory evaluation of quality parameters of noodles made from reconstituted flours, distilled water was used to isolate and separate fractions. Starch, gluten, water solubles, and tailings were isolated by the handwashing method described by Park et al (2005) with slight modifications. Flour (300 g) was mixed to the optimum level with 200 mL of distilled water and was rested for 15 min before the dough was kneaded in 200 mL of chilled distilled water in a 2 L beaker. The kneading and washing of the elastic mass was repeated with 800 mL of chilled distilled water (about 100 mL of chilled distilled water per repeat) until no starch leached into the water. The starch suspension was decanted and strained through a nylon sieve (200 μm mesh) to remove the small pieces of elastic mass, which were subsequently returned to the gluten mass, and a nylon sieve (75 μm mesh) was used to remove fine fiber and a small proportion of other unknown particles. The wet gluten mass was rested for 1 h. Prime starch, water solubles, and tailings contained in the starch suspension were isolated by repeated centrifugations (2, 000 $\times g$, 15 min). Large and small starch fractions were separated from prime starch by repeated suspension in six cycles (Park et al 2005). The sediment comprised mainly A-type starch granules, and the supernatant contained mainly B-type granules. The sediment and supernatant were centrifuged for 15 min (4, 000 $\times g$), and the large and small starch granule factions were assembled. All fractions were frozen, lyophilized, and ground with a household coffee mill to pass through a 149 μm mesh sieve.

TABLE I Composition of Reconstituted Flours[w]

Reconstituted Flours	Small Granules[x]	Large Granules[x]	Tailings	Gluten	Water Solubles
A[y]	9.6 (12.1)	61.5 (77.5)	8.3 (10.4)	15.6	5
B[z]	10.7 (13.5)	68.7 (86.5)	0	15.6	5
C (0)	0 (0)	79.4 (100)	0	15.6	5
D (5)	4.0 (5)	75.4 (95)	0	15.6	5
E (10)	8.0 (10)	71.5 (90)	0	15.6	5
F (15)	11.9 (15)	67.5 (85)	0	15.6	5
G (20)	15.9 (20)	63.5 (80)	0	15.6	5
H (25)	19.8 (25)	59.6 (75)	0	15.6	5
I (30)	23.8 (30)	55.6 (70)	0	15.6	5
J (35)	27.8 (35)	51.6 (65)	0	15.6	5
K (40)	31.8 (40)	47.6 (60)	0	15.6	5
L (45)	35.7 (45)	43.7 (55)	0	15.6	5
M (50)	39.7 (50)	39.7 (50)	0	15.6	5

(continued)

Reconstituted Flours	Small Granules[x]	Large Granules[x]	Tailings	Gluten	Water Solubles
N (60)	47.6 (60)	31.8 (40)	0	15.6	5
O (80)	63.5 (80)	15.9 (20)	0	15.6	5
P (100)	79.4 (100)	0 (0)	0	15.6	5

[w] Weight percentages of fractions (dry weight).

[x] Weight proportions of fractions in total solid, followed by weight proportions of starch granule fractions in total starch (in parentheses).

[y] Reconstituted flour of original components and proportions with tailings.

[z] Reconstituted flour of original proportions without tailings.

The sizes of large and small starch fractions were examined with a scanning electron microscope (Kim and Huber 2008), and starch granule size distributions were determined by a HELOS and RODOS laser diffraction particle size analyzer (Japan Laser Co., Tokyo, Japan).

Flour Reconstitution. For reconstituting the fractions, the original weights were maintained for gluten, water solubles, and starch fractions, with starch composites containing variable ratios of large and small starch granules. Only one of the reconstituted flours included the tailings fraction. The tailings contained mainly starch granules (Shinde et al 2003). Finally, the reconstituted flours were rehydrated to ≈14% moisture in a National fermentation cabinet (20℃ and 85% rh) (National Manufacturing, Lincoln, NE, U. S. A.) for 5 h with regular stirring every 15 min. Fractional proportions of reconstituted flours are shown in Table I.

Reconstituted Flour Quality. Flour protein content was determined with a Foss-Tecator 1241 near-infrared transmittance analyzer (Foss, Höganäs, Sweden). Flour moisture content and farinograph parameters were determined according to AACCI Approved Methods 44-16.01 and 54-21.01, respectively. Flour pasting parameters were determined with a rapid viscosity analyzer (RVA-3D Super, Newport Scientific, Warriewood, NSW, Australia). Flours (≈3.5 g) corrected to 14% moisture content were weighed into the canister and stirred in distilled water to achieve a weight of 28.5 g (Batey and Curtin 2000).

Amylose Content Determination. A 0.1 mg flour sample was dissolved in 1 mL of ethanol (95%) and 9 mL of NaOH (1N) solution and placed overnight in a refrigerator at 4℃. Distilled water was added to make up 100 mL in a volumetric flask containing the dissolved reconstituted flour. The amount of amylose was determined with an automated chemistry analyzer (FS 3100, OI Analytical, College Station, TX, U. S. A.) according to the operator's manual. Rice amylose concentrations of 1.5, 10.4, 16.2, 19.3, and 22.5% from the China National Rice Research Institute, Chinese Academy of Agricultural Sciences, were used to form the standard curve. An additional rice starch with 26.5% amylose was used as the control sample.

CSB Preparation and Quality Evaluation. CSB was prepared and evaluated according to the method of Chen et al (2007) with minor modifications. The optimum water addition was set at 80% of farinograph water absorption for flour samples. Flour (200 g) was mixed with yeast (SAF-instant, Lesaffre, Marcq-en-Barœul, France) slurry and water in a National mixer for 1-2 min. The dough was divided into two parts with similar weights and then sheeted by passing 10 times through a pair of rollers set with a gap of 9/32 in. (National Manufacturing). After each pass, the sheeted dough was folded along the side and rotated to 90° before the next pass through the rollers. The dough piece was gently shaped by hand to form a rounded piece with a smooth upper surface. The dough piece was placed on a platform and then rounded with a suitably sized bowl five times in a space of 40 cm diameter and then five times more in a space of 28-30 cm diameter. The rounded dough pieces were proofed for 20 min in a National fermentation cabinet (35℃, 85% rh) and steamed for 25 min in a steamer containing cold water. The CSB score included specific volume (weighting, 20), skin color (10), smoothness (10), shape (10), structure (15), and stress relaxation (35). Specific volume indicated volume/weight ratio of steamed bread. Skin color measurement was carried out with a Minolta CR 310 chromameter (Minolta Camera, Osaka, Japan). Stress relaxation was measured with a TA-XT2i texture analyzer (Stable Micro Systems, Surrey, Godalming, U.K.). Smoothness, shape, and structure of steamed bread were scored subjectively by the panelists. A high score for smoothness was given to very smooth skin, free of wrinkles, dimples, blisters, or gelatinized spots. A round shape and fine crumb with uniform porosity contributed to high scores for shape and structure, respectively.

RWN Preparation and Quality Evaluation. RWN were prepared and evaluated according to the method described by Zhang et al (2005). At least five trained panelists performed the sensory evaluation and evaluated six parameters, namely, color (weighting, 15), appearance (10), firmness (20), viscoelasticity (30), smoothness (15), and taste or flavor (10). The experiment was performed at room temperature (20-25℃) and 50-60% rh. For sensory evaluation of noodle quality, a well-known Chinese commercial flour (Hetao Xuehua flour) was used as a control.

Noodles (150 g) were cooked in 2 L of boiling distilled water for 6 min. The cooking water was collected and weighed. Cooking loss was determined by evaporating the cooking water to dryness in a preweighed glass beaker in an air oven at 110℃. The residue was weighed and reported as a percentage of the weight of dry noodles before cooking (Collado et al 2001).

Color Measurements of Noodle Sheets. The color of noodle sheets was assessed with a Minolta CR 310 chromameter equipped with a D65 illuminant using the CIE 1976 L^*, a^*, b^* color scale. Sheets after sheeting were measured on a Royal Australian Cereal Institute standard backing tile, with three measurements being made on each side of the noodle sheet.

Statistical Analysis. All tests were conducted at least twice. Analysis of variance was performed with SAS software version 9.2 (SAS Institute, Cary, NC, U.S.A.). Least square means were calculated for each parameter and used to test the significance of differences ($P<0.05$) between samples.

RESULTS AND DISCUSSION

Flour and Starch Fractionation and Reconstitution. Flour was fractionated, and the dry weight percentages of five fractions were as follows: small starch granules, 9.6%; large starch granules, 61.5%; tailings, 8.3%; gluten, 15.6%; and water solubles, 5.0%. The tailings consisted for the most part of the endosperm cell walls with some aleurone and bran fragments, having an approximate composition of 87-94% starch, 4% pentosans, 0.7% lipids, and 3.0% cellulosic components (MacMasters and Hilbert 1944). The water solubles contained essentially soluble proteins (albumin and globulin), soluble pentosans, and many other low- molecular-weight compounds (Sollars 1959; Miller and Hoseney 1997). Isolated prime starch and separated large and small starch granule fractions were examined to verify the efficiency of separation. The separation was considered successful based on data shown in Table II. According to the data from the laser diffraction particle size analyzer, the fraction with large starch granules consisted of 94.2% starch granules >10 μm in diameter, and the fraction with small granules contained 96.7% starch granules <10 μm in diameter.

Flours were reconstituted to the original composition of Zhongmai 175 with the tailings fraction (flour A) and without the tailings fraction (flour B), and 14 different starch composites, containing small starch granule proportions of 0, 5, 10, 15, 20, 25, 30, 35, 40, 45, 50, 60, 80, and 100%, were designated as flours C-P (Table I). The starch-interchanged flours contained (by weight of dry solids) 79.4% starch, 15.6% gluten, and 5.0% water solubles but not the tailings fraction.

TABLE II Volume Percent of A-and B-Type Starch Granules in Isolated Starch Granule Fractions

Starch Granule Fractions	A-Type (%)	B-Type (%)
Prime	83.91	16.09
Large	94.22	5.78
Small	3.35	96.65

Protein Characteristics of Reconstituted Flours. Farinograph parameters of the original and reconstituted flours are shown in Table III. Compared with the original flour, flour A had lower water absorption, development time, and stability. The reasons for this change might be that the freeze-drying method caused structural changes in the gluten, that the dehydration procedure altered the rheological properties (Bache and Donald 1998), or that the gluten fractions influenced mixing properties (Alamri et al 2010). With the exception of flour A, there was no significant difference between the protein contents of any of the flours. Flour B showed significantly lower water absorption and stability than flour A because flour A also contained the tailings fraction, which was rich in starch (Shinde et al 2003), pentosans (D'Appolonia and MacArthur 1975), and protein (Czuchajowska and Pomeranz 1993) with a high water-holding capacity (Shogren et al 1987). Water absorptions of flours C-J were not significantly changed with increases of 0-35% small starch granules. However, water absorptions of flours K-P showed significant increase with additions of small starch granules to beyond 40% of total starch. These results were in agreement with previous reports that small starch granules have higher water-binding capacity because of their high specific surface areas (Vasanthan and Bhatty 1996; Stoddard 1999). Development time and stability of reconstituted flours containing small starch granules at proportions of 0-15% were slightly lower than those with proportions of 20-100%, and no significant differences were observed within the two range groups. This observation was inconsistent with those of Park et al (2005) indicating that bake mixing time decreased with increases in small granules. Zhongmai 175 is classified as a weak gluten cultivar, with a smaller variation in dough gluten strength.

TABLE III

Farinograph Parameters of Original and Reconstituted Flours[y]

Flour Type	Water Absorption (%)	Development Time (min)	Stability (min)
Original	50.7b	1.8a	3.4a
A	48.9d	1.1bcd	2.6b
B	43.7m	0.7cde	1.8de
C (0)	45.9gh	0.5e	1.5e
D (5)	44.9ijk	0.5e	1.6e
E (10)	44.5kl	0.6de	1.6e
F (15)	44.5kl	0.7cde	1.9cde
G (20)	44.2lm	0.9bcde	2.3bc
H (25)	44.8jkl	1.0bcd	2.2bcd
I (30)	45.4hij	0.9bcde	2.3bc
J (35)	45.0ijk	1.2bc	2.3bc
K (40)	45.5hi	1.2bc	2.3bc
L (45)	46.4fg	1.3b	2.4bc
M (50)	47.0ef	1.2bc	2.3bc
N (60)	47.6e	1.1bcd	2.1cd
O (80)	49.7c	1.2bc	1.9cde
P (100)	52.6a	1.1bcd	1.8de
LSD[z]	0.7	0.5	0.5

[y] Means of two measurements. Means in the same column with the same letter are not significantly different.

[z] Least significant difference ($P = 0.05$).

Starch Properties of Reconstituted Flours. The RVA pasting property values of the original flour were higher than those of the reconstituted flours (Table

IV). This profile was probably because of slight damage to flour components during fractionation and loss of fractions during repeated starch washing. Although the starch fraction is the most important determinant of flour pasting potential (Morris et al 1997), flour pasting properties are also influenced by other flour components, including proteins, pentosans, α-amylase, phosphorous, and lipids (Mathewson and Pomeranz 1978; Olkku et al 1978; Tester and Morrison 1990; Sahlström et al 2003a; Geera et al 2006). Peak viscosity, trough, and final viscosity of reconstituted flours exhibited decreasing trends with increasing proportions of small granules, consistent with earlier results (Shinde et al 2003; Soh et al 2006; Ao and Jane 2007; Wei et al 2010). With higher contents of lipid-complexed amylose of B-type granules relative to A-type granules, B-type granules produced a greater hot paste stability that delayed the RVA pasting profiles and caused higher peak viscosities of A-type granules rather than B-type (Shinde et al 2003). However, breakdown and setback of reconstituted flours with small starch granules at proportions of 0-30% were smaller than that at proportions of 35-100%, except for setback in flour P with 100% small granules. This result differed from reports that B-type starch granules displayed lower breakdown and setback values than A-type (Fortuna et al 2000; Ao and Jane 2007; Kim and Huber 2010). A possible explanation was that some fractions in the reconstituted flours such as gluten and water-soluble components also had a significant influence on pasting behavior.

TABLE IV Amylose Contents and Pasting Properties of Original and Reconstituted Flours[z]

Flour Type	Amylose (%)	Rapid Viscosity Analyzer Parameter				
		Peak (cP)	Trough (cP)	Breakdown (cP)	Final Viscosity (cP)	Setback (cP)
Original	26.7bc	3 161a	2 358b	803a	3 822a	1 464a
A	23.2gh	2 700d	2 136def	565cd	3 351cde	1 216defg
B	26.3c	2 746d	2 231cd	516cd	3 340cde	1 109ghi
C (0)	27.6a	3 065b	2 476a	590bcd	3 585b	1 109ghi
D (5)	27.7a	2 895c	2 397ab	498d	3 430c	1 033i
E (10)	27.1b	2 839c	2 310bc	530cd	3 390cd	1 080hi
F (15)	26.5c	2 697d	2 188d	510cd	3 287def	1 099hi
G (20)	25.8d	2 694d	2 177de	517cd	3 326cde	1 149fgh
H (25)	25.4d	2 602ef	2 067ef	535cd	3 231ef	1 164efgh
I (30)	24.4e	2 631e	2 049f	582cd	3 242ef	1 193efgh
J (35)	24.2ef	2 554f	1 818g	737a	3 192fg	1 375ab
K (40)	23.7fg	2 466g	1 698h	768a	3 006hi	1 308bcd
L (45)	23.5g	2 447g	1 836g	611bc	3 106gh	1 271bcde
M (50)	23.3g	2 376h	1 625h	751a	2 961ij	1 336bc
N (60)	23.4g	2 309i	1 612h	697ab	2 870j	1 258cdef
O (80)	22.8hi	2 209j	1 431i	778a	2 696k	1 265bcde
P (100)	22.5i	2 054k	1 252j	802a	2 421m	1 169efgh
LSD	0.5	63	115	111	124	115

[z] Means of two measurements. Means in the same column with the same letter are not significantly different. LSD= least significant difference (P=0.05).

Amylose content of flour A (23.2%) was significantly lower than in the original flour (26.7%) and flour B (26.3%). This difference may also be because of slight flour component damage and loss of starch granules during the isolation process along with the inclusion of the tailings fraction containing mostly small starch granules (Shinde et al 2003). Amylose content of reconstituted flours significantly decreased as small starch granule content increased, agreeing with reports that A-type starch granules had a higher amylose content relative to B-type starch granules (Shinde et al 2003; Park et al 2004; VanHung and Morita 2005; Ao and Jane 2007; Kim and Huber 2010).

CSB Quality of Reconstituted Flour. Although there was no significant variation in CSB total score among flour A, flour B, and original flour, the three flours did show significant differences in CSB quality (Table V). Flour A had worse shape, smoothness, and crumb structure and better specific volume and stress relaxation than the original flour. Flour B had better smoothness and crumb structure, lower stress relaxation, and similar specific volume and shape compared with flour A. We speculate that some flour components were slightly damaged and that fine fiber, some soluble components, and starch granules were lost during fractionation. In addition, the tailings fraction containing mostly small starch granules (Shinde et al 2003) and near 20% protein in soft wheat (Czuchajowska and Pomeranz 1993) might have reduced skin color, smoothness, and crumb structure and enhanced stress relaxation.

Specific volume, skin color, and smoothness of flours L-P showed significant decreases with increasing small starch granule contents at proportions of 40-100%, but there was no significant difference at 0-35%. This result suggests that large starch granules positively affect specific volume, skin color, and smoothness of steamed bread. Flour with more than 60% large starch granules produced steamed bread with better specific volume, skin color, and smoothness. No significant trend was observed in stress relaxation of steamed bread, which may be because stress relaxation, indicating elasticity in steamed bread, was mainly affected by the protein properties (He et al 2003; Jin et al 2013). The shapes of steamed bread made from flours C, D, E, and P showed more flatness than breads made from flours F-O; the reason might be that flours C, D, and E, containing 90-100% large starch granules, would exude more amylose during steaming, which would enhance cross-linking of amylose and starch in the starch-gluten network (Martin and Hoseney 1991). In addition, an extra stiff starch-gluten matrix system would have an adverse effect on the stability of gas cells (Park et al 2005), and thus some gas cells rupturing would lead to flatter steamed bread. On the other hand, the flat shape of steamed bread made from flour P with 100% small starch granules may be the result of softer dough owing to higher water absorption by small granules. Steamed breads made from flours G-J with 20-35% small starch granules had higher crumb structure scores of 10.5 to 12.0 compared with other reconstituted flours. Flour J with 35% small starch granules performed well with the highest crumb structure score (12.0) and had a fine and well-distributed crumb grain (Fig. 1). Flour C with 100% large starch granules and flours K-N with 40-60% small starch granules showed mediocre crumb grain scores with fair crumb fineness. Flours O and P (80 and 100% small starch granules, respectively) had low crumb grain scores with more open and rough grain. This observation was completely consistent with results for conventional bread reported by Park et al (2005) and suggested that there was a range of optimum weight ratios of small to large starch granules for producing steamed breads with a good crumb grain score. The suggested optimum weight percentage of small starch granules from this investigation (35%) was in agreement with Park et al (2005), who concluded that ≈ 30% weight percentage of small starch granules produced breads with the best crumb grain.

TABLE V Quality Evaluation of Chinese Steamed Bread Made from Original and Reconstituted Wheat Flours[z]

Flour Type	Specific Volume (20)	Stress Relaxation (35)	Skin Color (10)	Shape (10)	Smoothness (10)	Crumb Structure (15)	Total Score (100)
Original	18. 5bc	28f	7. 5ab	8. 5a	7. 8ab	12. 0a	82. 3abc
A	20. 0a	35a	6. 0bc	6. 0cd	6. 0d	9. 0fg	82. 0abcd
B	20. 0a	29ef	8. 0ab	6. 0cd	8. 3a	10. 5cd	81. 8abcd
C (0)	20. 0a	28f	8. 0ab	5. 5d	7. 3bc	9. 0fg	77. 8def
D (5)	20. 0a	23g	8. 5a	6. 0cd	8. 0ab	9. 0fg	74. 5f
E (10)	20. 0a	24g	8. 0ab	6. 0cd	7. 3bc	10. 1cde	75. 4ef
F (15)	20. 0a	30def	8. 5a	6. 5bc	7. 5ab	9. 8def	82. 3abc
G (20)	20. 0a	28f	8. 0ab	6. 5bc	8. 3a	11. 6ab	82. 4abc
H (25)	20. 0a	28f	8. 0ab	6. 5bc	7. 8ab	10. 5cd	80. 8bcd
I (30)	19. 0b	33abc	7. 5ab	7. 0b	8. 0ab	10. 9bc	85. 4a
J (35)	18. 5b	30def	8. 0ab	7. 0b	7. 8ab	12. 0a	83. 3ab
K (40)	17. 5de	31cde	7. 5ab	7. 3b	7. 3bc	9. 0fg	79. 5bcde
L (45)	18. 0cd	31cde	5. 0cd	7. 3b	6. 3d	8. 3gh	75. 8ef
M (50)	17. 0e	33abc	4. 5cd	7. 3b	6. 5cd	9. 8efg	78. 0cdef
N (60)	16. 0f	34ab	4. 5cd	7. 3b	6. 5cd	9. 4ef	77. 6def
O (80)	15. 0g	35a	3. 0de	7. 0b	5. 8d	7. 9h	73. 6f
P (100)	13. 0h	32bcd	1. 0e	6cd	4. 5e	6. 8i	63. 3g
LSD	1. 0	2. 2	2. 1	0. 9	0. 8	1. 0	4

[z] Means of two measurements. Means in the same column with the same letter are not significantly different. Number in parentheses in heading indicates full score. LSD=least significant difference (P=0. 05).

Total scores of steamed bread made from reconstituted flours exhibited a quadratic curve trend as the proportion of small starch granules increased (Fig. 2). The total scores ranged from 63. 3 to 85. 4, but flours I and J (30 and 35% small granules) with scores of 85. 4 and 83. 3, respectively, had the highest total scores in com- parison with other flours. These results clearly indicated that wheat starch granule distribution played a significant role in determining processing quality of CSB and suggested that flours containing the optimum weight percentage of small starch granules of 30-35% will produce the best quality steamed bread.

RWN Quality of Reconstituted Flour. Noodle sheet color and eating and cooking quality parameters of RWN are shown in Table VI. RWN sensory evaluation parameters for appearance, taste, and flavor were not listed because the data revealed no significant differences. Flour A exhibited lower L^* and b^* values for noodle sheets, lower firmness, viscoelasticity, and smoothness in noodle sensory evaluation, higher a^* values for noodle sheets, and higher cooking loss than the original flour. This phenomenon indicated that reconstituted flour may produce significantly inferior noodle sheet color and noodle processing quality than original flour because of the negative effects of fractionation. Flour B had a higher L^* value and firmness, lower a^* and b^* values, and lower viscoelasticity than flour A, but there were no significant differences in smoothness and cooking loss between flours A and B. Hence, the lack of the tailings fraction in flour B may have positively contributed to noodle sheet color and noodle firmness and may have reduced noodle viscoelasticity owing to flour B having a higher amylose content than flour A (Table IV).

The color of noodle sheets and sensory evaluation parameters of cooked noodles made from flours C and D were not obtained because the noodle making was not successful owing to problems with the layered noodle sheets. It seemed likely that lower water absorption and higher amylose contents in flours C and D with 95-100% large starch granules (Table IV) reduced cross-linking of the starch-gluten network and that this reduced network formation may have been less able to hold flour components together during dough sheeting in reconstituted flours C and D (Baik and Lee 2003). Increases in L^* value of noodle sheets and decreases in a^* and b^* values were observed in flours G-P with increasing small starch granule proportions of 20-100%, indicating that small starch granules positively contribute to noodle sheet color. The L^* values of noodle sheets for flours E and F were higher than flours G-K and similar to flour L. The a^* and b^* values for flours E and F were lower than for flours G-K and similar to flours L and N, respectively. A possible explanation was that flours E and F with 85-90% large starch granules, having lower water retention capacity, may produce better noodle sheet color than flours G-K containing 60-80%. However, that was not the cause of the changing trend in noodle sheet color in flours G-P, in which increasing small starch granule content should improve noodle sheet color. Constant levels of water addition in noodle making would lead to insufficient water absorption in noodle sheets as small starch granule content increased. The firmness score of noodles made from reconstituted flours I-K with 30－40% small granules was higher than other reconstituted flours. However, an increasing trend in noodle firmness with increasing small starch granule content was observed. This observation was mainly because of moderate firmness of RWN giving a high sen-sory firmness score. The current result was in agreement with those of Chen et al (2003), who observed that potato starch noo-dles containing more small granules had firmer cutting properties.

Cooked spaghetti with 32-44% (volume percentage basis) B-type granules also had the highest firmness (Soh et al 2006), and smoothness increased significantly as small starch granule content increased. Decreasing amylose content with increasing small starch granules could enhance water-holding capacity, viscoelasticity, and smoothness of cooked noodles (Toyokawa et al 1989).

Cooking loss is an important parameter for cooking quality of RWN. Small variation in cooking loss was observed for RWN made from reconstituted flours. Cooking loss ranged from 3.6 to 4.2% for flours E-P, which indicated that starch granule size could not significantly affect RWN cooking quality. In summary, starch granule size distribution significantly affected RWN quality. Improved color, viscoelasticity, and smoothness of RWN were observed as the proportion of small starch granules increased. Flours with 30-40% small granules tended to produce RWN with the desired moderate firmness.

CONCLUSIONS

This study showed that variation in wheat starch granule distribution affected qualities of CSB and RWN made from reconstituted flours. An increase in small starch granule content increased farinograph water absorption and decreased the amylose content of the reconstituted flours. Peak viscosity, trough, and final viscosity of reconstituted flours decreased with increasing proportions of small granules. CSB made from reconstituted flour with 30-35% small starch granules exhibited the best crumb structure and the highest total score. Improvements in color, viscoelasticity, and smoothness of RWN were observed with increased small starch granule content, and reconstituted flours with 30-40% small starch granules produced the preferred moderate firmness of RWN.

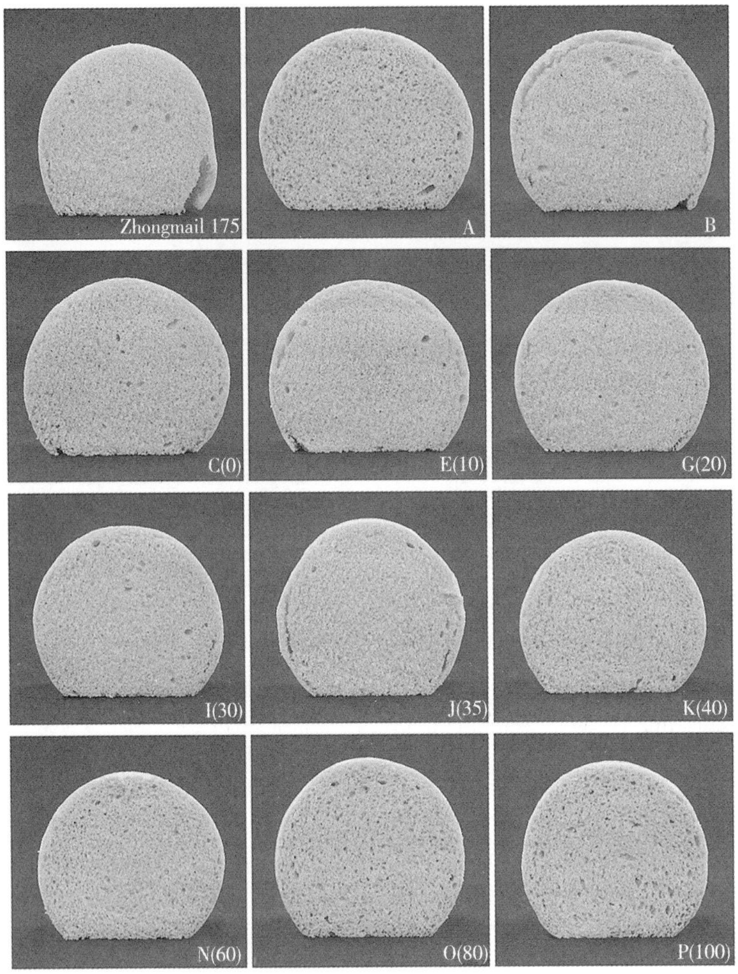

Fig. 1 Crumb structures of steamed bread made from original Zhongmai175 and reconstituted flours. Values in parentheses indicate the proportions of small starch granules in reconstituted flour samples. Composition of flours is described in Table I.

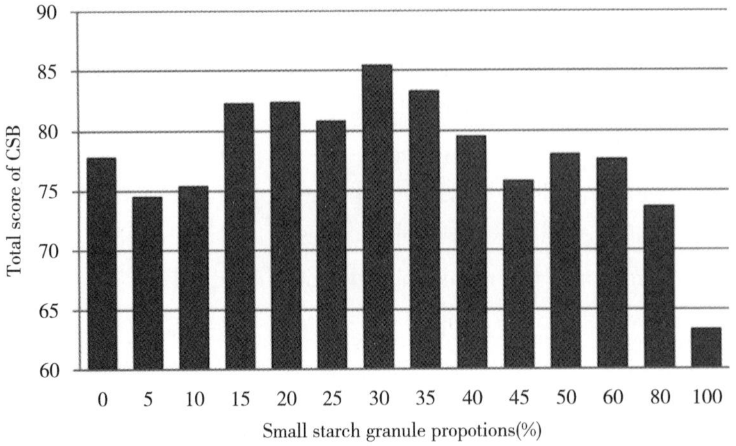

Fig. 2 Relationship between Chinese steamed bread (CSB) total scores and proportions of small starch granules in reconstituted flours.

TABLE VI Quality Evaluation of Raw White Noodles Made from Original and Reconstituted Wheat Flours[y]

Flour Type	Noodle Sheets			Sensory Evaluation Parameters			Cooking Loss (%)
	L^*	a^*	b^*	Firmness (20)	Viscoelasticity (30)	Smoothness (15)	
Original	84.13de	0.09m	20.90a	12.1e	24.0b	11.9b	3.7de
A	81.16j	1.97a	17.06b	10.1g	21.2e	10.5d	4.1abc
B	84.42cd	1.52de	13.05fg	11.2f	19.5f	10.6d	4.3a
C (0)[z]	…	…	…	…	…	…	…
D (5)[z]	…	…	…	…	…	…	…
E (10)	84.1def	1.49ef	14.26e	11.0f	19.7f	11.2c	4.2ab
F (15)	83.92ef	1.45fg	14.48e	13.3d	19.2f	9.8e	4.0bcd
G (20)	82.08i	1.77b	17.24b	13.1d	20.7e	10.5d	3.6e
H (25)	82.08i	1.73b	17.23b	13.2d	21.0e	11.4c	3.8de
I (30)	82.45i	1.65c	16.82b	14.7a	22.7d	11.3c	3.6e
J (35)	83.26gh	1.53de	15.90c	14.8a	23.3c	12.1b	4.0bcd
K (40)	83.02h	1.56d	16.02c	15.0a	22.5d	11.3c	3.9cde
L (45)	83.64fg	1.44fg	15.30d	14.2bc	22.7d	11.3c	3.6e
M (50)	84.06def	1.35hi	15.31d	14.6ab	22.7d	11.9b	3.7de
N (60)	84.45cd	1.30ij	14.49e	13.8c	23.7bc	12.7a	3.8de
O (80)	85.95b	1.10k	12.68g	13.0d	24.0bc	12.8a	4.2ab
P (100)	87.49a	0.89l	11.01h	12.2e	25.2a	12.8a	4.1abc
LSD	0.5	0.06	0.6	0.5	0.5	0.2	0.3

[y] Means of two measurements. Means in the same column with the same letter are not significantly different at $P=0.05$. Number in parentheses in heading indicates full score. LSD=least significant difference ($P=0.05$).

[z] No data because noodles from flours C (0) and D (5) were not successfully prepared because of problems with the layered noodle sheets.

❖ ACKNOWLEDGMENTS

The authors are grateful to Prof. R. A. McIntosh, Plant Breeding Insti- tute, University of Sydney, for critical review of this manuscript. This work was supported by the National Natural Science Foundation of China(31171547), National 863 Project(2012AA101105), International Collaboration Project from the Ministry of Agriculture (2011-G3), and China Agriculture Research System (CARS-3-1-3).

❖ LITERATURE CITED

AACC International. Approved Methods of Analysis, 11th Ed. Method 26-21.02. Experimental milling-Bühler method for hard wheat. Approved October 12, 1988; reapproved November 3, 1999. Method 44-16.01. Moisture-Air-oven (aluminum-plate) method. Approved April 13, 1961; reapproved November 3, 1999. Method 54-21.02. Rheological behavior of flour by farinograph: Constant flour weight procedure. Approved November 8, 1995; revised and approved January 6, 2011. Available online only. AACCI: St. Paul, MN.

Alamri, M., Manthey, F., Mergoum, M., Elias, E., and Khan, K. 2010. The effects of reconstituted semolina fractions on pasta processing and quality parameters and relationship to glutograph parameters. J. Food Technol. 8: 159-168.

Ao, Z., and Jane, J. 2007. Characterization and modeling of the A- and B-granule starches of wheat, triticale, and barley. Carbohydr. Polym. 67: 46-55.

Bache, I. C., and Donald, A. M. 1998. The structure

of the gluten network in dough: A study using environmental scanning electron microscopy. J. Cereal Sci. 28: 127-133.

Baik, B.-K., and Lee, M. -R. 2003. Effects of starch amylose content of wheat on textural properties of white salted noodles. Cereal Chem. 80: 304-309.

Batey, I. L., and Curtin, B. M. 2000. Effects on pasting viscosity of starch and flour from different operating conditions for the Rapid Visco Analyser. Cereal Chem. 77: 754-760.

Bechtel, D. B., and Wilson, J. D. 2003. Amyloplast formation and starch granule development in hard red winter wheat. Cereal Chem. 80: 175- 183.

Bechtel, D. B., Zayas, I., Kaleikau, L., and Pomeranz, Y. 1990. Size-distribution of wheat starch granules during endosperm development. Cereal Chem. 67: 59-63.

Black, C. K., Panozzo, J. F., Wright, C. L., and Lim, P. C. 2000. Survey of white salted noodle quality characteristics in wheat landraces. Cereal Chem. 77: 468-472.

Chen, F., He, Z. H., Chen, D. S., Zhang, C. L., Zhang, Y., and Xia, X. C. 2007. Influence of puroindoline alleles on milling performance and qualities of Chinese noodles, steamed bread and pan bread in spring wheats. J. Cereal Sci. 45: 59-66.

Chen, Z., Schols, H. A., and Voragen, A. G. J. 2003. Starch granule size strongly determines starch noodles processing and noodle quality. J. Food Sci. 68: 1584-1589.

Collado, L. S., Mabesa, L. B., Oates, C. G., and Corke, H. 2001. Bihon- type noodles from heat-moisture-treated sweet potato starch. J. Food Sci. 66: 604-609.

Crosbie, G. B. 1991. The relationship between starch swelling properties, paste viscosity and boiled noodle quality in wheat flours. J. Cereal Sci. 13: 145-150.

Crosbie, G. B. 1998. Wheat quality requirements of Asian foods. Euphytica 100: 155-156.

Czuchajowska, Z., and Pomeranz, Y. 1993. Protein concentrates and prime starch from wheat flours. Cereal Chem. 70: 701-706.

D' Appolonia, B. L., and MacArthur, L. A. 1975. Comparison of starch, pentosans, and sugars of some conventional height and semidwarf hard red spring wheat flours. Cereal Chem. 52: 230-239.

Edwards, N. M., Dexter, J. E., and Scanlon, M. G. 2002. Starch participation in durum dough linear viscoelastic properties. Cereal Chem. 79: 850-856.

Evers, A. D. 1973. The size distribution among starch granules in wheat endosperm. Starch/Stärke 25: 303-304.

FAO. 2014. FAOSTAT—Agriculture Database. Available at: http: // faostat. fao. org/site/339/default. aspx

Fortuna, T., Januszewska, R., Juszczak, L., Kielski, A., and Palasinski, M. 2000. The influence of starch pore characteristics on pasting behaviour. Int. J. Food Sci. Technol. 35: 285-291.

Ge, X. X., He, Z. H., Yang, J., and Zhang, Q. J. 2003. Polyphenol oxidase activities of Chinese winter wheat cultivars and correlations with quality characteristics. Acta Agron. Sin. 29: 481-485.

Geera, B. P., Nelson, J. E., Souza, E., and Huber, K. C. 2006. Composition and properties of A- and B-type starch granules of wild-type, partial waxy, and waxy soft wheat. Cereal Chem. 83: 551-557.

He, Z. H., Liu, A. H., Peña, R. J., and Tajaram, S. 2003. Suitability of Chinese wheat cultivars for production of northern style Chinese steamed bread. Euphytica 131: 155-163.

He, Z. H., Liu, L., Xia, X. C., Liu, J. J., and Peña, R. J. 2005. Composition of HMW and LMW glutenin subunits and their effects on dough properties, pan bread, and noodle quality of Chinese bread wheats. Ce- real Chem. 82: 345-350.

He, Z. H., Xia, X. C., and Bonjean, A. P. A. 2010a. Wheat improvement in China. Pages 51-68 in: Cereals in China. Z. H. He and A. P. A. Bonjean, eds. CIMMYT: Mexico, D. F.

He, Z. H., Xia, X. C., and Zhang, Y. 2010b. Breeding noodle wheat in China. Pages 1-24 in:

Asian Noodles: Science, Technology and Processing. G. G. Hou and J. W. Johnson, eds. Wiley: Hoboken, NJ.

Huang, S., Yun, S. H., Quail, K., and Moss, T. 1996. Establishment of flour quality guidelines for northern style Chinese steamed bread. J. Cereal Sci. 24: 179-185.

Jin, H., Zhang, Y., Li, G. Y., Mu, P. Y., Fan, Z., Xia, X. C., and He, Z. H. 2013. Effects of allelic variation of HMW-GS and LMW-GS on mixograph properties and Chinese noodle and steamed bread qualities in a set of Aroona near-isogenic wheat lines. J. Cereal Sci. 57: 146-152.

Kim, H. S., and Huber, K. C. 2008. Channels within soft wheat starch A- and B-type granules. J. Cereal Sci. 48: 159-172.

Kim, H. S., and Huber, K. C. 2010. Physicochemical properties and amylopectin fine structures of A- and B-type granules of waxy and normal soft wheat starch. J. Cereal Sci. 51: 256-264.

Liu, A. H., He, Z. H., Wang, G. R., Wang, D. S., Zhang, Y., and Zhou, G. Y. 2000. Investigation of wheat flour quality for northern style Chinese steamed bread. J. Chin. Cereals Oils Assn. 15: 10-14.

Liu, J. J., He, Z. H., Yang, J., Xu, Z. H., Liu, A. F., and Zhao, Z. D. 2003a. Variation of starch properties in wheat cultivars and their relationship with dry white Chinese noodle quality. Sci. Agric. Sin. 36: 7-12.

Liu, J. J., He, Z. H., Zhao, Z. D., Peña, R. J., and Rajaram, S. 2003b. Wheat quality traits and quality parameters of cooked dry white Chinese noodles. Euphytica 131: 147-154.

MacMasters, M. M., and Hilbert, G. E. 1944. The composition of the "amylodextrin" fraction of wheat flour. Cereal Chem. 21: 548-555.

Martin, M. L., and Hoseney, R. C. 1991. A mechanism of bread firming. II. Role of starch hydrolyzing enzymes. Cereal Chem. 68: 503-507.

Mathewson, P. R., and Pomeranz, Y. 1978. Hot paste viscosity and alpha-amylase susceptibility of hard red winter wheat flour. J. Food Sci. 43: 60-63.

Miller, R. A., and Hoseney, R. C. 1997. Factors in hard wheat flour re-sponsible for reduced cookie spread. Cereal Chem. 74: 330-336.

Morris, C. F., King, G. E., and Rubenthaler, G. L. 1997. Contribution of wheat flour fractions to peak hot paste viscosity. Cereal Chem. 74: 147-153.

Olkku, J., Fletcher, S. W., III, and Rha, C. 1978. Studies on wheat starch and wheat flour model paste systems. J. Food Sci. 43: 52-59.

Park, S. H., Wilson, J. D., Chung, O. K., and Seib, P. A. 2004. Size distri- bution and properties of wheat starch granules in relation to crumb grain score of pup-loaf bread. Cereal Chem. 81: 699-704.

Park, S.-H., Chung, O. K., and Seib, P. A. 2005. Effects of varying weight ratios of large and small wheat starch granules on experimental straight-dough bread. Cereal Chem. 82: 166-172.

Parker, M. L. 1985. The relationship between A-type and B-type starch granules in the developing endosperm of wheat. J. Cereal Sci. 3: 271-278.

Peng, M., Gao, M., Abdel-Aal, E.-S. M., Hucl, P., and Chibbar, R. N. 1999. Separation and characterization of A- and B-type starch granules in wheat endosperm. Cereal Chem. 76: 375-379.

Raeker, M. Ö., Gaines, C. S., Finney, P. L., and Donelson, T. 1998. Gran- ule size distribution and chemical composition of starches from 12 soft wheat cultivars. Cereal Chem. 75: 721-728.

Sahlström, S., Bråthen, E., Lea, P., and Autio, K. 1998. Influence of starch granule size distribution on bread characteristics. J. Cereal Sci. 28: 157-164.

Sahlström, S., Bævre, A. B., and Bråthen, E. 2003a. Impact of starch properties on hearth bread characteristics. I. Starch in wheat flour. J. Cereal Sci. 37: 275-284.

Sahlström, S., Bævre, A. B., and Bråthen, E. 2003b. Impact of starch properties on hearth bread characteristics. II. Purified A- and B-granule fractions. J. Cereal Sci. 37: 285-293.

Shinde, S. V., Nelson, J. E., and Huber, K. C. 2003. Soft wheat starch pasting behavior in relation to A- and B-type granule content and com- position. Cereal Chem. 80：91-98.

Shogren，M. D.，Hashimoto，S.，and Pomeranz，Y. 1987. Cereal pentosans：Their estimation and significance. II. Pentosans and breadmaking characteristics of hard red winter wheat flours. Cereal Chem. 64：35-38.

Soh，H. N.，Sissons，M. J.，and Turner，M. A. 2006. Effect of starch granule size distribution and elevated amylose content on durum dough rheology and spaghetti cooking quality. Cereal Chem. 83：513-519.

Sollars，W. F. 1959. Effects of the water-soluble constituents of wheat flour on cookie diameter. Cereal Chem. 36：498-513.

Soulaka，A. B.，and Morrison，W. R. 1985. The bread baking quality of six wheat starches differing in composition and physical properties. J. Sci. Food Agric. 36：719-727.

Stoddard，F. L. 1999. Survey of starch particle-size distribution in wheat and related species. Cereal Chem. 76：145-149.

Tester，R. F.，and Morrison，W. R. 1990. Swelling and gelatinization of cereal starches. I. Effects of amylopectin，amylose，and lipids. Cereal Chem. 67：551-557.

Toyokawa，H.，Rubenthaler，G. L.，Powers，J. R.，and Schanus，E. G. 1989. Japanese noodle qualities. II. Starch components. Cereal Chem. 66：387-391.

VanHung，P.，and Morita，N. 2005. Physicochemical properties of hy- droxypropylated and cross-linked starches from A-type and B-type wheat starch granules. Carbohydr. Polym. 59：239-246.

Vasanthan，T.，and Bhatty，R. S. 1996. Physicochemical properties of small- and large-granule starches of waxy，regular，and high-amylose barleys. Cereal Chem. 73：199-207.

Wei，C. X.，Zhang，J.，Chen，Y. F.，Zhou，W. D.，Xu，B.，Wang，Y. P.，and Chen，J. M. 2010. Physicochemical properties and development of wheat large and small starch granules during endosperm development. Acta Physiol. Plant. 32：905-916.

Yin，Y. A.，Qi，J. C.，Li，W. H.，Cao，L. P.，and Wang，Z. B. 2012. Formation and developmental characteristics of A- and B-type starch granules in wheat endosperm. J. Integr. Agric. 11：73-81.

Zhang，Y.，Nagamine，T.，He，Z. H.，Ge，X. X.，Yoshida，H.，and Peña，R. J. 2005. Variation in quality traits in common wheat as related to Chinese fresh white noodle quality. Euphytica 141：113-120.

Mineral element concentrations in grains of Chinese wheat cultivars

Yong Zhang • Qichao Song • Jun Yan • Jianwei Tang • Rongrong Zhao • Yueqiang Zhang • Zhonghu He • Chunqin Zou • Ivan Ortiz-Monasterio

Y. Zhang J. Tang Z. He
Institute of Crop Science, National Wheat Improvement Centre/The National Key Facility for Crop Gene Resources and Genetic Improvement, Chinese Academy of Agriculture Sciences (CAAS), 12 Zhongguancun South Street, Beijing 100081, China

Q. Song R. Zhao Y. Zhang C. Zou
Department of Plant Nutrition, China Agricultural University, 2 West Yuanmingyuan Road, Haidian District, Beijing 100193, China

J. Yan
Cotton Research Institute, Chinese Academy of Agricultural Sciences (CAAS), Huanghedadao, Anyang 455000, Henan Province, China

Z. He
CIMMYT-China Office, c/o CAAS, 12 Zhongguancun South Street, Beijing 100081, China
e-mail: zhhe@public3. bta. net. cn; zhangyongzhy@263. net

I. Ortiz-Monasterio
CIMMYT, Apdo. Postal 6-641, 066000 Mexico, DF, Mexico

Abstract: Investigations on concentration of mineral elements including Fe and Zn in wheat grains are important for human health. Two hundreds and sixty- five cultivars and advanced lines were collected and sown at Anyang experimental station of the Institute of Crop Science of the Chinese Academy of Agriculture Sciences in season 2005—2006 to evaluate the genetic variation of major mineral element concentrations in wheat grain. Twenty-four selected cultivars were also planted at seven representative locations in seasons 2005—2006 and 2006—2007 to evaluate the effects of genotype, environment, and genotype by environment interaction on mineral element concentrations. The 265 genotypes displayed a large variation for all mineral elements investigated including Fe and Zn, ranging from 28. 0 to 65. 4mg • kg^{-1} and 21. 4 to 58. 2mg • kg^{-1} for Fe and Zn, with mean values of 39. 2 and 32. 3mg • kg^{-1}, respectively. Jimai 26, Henong 326, and Jingdong 8 displayed high Fe and Zn concentrations, and Jimai 26 and Henong 326 also displayed high concentrations of Cu, Mg, K, P, and protein content. Jingdong 8 is the most promising leading cultivar for increasing Fe and Zn concentrations. All mineral element concentrations including Fe and Zn were largely influenced by environment effects. Production of high Fe concentra-

Published in Euphytica, 2010, 174: 303-313.

tion can be best secured at Jiaozuo and Jinan, and high Zn concentration can be best secured at Jinan and Xuzhou, since samples from these locations in the two seasons are characterized by high Fe or Zn concentration, compared with the other locations. High and significant genotype by environment interaction effects on all mineral element concentrations were also observed, with ratios of genotype by environment to genotype variances all larger than 1.20. Grain Fe concentration was highly significant and positively correlated with that of Zn, indicating a high possibility to combine high Fe and Zn traits in wheat breeding. It also indicated strong positive correlations between concentrations of Fe, Zn, and protein content.

Keywords: *Triticum aestivum* • Grain • Mineral element • Genotype by environment

Introduction

Nutritional problems related to diets of cereal suchas wheat were observed throughout the world including China, and the consequences of malnutrition create immense economic and social costs (Bouis et al. 2000; Chen 2004). A recent UN report (United Nations System Standing Committee on Nutrition 2004) has even indicated that more than half of the world's population suffers micronutrient undernourishment. It is estimated that two billion people worldwide suffer from iron deficiency, especially children and women (Stoltzfus and Dreyfuss 1998; Welch and Graham 2004), and the mineral deficiencies in the vulnerable groups has led to concerns of the Chinese government (Chen 2004). The anemia prevalence in China is about 25% on average, and can be as high as 60% for children and 35% for pregnant women in rural areas, largely due to dietary Fe deficiency. Children suffering from Fe deficiency have poor attention spans, impaired fine motor skills and less capacity for memory (Walter et al. 1997). Fe deficiency in pregnant women may cause irreversible damage to fetal brain development leading to unfulfilled intellectual development in their babies (Gordon 1997).

The situation has led to the Fe fortification program in wheat flour in China by the Joint FAO/ WHO Expert Committee on Food Additives (Hurrell et al. 1992; Chen 2004). However, fortification efforts are highly dependent on funding, and the scope is usually restricted to urban areas. Moreover, the benefits will disappear if the investment is not sustained. Bio-fortification through plant breeding, by contrast, has multiplicative advantages (Bouis et al. 2000; Calderini and Ortiz-Monasterio 2003; Ortiz- Monasterio et al. 2007). The benefits reach the total population in both urban and rural areas, and moreover, it will not require continued investment once the bio-fortified cultivars have been developed, and do not disappear after initial successful investment and research, as long as a domestic agricultural research infrastructure is maintained (Bouis et al. 2000). Therefore due to these reasons, the Consultative Group on International Agricultural Research (CGIAR) has developed a bio-fortification challenge program also known as Harvest-Plus with the objective of developing bio-fortified (with high micronutrient concentrations) staple crops such as common wheat (*Triticum aestivum* L.) through plant breeding (Bouis et al. 2000; Calderini and Ortiz-Monasterio 2003), with a hope to elevate the current level of 35 and 31 mg kg^{-1} to 57 and 41 mg kg^{-1} for Fe and Zn, respectively, based on personal daily requirement of Fe and Zn, and the information of bioavailability provided by the WHO (2006).

Common wheat is one of the most important staple food crops and the cheapest source of calories and protein for local inhabitants in China, in which the Northern China Winter Wheat Region plays the most important role, contributing for about 70% of the national wheat production and 75% or more of the commercial wheat (He et al. 2001; Zhuang 2003). Its wheat is traded across the nation, adding to local consumption in the form of various kinds of noodles and steamed bread as the main end products, especially in the rural areas. Therefore, the objectives of this study were to (1) determine the levels of mineral element concentrations in grain of current leading wheat cultivars and advanced lines from the Northern China Winter Wheat Zone for identifying promising lines with high Fe and Zn concentrations, (2) assess the effects of genotype by environment interaction on these traits, and (3) analyze the relationships among these traits and kernel charac-teristic traits including thousand kernel weight and protein content, the two very important traits for grain yield and end-use quality. The information generated from this study can be very important for Chinese wheat breeding programs, and also has potential application in other wheat producing countries since major wheat producing countries including India, Pakistan, and Turkey are also involved in the Harvest-Plus program.

Materials and methods

Wheat samples and field experiments

Two hundreds and sixty-five genotypes, including leading cultivars and advanced lines recommendedby major breeding programs in the Northern China Winter Wheat Region, were collected and grown at 1 row 1 m length plot in irrigated field trials with 2 replications at Anyang in Henan province in the 2005- 2006 crop season to investigate the current status of Fe/Zn in terms of genotypes. These genotypes repre- sent the current wheat breeding status of Chinese wheat genotypes grown in the Northern China Winter Wheat Region. Henan Province is one of the major wheat producers in this region.

In addition, knowledge of the effect of genotype by environment interaction on mineral elementconcentrations is required for the design and implementation of an efficient and economic selection for high Fe and Zn cultivars. Therefore, 24 leading winter and facultative wheat cultivars with medium to high Fe/Zn concentrations selected on the basis of Zhang et al. (2007) were sown with 2 replications at seven representative locations which cover a geographic region ranging from latitude 30N-40N and longitude 113E-120E, in the two successive wheat seasons from 2005 to 2007, to determine the effects of genotype, environment, and their interaction on mineral element concentrations.

Each trial was planted on the locally recommended date at a seeding rate of 130 kg • ha^{-1}. Genotypes were sown in a latinized alpha lattice design (Barreto et al. 1997), and one local check was also given in every 10 genotypes. Each plot consisting of 2 rows 2 m long with 0.20 m apart between rows was given standard fungicide and hand-weeded protection to ensure optimal grain yield. All trials were irrigated by flooding. Field management and timing of management practices including fertilization generally matched local commercial production practices. Grain samples were harvested and cleaned by hand to avoid potential contamination of mineral element concentrations.

Mineral element concentration and kernel characteristic trait testing

Thousand kernel weight (TKW) was determinedas an average of three samples of 200 seeds. Protein content (14% moisture basis, mb) was determined with a Near Infrared Transmittance (NIT) analyzer Foss-Tecator 1241 (Foss, Höganäs, Sweden), calibrated by the Kjeldahl method (AACC approved, AACC 2000). Mineral element concentrations of grain were measured using inductively coupled plasma atomic e-mission spectrometry (ICP-AES, OPTIMA 3300 DV) to quantify aqueous constituents following microwave digestion with $HNO_3-H_2O_2$ solution.

Statistical analyses

Prior to performing the statistical analyses, the data was checked for unusually high aluminum (Al) values which could indicate contamination of samples with soil or dust, and no samples were omitted from further analyses based on this precaution. All trials were separately analyzed by fitting an appropriate spatial model with rows and columns (Gilmour et al. 1997). The best linear unbiased predictions from the best-fit model were used as raw data for all subsequent analyses. Means, standard deviations, and ranges were determined using PROC MEANS in the Statistical Analysis System (SAS Institute 2000). Analysis of variance was conducted by PROC MIXED with genotype as fixed effect, while environment, the combination of location and year, as random for the dataset from the 24 genotypes sown at seven locations in two successive wheat seasons. Genotypic least square means across the 14 environments for all mineral element concentrations and kernel characteristic traits including thousand kernel weight and protein content were calculated and the differences among genotypes and environments were determined using the Tukey test (Zhang et al. 2004). Pearson's linear correlation coefficients among all the traits were obtained by PROC CORR (Zhang et al. 2008).

Results

Grain composition of Chinese winter and facultative wheats

The mean and range of mineral element concentrations in grain of the 265 genotypes are presentedin Table 1. The amount of all elements in grain showed a large variation among genotypes, with the highest concentration for all elements almost doubled than that of the lowest. The genotypes showed almost the same mean value and range of Fe and Zn concentration, compared with the report of Zhang et al. (2007), ranging from 28.0 to 60.2mg • kg^{-1} for Fe with mean of 39.2mg • kg^{-1}, and from 21.4 to 58.2mg • kg^{-1} for Zn with mean of 32.3mg • kg^{-1}. The thousand kernel weight and protein content of the samples ranged from 28.0 to 48.7 g and from 10.1 to 15.6%, with mean of 34.7 g and 12.6%.

In the 10 selected genotypes with the highest mineral element concentrations (Table 2), Jimai 26 and Henong 326 performed the highest concentrations for Fe (60.2 and 58.1mg • kg^{-1}), Zn (51.9 and 58.2mg • kg^{-1}), Cu (12.1 and 11.2mg • kg^{-1}), Mg (2 165 and 2 121mg • kg^{-1}), K (6 043 and 6 053mg • kg^{-1}), and protein content (15.3 and 15.6%) (Table 2), among which Henong 326 also performed the second highest concentration of Mg (70.4mg • kg^{-1}) following Nongda 3383 (70.6mg • kg^{-1}), Jimai 26 performed the highest con- centration of K. Xumai 856 and Xiaoyan 93166 had the highest concentration of Ca, Lankao 4 and Lu 954072 had the highest concentration of P.

For the 10 genotypes with the highest Fe concentration, five including Henong 326, Jimai 26, Nongda 3197, Jingdong 8, and Nongda 179 were among the genotypes with the highest Zn concentration, three including Henong 326, Jimai 26, Nongda 179 were among the genotypes with the highest P concentration, four including Henong 326, Jimai 26, Nongda 3197, and Nongda 179 were among the genotypes with the highest protein content. Among the 10 genotypes with the highest Zn concentration, five including Henong 326, Jimai 26, Zhongmai 175, Nongda 179, and Lumai 23 were among the geno- types with the highest P concentration, four including Henong 326, Jimai 26, Nongda 3197, and Nongda 179 were among the genotypes with the highest protein content. Among the 10 genotypes with the highest P concentration, four including Henong 326, Jimai 26, CA 0204, and Nongda 179 were among the genotypes with the highest protein content.

Genotype by environment interaction

Genotype, environment, and their interaction all had highly significant effects on all mineral element concentrations and kernel characteristic traits (Table 3). The environment effects were much larger than those of genotype for all traits except for thousand kernel weight and protein content, on which there were larger effect of genotype. Therefore, the environment effect were the predominant source of variation for Fe, Zn, Mn, Cu, Ca, Mg, K, and P concentrations, with the ratios of environment variance and genotype variance all larger than 2.17, indicating large differences between environments for these traits. All the traits were also heavily influenced by the effects of genotype by environment interaction except for thousand kernel weight and protein content, with the ratios of genotype by environment interaction variance and genotype variance all larger than 1.20, and the highest being 3.14 for Ca.

Jingdong 6, Jingdong 8, Zhongyou 16, Youxuan 9, and Liken 2 had high Fe concentration (C44.7mg • kg^{-1}) (Table 4). Jingdong 6, Jingdong 8, Zhong-zuo 8131-1, Youxuan 9, and Jimai 24 had high Zn concentration (C36.6 mg • kg^{-1}). Zhongzuo 8131-1, Zhongyou 16, Liken 2, and Jinmai 33 had high Mn concentration (C39.4mg • kg^{-1}). Jing 411 (7.15 mg • kg^{-1}) and Jimai 26 (7.24mg • kg^{-1}) had high Cu concentration. Zhongyou 14, Youxuan 9, Youxuan 14, Jinmai 33, and Linfen 615 had high Ca concentration (C516mg • kg^{-1}). Zhongzuo 8131-1 (1 569 mg • kg^{-1}), Zhongyou 16 (1 543mg • kg^{-1}), and Youxuan 14 (1 544mg • kg^{-1}) had high Mg concentration. Jimai 26 (4 431mg • kg^{-1}) had high K concentration. Zhongzuo 8131-1 (3 890mg • kg^{-1}) had high P concentration. Jingdong 6 (46.8 g) and Zhongyou 9507 (47.8 g) had the biggest grain size. Zhongzuo 8131-1 (15.6%) had the highest protein content.

A wide range of variation in all mineral element concentrations and kernel characteristic traits were observed. There were significant differences among samples collected from different environments (Table5). Among the 14 testing environments, samples from Anyang-06 (means from Anyang 2006 environment) (46.6mg • kg^{-1}), Jiaozuo-06 (51.8mg • kg^{-1}), Jiaozuo-07 (46.2mg • kg^{-1}), and Jinan-07 (51.0mg • kg^{-1}) were characterized by high Fe concentrations, followed by Beijing-06 and Jinan- 06, whereas samples from Beijing-07 (35.5mg • kg^{-1}), Shijiazhuang-06 (38.0 mg • kg^{-1}), Shijiazhuang-07 (33.7mg • kg^{-1}), Xuzhou-06 (35.8mg • kg^{-1}), and Xuzhou-07 (32.8 mg • kg^{-1}) were characterized by low Fe concentration. Samples from Xuzhou-06 (51.7mg • kg^{-1}) were characterized by high Zn concentration, followed by Beijing-06, Anyang-06, Jiaozuo-06, and Jinan-06, whereas samples from Shijiazhuang-06 (24.6mg • kg^{-1}), Shijiazhuang-07 (24.2mg • kg^{-1}), Anyang-07 (27.5mg • kg^{-1}), and Zhengzhou-06 (21.0mg • kg^{-1}) were characterized by low Zn concentration. Samples from Anyang- 06 (52.9mg • kg^{-1}), Anyang-07 (46.9mg • kg^{-1}), Jiaozuo-06 (52.4mg • kg^{-1}) Jiaozuo-07 (47.8mg • kg^{-1}), Zhengzhou-07 (40.0mg • kg^{-1}), and Jinan-06 (42.0mg • kg^{-1}) were characterized by high Mn concentration, whereas samples from Zhengzhou-06 (28.2mg • kg^{-1}), Xuzhou-06 (19.1 mg • kg^{-1}), and Xuzhou-07 (19.8mg • kg^{-1}) were characterized by low Mn concentration. Samples from Anyang-06 (8.49mg • kg^{-1}), Jinan-06 (8.25mg • kg^{-1}) and Xuzhou-06 (8.25mg • kg^{-1}) were characterized by high Cu concentration, whereas samples from Beijing-07 (5.48mg • kg^{-1}), Shijiazhuang-07 (5.09 mg • kg^{-1}), Anyang-07 (5.20mg • kg^{-1}), Jiaozuo-07 (5.57mg • kg^{-1}), and Zhengzhou-07 (5.46mg • kg^{-1}) were characterized by low Cu concentration.

Samples from Beijing-06 (573mg • kg^{-1}), Beijing- 07 (530mg • kg^{-1}), Jiaozuo-06 (536mg • kg^{-1}), Zhengzhou-06 (522mg • kg^{-1}), hengzhou-07 (529mg • kg^{-1}), Jinan-06 (539 mg • kg^{-1}), and Jinan-07 (518mg • kg^{-1}) were characterized by high Ca concentration, whereas samples from Anyang-07 (380mg • kg^{-1}) were characterized by low Ca concentration. Samples from Shijiazhuang-07 (1663mg • kg^{-1}) were characterized by high Mg concentration, followed by Anyang-07, Jiaozuo-06, Jiaozuo-07, Jinan-06, and Jinan-07, whereas samples from Beijing-06 (1 393mg • kg^{-1}), Anyang-06 (1 393mg • kg^{-1}), Zhengzhou-07 (1 327mg • kg^{-1}), Xuzhou-06 (1 351mg • kg^{-1}), and Xuzhou-07 (1 398 mg • kg^{-1}) were characterized by low Mg concentration. Samples from Beijing-06 (4 913mg • kg^{-1}) and Beijing-07 (4 913mg • kg^{-1}) were characterized by high K concentration, followed by Shijiazhuang-07, Anyang-06, Anyang-07, and Jinan-07, whereas samples from Jiaozuo-06 (3 616mg • kg^{-1}), Zhengzhou-06 (3 652 mg • kg^{-1}), Zhengzhou-07 (3 590mg • kg^{-1}), and Xuzhou-07 (3 320mg • kg^{-1}) were characterized by low K concentration. Samples from Anyang-06 (4 063mg • kg^{-1}) and Zhengzhou-06 (4 102mg • kg^{-1}) were characterized by high P concentration, whereas samples from Zhengzhou-07 (2 863mg · kg^{-1}) and Xuzhou-07 (2 669mg · kg^{-1}) were characterized by low P concentration.

Samples from Jiaozuo-07 (47.3 g), Xuzhou-06 (46.3 g), and Xuzhou-06 (45.0 g) were characterized by large grain size, whereas samples from Beijing-06 (33.0 g), Beijing-07 (35.1 g), and Anyang-06 (32.8 g) were characterized by small grain size. Samples from Jiaozuo-07 (14.4%) and Zhengzhou-07 (14.5%) were characterized by high protein content, followed by Beijing-06, Beijing-07, Shijiazhuang-07, Anyang-06, Anyang-07, and Jinan-06, whereas samples from Xuzhou-06 (12.3%) and Xuzhou-07 (12.6%) were characterized by low protein content.

Correlation among mineral element concentrations and kernel characteristic traits

Associations among mineral element concentrations and kernel characteristic traits including thousand kernel weight and protein content were presentedin Table 6. Significant and positive correlations among Fe, Zn, Mn, Mg, P were observed, among which Fe concentration was strongly and positively correlated with Zn ($r = 0.75$, $P < 0.001$), and high significantly positive correlations among Mg, P, and protein content were also observed, with r all larger than 0.74 ($P < 0.001$). There were significant and positive correlations between Fe, Zn, Mn, and protein content, with r all larger than 0.54. Significant and positive correlations between Cu and Mg ($r = 0.54$, $P < 0.01$), P ($r = 0.57$, $P < 0.01$), and significant and negative correlations between Cu and Ca, between K and Fe, Zn, Mn, P, thousand kernel weight, protein content were also observed, with r all lower than -0.41. China, and the consequences of malnutrition create immense economic and social costs (Chen 2004). A unique opportunity thus exists for agriculture to invest in developing more nutrient-dense staple food crops that could help to reduce not only energy shortage but also malnutrition (Underwood 2000). Welch and Graham (1999) pointed out that although the world's food supply in recent years has been somewhat sufficient, it does not promote an adequate nutritional balance, and thus, a new balanced nutrition paradigm for crop production through breeding staple crops with high nutrient concentration in the grain has been proposed as a low cost, sustainable strategy for reducing mineral deficiencies in humans (Welch and Graham 1999, 2000). To explore this possibility, the International Maize and Wheat Improvement Center (CIMMYT) and other international centers and research institutions have been participating in the CGIAR Challenge Program for micronutrient since 1995, to study the feasibility of developing micronutrient-dense cultivars such as wheat through breeding. In attempting to enhance the micronutrient levels in wheat through conventional plant breeding, it is very important to identify genetic resources with high levels of the targeted micronutrients that could be exploited in the future breeding programs (Bouis et al. 2000), and to gain a better understanding of genotype and environment effects (Ortiz-Monasterio et al. 2007).

Wide variation among genotypes was pronounced for all traits in this study. The genetic variability of Fe and Zn in wheat grain ranged from 28.0 to 60.2mg · kg^{-1} and from 21.4 to 58.2mg · kg^{-1}, with an average of 39.2mg · kg^{-1} for Fe and 32.2mg · kg^{-1} for Zn, respectively. Jimai 26, Hnong 326, Han3475, and Jingdong 8 displayed high Fe concentration, with value all more than 50mg · kg^{-1}; Jimai 26, Henong 326, Nongda 3197, and Jingdong 8 displayed high Zn concentration, with value all more than 46.1mg · kg^{-1}; among which Jimai 26 and Henong 326 also displayed high concentrations of Cu, Mg, K, P, and protein content. These cultivars could be used as crossing parents in breeding programs targeting for high nutrient concentrations. The high concentrations of Fe and Zn for Jingdong 8, the leading cultivar as local check in the northern part of Northern China Winter Wheat Region with high yield potential, good resistance to powdery mildew, and good tolerance to high temperature during grain filling (Zhuang 2003), were also confirmed by the multi-environment trial. This made it the most promising cultivar, since the selection of high Fe and Zn wheat cultivar must have good agronomic performance and broad adaptation. Zhongyou 9507, Zhongyou 16, and Liken 2, which were all reselected from Zhongzou 8131-1 (Zhuang 2003), had high concentrations of Fe and Zn. These cultivars also performed outstanding pan bread making quality and were recommended as nutrition quality donor for Chinese wheat breeding programs. Moreover, these lines, together with some contrasting lines for Fe and Zn concentrations, may be useful for the development of inexpensive molecular markers that can be used effectively to assist breeding programs targeted on biofortified wheat cultivars potentially by QTL mapping through populations, or possibility by association mapping which is becoming more and more popular and has been used successfully for identify lycopene epsilon cyclase (lcyE) locus to alter flux down alpha-carotene versus beta-carotene branches of the carotenoid pathway among diverse maize inbred lines by Harjes et al. (2008).

Table 1 Mean, standard deviation, and range of grain characteristics and mineral element concentrations in 265 Chinese leading cultivars and advanced lines

Trait	Mean	Std	Range
Fe	39.2	4.1	28.0—60.2
Zn	32.3	4.7	21.4—58.2
Mn	48.8	6.9	32.2—70.6
Cu	7.39	0.89	5.3—12.1
Ca	473	58	317—653
Mg	1 519	111	1 215—2 165
K	4 847	478	3 315—7 061
P	4 179	348	3 134—6 353
TKW	34.7	4.8	28.0—48.7
PRO	14.7	1.1	10.1—15.6

[a]Fe, iron concentration (mg · kg^{-1}); Zn, zinc concentration (mg · kg^{-1}); Mn, manganese concentration (mg · kg^{-1}); Cu, copper concentration (mg · kg^{-1}); Mg, magnesium concentration (mg · kg^{-1}); K, kalium concentration (mg · kg^{-1}); Ca, calcium concentration (mg · kg^{-1}); P, total phosphorus concentration (mg · kg^{-1}); TKW, thousand kernel weight (g); PRO, protein content (%)

Discussion

Plant breeding and improved crop managementhave successfully increased grain yield of common wheat during the 20th century worldwide (Calderini and Slafer 1998; Zhuang 2003; Zhou et al. 2007) and therefore, effectively reduced food shortages. However, nutritional problems related to diets of cereal were observed throughout the world including

Based on personal daily requirement of Fe and Zn, and the information of bioavailability provided by the WHO (2006), the concentrations of Fe and Zn have been proposed to be elevated to 57 and 41mg · kg^{-1}, respectively. However, only Jimai 26 had high Fe concentration more than 60mg · kg^{-1}, whereas tens of cultivars had higher Zn concentration more than 40 mg · kg^{-1} in this study. Therefore, there is sufficient genetic variability to develop wheat cultivars with increased Zn levels in the grain and promising genetic variability for Fe, but due to the lower bioavailability of Fe when compared with Zn, target levels for Fe are significantly higher and meeting them will be more challenging.

Though crucial, screening for high Fe and Zn concentration of the current germplasm is only an initial step. A better understanding of the effects of genotype, environment and their interaction are required to make breeding efforts more efficient, since both genotype and environment contributed to the wide range of variation in nutrient element concentrations in this paper, which is in agreement with the previous study (Morgounov et al. 2007). Success in crop improvement through plant breeding strategies depends on the existence of genetic variation for the target traits in the gene pool available to the breeder. When breeding for Fe and Zn concentration in the grain, the task would be further complicated by the fact that grain micronutrient concentrations depend largely on environmental effect and condition in this study, which is in agreement with the previous report (Feil et al.

Table 2 Mean and standard deviation of grain characteristic traits and mineral element concentrations for the ten cultivars and advanced lines with the highest value

Genotype	Fe[a]	Genotype	Zn	Genotype	Mn	Genotype	Cu	Genotype	Ca
Jimai 26	60. 2±10. 3	Henong 326	58. 2±7. 3	Nongda 3383	70. 6±4. 1	Jimai 26	12. 1±1. 1	Xumai 856	653±48
Henong 326	58. 1±13. 0	Jimai 26	51. 9±4. 1	Henong 326	70. 4±2. 3	Hnong 326	11. 2±1. 1	Xiaoyan 93166	650±48
Han 3475	52. 7±9. 3	Nongda 3197	47. 8±3. 5	Tai'shan 9818	70. 4±1. 8	Nongda 3383	10. 5±0. 7	CA 0399	625±57
Jingdong 8	51. 8±6. 3	Jingdong 8	46. 1±2. 5	Zheng9405	69. 9±4. 4	Nongda 3197	9. 9±0. 5	Jimai 47	622±54
Zhoumai 3	49. 4±7. 9	Zhongmai 175	45. 8±2. 7	Lumai 23	68. 7±2. 5	Zhongmai 175	9. 7±0. 4	Zheng9405	601±46
Jimai 3	49. 4±4. 6	Nongda 179	45. 6±2. 8	Lankao 2	67. 5±4. 2	CA0178	9. 5±0. 4	CA 0368	598±23
Nongda 3197	47. 1±3. 0	Lumai 23	45. 5±3. 4	Lankao 906	67. 3±5. 2	Nongda 179	9. 4±0. 5	Guanfeng 2	595±43
Lankao 2	46. 5±4. 9	Linfen 139	43. 3±3. 6	Zheng9409	65. 3±3. 9	Lu 954072	9. 2±0. 7	Zheng7898	591±53
Nongda 179	46. 4±2. 5	Weimai 8	42. 4±3. 0	Lu 954072	63. 1±2. 2	Lu 393	9. 1±0. 4	Ji'ai1	584±28
Nongda 393	46. 0±2. 3	Nongda 123	42. 0±1. 4	Nongda 179	61. 7±4. 0	Jing 411	9. 1±0. 1	Lankao 906	581±45
Genotype	**Mg**	**Genotype**	**K**	**Genotype**	**P**	**Genotype**	**TKW**	**Genotype**	**PRO**
Jimai 26	2 165±201	Jimai 26	7 061±482	Henong 326	6 353±847	Lankao 4	48. 7±3. 8	Henong 326	15. 6±1. 1
Henong 326	2 121±217	Zhongmai 175	6 385±403	Jimai 26	6 043±637	Lu 954072	48. 4±2. 1	Jimai 26	15. 3±1. 7
Nongda 179	1 865±97	CA 0204	6 112±339	Zhongmai 175	5 514±443	Nongda 3383	47. 5±1. 6	Shaanyou 225	15. 2±0. 8
CA 0368	1 812±71	Zhoumai 13	6 111±314	Lankao 906	5 129±406	Linmai 2	46. 5±3. 7	Jimai 21	15. 1±0. 9
Zhongmai 175	1 789±73	CA 0203	6 038±310	CA 0203	5 117±272	Weimai 8	44. 5±3. 8	Nongda 3197	15. 1±1. 0
CA 0204	1 783±63	Nongda 393	6 024±286	CA 0349	5 109±255	Zheng9409	44. 5±2. 2	Yannong 15	15. 0±1. 1
CA 0178	1 764±68	Ji'ai 1	6 007±270	Nongda 179	5 083±272	Jingdong 8	43. 8±1. 3	Jinan 17	14. 9±1. 1
Xiaoyan 93166	1 748±100	Zhoumai 14	5 971±264	CA 0204	5 044±258	Yanfu 188	42. 1±0. 8	CA 0204	14. 8±1. 1
CA 0349	1 747±47	Henong 326	5 970±319	Lumai 23	4 959±335	Tai'shan 9818	42. 1±2. 3	Nongda 179	14. 6±0. 8
Nongda 3383	1 747±55	Xiaoyan 93166	5 880±278	Lu 954072	4 920±313	Zheng7898	41. 8±0. 6	CA 0399	14. 5±0. 8

[a]Fe, iron concentration ($mg \cdot kg^{-1}$); Zn, zinc concentration ($mg \cdot kg^{-1}$); Mn, manganese concentration ($mg \cdot kg^{-1}$); Cu, copper concentration ($mg \cdot kg^{-1}$); Mg, magnesium concentration ($mg \cdot kg^{-1}$); K, kalium concentration ($mg \cdot kg^{-1}$); Ca, calcium concentration ($mg \cdot kg^{-1}$); P, total phosphorus concentration ($mg \cdot kg^{-1}$); TKW, thousand kernel weight (g); PRO, protein content (%)

Table 3 Sum of square for mineral element concentrations and kernel characteristic traits for 24 cultivars across seven locations over 2 years

Source	df	Fe[a]	Zn	Mn	Cu	Ca	Mg	K	P	TKW	PRO
Genotype (G)[b]	23	3 343***	1 545***	1 933***	69***	392 550***	2 327 427***	28 965 029***	12 215 400***	12 782***	382***

(continued)

Source	df	Fe[a]	Zn	Mn	Cu	Ca	Mg	K	P	TKW	PRO
Environment (E)	13	22 806***	49 403***	70 967***	906***	2 095 844***	5 039 971***	102 743 267***	120 011 275***	7 282***	346***
G×E	299	8 415***	4 262***	3 692***	111***	1 231 805***	2 818 975***	34 776 297***	26 161 372***	5 203***	301***
Rep (E)	14	45	13	13	0.35	1 059***	7 223	347 857	43 808	18	1.5
Error	319	2 336	1 713	1 312	27	202 959	723 055	14 037 349	7 509 984	1 711	82
E/G		6.82	31.97	36.72	13.14	5.34	2.17	3.55	9.82	0.57	0.91
G×E/G		2.52	2.76	1.91	1.60	3.14	1.21	1.20	2.14	0.41	0.79

*, **, and *** indicate significant at $P=0.05$, 0.01, and 0.001 levels, respectively

[a] Fe, iron concentration ($mg \cdot kg^{-1}$); Zn, zinc concentration ($mg \cdot kg^{-1}$); Mn, manganese concentration ($mg \cdot kg^{-1}$); Cu, copper concentration ($mg \cdot kg^{-1}$); Mg, magnesium concentration ($mg \cdot kg^{-1}$); K, kalium concentration ($mg \cdot kg^{-1}$); Ca, calcium concentration ($mg \cdot kg^{-1}$); P, total phosphorus concentration ($mg \cdot kg^{-1}$); TKW, thousand kernel weight (g); PRO, protein content (%)

[b] G, genotype; E, environment; G×E, genotype by environment interaction; Rep (E), Replication nested within environment; E/G, the ratio of environment variance with genotype variance; G×E/G, the ratio of genotype and environment interaction variance with genotype variance

Table 4 Mean of major mineral element concentrations and kernel characteristic traits for 24 cultivars across seven locations over 2 years

Genotype	Fe[a]	Zn	Mn	Cu	Ca	Mg	K	P	TKW	PRO
Jingdong 6	44.7abc[b]	37.1abc	36.2ijk	6.97cdefg	458i	1 411g	3 656k	3 535defghi	46.8ab	13.6d
Jingdong 8	45.4ab	37.3ab	37.2fghij	7.00bcdef	494def	1401gh	3707jk	3505efghij	45.2c	13.6d
Zhongmai 9	40.9ghij	35.2efgh	37.9defg	6.27jk	484fgh	1 399ghi	3 980ef	3 481ghijk	44.6cd	13.2ghij
Jing 411	41.0ghij	34.0hi	37.0ghij	7.15ab	448ij	1459def	3853ghi	3446ijkl	43.0e	13.1hijk
Zhongzuo 8131-1	44.0bcd	37.7a	39.8ab	7.07bcd	475h	1 569a	3 650k	3 890a	43.6de	15.6a
Zhongyou 9507	40.8hij	34.1hi	38.2cdefg	6.90defg	445ij	1 509c	3 826hij	3 634bc	47.8a	14.1c
Zhongyou 9507ai	38.7k	32.4j	32.3l	7.00bcde	480fgh	1 467def	4 303b	3 378lmn	37.4jk	12.9jk
Zhongyou 8	41.5fghi	35.efgh	37.9defg	6.62i	474h	1 479d	3 873fghi	3 589bcdef	45.7bc	13.9c
Zhongyou 14	43.2cde	36.0bcdef	38.3cdef	6.59i	527a	1 540b	3 753hijk	3 623bcd	37.4jk	14.6b
Zhongyou 16	46.3a	36.0bcdef	39.8ab	6.84efgh	502cde	1 543ab	3 760hijk	3 652b	38.8ghi	14.5b
Youxuan 9	44.7abc	37.0abcd	38.9bcd	6.82fgh	517ab	1 529bc	3 794hij	3 597bcde	38.0jki	14.5b
Youxuan 14	43.1cdef	35.9bcdef	38.7bcde	6.70hi	526a	1 544ab	3 754hijk	3 568bcdefg	36.8jk	14.6b
Zhongyou 9843	38.5k	32.5j	35.6k	6.36j	486fgh	1 379hi	4 153cd	3 352mno	39.9fg	13.5def
Zhongyou 9844	39.6jk	32.ij	37.2fghi	6.35j	491defg	1387ghi	4192c	3401klm	40.9f	13.4defg

(continued)

Genotype	Fe[a]	Zn	Mn	Cu	Ca	Mg	K	P	TKW	PRO
Jimai 24	40. 8ijh	36. 6abcde	38. 3cdef	6. 80gh	481fgh	1482d	3954efg	3533defghi	40. 2fg	14. 5fghi
Jimai 26	42. 5defg	35. 7cdef	36. 0jk	7. 24a	438j	1535bc	4431a	3517efghi	36. 6k	13. 3efgh
Liken 2	45. 4ab	35. 6cdef	40. 4a	6. 84efgh	503bcd	1528bc	3808hij	3552cdefgh	39. 1ghi	13. 3b
Yumai 28	38. 4k	35. 2efgh	36. 2ijk	7. 10abc	459i	1449ef	3743ijk	3498fghij	39. 5gh	13. 4efgh
Jinmai 45	43. 9bcd	35. 8cdef	35. 3k	6. 13k	479fgh	1399ghi	4050de	3304nop	41. 2f	13. 5def
Jinmai 33	43. 1cdef	35. 1fgh	39. 4abc	6. 55i	516abc	1466def	3796hij	3414jklm	44. 9cd	13. 3efgh
Linfeng 615	41. 5fghi	34. 2ghi	36. 4hijk	6. 36j	516abc	1371i	4073de	3236p	39. 1ghi	13. 0ijk
Xiaoyan 6	42. 0efgh	33. 4ij	38. 2cdefg	6. 55i	477gh	1474de	3882fgh	3352mno	38. 2hij	13. 6de
Shaan 160	40. 3j	33. 3ij	37. 6efgh	6. 34j	488efgh	1443f	4312b	3460hijkl	37. 8ijk	13. 5def
Shaan 225	39. 7jk	34. 2ghi	37. 7defg	6. 15k	480fgh	1414g	3982ef	3277op	41. 1f	12. 9k

[a]Fe, iron concentration (mg • kg^{-1}); Zn, zinc concentration (mg • kg^{-1}); Mn, manganese concentration (mg • kg^{-1}); Cu, copper concentration (mg • kg^{-1}); Mg, magnesium concentration (mg • kg^{-1}); K, kalium concentration (mg • kg^{-1}); Ca, calcium concentration (mg • kg^{-1}); P, total phosphorus concentration (mg • kg^{-1}); TKW, thousand kernel weight (g); PRO, protein content (%)

[b] Values within column with the same letter are not significantly different at $P=0.05$ by Tukey test

Table 5 Mean of major mineral element concentrations and kernel characteristic traits for 14 environments with 24 cultivars sown over 2 years

Environment	Fe[a]	Zn	Mn	Cu	Ca	Mg	K	P	TKW	PRO
Beijing-06	44.4cd[b]	43.1b	37.3g	7.73c	573a	1393e	4913a	3604d	33.0i	14.0c
Beijing-07	35.5g	35.2e	31.0j	5.48i	530bc	1486d	4458b	3076g	35.1h	14.2b
Shijiazhuang-06	38.0f	24.6i	35.1h	7.04e	459f	1409e	3717g	3862b	39.9g	13.3f
Shijiazhuang-07	33.7h	24.2i	33.7i	5.09j	418g	1663a	4151d	3402f	44.9c	13.9c
Anyang-06	46.6b	41.8c	52.9a	8.49a	422g	1393e	4285c	4063a	32.8i	13.8d
Anyang-07	42.6e	27.5h	46.9c	5.20j	380h	1515c	4004e	3514e	43.0d	14.1bc
Jiaozuo-06	51.8a	41.2c	52.4a	7.35d	536b	1535b	3616hi	3678c	39.9g	13.5ef
Jiaozuo-07	46.2b	28.6g	47.8b	5.57i	424g	1536b	3703gh	3577d	47.3a	14.4a
Zhengzhou-06	42.6e	21.0j	28.2k	6.70g	522cd	1497cd	3652ghi	4102a	40.8fg	13.6de
Zhengzhou-07	43.4de	32.8f	40.0e	5.46i	529bc	1327g	3590i	2863h	41.5ef	14.5a
Jinan-06	44.9c	39.1d	42.0d	8.25b	539b	1500cd	3816f	3896b	41.1f	14.1bc
Jinan-07	51.0a	42.0c	38.7f	6.89f	518d	1509c	4016e	3025g	42.2de	13.6de
Xuzhou-06	35.8g	51.7a	19.1l	8.25b	489e	1351f	3831f	3485e	46.3b	12.3h
Xuzhou-07	32.8h	38.7d	19.8l	6.23h	462f	1398e	3320j	2669i	45.0c	12.6g

[a] Fe, iron concentration ($mg \cdot kg^{-1}$); Zn, zinc concentration ($mg \cdot kg^{-1}$); Mn, manganese concentration ($mg \cdot kg^{-1}$); Cu, copper concentration ($mg \cdot kg^{-1}$); Mg, magnesium concentration ($mg \cdot kg^{-1}$); K, kalium concentration ($mg \cdot kg^{-1}$); Ca, calcium concentration ($mg \cdot kg^{-1}$); P, total phosphorus concentration ($mg \cdot kg^{-1}$); TKW, thousand kernel weight (g); PRO, protein content (%)

[b] Values within column with the same letter are not significantly different at $P=0.05$ by Tukey test

Table 6 Correlations among mineral element concentration and kernel characteristic traits for 24 cultivars across seven locations over 2 years

	Zn	Mn	Cu	Ca	Mg	K	P	TKW	PRO
Fe[a]	0.75***	0.53*	0.24	0.33	0.43*	−0.54*	0.45*	0.09	0.58**
Zn		0.46*	0.38	0.13	0.42*	−0.64**	0.63**	0.21	0.54*
Mn			0.00	0.37	0.52*	−0.58**	0.55*	0.12	0.63**
Cu				−0.41*	0.54**	−0.28	0.57**	0.05	0.25
Ca					0.09	−0.21	−0.04	−0.36	0.29
Mg						−0.32	0.76***	−0.27	0.74***
K							−0.53**	−0.42*	−0.55**
P								0.19	0.83***
TKW									−0.03

*, **, and *** indicate correlation coefficient significantly different from zero at $P=0.05$, 0.01, and 0.001 levels, respectively

[a] Fe, iron concentration ($mg \cdot kg^{-1}$); Zn, zinc concentration ($mg \cdot kg^{-1}$); Mn, manganese concentration ($mg \cdot kg^{-1}$); Cu, copper concentration ($mg \cdot kg^{-1}$); Mg, magnesium concentration ($mg \cdot kg^{-1}$); K, kalium concentration ($mg \cdot kg^{-1}$); Ca, calcium concentration ($mg \cdot kg^{-1}$); P, total phosphorus concentration ($mg \cdot kg^{-1}$); TKW, thousand kernel weight (g); PRO, protein content (%)

2005; Morgounov et al. 2007). Therefore, the responses of cultivars to production environments need to be well characterized. A thorough understanding of the variation among cultivars in their response to environment would further improve the probability of predicting and identifying cultivars with high Fe or Zn concentration (Dong et al. 1995; Garnett and graham 2005). Grain nutrient concentration could be substantially improved through integrating knowledge of geographic cultivar production with key environmental variables that relate to nutrient concentration like pH, organic matter and soil texture (Tisdale and Nelson 1975; Romheld and Marschner 1986). It also indicated that selection of the production environment will have a large impact on nutrient concentrations. Production of high Fe concentration was best observed at Jiaozuo and Jinan, and high Zn concentration was best expressed at Jinan and Xuzhou in this study, due to that samples from the locations in the two seasons were all characterized by high Fe or Zn concentration, compared with the other locations.

Genotype by environment interaction hasan important effect on all nutrient concentrations. Therefore, the importance of genotype by environment interaction should not be overlooked. A better strategy to characterize the environmental response of cultivars or advanced breeding lines should be developed to accommodate the significant genotype by environment interactions. It would seem important to clearly define target areas and to subsequently devote resources to screening of genotypes in a variety of these target areas, and efforts to understand genotype by environment interaction need to be included in the breeding programs.

Highly significant and positive correlations between protein content, Fe and Zn concentrations were also observed, therefore, suggesting the possibility to combine high Fe and Zn traits during wheat breeding. In agreement with this result, Distelfeld et al. (2004) and Uauy et al. (2006) has showed that a locus on the short arm of chromosome 6B (*Gpc-B1*) in wheat was effective in increasing the accumulation of protein content, Fe, and Zn in grain, due to an increased remobilization of nutrients from leaves to the developing grains under an accelerated senescence caused by a NAC transcription factor.

Acknowledgments

Financial support was supported inpart by a grant from the Harvest-Plus China (#8017) program, National Basic Research Program (2009CB118300), the international collaboration project on wheat improvement from Ministry of Agriculture of the People's Republic of China (2006-G2), and Core Research Budget of the Non-profit Governmental Research Institutions (ICS, CAAS).

References

AACC Approved methods of the AACC (2000) 10 edn. American Association of Cereal Chemists, St. Paul, MN Barreto HJ, Edmeades GO, Chapman SC, Crossa J (1997) The alpha lattice design in plant breeding and agronomy: Generation and analysis. In: Edmeades GO, Banziger M, Mickelson HR, Pena-Valdivia CB (eds) Developing drought- and low N-tolerant maize: proceedings of a symposium. CIMMYT, Mexico, DF, pp 25-29.

Bouis HE, Graham RD, Welch RM (2000) The Consultative Group on International Agriculture Research (CGIAR) Micronutrients Project: justification and objectives. UNU Food Nutr Bull 21: 374-381.

Calderini DF, Slafer GA (1998) Changes in yield andyield stability in wheat during the 20th century. Field Crops Res 57: 335-347.

Calderini DF, Ortiz-Monasterio I (2003) Grain position affects grain macronutrient and micronutrient concentrations in Wheat. Crop Sci 43: 141-151.

Chen CM (2004) Ten year tracking nutritional status in China. People's Medical Publishing House, Beijing (in Chinese).

Distelfeld A, Uauy C, Olmos S, Schlatter AR, Dubcovsky J, Fahima T (2004) Microcolinearity between a 2-cM region encompassing the grain protein content locus *Gpc-6B1* on wheat chromosome 6B and a 350-kb region on rice chromosome 2. Funct Integr Genemics 4: 59-66.

Dong B, Rengel Z, Graham RD (1995) Root morphology of wheat genotypes differing in zinc efficiency. J Plant Nutr 18: 2761-2773.

Feil B, Moser SB, Jampatong S, Stamp P (2005) Mineral composition of the grains of tropical maize varieties as affected by pre-anthesis drought and rate of nitrogen fertilization. Crop Sci 45: 516-523.

Garnett TP, Graham RD (2005) Distribution and remobilization of iron and copper in wheat. Ann Bot 95: 817-826.

Gilmour AR, Cullis BR, Verbyla AP (1997) Accounting for natural and extraneous variation in the analysis of field experiments. J Agric Biol Environ Stat 2: 269-293.

Gordon N (1997) Nutrition and cognitive function. Brain Dev 19: 165-170.

Harjes CE, Rocheford TR, Bai L, Brutnell TP, Kandianis CB, Sowinski SG, Stapleton AE, Vallabhaneni R, Williams M, Wurtzel ET, Yan J, Buckler ES (2008) Natural genetic variation in lycopene epsilon cyclase tapped for maize biofortification. Science 319: 330-333.

He ZH, Rajaram S, Xin ZY, Huang G (eds) (2001) A history of wheat breeding in China. CIMMYT, Mexico, DF, pp 1-14.

Hurrell RF, Juillerat MA, Reddy MB, Lynch SR, Dassenko SA, Cook JD (1992) Soy protein, phytate and iron absorption in man. Am J Clin Nutr 56: 573-578.

Morgounov A, Gomez-Becerra HF, Abugalieva A, Dzhunus- ova M, Yessimbekova M, Muminjanov H, Zelenskiy Y, Ozturk L, Cakmak I (2007) Iron and zinc grain density in common wheat grown in Central Asia. Euphytica 155: 193-203.

Ortiz-Monasterio I, Palacios-Rojas N, Meng E, Pixle K, Trethowan R, Pena RJ (2007) Enhancing the mineral and vitamin content of wheat and maize through plant breeding. J Cereal Sci 46: 293-307.

Romheld V, Marschner H (1986) Evidence for a specific uptake system for iron phytosidorphore in roots of grasses. Plant Physiol 80: 175-180.

SAS Institute SAS User ' s Guide (2000) Statistics. SAS Institute, Inc., Cary, NC.

Stoltzfus RJ, Dreyfuss ML (1998) Guidelines for the useof iron supplements to prevent and treat iron deficiency anemia. ILSI Press, Washington, DC.

Tisdale SL, Nelson WL (1975) Soil fertility and fertilizer, 3rd edn. Macmillian Publishing Co., Inc., USA. Uauy C, Distelfeld A, Fahima T, Blechl A, Dubcovsky J (2006) A NAC gene regulating senescence improves grain protein, zinc, and iron content in wheat. Science 314: 1298-1301.

Underwood BA (2000) Overcoming micronutrient deficiencies in developing countries: Is there a role for agriculture? UNU Food Nutr Bull 21: 356-360.

United Nations System Standing Committee on Nutrition (SCN) (2004) Nutrition for improved development out- comes. Fifth Report on the World Nutrition Situation. SCN, Geneva.

Walter T, Peirano P, Roncagliolo M (1997) Effect ofiron deficiency anemia on cognitive skills and neuromaturation in infancy and childhood. In: Fischer PWF, L'Abbe', MR, Cockell KA, Gibson RS (eds) Trace elements in man and animals, vol 9. Proceedings of the ninth international symposium on trace elements in man and animals. National Research Council of Canada, Ottawa, pp 217-219.

Welch RM, Graham RD (1999) A new paradigm for world agriculture: meeting human needs productive, sustainable, nutritious. Field Crops Res 60: 1-10.

Welch RM, Graham RD (2000) A new paradigm for world agriculture: productive, sustainable, nutritious, healthful food systems. UNU Food Nutr Bull 21: 361-366.

Welch RM, Graham RD (2004) Breeding for micronutrientsin staple food crops from a human nutrition perspective. J Exp Bot 55: 353-364.

WHO (2006) Guidelines on food fortification with micronu- trients. World Health Organization, Geneva.

Zhang Y, He ZH, Ye GY, Zhang AM, van Ginkel M (2004) Effect of environment and genotype on bread-making quality of spring-sown spring wheat cultivars in China. Euphytica 139: 75-83.

Zhang Y, Wang DS, Zhang Y, He ZH (2007) Variation of major mineral elements concentration and their relation- ships in grain of Chinese wheats. Scientia Agricultura Sinica 40: 1871-1876 (in Chinese).

Zhang Y, Zhang QJ, He ZH, Qian SH, Zhang Y, Pena RJ, Ye GY (2008) Solvent retention capacities as indirect selection criteria for sugar snap cookie quality in Chinese soft wheat genotypes. Aust J Agric Res 59: 911-917.

Zhou Y, He ZH, Sui XX, Xia XC, Zhang XK, ZhangGS (2007) Genetic improvement of grain yield and associated traits in the northern China winter wheat region from 1960 to 2000. Crop Sci 47: 245-253.

Zhuang QS (2003) Chinese wheat improvement and pedigree analysis. China Agricultural Press, Beijing (in Chinese).

QTL Mapping for Grain Zinc and Iron Concentrations in Bread Wheat

Yue Wang[1], Xiaoting Xu[1], Yuanfeng Hao[1], Yelun Zhang[2], Yuping Liu[2], Zongjun Pu[3], Yubing Tian[1], Dengan Xu[1], Xianchun Xia[1], Zhonghu He[1,4] and Yong Zhang[1]

[1] National Wheat Improvement Centre, Institute of Crop Sciences, Chinese Academy of Agricultural Sciences, Beijing, China, [2] Hebei Laboratory of Crop Genetics and Breeding, Institute of Cereal and Oil Crops, Hebei Academy of Agricultural and Forestry Sciences, Shijiazhuang, China, [3] Institute of Crop Sciences, Sichuan Academy of Agricultural Sciences, Chengdu, China, [4] International Maize and Wheat Improvement Center (CIMMYT) China Office, Chinese Academy of Agricultural Sciences, Beijing, China

Abstract: Deficiency of micronutrient elements, such as zinc (Zn) and iron (Fe), is called "hidden hunger," and biofortification is the most effective way to overcome the problem. In this study, a high-density Affymetrix 50K single-nucleotide polymorphism (SNP) array was used to map quantitative trait loci (QTL) for grain Zn (GZn) and grain Fe (GFe) concentrations in 254 recombinant inbred lines (RILs) from a cross Jingdong 8/Bainong AK58 in nine environments. There was a wide range of variation in GZn and GFe concentrations among the RILs, with the largest effect contributed by the line ×environment interaction, followed by line and environmental effects. The broad sense heritabilities of GZn and GFe were 0.36 ± 0.03 and 0.39 ± 0.03, respectively. Seven QTL for GZn on chromosomes 1DS, 2AS, 3BS, 4DS, 6AS, 6DL, and 7BL accounted for 2.2-25.1% of the phenotypic variances, and four QTL for GFe on chromosomes 3BL, 4DS, 6AS, and 7BL explained 2.3-30.4% of the phenotypic variances. QTL on chromosomes 4DS, 6AS, and 7BL might have pleiotropic effects on both GZn and GFe that were validated on a germplasm panel. Closely linked SNP markers were converted to high-throughput KASP markers, providing valuable tools for selection of improved Zn and Fe bio-fortification in breeding.

Keywords: *Triticum aestivum*, mineral biofortification, quantitative trait locus, 50K SNP array, KASP marker

INTRODUCTION

Wheat provides the starch, protein, and mineral nutrition needs for 35-40% of the world population[1]. Mineral nutrition is crucial for a healthy diet. Over 17% of people suffer from malnutrition worldwide due to lack of mineral nutrition and more than 100 000 children under the age of five die from zinc (Zn) deficiency annually[2-4]. The CIMMYT Harvest-Plus program initiated in the early 21st century aimed to address the "hidden hunger" issue by increasing micronutrient concentrations in staple food grains by plant breeding[5]. Zn and Fe deficiency were identified as major causes of malnutrition, especially in underdeveloped regions where cereal grains make up most of the food[6].

Published in Frontiers in Nutrition, 2021, 8: 680391.

Zn is a crucial cofactor in many enzymes and regulatory proteins, such as carbonic anhydrase, alkaline phosphatase, and DNA polymerase enzyme synthesis[7]. Zn deficiency, first reported in 1961, affects the immune system, taste perception, site, and sexual function[4]. Fe deficiency in humans most commonly leads to nutritional anemia in women and children[8]. Therefore, it is very important to improve the nutritional quality of wheat by enhancing the Zn(GZn) and Fe(GFe) concentrations in grain[9,10] . environment. Each plot comprised a 1-m row with an inter-row spacing of 20 cm, and a parental check was sown every 30 plots. Standard agronomic practices were applied at each location, along with a soil application of 25 kg/ha $ZnSO_4 \cdot 7H_2O$ in all fields except Beijing.

Grain samples were hand-harvested and cleaned to avoid potential contamination of mineral elements. Micronutrient analysis of grain samples collected from the 2016-2017 cropping season was performed at the Institute of Quality Standards and Testing Technology for Agro-products of CAAS using inductively coupled plasma atomic emission spectrometry (ICP- AES, OPTIMA 3300 DV) after samples were digested in a microwave system with HNO_3-H_2O_2 solution[20]. For grain samples from the 2017-2018 and 2018-2019 cropping seasons and the germplasm panel, a "bench-top," nondestructive, energy- dispersive X-ray fluorescence spectrometry (EDXRF) instrument (model X-Supreme 8 000, Oxford Instruments plc, Chengdu) was used to measure GZn and GFe, following the standard method for high-throughput screening of micronutrients in whole wheat grain[21]. Bio-fortification in wheat breeding demands identification of genetic resources with high GZn and GFe[9]. Wide ranges in variation in GZn and GFe have been reported in bread wheat[11-13] and its cultivated and wild relatives[12,14,15]. Quantitative trait locus (QTL) mapping was used to identify genetic loci affecting GZn and GFe in biparental mapping populations, including recombinant inbred lines (RILs)[16-18]. Genome-wide association studies (GWAS) with high-density single-nucleotide polymorphism (SNP) arrays were also used; for example, Alomari et al.[19] performed a GWAS for GZn concentration in 369 European wheats using the 90K and 35K SNP arrays and detected 40 marker-trait associations on chromosomes 2A, 3A, 3B, 4A, 4D, 5A, 5B, 5D, 6D, 7A, 7B, and 7D and 10 candidate genes on chromosomes 3B and 5A. With wide application of molecular markers, such as SSR, DArT, and SNPs, increasing numbers of QTLs for GZn and GFe were detected, including 35 and 32 QTL for GZn and GFe in the A genome, 37 and 30 in the B genome, and 15 and 12 in the D genome, respectively (Supplementary Table 1). The GZn QTL in homoeologous groups 1 to 7 were 9, 10, 13, 11, 13, 12, and 19, respectively, whereas the corresponding numbers of GFe QTL were 6, 17, 10, 8, 15, 7, and 11. QTL pleiotropic for GZn and GFe were identified in homoeologous group 3, 4, 5, and 7 chromosomes. Cultivar Jingdong 8, with high yield and resistance to stripe rust, leaf rust, and powdery mildew, was released in the early 1990s in the China Northern Winter Wheat Region. It was used widely as a parent in breeding and was verified to have high GZn and GFe levels across environments[13]. Bainong AK58, a high yielding cultivar in the Southern Yellow-Huai Valley Winter Wheat Region, has wide adaptability and good resistance to stripe rust, powdery mildew, and lodging, but has lower GZn and GFe. The main goals of the present study were to (1) identify QTL for GZn and GFe in the Jingdong 8/Bainong AK58 RIL population using inclusive composite interval mapping, and (2) develop and validate breeder-friendly markers for marker-assisted selection (MAS) for Zn and Fe biofortification in wheat breeding programs.

MATERIALS AND METHODS

Plant Materials

Two hundred fifty-four F6 RILs developed from Jingdong 8/Bainong AK58 cross were used for QTL mapping of GZn and GFe concentrations. A germplasm panel, including 145 cultivars/lines with a wide range of variation in GZn and GFe from the Chinese wheat germplasm bank[13], were used for validation of QTL for GZn and GFe identified in the RIL population.

Field Trials and Phenotyping

The field trials were conducted at the wheat breeding station of the Institute of Crop Sciences (ICS, CAAS) located at Gaoyi (37° 33′ N, 114° 26′ E) and Shijiazhuang (37°27′N, 113°30′E) in Hebei province and Beijing (39°56′N, 116°20′E) during 2016 to 2019 cropping seasons. The parents and RILs were planted in randomized complete blocks with two replications in each environment. Each plot comprised a 1-m row with an inter-row spacing of 20 cm, and a parental check was sown every 30 plots. Standard agronomic practices were applied at each location, along with a soil application of 25 kg/ha $ZnSO_4 \cdot 7H_2O$ in all fields except Beijing.

Grain samples were hand-harvested and cleaned to avoid potential contamination of mineral elements. Micronutrient analysis of grain samples collected from the 2016-2017 cropping season was performed at the Institute of Quality Standards and Testing Technology for Agro-products of CAAS using inductively coupled plasma atomic emission spectrometry (ICPAES, OPTIMA 3300 DV) after samples were digested in a microwave system with HNO_3-H_2O_2 solution[20]. For grain samples from the 2017-2018 and 2018-2019 cropping seasons and the germplasm panel, a "bench-top," nondestructive, energy dispersive X-ray fluorescence spectrometry (EDXRF) instrument (model X-Supreme 8000, Oxford Instruments plc, Chengdu) was used to measure GZn and GFe, following the standard method for high-throughput screening of micronutrients in whole wheat grain[21].

Statistical Analysis

Analysis of variance (ANOVA) was performed by PROC MIXED with method type3 and all effects were treated as fixed in SAS 9.4 software (SAS Institute, Cary, NC). Variance and covariance components for genotype and genotype by environment interaction effects were estimated using PROC MIXED, assuming all effects as random. A similar model was also performed by PROC MIXED with genotype effect as fixed, while environment, replication nested in environment, and interactions involving environment as random, to estimate best linear unbiased estimate (BLUE). Broad-sense heritabilities ($H_b{}^2$) on the basis of BLUE value were estimated using the following equation and standard errors were calculated following Holland et al.[22]:

$$H_b^2 = \frac{\sigma_g^2}{(\sigma_g^2 + \frac{\sigma_{ge}^2}{e} + \frac{\sigma_\varepsilon^2}{re})}$$

where σ_g^2 represents the variance of genotypes, σ_{ge}^2 and σ_ε^2 represent the variances of genotype × environment interaction and error, and e and r represent environments and number of replicates per environment, respectively. Phenotypic and genotypic correlations and their standard errors were estimated after Becker[23]. Student's t test was performed by PROC TTEST.

SNP Genotyping and QTL Analysis

Genomic DNA extracted from fresh seedling leaves of RILs and parents by CTAB method[24] were used for genotyping by the wheat 50K SNP Array. The wheat 50K SNP Array was developed in collaboration by CAAS and Capital-Bio, Beijing, China (https://www.capitalbiotech.com/). Linkage analysis was performed with JoinMap v4.0 using the regression mapping algorithm[25]. QTL analysis was performed by inclusive composite interval mapping with the ICIM-ADD function using QTL IciMapping v4.1 (http://www.isbreeding.net). Phenotypic values of RILs averaged from two replicates in each environment and BLUE value across nine environments were used for analyses. QTL detection was done using a logarithm of odds (LOD) threshold of 2.5. Pleiotropic QTL were analyzed using the module JZmapqtl of multi-trait composite interval mapping (MCIM) in Windows QTL Cartographer v2.5[26]. QTL pyramids were plotted using ggplot2 in R[27]. Physical maps for the positional comparisons of GZn and GFe QTL with previous reports were exhibited using MapChart v2.3[28].

Conversion of SNPs to KASP Markers

Kompetitive Allele Specific PCR (KASP) markers were developed from SNPs tightly linked with the targeted QTL, each including two competitive allele-specific forward primers and one common reverse primer. Each forward primer incorporated an additional tail sequence that corresponds to only one of the two universal fluorescence resonance energy transfers. Primers were designed from information in the PolyMarker website (http://polymaker.tgac.ac.uk/). PCR procedures and conditions followed Chandra et al. [29]. Gel-free fluorescence signal scanning and allele separation were conducted by microplate reader (Multiscan Spectrum BioTek, Synegy/H1) with Klustercaller 2.24.0.11 software (LGC, Hoddesdon, UK)[30].

RESULTS

Phenotypic Evaluation

ANOVAshowed that GZn and GFe were significantly influenced by lines, environments, and line by environment interaction effects, with line by environment interaction effects contributing the highest variation, followed by line and environment effects (Table 1). The broad-sense heritabilities of GZn and GFe were 0.36 ± 0.03 and 0.39 ± 0.03, respectively. Jingdong 8 accumulated significantly higher GZn and GFe than Bainong AK58. Wide-ranging continuous variation among the RILs suggests polygenic inheritance (Table 2, Figure 1). Significant and positive correlations of GZn ($r = 0.25$-0.67, $P < 0.01$) and GFe ($r = 0.26$-0.70, $P < 0.01$) were observed across the nine environments (Table 3). Additionally, positive phenotypic and genotypic correlations between GZn and GFe ($r = 0.78 \pm 0.01$ and 0.81 ± 0.03, $P < 0.001$) (Figure 2), indicated that GZn and GFe were, to some degree, simultaneously accumulated.

Linkage Map Construction

Among 54 680 SNP markers in the 50K SNP array, 20 060 were polymorphic after removal of markers that were monomorphic, absent in more than 20% of assays, and minor allele frequency was < 30%. A high-density linkage map spanning 3 423 cM and including all 21 chromosomes was constructed using 3328 representative SNP markers of each bin. The average chromosome length was 163 cM, ranging from 116.72 cM (1B) to 237.40 cM (5A) (Supplementary Table 2).

QTL Mapping of GZn and GFe using ICIM and MCIM

Seven QTL for GZn were mapped on chromosomes 1DS, 2AS, 3BS, 4DS, 6AS, 6DL, and 7BL, explaining 2.2-25.1% of the phenotypic variances (Table 4, Supplementary Table 3, and Figure 3), with five favorable alleles coming from Jingdong 8, and with the other two, i.e., *QZn.caas-1DS* and *QZn.caas-3BS*, coming from Bainong AK58. Four QTL for GFe were detected on chromosomes 3BL, 4DS, 6AS, and 7BL, explaining 2.3-30.4% of the phenotypic variances (Table 4, Supplemenatry Table 3, and Figure 3), with all superior alleles coming from Jingdong 8. Among these QTL, three were identified for both GZn and GFe at the same or overlapping location on chromosomes 4DS, 6AS, and 7BL. Three chromosomal intervals were detected using MCIM including 4DS, 6AS, and 7BL, corresponding to colocalized QTL for GZn and GFe by ICIM-ADD (Table 5). Two intervals on chromosomes 4DS and 6AS were detected in most environments for GZn and GFe, while the one on chromosome 7BL was found in most environments for GZn but only one environment for GFe. QTL Pyramids and Validation It indicated that superior alleles of pleiotropic QTL on 4DS, 6AS, and 7BL were all from Jingdong 8. Accumulation effect of the three co-localization QTL for GZn and GFe was calculated based on the closely linked markers. The average concentration of GZn increased from 37.79 to 44.43 mg/kg and that of GFe increased from 41.02 to 50.37 mg/kg, with lines containing zero to three favorable alleles (Supplementary Figure 1).

Flanking SNPs closely linked to the QTL on chromosomes 4DS and 7BL and a SNP near QTL region of 6AS were converted to KASP markers and validated in the germplasm panel (Tables 6, 7). Cultivars with the same superior allele as Jingdong 8 had significantly higher GZn and GFe than those with the inferior allele from Bainong AK58 for all QTL, except for QFe. caas-6AS. The difference between the superior and inferior allele of the QTL on chromosomes 4DS, 6AS, and 7BL was 1.7, 2.8, and 3.5 mg/kg for GZn and 1.4, 1.0, and 4.7 mg/kg for GFe, respectively (Table 7).

TABLE 1 Analysis of variance of GZn and GFe in 254 RILs derived from the cross Jingdong 8/Bainong AK58 grown in nine environments

Source of variation	DF	Sum square	
		Zn	Fe
Line	253	39 148**	50 429**
Environment (Env)	8	67 264**	24 847**
Line×Env	2 024	74 960**	91 715**
Rep (Env)	9	3 385**	1 658**
Error	2021	46 076	49 910
Heritability		0.36 ±0.03	0.39 ±0.03

***Significant at P<0.01.*

TABLE 2 Mean and range of GZn and GFe (mg/kg) in the Jingdong 8/Bainong AK58 RIL population among nine environments

Trait	Environment	Parents		RILs	
		Jingdong 8	Bainong AK58	Range	Mean±SD
Zn (mg/kg)	E1	42.1	35.1	25.4-52.6	38.9±4.6
	E2	41.5	34.1	25.2-56.6	39.1±5.6
	E3	53.4	46.4	29.5-60.7	43.5±5.7
	E4	41.3	34.5	28.7-52.6	38.0±4.1
	E5	52.1	44.1	33.5-62.2	46.4±4.7
	E6	46.2	36.9	28.9-54.3	41.3±5.3
	E7	40.7	33.7	25.7-49.0	34.6±3.9
	E8	55.0	42.3	34.6-62.5	47.6±6.0
	E9	48.6	34.7	27.0-57.9	40.0±5.8
Fe (mg/kg)	E1	51.0	43.0	32.8-62.6	47.3±5.2
	E2	53.5	40.9	34.5-68.9	48.0±6.5
	E3	53.2	42.8	34.2-64.0	48.5±6.3
	E4	46.6	40.2	35.3-52.3	42.2±3.2
	E5	49.8	42.3	37.0-59.5	44.9±3.8
	E6	49.2	34.5	31.1-65.1	42.7±5.5
	E7	54.2	37.8	32.0-67.2	45.0±6.7
	E8	56.6	39.7	33.9-69.2	49.2±6.2
	E9	55.0	38.0	28.0-63.9	47.1±6.4

E1-E9, Shijiazhuang 2016-2017, Gaoyi 2016-2017, Beijing 2016-2017, Shijiazhuang 2017-2018, Gaoyi 2017-2018, Beijing 2017-2018, Shijiazhuang 2018-2019, Gaoyi 2018-2019, and Beijing 2018-2019.

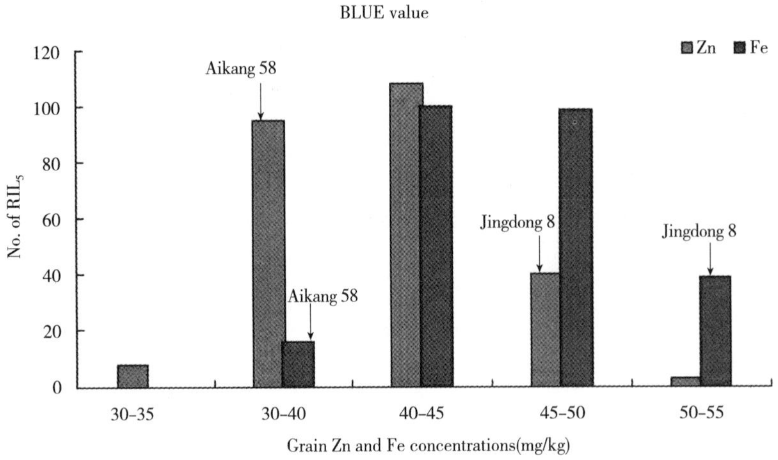

FIGURE. 1 Frequency distributions of GZn and GFe based on BLUE value across nine environments for 254 RILs in the Jingdong 8/Bainong AK58 population

TABLE 3 Pearson correlation coefficients of GZn and GFe in the Jingdong 8/Bainong AK58 RIL population among nine environments

Environment	E1	E2	E3	E4	E5	E6	E7	E8	E9
E1		0.45***	0.32***	0.40***	0.37***	0.48***	0.36***	0.47***	0.48***
E2	0.70***		0.25***	0.41***	0.46***	0.52***	0.27***	0.43***	0.39***
E3	0.52***	0.52***		0.40***	0.33***	0.40***	0.34***	0.34***	0.36***
E4	0.44***	0.47***	0.47***		0.49***	0.57***	0.40***	0.52***	0.51***
E5	0.40***	0.46***	0.42***	0.40***		0.50***	0.31***	0.58***	0.46***
E6	0.51***	0.51***	0.44***	0.53***	0.51***		0.50***	0.67***	0.60***
E7	0.53***	0.54***	0.46***	0.55***	0.51***	0.63***		0.47***	0.43***
E8	0.41***	0.47***	0.45***	0.51***	0.48***	0.53***	0.62***		0.55***
E9	0.26***	0.28***	0.31***	0.38***	0.32***	0.38***	0.41***	0.39***	

****Significant at* $P<0.001$.

Upper right triangle: Correlation coefficients between environments for GZn. Lower left triangle: Correlation coefficients between environments for GFe.

E1-E9, Shijiazhuang 2016-2017, Gaoyi 2016-2017, Beijing 2016-2017, Shijiazhuang 2017-2018, Gaoyi 2017-2018, Beijing 2017-2018, Shijiazhuang 2018-2019, Gaoyi 2018-2019, and Beijing 2018-2019.

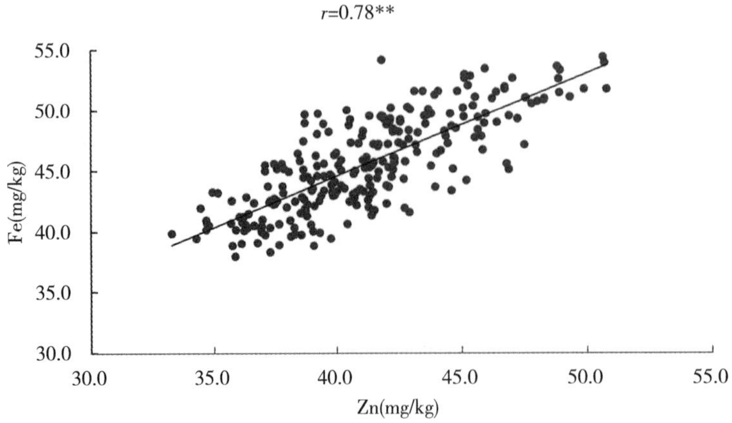

FIGURE. 2 Phenotypic correlation of GZn with GFe based on BLUEs across nine environments in the Jingdong 8/Bainong AK58 RIL population

TABLE 4 QTL for GZn and GFe identified by inclusive composite interval mapping in the Jingdong 8/Bainong AK58 RIL population

Trait	QTL	Environment	Physical interval[a]	Marker interval	LOD[b]	PVE (%)[c]	Add[d]
Zn	*QZn. caas-1DS*	E3	32. 5-38. 8	*AX-95235028-AX-94939596*	3. 0	3. 5	1. 1
		E6			2. 7	3. 4	0. 9
		E8			6. 0	6. 0	1. 5
	QZn. caas-2AS	E5	46. 1-48. 4	*AX-94592263-AX-108732889*	4. 1	2. 2	−1. 1
		E8			9. 2	9. 3	−1. 8
	QZn. caas-3BS	E2	42. 5-59. 1	*AX-110975262-AX-109911679*	3. 7	5. 7	1. 3
		E4			4. 8	5. 5	1. 0
	QZn. caas-4DS	E1	16. 0-19. 5	*AX-89593703-AX-89445201*	10. 8	12. 1	−1. 8
		E2			4. 6	7. 2	−1. 5
		E4			9. 0	10. 7	−1. 4
		E5			4. 4	2. 4	−1. 1
		E6			17. 1	25. 1	−2. 5
		E7			3. 5	4. 9	−0. 9
		E8			13. 1	14. 3	−2. 3
		E9			8. 5	11. 2	−1. 9
	QZn. caas-6AS	E4	77. 1-100. 3	*AX-108951317-AX-110968221*	4. 4	5. 2	−1. 0
		E6			3. 7	4. 8	−1. 1
	QZn. caas-6DL	E3	454. 1-459. 4	*AX-109058428-AX-111841126*	7. 3	8. 5	−1. 7
		E4			3. 0	3. 5	−0. 8
	QZn. caas-7BL	E1	721. 8-725. 4	*AX-95658138-AX-89745787*	4. 2	4. 3	−1. 0
		E3			5. 4	6. 4	−1. 5
		E6			5. 4	6. 9	−1. 3
		E8			6. 3	6. 3	−1. 5
		E9			5. 0	6. 2	−1. 4
Fe	*QFe. caas-3BL*	E5	764. 7-822. 9	*AX-111016352-AX-94835626*	3. 1	5. 8	−0. 9
		E6			2. 8	2. 9	−0. 9
	QFe. caas-4DS	E1	16. 0-17. 1	*AX-89593703-AX-89398511*	16. 8	20. 4	−2. 5
		E2			18. 7	27. 0	−3. 4
		E3			12. 3	19. 4	−2. 7
		E4			24. 1	24. 3	−1. 9
		E5			6. 2	9. 0	−1. 1
		E6			20. 6	25. 6	−2. 7
		E7			20. 8	30. 4	−3. 4
		E8			11. 2	5. 5	−2. 3
		E9			4. 6	2. 3	−1. 7
	QFe. caas-6AS	E1	77. 1-106. 9	*AX-108951317-AX-109304443*	4. 7	5. 1	−1. 2
		E6			7. 0	7. 5	−1. 5
	QFe. caas-7BL	E1	718. 5-725. 4	*AX-95631535-AX-89745787*	2. 8	2. 9	−0. 9
		E6			6. 2	6. 9	−1. 4

[a]Physical interval; Mb, according to IWGSC RefSeq v1. 0 (31), http: //www. wheatgenome. org/.

[b]LOD; likelihood of odds ratio for genetic effects.

[c]PVE; percentage of phenotypic variance explained by individual QTL.

[d]Add; Additive effect of QTL; negative values indicate that the superior allele came from Jingdong 8, whereas positive values indicate that the superior allele was from Bainong AK58. E1-E9, Shijiazhuang 2016-2017, Gaoyi 2016-2017, Beijing 2016-2017, Shijiazhuang 2017-2018, Gaoyi 2017-2018, Beijing 2017-2018, Shijiazhuang 2018-2019, Gaoyi 2018-2019, and Beijing 2018-2019.

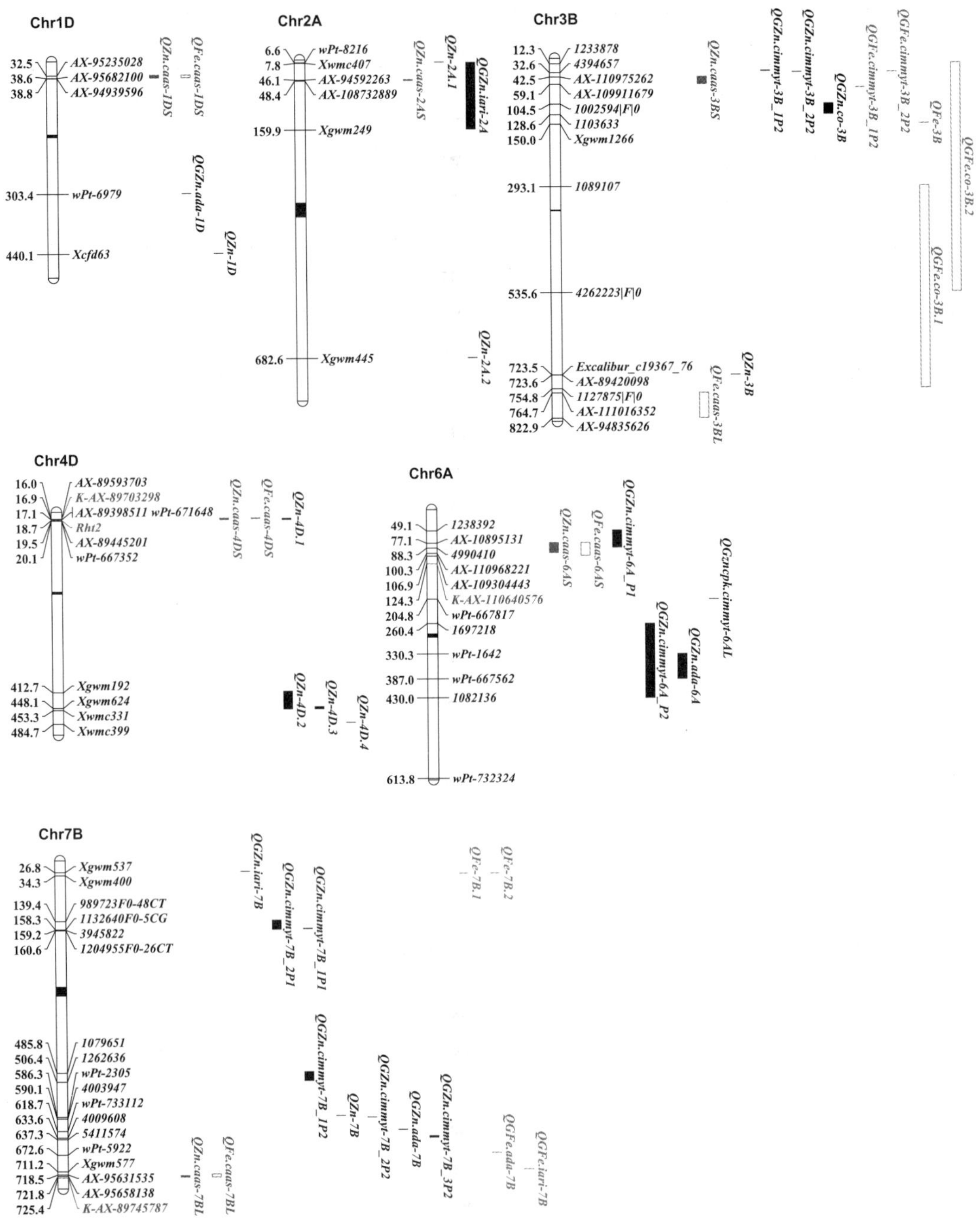

FIGURE. 3 Physical maps for the positional comparisons of GZn and GFe QTL reported on chromosomes 1DS, 2AS, 3BL, 4DS, 6AS, and 7BL with those identified in the present study. QTL linked markers are shown on the right, physical positions are shown on the left, and centromere is shown in the middle (black bar). KASP markers developed in the present study were shown in red. QTL for GZn and GFe in the present study were in red; QTL for GZn in previous studies were in blue; QTL for GFe in previous studies were in purple

TABLE 5 Chromosomal intervals for GZn and GFe identified by multi-trait composite interval mapping (MCIM)

Chromosomes	Flanking markers	Physical position (Mb)	Traits (Environment)
4DS	*AX-89593703-AX-89398511*	16.0-17.1	GZn (E1, E2, E4, E6, E7, E8, E9, BLUE value) GFe (E1, E2, E4, E5, E6, E7, E8, E9, BLUE value)
6AS	*AX-108951317-AX-110968221*	77.1-100.3	GZn (E1, E2, E4, E6, E7, BLUE value) GFe (E1, E2, E4, E6, BLUE value)

(continued)

Chromosomes	Flanking markers	Physical position (Mb)	Traits (Environment)
7BL	*AX-95658138-AX-89745787*	721.8-725.4	GZn (E1, E3, E6, E7, E8, E9, BLUE value) GFe (E6)

TABLE 6 Kompetitive allele specific PCR (KASP) markers converted from single nucleotide polymorphisms (SNPs) tightly linked to identified QTL on three chromosomes

Chromosome	SNP name	Physical position (Mb)	KASP primer	Primer sequence
4DS	*AX*-89703298	16.9	*K-AX-89703298*	5'-GAAGGTGACCAAGTTCATGCTCTAACCATTGGATAGGGCGAC-3'
				5'-GAAGGTCGGAGTCAACGGATTCTAACCATTGGATAGGGCGAA-3'
				5'-CCCAGCTTCAGCCCATGA-3'
6AS	*AX-110640576*	124.3	*K-AX-110640576*	5'-GAAGGTGACCAAGTTCATGCTCACAGATGTTCTCCACTCTCTG-3'
				5'-GAAGGTCGGAGTCAACGGATTCACAGATGTTCTCCACTCTCTC-3'
				5'-CCCTCCAAGGTCCATGGGT-3'
7BL	*AX-89745787*	725.4	*K-AX-89745787*	5'-GAAGGTGACCAAGTTCATGCTGGAGGACATTGTGCAACCG-3'
				5'-GAAGGTCGGAGTCAACGGATTGGAGGACATTGTGCAACCT-3'
				5'-AGGATTGGTTCTGCAATCCA-3'

TABLE 7 Mean values of GZn and GFe for genotype classes in the germplasm panel

Trait	QTL	Marker	Genotype	Number	Mean (mg/kg)	*T* value
GZn	*QZn. caas-4DS*	*K-AX-89703298*	CC	79	32.4	−2.28*
			AA	66	30.7	
	QZn. caas-6AS	*K-AX-110640576*	GG	19	34.0	−2.54*
			CC	126	31.2	
	QZn. caas-7BL	*K-AX-89745787*	GG	11	34.7	−2.41*
			TT	134	31.4	
GFe	*QFe. caas-4DS*	*K-AX-89703298*	CC	79	39.4	−2.58*
			AA	66	38.0	
	QFe. caas-6AS	*K-AX-110640576*	GG	19	39.6	−1.18
			CC	126	38.6	
	QFe. caas-7BL	*K-AX-89745787*	GG	11	43.1	−2.55*
			TT	134	38.4	

* *Significant at* $P<0.05$.

DISCUSSION

Comparisons With Previous Reports

In this study, QTL for GZn and GFe were mapped on chromosomes1D, 2A, 3B, 4D, 6A, 6D, and 7B, and on chromosomes 3B, 4D, 6A, and 7B, respectively. Previously identified QTL are summarized in Supplementary Table 1 and partly shown in FIGURE 3. In addition to consensus maps, the IWGSC RefSeq v1.0 Chinese Spring reference sequence[31] was used for comparisons of QTL identified in different studies.

QZn. caas-1DS

QZn. caas-1DS, flanked by SNP markers *AX-95235028* and *AX-94939596* at 32.5-38.8 Mb, was detected in three environments. Velu et al. [18] identified *QGZn. ada-1D* linked with a DArT marker *wPt-6979* at 303.4 Mb. Gorafiet al.[32] detected a QTL linked with SSR marker *Xcfd63* at physical position 440 Mb. The present QTL appears to be new.

QZn. caas-2AS

QZn. caas-2AS, flanked by *AX-94592263* and *AX-108732889* at physical positions of 46.1 and 48.4 Mb, was identified in two environments. Peleg et al.[33] identified *QZn-2A.1* and *QZn-2A.2* linked with *wPt-8216* and *Xgwm445* at 6.6 and 682.6 Mb, respectively. Krishnappa et al.[34] mapped *QGZn. iari-2A* flanking by *Xwmc407* and*Xgwm249* at physical position 28.2 and 159.9 Mb, respectively. *QZn. caas-2AS* detected in the present study was located within the region of *QGZn. iari-2A*; therefore, these two QTL may be the same.

QZn. caas-3BS

QZn. caas-3BS, flanked by *AX-110975262* and *AX-109911679* at physical positions of 42.5 and 59.1 Mb, was detected in two environments. Crespo-Herrera et al.[17] identified two QTL for GZn on this chromosome. QGZn. cimmyt-3B _ 2P2 was at the physical position 32.6 Mb linked with DArT markers 4394657, and QGZn. cimmyt-3B _ 1P2 flanked by 3533713 and 1007339 ismuch more near the distal end of 3BS than QGZn. cimmyt-3B _ 2P2 on the basis of the genetic map, although both markers were not on chromosome 3B with the result of blast. Furthermore, Liu et al.[35] mapped QGZn. co-3B flanked by DArT markers 1002594 | F | 0 and 1103633 at physical positions of 104.5 and 128.6Mb, respectively. Alomari et al.[19] identified a locus for GZn on chromosome 3BL, linked with *AX-89420098* at 723.5Mb. Thus, the previous QTL were around 10Mb from *QZn. caas-3BS*, indicating that *QZn. caas-3BS* is likely a new QTL.

QFe. caas-3BL

QFe. caas-3BL, flanked by *AX-111016352* and *AX-94835626* at physical positions of 764.7 and 822.9 Mb, was detected in two environments. Crespo-Herrera et al.[17] identified two QTL for GFe that were at the similar position as QTL for GZn as mentioned previously, both of which were on the short arm of chromosome 3B. Peleg et al.[33] mapped a QTL on chromosome 3B, closely linked with *Xgwm1266* at physical position 150 Mb. Liu et al.[35] identified *QGFe. co- 3B.1* and *QGFe. co-3B.2* flanked by DArT markers 1089107 and 1127875 | F | 0, 1233878-4262223 | F | 0 at physical positions 37.2-754.8 and 12.3-536.6 Mb, respectively. These five QTL were at least 10 Mb distant from *QFe. caas-3BL*. Therefore, *QFe. caas-3BL* is likely a new QTL for GFe.

QZn. caas-4DS* and *QFe. caas-4DS

QZn. caas-4DS and *QFe. caas-4DS*, flanked by *AX-89593703* and *AX-89445201* at physical positions of 16.0 and 19.5 Mb were detected in eight and nine environments, respectively. Pu et al.[36] identified a QTL for GZn at the same position, flanked by *wPt-671648* and *wPt-667352* located between 17.1 and 20.1 Mb on chromosome 4D, with reduced height gene *Rht2* (*Rht-D1b*) located in this region. Using a limited number of isogenic lines, Graham et al.[37] found that lower GZn and GFe in wheat was associated with reduced height genes. Velu et al.[38] verified this association using nine bread wheat (*Triticum aestivum*) and six durum (*T. turgidum*) isogenic line pairs differing at the *Rht1* (*Rht- B1*) locus and one bread wheat pair differing at the *Rht2* locus, indicating that the presence of reduced height genes decreased GZn by 1.9 to 10.0 ppm and GFe by 1.0 to 14.4 ppm. In this study, Bainong AK58 carried *Rht2* (*Rht-D1b*), while Jingdong 8 had *rht2* (*Rht-D1a*)[39]. A gene-specific KASP marker *K-AX-86170701* was identified for *Rht2*[40], and lines with allele from Bainong AK58 had significantly lower GZn and GFe than that with allele from Jingdong 8(Supplementary Figure 2). Therefore, it was possible that the lower concentrations of Zn and Fe in Bainong AK58 was associated with the *Rht2* allele.

QZn. caas-6AS* and *QFe. caas-6AS

QZn. caas-6AS and *QFe. caas-6AS*, flanked by *AX-108951317* and *AX-110968221* at physical positions of 77.1 and 106.9 Mb, were detected in four environments. No QTL for GFe was detected on chromosome 6AS previously, while two QTL for GZn were reported. Crespo-Herrera et al.[17] identified *QGZn. cimmyt- 6A _ P1*, linked with *1238392* and *4990410* at physical positions of 49.1 and 88.2 Mb. Hao et al.[16] mapped *QGZn. cimmyt-6AL* at 204.8 Mb with nearest marker *wPt-667817*. The present QTL was somewhat near the *QGZn. cimmyt-6A _ P1*, indicating that they might be the same.

QZn. caas-6DL

QZn. caas-6DL, flanked by *AX-109058428* and *AX-11184112* at physical positions of 454.1 and 459.4 Mb, was detected in two environments. It is likely a new one since no previous QTL for GZn was mapped on this chromosome.

QZn. caas-7BL and *QFe. caas-7BL*

Markers *AX-95631535* and *AX-89745787* at positions 718.5 and 725.4 Mb flanking *QZn. caas-7BL* and *QFe. caas-7BL* are distally located on chromosome 7BL. Several QTL were previously identified on this chromosome. Krishnappa et al.[34] mapped *QGZn. iari-7B* with closest marker Xgwm537 at 26.8 Mb. Peleg et al.[33] detected *QZn-7B* linked with *wPt-2305* at 586.3 Mb. Crespo-Herrera et al.[17] identified five QTL, including *QGZn. cimmyt-7B _ 2P1*, *QGZn. cimmyt-7B _ 1P1*, *QGZn. cimmyt-7B _ 1P2*, *QGZn. cimmyt-7B _ 2P2*, and *QGZn. cimmyt-7B _ 3P2* at physical positions of 139.4-160.6, 158.3-159.2, 485.8-506.4, 590.1, and 633.6-637.3 Mb, respectively. Velu et al.[18] reported *QGZn. ada-7B*, which was located at around 618.7 Mb with closely linked marker *wPt-733112*. All these eight QTL were well proximal (>80 Mb) from the QTL in this study, indicating that *QZn. caas-7BL* was reported for the first time. In addition, four QTL for GFe were mapped on chromosome 7B, among which two of them were at the same physical position of 34.3 Mb (*QFe-7B.1* and *QFe-7B.2*), and the other two were at 672.6 Mb (*QGFe. ada-7B*) and 711.2 Mb (*QGFe. iari-7B*), respectively[18,28,29,41]. *QGFe. iari-7B* and *QFe. caas-7BL* might be the same, since their distance is <10 Mb.

Pleiotropic Effects of QTL

The co-localization QTL for GZn and GFe on chromosomes 4DS, 6AS, and 7BL might be pleiotropic QTL based on the same or overlapping region detected using MCIM, in agreement with the significant positive phenotypic and genotypic correlations ($r = 0.78 \pm 0.01$ and 0.81 ± 0.03, $P < 0.01$) between GZn and GFe. Gorafi et al.[32] identified a significant and positive phenotypic correlation between GZn and GFe ($r = 0.78$) and a pleiotropic QTL on chromosome 5D; significant and positive correlations between GZn and GFe were also found in other studies[11,42]. It has been reported that some transporters, chelators, and genes regulated GZn and GFe simultaneously in a high frequency[10,43]. These findings indicated that Zn and Fe could be improved simultaneously in breeding programs targeting mineral biofortification.

Potential Implication in Wheat Breeding

MAS has been applied in crop breeding for more than two decades. Therefore, it would be effective for traits that were controlled by low numbers of major QTL[37]. The phenotypic analysis on GZn and GFe was time-consuming and laborious, indicating that identification of molecular markers linked to GZn and GFe would be of interest for improvement of nutritional quality in wheat. In the present study, pleiotropic QTL on chromosomes 4D, 6A, and 7B were detected in multiple environments. SNP markers linked to some of these QTL were converted to KASP markers, and the QTL were verified in a germplasm panel, indicating potential application in wheat breeding programs.

ACKNOWLEDGMENTS

The authors are very grateful to Prof. R. A. McIntosh, Plant Breeding Institute, University of Sydney, Australia, for reviewing this manuscript.

REFERENCES

[1] Sharma S, Chunduri V, Kumar A, Kumar R, Khare P, Kondepudi KK, et al. Anthocyanin bio-fortified colored wheat: nutritional and functional characterization. *PLoS ONE*. (2018) 13: e0194367. doi: 10.1371/

journal. pone. 0194367

[2] Shah D, Sachdev HPS. Zinc deficiency in pregnancy and fetal outcome. *Nutr Rev.* (2006) 64: 15-30. doi: 10.1111/j.1753-4887.2006.tb00169.x

[3] Wessells KR, Brown KH. Estimating the global prevalence of zinc deficiency: results based on zinc availability in national food supplies and the prevalence of stunting. *PLoS ONE.* (2012) 7: e50568. doi: 10.1371/journal.pone.00 50568

[4] Black RE, Victora CG, Walker SP, Bhutta ZA, Christian P, de Onis M, et al. Maternal and child undernutrition and overweight in low-income and middle-income countries. *Lancet.* (2013) 382: 427-51. doi: 10.1016/S0140-6736 (13) 60937-X

[5] Bouis HE, Welch RM. Biofortification—A sustainable agricultural strategy for reducing micronutrient malnutrition in the global south. *Crop Sci.* (2010) 50: S20-32. doi: 10.2135/cropsci2009.09.0531

[6] Sands DC, Morris CE, Dratz EA, Pilgeram AL. Elevating optimal human nutrition to a central goal of plant breeding and production of plant-based foods. *Plant Sci.* (2009)177:377-89. doi:10.1016/j. plantsci. 2009.07.011

[7] Deshpande J, Joshi M, Giri P. Zinc: the trace element of major importance in human nutrition and health. *Int J Med Sci.* (2013) 2: 1-6. doi: 10.5455/ijmsph.2013.2.1-6

[8] Stoltzfus RJ. Defining iron-deficiency anemia in public health terms: a time for reflection. *J Nutr.* (2001) 131: 565S—7. doi: 10.1093/jn/131.2.565S

[9] Crespo-Herrera LA, Velu G, Singh RP. Quantitative trait loci mapping reveals pleiotropic effect for grain iron and zinc concentrations in wheat. *Ann Appl Biol.* (2016) 169: 27-35. doi: 10.1111/aab.12276

[10] Tong J, Sun M, Wang Y, Zhang Y, Rasheed A, Li M, et al. Dissection of molecular processes and genetic architecture underlying iron and zinc homeostasis for biofortification: from model plants to common wheat. *Int J Mol. Sci.* (2020) 21: 9280. doi: 10.3390/ijms21239280

[11] Morgounov A, Gómez-BecerraHF, Abugalieva A, Dzhunusova M, Yessimbekova M, Muminjanov H, et al. Iron and zinc grain density in common wheat grown in Central Asia. *Euphytica.* (2007) 155: 193-203. doi: 10.1007/s10681-006-9321-2

[12] Tiwari VK, Rawat N, ChhunejaP, Neelam K, Aggarwal R, Randhawa GS, et al. Mapping of quantitative trait loci for grain iron and zinc concentration in diploid A genome wheat. *J Hered.* (2009) 100: 771- 6. doi: 10.1093/jhered/esp030

[13] ZhangY, Song Q, Yan J, Tang J, Zhao R, Zhang Y, et al. Mineral element concentrations in grains of Chinese wheat cultivars. *Euphytica.* (2010) 174: 303-13. doi: 10.1007/s10681-009-0082-6

[14] Srinivasa J, Arun B, Mishra VK, Singh GP, Velu G, Babu R, et al. Zinc and iron concentration QTL mapped in a *Triticum spelta* × *T. aestivum* cross. *Theor Appl Genet.* (2014) 127: 1643-51. doi: 10.1007/s00122-014-2327-6

[15] Velu G, Ortiz-Monasterio I, Cakmak I, HaoY, Singh RP. Biofortification strategies to increase grain zinc and iron concentrations in wheat. *J Cereal Sci.* (2014) 59: 365-72. doi: 10.1016/j.jcs.2013.09.001

[16] HaoY, Velu G, Peña RJ, Singh S, Singh RP. Genetic loci associated with high grain zinc concentration and pleiotropic effect on kernel weight in wheat (*Triticum aestivum* L.). *Mol Breeding.* (2014) 34: 1893-902. doi: 10.1007/s11032-014-0147-7

[17] Crespo-Herrera LA, GovindanV, Stangoulis J, Hao Y, Singh RP. QTL Mapping of grain Zn and Fe concentrations in two hexaploid wheat RIL populations with ample transgressive segregation. *Front Plant Sci.* (2017) 8: 1800. doi: 10.3389/fpls.2017. 01800

[18] Velu G, Tutus Y, Gomez-Becerra HF, Hao Y, Demir L, Kara R, et al. QTL mapping for grain zinc and iron concentrations and zinc efficiency in a tetraploid and hexaploid wheat mapping populations. *Plant Soil.* (2017) 411: 81-99. doi: 10.1007/s11104-016-3025-8

[19] Alomari DZ, Eggert K, von Wirén N, Alqudah AM, Polley A, PlieskeJ, et al. Identifying candidate genes for enhancing grain Zn concentration in wheat. *Front Plant Sci.* (2018) 9: 1313. doi: 10.3389/fpls.2018.01313

[20] Zarcinas BA, Cartwright B, Spouncer LR. Nitric acid digestion and multi-element analysis of plant material by inductively coupled plasma spectrometry. *Soil Sci Plant Anal.* (1987) 18: 131-46. doi: 10.1080/00103628709367806

[21] Paltridge NG, MilhamPJ, Ortiz-Monasterio JI, Velu G, Yasmin Z, Palmer LJ, et al. Energy-dispersive X-ray fluorescence spectrometry as a tool for zinc, iron and selenium analysis in whole grain wheat. *Plant Soil.* (2012) 361: 261-9. doi: 10.1007/s11104-012-1423-0

[22] Holland JB, Nyquist WE, Cervantes-Matinez CT. Estimating and interpreting heritability for plant

breeding: an update. *Plant Breed Rev.* (2003) 22: 9-111. doi: 10.1002/9780470650202.ch2

[23] Becker WA. *Manual of Quantitative Genetics*, 4th ed. Pullman, WA: Academic Enterprises (1984)

[24] DoyleJ, Doyle JL. A rapid DNA isolation procedure from small quantities of fresh leaf tissues. *Phytochem Bull.* (1987) 19: 11-5

[25] Stam P (1993) Construction of integrated genetic linkage maps by means of a new computer package: join Map. *Plant J.* (1993) 3: 739-744. doi: 10.1111/j.1365-313X.1993.00739.x

[26] JiangCJ, Zeng ZB. Multiple trait analysis of genetic mapping for quantitative trait loci. *Genetics.* (1995) 140: 1111-27. doi: 10.1093/genetics/140.3.1111

[27] Wickham H. *ggplot2: Elegant Graphics for Data Analysis.* New York, NY: Springer-Verlag

[28] Voorrips RE. MapChart: software for the graphical presentation of linkage maps and QTLs. *J Hered.* (2002) 93: 77-78. doi: 10.1093/jhered/93.1.77

[29] Chandra S, SinghD, Pathak J, Kumari S, Kumar M, Poddar R, et al. SNP discovery from next-generation transcriptome sequencing data and their validation using KASP assay in wheat (*Triticum aestivum* L.). *Mol Breeding.* (2017) 37: 92. doi: 10.1007/s11032-017-0696-7

[30] Jia A, RenY, Gao F, Yin G, Liu J, Guo L, et al. Mapping and validation of a new QTL for adult-plant resistance to powdery mildew in Chinese elite bread wheat line Zhou8425B. *Theor Appl Genet.* (2018) 131: 1063- 71. doi: 10.1007/s00122-018-3058-x

[31] Appels R, Eversole K, Stein N, Feuillet C, Keller B, RogersJ, et al. Shifting the limits in wheat research and breeding using a fully annotated reference genome. *Science.* (2018) 361: eaar7191. doi: 10.1126/science.aar7191

[32] Gorafi YSA, IshiiT, Kim J-S, Elbashir AAE, Tsujimoto H. Genetic variation and association mapping of grain iron and zinc contents in synthetic hexaploid wheat germplasm. *Plant Genet Resour.* (2018) 16: 9-17. doi: 10.1017/S1479262116000265

[33] Peleg Z, Cakmak I, Ozturk L, Yazici A, Jun Y, Budak H, et al. Quantitative trait loci conferring grain mineral nutrient concentrations in durum wheat × wild emmer wheat RIL population. *Theor Appl Genet.* (2009) 119: 353-69. doi: 10.1007/s00122-009-1044-z

[34] Krishnappa G, Singh AM, Chaudhary S, Ahlawat AK, Singh SK, Shukla RB, et al. Molecular mapping of the grain iron and zinc concentration, protein content and thousand kernel weight in wheat (*Triticum aestivum* L.). *PLoS ONE.* (2017) 12: e0174972. doi: 10.1371/journal.pone.0174972

[35] Liu J, Wu B, Singh RP, Velu G. QTL mapping for micronutrients concentration and yield component traits in a hexaploid wheat mapping population. *J Cereal Sci.* (2019) 88: 57-64. doi: 10.1016/j.jcs.2019.05.008

[36] Pu ZE, Yu M, He QY, Chen GY, Wang JR, Liu YX, et al. Quantitative trait loci associated with micronutrient concentrations in two recombinant inbred wheat lines. *J Integr Agric.* (2014) 13: 2322-9. doi: 10.1016/S2095-3119 (13) 60640-1

[37] Graham R, SenadhiraD, Beebe S, Iglesias C, Monasterio I. Breeding for micronutrient density in edible portions of staple food crops: conventional approaches. *Field Crops Res.* (1999) 60: 57-80. doi: 10.1016/S0378-4290 (98) 00133-6

[38] Velu G, Singh RP, Huerta J, Guzmán C. Genetic impact of Rht dwarfing genes on grain micronutrients concentration in wheat. Field Crops Res. (2017) 214: 373-7. doi: 10.1016/j.fcr.2017.09.030

[39] Li X, Xia X, XiaoY, He Z, Wang D, Trethowan R et al. QTL mapping for plant height and yield components in common wheat under water-limited and full irrigation environments. *Crop Pasture Sci.* (2015) 66: 660-70. doi: 10.1071/CP14236

[40] Rasheed A, Wen W, GaoF, Zhai S, Jin H, Liu J, et al. Development and validation of KASP assays for genes underpinning key economic traits in bread wheat. *Theor Appl Genet.* (2016) 129: 1843-60. doi: 10.1007/s00122-016-2743-x

[41] Shi RL, TongYP, Jing RL, Zhang FS, Zou CQ. Characterization of quantitative trait loci for grain minerals in hexaploid wheat (*Triticum aestivum* L.). *J Integr Agric.* (2013) 12: 1512-21. doi: 10.1016/S2095-3119 (13) 6 0559-6

[42] ZhaoFJ, Su YH, Dunham SJ, Rakszegi M, Bedo Z, McGrath SP, et al. Variation in mineral micronutrient concentrations in grain of wheat lines of diverse origin. *J Cereal Sci.* (2009) 49: 290-5. doi: 10.1016/j.jcs.2008.11.007

[43] ShanmugamV, Lo JC and Yeh KC. Control of Zn uptake in *Arabidopsis halleri*: a balance between Zn and Fe. *Front Plant Sci.* (2013) 4: 281. doi: 10.3389/fpls.2013.00281

Determination of phenolic acid concentrations in wheat flours produced at different extraction rates

Lan Wang[a], Yang Yao[a], Zhonghu He[a,b], Desen Wang[a], Aihua Liu[c], Yong Zhang[a]

[a] *Institute of Crop Science, National Wheat Improvement Centre, Chinese Academy of Agricultural Sciences (CAAS), 12 Zhongguancun South Street, Beijing 100081, China*

[b] *CIMMYT China Office, C/O CAAS, Beijing 100081, China*

[c] *Qingdao Entryeexit Inspection and Quarantine Bureau, Qingdao 266000, Shandong province, China*

Abstract: High-performance liquid chromatography (HPLC) was used to determine the distribution of phenolic acids in wheat flours produced from five milling extraction rates ranging from 60% to 100% in four cultivars sown in two locations in the 2008—2009 season. Considerable variation was observed in free and bound phenolic acids, and their components in flours with different extraction rates. Most phenolic acids, including the component ferulic, were present in the bound form (94.0%). Ferulic (51.0%) was the predominant phenolic acid in wheat grain, and caffeic (22.8%) and *p*-coumaric (17.6%) acids were abundant. The phenolic acids and their components were all significantly influenced by effects of cultivar, milling, location, and cultivar×milling interaction, with milling effect being the predominant. The proportions of phenolic compounds varied considerably among milling extractions and cultivars, and their levels depended on both initial grain concentrations and on selection of milling extraction that was incorporated into the final product. The grain phenolic acid concentrations determined ranged from 54 mg • g^{-1} in flour produced at 60% extraction rate to 695 mg • g^{-1} in flour produced at 100% extraction rate, indicating their higher concentrations in bran associated with cell wall materials. Therefore, wholemeal wheat products maximize health benefits and are strongly recommended for use in food processing.

Keywords: Common wheat, Flour milling, Phenolic acid, Phenolic components

1 Introduction

Common wheat (*Triticum aestivum* L.) is an important component of the human diet, and is used in the production of many food products, including bread, noodles, steamed bread, and cakes, providing energy based on the high contents of protein and carbohydrate. Wheat products contain high levels of antioxidants, which confer protection against cancer and heart diseases mostly coming from phenolics (Adom and Liu, 2002; Adom et al., 2005; Jacobs et al., 1998; Ward et al., 2008). Thus, wheat derived phenolics hold great promise for the provision of health benefits (Jacobs et al., 1998; Nicodemus et al., 2001; Ward et al., 2008). Phenolic acids represent the most common form of phenolic compounds in wheat and make up one of the major and most complex

Published in Journal of Cereal Science, 2013, 57: 67-72.

groups of phytochemicals in the cereal grain (Mattila et al., 2005). They are derivatives of either hydroxycinnamic acid or hydroxybenzoic acid and can be divided into two groups (Saulnier et al., 2007), with hydroxybenzoic acid derivatives including *p*-hydroxybenzoic, protocatechuic, vannilic, syringic, and gallic acids, and hydroxycinnamic acid derivatives including *p*-coumaric, caffeic, ferulic, and sinapic acids (Adom and Liu, 2002). These acids are present mainly in bound forms, linked to cell wall structural components such as cellulose, lignin, and proteins through ester bonds (Parker et al., 2005). The bound phenolics are considered to have greater antioxidant capacity, because they can escape from upper gastrointestinal digestion along with cell wall materials, and are absorbed into blood plasma during digestion by intestinal microflora (Adom and Liu, 2002).

Previous studies indicated there were highly significant genotypic differences in phenolic acids among wheat genotypesfrom Europe, Canada, the U.S.A., and China (Li et al., 2008; Verma et al., 2009; Ward et al., 2008; Zhang et al., 2012). Most studies, however, were focused on the bran fractions (Zhou and Yu, 2004) or endosperm fractions alone (Yu et al., 2004), with little attempt to analyze combinations of whole grain and milled extractions (Adom et al., 2005; Beta et al., 2005; Skrabanja et al., 2004). Zhou and Yu (2004) examined the effects of cultivars and growing conditions on phenolic concentrations and antioxidant activities of bran extracts of a wheat cultivar. They found antioxidant activity and phenolic concentration in wheat endosperm were significantly influenced by growing locations. Skrabanja et al. (2004) and Tang et al. (2008) presented data of milled fractions on nutrients such as Fe and Zn concentrations, but did not include phytochemical profiles of phenolic acids. All these reports indicate that the milling process is crucial for nutrient retention in wheat grain, and that distribution of phytochemical profiles and antioxidant activities, including phenolic acids, might vary among cultivars and locations. However, no information is available on the effects of milling, combined with genotype and location effects on phenolic acids.

With the increasing popularity of functional foods, it is crucialto understand the distribution of phenolic compounds in different milling extractions which may have important implications in ensuring their health benefits. Most wheat products in China are used for steamed bread and noodle flours produced at around 60% milling extraction (Ye et al., 2009), but the use of wholemeal foods is rapidly increasing as health concerns become more important (Chen, 2004). The objectives of this study were, therefore, to investigate the distribution of phenolic acids and their major components in flour from different extraction rates, and to determine whether growing locations and cultivars have any effect on the distribution of phenolic acis in different milling extractions.

2 Materials and methods

2.1 Wheat samples and sample preparation

Four cultivars, namelyJingdong 8, Beijing 0045, Zhongmai 175 and Zhongyou 206 from the Northern China Winter Wheat Plain were sown in the 2008—2009 cropping season at the Institute of Crop Science, Chinese Academy of Agricultural Sciences, Beijing (lat. 39° 92′ N, long. 116° 23′ E, 43.5 m above sea level, with loam soil type) and Gaobeidian (lat. 39° 50′N, long. 115°47′E, 25.5 m above sea level, with sandy loam soil type) in Hebei Province. Jingdong 8, Beijing 0045, and Zhongmai 175 are current leading cultivars, all with annual planting areas of more than 70 000 ha. Jingdong 8 was the check in the regional trials from 1998 to 2010, and Zhongmai 175 is the current check used in the regional trials. Zhongyou 206 has superior pan bread making quality. The two locations are representative sites within the winter wheat zone. A completely randomized block design with two replications was used, and each plot consisted of 12 9.0 m rows spaced 0.2 m apart. Seeding rate was about 240 kg • ha^{-1} and field management was according to local practices. Grain samples harvested from each replication were cleaned prior to conditioning and milling. Grain hardness was determined on 300-kernel samples with a Perten 4100 Single Kernel Characterization System (SKCS, Perten Instruments North America Inc., Reno, NV). Hard and soft wheats were conditioned to 16% and 14% moisture content, respectively, following AACC approved method 26-10 (AACC International, 2000). A Buhler experimental mill (MLU 220, Uzvil, Switzerland) was used to obtainflour samples at five extraction rates, namely 60%, 70%, 80%, 90%, 100% for each grain sample, according to AACC approved method 26-21A. Each milling extraction was thoroughly mixed, and cooled immediately and stored at 20 ℃ for analysis to protect bioactive components from degradation.

2.2 Analytical methods

Free and bound phenolic acid extractions were carried out as previously described (Zhang et al., 2012), following modified methods of Adom et al. (2003, 2005) and Mattila et al. (2005). Free phenolic acids were extracted using an 80：20 chilled ethanol/water solvent mixture. Insoluble bound phenolic acids were released via alkaline hydrolysis (6 M NaOH, 16 h) of the residue from the initial ethanol/water extraction, followed by acid hydrolysis. Both phenolic acid fractions extracted by cold diethyl ether (DE) and ethyl acetate (EA, 1：1 v/v) after alkaline and acid hydrolysis were acidified to pH 2 to enable extraction into organic solvent. Individual phenolic acids in the flour extracts were analyzed by an Agilent 1100 Series high-performance liquid chromatograph (HPLC), equipped with a UV detector at 280 nm and an Agilent TC-C18 (250×4.60 mm, 5 mm) column (Irmak et al., 2008). The injection volume was 20 mL, and the flow rate was set to 1 mL • min^{-1}. A gradient elution program was utilized, as previously described (Zhang et al., 2012), and incorporated a mobile phase of water with 0.05% trifluoroacetic acid (solvent A) and 30% acetonitrile, 10% methanol, 59.95% water and 0.05% trifluoroacetic acid (solvent B). Identification and quantification of phenolic acids in samples were performed by comparison with chromatographic retention times and areas of external standards. Calibration curves of phenolic acid standards were constructed using authentic standards that had undergone the same extraction procedure to ensure that losses due to the extraction were accounted for. The authentic standards used for peak identification of ferulic, syringic, chlorogenic, caffeic, vanillic, *p*-coumaric and gentisic acids, were purchased from Sigma and Aldrich (Sigma-Aldrich Corporation, St. Louis, MO) and used without further purification (97% and higher purity). Standard compounds were prepare as stock solutions at 2 mg • mL^{-1} in 80% ethanol. The stock solutions were stored in darkness at 18 ℃ and remained stable for at least 3 months. All samples were analyzed in duplicate unless otherwise stated, and concentrations of individual phenolic acids were expressed in micrograms per gram (mg • g^{-1}) of dry matter (dm).

2.3 Statistical analysis

Means, standard deviations, and ranges were determined using PROC MEANS in the Statistical Analysis System (SAS Institute, 2000). Analysis of variance was conducted by PROC GLM, treating cultivars and locations as random effects, and milling extractions as fixed. Fisher's F-protected least significant difference (LSD) method was used to separate means. PROC CORR was used to analyze the correlations among phenolic acids.

3 Results

The concentration of free phenolic acids in flour samples was17 mg • g^{-1} of dm, making up only 6.1% of the grain phenolic acids determined (Table 1). Syringic accounted for 55.9% of the free phenolic acids, while ferulic accounted for 34.6% of the total of this class. There were no detectable amounts of the free phenolic acid components caffeic, chlorogenic and gentisic. The concentration of bound phenolic acids was 262 mg • g^{-1} of dm, making up 93.9% of the grain phenolic acids determined, among which ferulic accounted for 52.0% of it, while caffeic and *p*-coumaric accounted for 24.3% and 18.1% of it. Wide ranges of variation among flour samples with five extraction rates of the four cultivars sown in two locations were observed on concentrations of free and bound phenolic acids, and all their components.

Analysis of variance indicated that most of the parameters were significantly influenced bycultivar, location, milling, and their interaction effects (Table 2). Milling variance was predominant for all parameters including free and bound phenolic acids, and their components, except for free *p*-coumaric which was mainly influenced by cultivar milling interaction variance, followed by cultivar milling interaction variances. Cultivar was important relative to other sources of variances for all phenolic acids and their components, location was important relative to other sources of variances for bound phenolic acids and their components ferulic and caffeic, and cultivar location interaction was important relative to other sources of variances for free and bound phenolic acids, and their component ferulic. Significant variances of location milling interaction and cultivar location milling interaction were also observed for bound gentisic and *p*-coumaric concentrations.

3.1 Distribution of phenolic acids in milling extractions

The amounts of free phenolic acids varied significantly among flour samples from different extraction rates (Table 3). The mean concentrations of free phenolic acids ranged from 5 mg • g^{-1} of dm to 37 mg • g^{-1} of dm, with flour samples produced at 100% extraction rate having the highest amounts, while flour sample from 60% extraction having the lowest amounts. Syringic represent the most abundant class, contributing from 29.8% in flour at 60% extraction rate to 36.8% in flour at 100% extraction rate of the total free phenolic acid among cultivars. The amounts of free phenolic acid increased slowly in flour at extraction rates from 60% to 80%, but sharply from 80% to 100%, and the same trend was observed for ferulic, syringic, and *p*-coumaric concentrations.

The amounts of bound phenolic acid varied significantly among flour samples from different extraction rates (Table 3). The mean concentrations ranged from 50 mg • g^{-1} of dm to 657 mg • g^{-1} of dm, with flour at 100% extraction rate having the highest amounts, and flour at 60% extraction rate having the lowest amounts. Ferulic represented the most abundant class, contributing from 46.0% in flour at 60% extraction rate to 55.8% in flour at 100% extraction rate of the total bound phenolic acids among cultivars, followed bycaffeic and *p*-coumaric. The amounts of bound phenolic acids increased slowly in flour at extraction rate from 60% to 80%, but sharply from 80% to 100%, and the same trend was observed for ferulic, caffeic, chlorogenic, and *p*-coumaric. The concentration of syringic increased in flour at extraction rates from 60% to 90%, but decreased in flour at extraction rates from 90% to 100%. The concentration of gentisic decreased in flour at extraction rates from 60% to 70%, but increased in flour at extraction rates from 70% to 100%.

Table 1 Mean and range of phenolic acids (μg • g^{-1} dry matter) in four wheat cultivars at five milling extractions from two locations in 2008—2009 season

Class	Parameter	Mean	Range			
			Overall	Milling	Cultivar	Location
Free	ferulic	6	0-44	1-8	6-26	13-15
	caffeic	—	—	—	—	—
	chlorogenic	—	—	—	—	—
	syringic	9	0-38	3-19	8-29	19-19
	gentisic	—	—	—	—	—

(continued)

Class	Parameter	Mean	Range			
			Overall	Milling	Cultivar	Location
	p-coumaric	2	0-17	1-4	1-12	3-6
	Subtotal	17	1-58	5-37	27-48	37-37
Bound	ferulic	136	5-466	23-367	282-452	352-383
	caffeic	64	0-264	14-147	74-241	144-149
	chlorogenic	6	0-32	1-13	7-20	13-14
	syringic	7	0-41	2-14	3-14	7-9
	gentisic	2	0-9	1-3	1-6	2-4
	p-coumaric	47	1-216	8-119	83-208	117-122
	Subtotal	262	16-803	50-657	545-783	643-672

—value below the limit of detection (1μg · g^{-1}).

3.2 Impact of cultivars on phenolic acids

Based on the average data of flour samples produced from 100% extraction rate from two locations, Zhongyou 206 had the highest free phenolic acid concentration (48 mg · g^{-1} of dm), followed by Beijing 0045 (36 mg · g^{-1} of dm) and Zhongmai 175 (37 mg · g^{-1} of dm), while Jingdong 8 had the lowest concentrations (27 mg · g^{-1} of dm) (Table 4). Beijing 0045 had high ferulic, but low syringic and *p*-coumaric concentrations, while Zhongmai 175 had medium ferulic, syringic, and *p*-coumaric concentrations. Zhongyou 206 had high syringic and *p*-coumaric concentrations, but low ferulic concentration and Jingdong 8 had medium syringic concentration, but low ferulic and *p*-coumaric concentrations.

Zhongmai 175 had the highest bound phenolic acid concentration (783 mg · g^{-1} of dm), followed by Jingdong 8 (703 mg · g^{-1} of dm), while Beijing 0045 (545 mg · g^{-1} of dm) and Zhongyou 206 (599 mg · g^{-1} of dm) had low concentrations (Table 4). Beijing 0045 had medium concentrations of ferulic and gentisic, but low concentrations of caffeic, chlorogenic, syringic, and *p*-coumaric; while Zhongmai 175 had high ferulic and cholorogenic concentrations, medium to high caffeic and syringic concentrations, but medium to low p-coumaric and low gentisic concentrations. Zhongyou 206 had high syringic, gentisic, and *p*-coumaric concentrations, but medium chlorogenic concentration, low ferulic and caffeic concentrations and Jingdong 8 had high caffeic concentration, medium ferulic, chlorogenic, and gentisic concentrations, but low syringic and *p*-coumaric concentrations.

3.3 Effect of cultivar by milling interaction on phenolic acids

Significant difference in free phenolic acids (Table 5) and the components (data not shown) among cultivars were observed only in flour samples at 100% extraction rate, while there were no significant differences among cultivars in flour at extraction rates from 60% to 90%. Beijing 0045 had the lowest free phenolic acid concentration in flour at 60% extraction rate, but the highest at 80% extraction rate, while medium at 70%, 90%, and 100% extraction rates; Zhongmai 175 had the highest free phenolic acid concentration in flour at 70% extraction rate, medium at 80%, 90%, and 100% extraction rates, while low at 60% extraction rate; Zhongyou 206 had the highest free phenolic acid concentration in flour at 60%, 90%, and 100% extraction rates, while medium at 80% extraction rate, and the lowest at 70% extraction rate; Jingdong 8 had medium to high free phenolic acid concentration in flour at 60% and 70% extraction rates, while low at 80%, 90%, and 100% extraction rates.

Significant differences in bound phenolic acids (Table 5) and the components (data not shown) among cultivars were observed in flour samples at 60%, 70%, 80%, and 100% extraction rates, while there were no significant difference among cultivars in flour at 90% extraction rate. Beijing 0045 had the highest bound phenolic acid concentration in flour at 60%, 80%, and 90% extraction rates, medium at 70% extraction rate, while low at 100% extraction rate; Zhongmai 175 had the highest bound phenolic acid concentration in flour at 70% and 100% extraction rates, medium at 60% and 80% extraction rates, but low at 90% extraction rate; Zhongyou 206 had medium to low bound phenolic acid concentration in flour at 70%, 80%, 90%, and 100% extraction rates, while low at 60% extraction rate; Jingdong 8 had medium to high bound phenolic acid concentration in flour at 100% extraction rate, medium to low at 60% and 90% extraction rates, but low at 70% and 80% extraction rates.

3.4 Effect of growing location on phenolic acids

Comparison of grain samples produced at 100% extraction rate from two locations showed that free phenolic acids displayed some variation in concentration, and significant difference in *p*-coumaric concentration was observed, while there was no significant difference in ferulic and syringic concentrations between samples from the two locations (data not shown). The samples from Beijing generally displayed higher concentrations of free phenolic acid components including syringic and *p*-coumaric, but lower ferulic than those grown in Gaobeidian.

Significant difference on ferulic, caffeic, and total amounts of bound phenolic acids was observed, while there was no significant difference in chlorogenic, syringic, gentisic, and *p*-coumaric concentrations between samples from the two locations. The samples grown in Beijing generally displayed higher concentrations of chlorogenic, syringic, and *p*-coumaric, but lower ferulic, caffeic, and gentisic than those grown in Gaobeidian.

3.5 Relationship of phenolic acids in whole grain and those in milling extractions

As shown in Table 6, correlation coefficients of free phenolic acids as well as the components between whole grain and milling extraction of the outer bran were higher than that in milling extraction of the inner endosperm. Whole grain free phenolic acid concentration was significant and positively correlated with those in milling extraction produced at 90% rate ($r=0.89$, $P<0.01$). For free ferulic concentrations, except for the milling extraction at 60% rate, significant and positive correlations between whole grain and all the other three milling extractions from 70% to 90% rate were observed, with r ranging from 0.90 ($P<0.01$) to 0.98 ($P<0.001$). Significant and positive correlations between free syringic ($r=0.84$, $P<0.01$), *p*-coumaric ($r=0.98$, $P<0.001$) in milling extraction both at 90% rate and those in whole grain were also observed. For bound phenolic acid concentrations, only caffeic in milling extraction at 90% rate ($r=0.71$, $P<0.05$), syringic in milling extraction at 60% ($r=0.87$, $P<0.01$) and 70% ($r=0.79$, $P<0.05$) rates were significant and positively correlated with those in whole grain, while significant and negative correlation of gentisic in milling extraction at 70% rate ($r=0.73$, $P<0.05$) and those in whole grain was observed.

Table 2 Sum of squares from a combined analysis of variance for phenolic acid concentrations (μg · g^{-1} dry matter) in flours produced at five extraction rates from four wheat cultivars grown at two locations in 2008—2009 season

Parameter	Source Df	Cultivar (C)	Milling (M)	Location (L)	C* M	C* L	M* L	C* M* L	Rep (L)	Error
		3	4	1	12	3	4	12	2	38
Free	Ferulic	916***	1 606***	8	729**	305**	11	200	44	793
	Syringic	731***	2 849***	12	548***	157***	17	84	16	267

(continued)

Parameter	Source Df	Cultivar (C)	Milling (M)	Location (L)	C*M	C*L	M*L	C*M*L	Rep (L)	Error
		3	4	1	12	3	4	12	2	38
	p-Coumaric	125***	199***	15***	235***	6	13*	20	3	31
	Subtotal	429**	11 496***	20	683*	395**	16	211	81	957
Bound	Ferulic	36 152***	1 300 110***	14 128***	74 472***	14 746***	2 973	10 074	2 210	26 805
	Caffeic	30 438***	183 459***	1 186**	65 496***	2 754**	1 456*	2 510	99	5 203
	Chlorogenic	278***	1 471***	1	395***	95**	12	195**	19	209
	Syringic	687***	1 414***	65	930**	402**	122	161	0	866
	Gentisic	13***	73***	1	38***	4*	11***	17***	1	13
	p-Coumaric	19 311***	163 895***	900**	36 775***	1 864**	3 348***	5 727***	612	4 410
	Subtotal	30 000***	4 110 231***	14 926**	132 607***	34 469***	4 269	18 611	4 556	47 523

*, **, and *** indicate significant at $P=0.05$, 0.01, and 0.001, respectively.

Table 3 Phenolic acid concentrations (μg · g^{-1} dry matter) in flours produced at five extraction rates from four wheat cultivars grown at two locations in 2008—2009 season

Parameter		60%	70%	80%	90%	100%
Free	Ferulic	1c	2c	4c	8b	14a
	Syringic	3d	5cd	6c	13b	19a
	p-Coumaric	1c	1c	1c	2b	4a
	Subtotal	5d	8d	11c	23b	37a
Bound	Ferulic	23d	39d	74c	178b	367a
	Caffeic	14e	25d	49c	83b	147a
	Chlorogenic	1c	3c	4c	7b	13a
	Syringic	2c	4c	9b	14a	8b
	Gentisic	2b	1c	1c	2b	3a
	p-Coumaric	8c	12c	15c	83b	119a
	Subtotal	50e	84d	152c	367b	657a

Values with different letters in the same row are significantly different from each other at $P=0.05$.

Table 4 Phenolic acid concentrations (μg · g^{-1} dry matter) of the wheat cultivars grown at two locations in 2008—2009 season

Parameter		Beijing 0045	Zhongmai 175	Zhongyou 206	Jingdong 8
Free	Ferulic	26a	15ab	7b	6b
	Syringic	8c	18b	29a	20b
	p-Coumaric	1c	4b	12a	1c
	Subtotal	36ab	37ab	48a	27b
Bound	Ferulic	375b	452a	282c	360b
	Caffeic	74c	195a	76c	241b
	Chlorogenic	7b	20a	14ab	13ab
	Syringic	4b	12a	14a	3b
	Gentisic	4ab	1c	6a	3bc

(continued)

Parameter		Beijing 0045	Zhongmai 175	Zhongyou 206	Jingdong 8
	p-Coumaric	83b	103b	208a	83b
	Subtotal	545b	783a	599b	703a

Values with different letters in the same row are significantly different from each other at $P=0.05$.

Table 5 Free and bound phenolic acid concentrations ($\mu g \cdot g^{-1}$ dry matter) in flours produced at five extraction rates from four wheat cultivars grown at two locations in 2008—2009 season

Extraction	Cultivar	Free	Bound
60%	Beijing 0045	3a	67a
	Zhongmai 175	4a	56ab
	Zhongyou 206	6a	32b
	Jingdong 8	5a	42ab
70%	Beijing 0045	7a	83b
	Zhongmai 175	8a	121a
	Zhongyou 206	7a	71b
	Jingdong 8	8a	58b
80%	Beijing 0045	13a	173a
	Zhongmai 175	11a	168a
	Zhongyou 206	11a	155ab
	Jingdong 8	9a	108b
90%	Beijing 0045	24a	395a
	Zhongmai 175	23a	346a
	Zhongyou 206	28a	369a
	Jingdong 8	18a	359a
100%	Beijing 0045	36ab	545b
	Zhongmai 175	37ab	783a
	Zhongyou 206	48a	599b
	Jingdong 8	27b	703a

Values with different letters in the same column are significantly different from each other at $P=0.05$.

Table 6 Correlations of phenolic acid concentrations ($\mu g \cdot g^{-1}$ dry matter) including the components in milling extractions and that in whole grain for four wheat cultivars grown at two locations in 2008—2009 season

Class	Parameter	60%	70%	80%	90%
Free	Ferulic	0.59	0.90**	0.91**	0.98***
	Syringic	0.60	0.69	0.53	0.84**
	p-Coumaric	0.02	−0.15	−0.18	0.98***
	Subtotal	0.54	0.59	0.70	0.89**
Bound	Ferulic	0.33	0.64	0.49	0.11
	Caffeic	0.18	−0.12	0.20	0.71*
	Chlorogenic	−0.20	0.03	−0.01	0.31
	Syringic	0.87**	0.79*	0.30	0.45

(continued)

Class	Parameter	60%	70%	80%	90%
	Gentisic	−0.18	−0.73*	−0.04	0.60
	p-Coumaric	−0.32	−0.18	−0.03	0.58
	Subtotal	0.06	0.48	0.32	0.42

*, **, and *** indicate correlation coefficients significantly different from zero at $P=0.05$, 0.01, and 0.001, respectively.

4 Discussion

Increasing numbers of reports have emphasized the positive contribution of consumption of grain products to reduce the incidence of chronic diseases, such as cardiovascular disease (Jacobs et al., 1998), diabetes (Meyer et al., 2000), and cancer (Nicodemus et al., 2001). These health benefits are attributed in part to the unique phytochemical content of grains, mostly phenolic acids, which occur in the endosperm, germ, and bran of whole wheat grain (Adom and Liu, 2002).

The present study showed that most of the phenolic acidsin wheat occur in the bound form, presumably attached to the cell walls. Free phenolic acids made up the least abundant class, contributing only about 6% of the grain phenolic acids determined, in agreement with Ward et al. (2008), Fernandez-Orozco et al. (2010) and Zhang et al. (2012). The average percentage contributions to total free and bound phenolic acid concentrations varied with components and cultivars. Syringic acid represented the most abundant class of free phenolic acids, contributing from 29.8% to 36.8% of the total concentrations among cultivars; whereas ferulic represented the most abundant class of the total bound phenolic acid, contributing from 46.0% to 55.8% of the concentrations among cultivars. Therefore, the free phenolic extracts of the grain were mostly contributed by syringic rather than ferulic, whereas ferulic acid primarily existed in bound form (Hung et al., 2009). Moreover, the bound form contributed 95.8% of the grain ferulic acid concentration determined, similar to the previous results (Adom and Liu, 2002; Zhang et al., 2012). Ferulic acid was, as expected, the dominant component in the bound fraction, consistent with previous reports by Rybka et al. (1993), Verma et al. (2009), and Zhang et al. (2012). The second most abundant phenolic acids in whole grain were caffeic and *p*-coumaric, largely in agreement with Sosulski et al. (1982) and Li et al. (2008).

Overall, the concentration of the phenolic acids and their components obtained in the cultivars here was in the range of previous ones (Adom et al., 2003, 2005; Li et al., 2008; Zhang et al., 2012), and the difference among cultivars was thought to have a genetic basis, demonstrating the extent of genetic potential for exploitation by breeding, which was very important for breeders, consistent with our previous report (Zhang et al., 2012).

The inconsistent relationship between whole grain and milling extractions in concentrations of phenolic acids and components indicated that milling had a great influence, and that cultivar × milling interaction was very important in regard to all phenolic acid components. There were several reports on the concentrations of phenolics in different parts of cereal grains including wheat, rye, and oat (Andreasen et al., 2000; Peterson, 2001; Rybka et al., 1993). The grain phenolic acid concentrations determined in this study ranged from 54 $\mu g \cdot g^{-1}$ in flour at 60% extraction rate to 694 $\mu g \cdot g^{-1}$ in flour at 100% extraction rate, indicating higher concentrations of phenolic acids in bran. Moreover, the highly significant effects of cultivar × milling interaction variance indicated that the proportions of phenolic compounds varied among milling extractions and cultivars. The levels of phenolic acids depended both on initial grain concentrations of cultivars and on choice of the milling extraction, and the loss of nutrient concentrations was quite different among phenolic acid components during milling processing. Therefore, much more attention should be given to milling procedures in regard to improving nutrition quality. Typically, bright white color is preferred with low extraction rate (Ye et al., 2009), which leads to a loss of beneficial compounds.

Clearly, there was an uneven distribution of phenolic compounds in the grain, with most in bran, as reported by Baublis et al. (2002). The total phenolic acid concentrations of bound type extracts were significantly higher than those of the free type extracts in every milling extraction. Therefore, greater health benefits will be obtained by consumption of the maximum outer layer extractions of the grain, and wholemeal is strongly recommended. This is especially important in countries, such as China, that do not have extensive nutrition fortification programs. Wholemeal wheat products have physicochemical and nutritional properties that differ from those of traditionally available milled extractions, especially in regard to pan bread and steamed bread (Dexter and Wood, 1996; Tang et al., 2008).

However, when milling extraction rates increased from 60% to 90%, differences in free phenolic acid concentrations were not significant among cultivars, and significant differences among cultivars occurred only in samples produced at 100% extraction rates, while significant differences in bound phenolic acid concentrations among cultivars were observed in samples at all extraction rates except 90%. Although both free and bound phenolic acids are primarily concentrated in bran, considerable portions are also found in starchy endosperm, especially the free phenolic acid components. The results indicated that the outermost pericarp is less rich in bound syringic acid than the aleurone layer, similar to the distribution of folate in wheat grains (Ward et al., 2008; Piironen et al., 2009). Generally, differences in free and bound phenolic acid concentrations between cultivars decreased progressively as milling progressed from the outer layers to very low extraction rates.

Acknowledgement

This study was supported by the Core Research Budget of the Non-profit Governmental Research Institutions (ICS, CAAS), National 863 Program (2012AA101105), and an international collaboration project on wheat improvement from the Chinese Ministry of Agriculture (2011-G3).

References

AACC International, 2000. In: Approved Methods of the AACC, tenth ed. AmericanAssociation of Cereal Chemists, St. Paul, MN.

Adom, K. K., Liu, R. H., 2002. Antioxidant activity of grains. Journal of Agricultural and Food Chemistry 50, 6182e6187.

Adom, K. K., Sorrells, M. E., Liu, R. H., 2003. Phytochemical profiles and antioxidant activity of wheat varieties. Journal of Agricultural and Food Chemistry 51, 7825e7834.

Adom, K. K., Sorrells, M. E., Liu, R. H., 2005. Phytochemicals and antioxidant activity of milled fractions of different wheat varieties. Journal of Agricultural and Food Chemistry 53, 2297e2306.

Andreasen, M. F., Christensen, L. P., Meyer, A. S., Hansen, Å., 2000. Ferulic acid dehydrodimers in rye (*Secale cereale* L.). Journal of Cereal Science 31, 303e307.

Baublis, A. J., Lu, C., Clydesdale, F. M., Decker, E. A., 2002. Potential of wheat-based breakfast cereals as a source of dietary antioxidants. Journal of the American College of Nutrition 19, 308Se311S.

Beta, T., Nam, S., Dexter, J. E., Sapirstein, H. D., 2005. Phenolic content and antioxidant activity of pearled wheat and roller milled fractions. Cereal Chemistry 82, 390e393.

Chen, C. M., 2004. Ten Year Tracking Nutritional Status in China. People's MedicalPublishing House, Beijing.

Dexter, J. E., Wood, P. J., 1996. Recent applications ofdebranning of wheat before milling. Trends in Food Science and Technology 7, 35e41.

Fernandez-Orozco, R., Li, L., Harflett, C., Shewry,

P. R., Ward, J. L., 2010. Effects of environment and genotype on phenolic acids in wheat in the HEALTH-GRAIN diversity Screen. Journal of Agricultural and Food Chemistry 58, 9341e9352.

Hung, P. V., Maeda, T., Miyatake, K., Morita, N., 2009. Total phenolic compounds and antioxidant capacity of wheat graded flours by polishing method. Food Research International 42, 185e190.

Irmak, S., Jonnala, R. S., MacRitchie, F., 2008. Effect of genetic variation on phenolic acid and policosanol contents of Pegaso wheat lines. Journal of Cereal Science 48, 20e26.

Jacobs, D. R., Meyer, K. A., Kushi, L. H., Folsom, A. R., 1998. Whole grain intake may reduce risk of coronary heart disease death in postmenopausal women: the Iowa Women' s Health Study. American Journal of Clinical Nutrition 68, 248e257.

Li, L., Shewry, P., Ward, J. L., 2008. Phenolic acids in wheat varieties in the HEALTHGRAIN diversity screen. Journal of Agricultural and Food Chemistry 56, 9732e9739.

Mattila, P., Pihlava, J. M., Hellström, J., 2005. Contents of phenolic acids, alkyl- and alkenylresorcinols, and avenanthramides in commercial grain products. Journal of Agricultural and Food Chemistry 53, 8290e8295.

Meyer, K. A., Kushi, L. H., Jacob, D. J., Slavin, J., Sellers, T. A., Folsom, A. R., 2000. Carbohydrates, dietary fiber, incident type 2 diabetes mellitus in older women. American Journal of Clinical Nutrition 71, 921e930.

Nicodemus, K. K., Jacobs, D. R., Folsom, A. R., 2001. Whole and refined grain intake and risk of incident postmenopausal breast cancer. Cancer Causes and Control 12, 917e925.

Parker, M. L., Ng, A., Waldron, K. W., 2005. The phenolic acid and polysaccharide composition of cell walls bran layers of mature wheat (*Triticum aestivum* L. cv. Avalon) grains. Journal of Agricultural and Food Chemistry 85, 2539e2547.

Peterson, D. M., 2001. Oat antioxidants. Journal of Cereal Science 33, 115e129.

Piironen, V., Lampi, A. M., Ekholm, P., Salmenkallio-Marttila, M., Liukkonen, K. H., 2009. Micronutrients and phytochemicals in wheat grain. In: Khan, K., Shewry, P. R. (Eds.), Wheat: Chemistry and Technology, fourth ed. AACC, St. Paul, MN, pp. 179e222.

Rybka, K., Sitarski, J., Raczynska-Bojanowska, K., 1993. Ferulic acid in rye and wheat grain and dietary fiber. Cereal Chemistry 70, 55e59.

SAS Institute SAS User' s Guide, 2000. Statistics. SAS Institute, Inc., Cary, NC. Saulnier, L., Sado, P. E., Branlard, G., Charmet, G., Guillon, F., 2007. Wheat arabi-noxylans: exploiting variation in amount and composition to develop enhanced varieties. Journal of Cereal Science 46, 261e281.

Skrabanja, V., Kreft, I., Golob, T., Modic, M., Ikeda, S., Ikeda, K., Kreft, S., Bonafaccia, G., Knapp, M., Kosmelj, K., 2004. Nutrient content in buckwheat milling fractions. Cereal Chemistry 81, 173e176.

Sosulski, F., Krygier, K., Hogge, L., 1982. Free, esterified, and insoluble bound phenolic acids. 3. Composition of phenolic acids in cereal and potato flours. Journal of Agricultural and Food Chemistry 30, 337e340.

Tang, J. W., Zou, C. Q., He, Z. H., Shi, R. L., Monasterio, I., Qu, Y. Y., Zhang, Y., 2008. Mineral element distributions in milling fractions of Chinese wheats. Journal of Cereal Science 48, 821e828.

Verma, B., Hucl, P., Chibbar, R. N., 2009. Phenolic acid composition and antioxidant capacity of acid and alkali hydrolysed wheat bran fractions. Food Chemistry 116, 947e954.

Ward, J. L., Poutanen, K., Gebruers, K., Piironen, V., Lampi, A. M., Nyström, L., Andersson, A. A. M., Åman, P., Boros, D., Rakszegie, M., Bedõ, Z., Shewry, P. R., 2008. The HEALTHGRAIN cereal diversity screen: concept, results and prospects. Journal of Agricultural and Food Chemistry 56, 9699e9709.

Ye, Y. L., Zhang, Y., Yan, J., Zhang, Y., He, Z. H., Huang, S. D., Quail, K. J., 2009. Effects of flour extraction rate, added water, and salt on color and texture of Chinese white noodles. Cereal Chemistry 86, 477e485.

Yu, L., Haley, S., Perret, J., Harris, M., 2004. Comparison of wheat flours grown atdifferent locations for their antioxidant properties. Food Chemistry 86, 11e16.

Zhang, Y., Wang, L., Yao, Y., Yan, J., He, Z. H., 2012. Phenolic acid profiles of Chinese wheat cultivars. Journal of Cereal Science 56, 629e635.

Zhou, K., Yu, L., 2004. Antioxidant properties of bran extracts from Trego wheat grown at different locations. Journal of Agricultural and Food Chemistry 52, 1112e1117.

Carotenoids in Staple Cereals: Metabolism, Regulation, and Genetic Manipulation

Shengnan Zhai[1], Xianchun Xia[1*] and Zhonghu He[1,2]

[1] *National Wheat Improvement Center, Institute of Crop Science, Chinese Academy of Agricultural Sciences, Beijing, China*, [2] *International Maize and Wheat Improvement Center, Chinese Academy of Agricultural Sciences, Beijing, China*

Abstract: Carotenoids play a critical role in animal and human health. Animals and humans are unable to synthesize carotenoids *de novo*, and therefore rely upon diet as sources of these compounds. However, major staple cereals often contain only small amounts of carotenoids in their grains. Consequently, there is considerable interest in genetic manipulation of carotenoid content in cereal grain. In this review, we focus on carotenoid metabolism and regulation in non-green plant tissues, as well as genetic manipulation in staple cereals such as rice, maize, and wheat. Significant progress has been made in three aspects: (1) seven carotenogenes play vital roles in carotenoid regulation in non-green plant tissues, including 1-deoxyxylulose-5-phosphate synthase influencing isoprenoid precursor supply, phytoene synthase, β-cyclase, and ε-cyclase controlling biosynthesis, 1-hydroxy-2-methyl-2-(*E*)-butenyl 4-diphosphate reductase and carotenoid cleavage dioxygenases responsible for degradation, and orange gene conditioning sequestration sink; (2) provitamin A-biofortified crops, such as rice and maize, were developed by either metabolic engineering or marker-assisted breeding; (3) quantitative trait loci for carotenoid content on chromosomes 3B, 7A, and 7B were consistently identified, eight carotenogenes including 23 loci were detected, and 10 gene-specific markers for carotenoid accumulation were developed and applied in wheat improvement. A comprehensive and deeper understanding of the regulatory mechanisms of carotenoid metabolism in crops will be beneficial in improving our precision in improving carotenoid contents. Genomic selection and gene editing are emerging as transformative technologies for provitamin A biofortification.

Keywords: carotenoid metabolism, carotenoid regulation, marker-assisted breeding, metabolic engineering, provitamin A biofortification, *Triticum*

Abbreviations: ABA, abscisic acid; b*, flour yellow color; CCDs, carotenoid cleavage dioxygenases; CrtI, carotene desaturase; CRTISO, carotene isomerase; DXS, 1-deoxyxylulose-5-phosphate synthase; GGPP, geranylgeranyl diphosphate; LCYB, β-cyclase; LCYE, ε-cyclase; LOX, lipoxygenases; MEP, 2-C-methyl-D-erythritol 4-phosphate; PDS, phytoene desaturase; PSY, phytoene synthase; QTL, quantitative trait loci; YPC, yellow pigment content; ZDS, ζ-carotene desaturase.

Published in Frontiers in Plant Science, 2016, 7: 1197.

INTRODUCTION

Carotenoids are mainly C40 isoprenoids comprising a large family with more than 700 members that are widely distributed in plants, algae, fungi, and bacteria (Khoo et al., 2011). In plants, they perform a multitude of functions involving the photosynthetic apparatus, photoprotection, and precursors to phytohormones such as ABA and strigolactones (Niyogi, 2000; Cazzonelli and Pogson, 2010). In addition, carotenoids provide color and aroma to flowers and fruits for attracting insects and other organisms for pollination and seed dispersal, and protect the seed from deterioration (Walter et al., 2010; Moise et al., 2013). Very recently, carotenoid derivatives were found in association with response to environmental stresses, such as photoxidative stress (Havaux, 2014).

Carotenoids also play a critical role in animal and human health. In animals, they can improve sexual behavior and reproduction, and protect animals from predation as well as parasitism (McGraw and Toomey, 2010). For humans, the most important function of carotenoids is as a dietary source of provitamin A (mainly α-carotene, β-carotene, zeaxanthin, and β-cryptoxanthin; Giuliano et al., 2008). Vitamin A deficiency (VAD) is the leading cause of preventable blindness in children and increases the risk of disease and death from severe infections. For pregnant women, VAD may cause night blindness and increase the risk of maternal mortality. The World Health Organization has estimated that 250 000-500 000 vitamin A-deficient children became blind each year, with half of them dying from loss of eyesight within 12 months①. In addition, carotenoids as antioxidants have a protective function in reducing the risk of age-related macular degeneration (ARMD), cancer, cardiovascular diseases, and other chronic diseases (Fraser and Bramley, 2004). Carotenoids are also used commercially as feed additives to enhance pigmentation of fish and eggs, colorizing agents for human food, cosmetics, and pharmaceutical products (Sandmann, 2001). Thus, understanding the regulatory mechanisms of carotenoids is a very important scientific pursuit and biofortification of staple foodstuffs for health benefits has become an important issue in food production.

Because animals and humans are unable tosynthesize carotenoids *de novo* they rely upon diet as the source of these compounds. However, most staple cereals, such as rice (*Oryza sativa*), wheat (*Triticum aestivum*), and maize (*Zea mays*), contain very little amounts of carotenoids in their grains. Therefore, the genetic manipulation of carotenoid accumulation in staple cereal grains should be a powerful means to combat vitamin A deficiency, and especially important for developing countries where people frequently rely on a single crop for sustenance. For better genetic manipulation of carotenoid content within cereal grains there is a particular interest in the regulatory mechanisms of carotenoid biosynthesis in non-green plant tissues (Farré et al., 2011). Various lines of evidence show that key nodes in the MEP pathway, carotenoid metabolism, and sequestration sink play vital roles in regulation of carotenoid biosynthesis.

In this review, we focus on carotenoid metabolism and regulation in non-green plant tissues, as well as genetic manipulation in staple cereals including rice, maize, and wheat. Compared with maize and rice (Harjes et al., 2008; Yan et al., 2010; Breitenbach et al., 2014; Bai et al., 2016), carotenoid biosynthesis in wheat has received much less attention. Therefore, a comprehensive overview of carotenoid biosynthesis in wheat was undertaken to provide a platform of understanding of carotenoid biosynthesis as wheat supplies significant amounts of dietary carbohydrate and protein for over 60% of the world population, and is also an important source of carotenoids in human diets (Shewry, 2009). In addition to cereals, the extensive literature on carotenoid biosynthesis in bacteria or other

① http://www.who.int/nutrition/topics/vad/en/

plants is also discussed, as it contributes to a better understanding of the pathway in cereals.

CAROTENOID METABOLISM

Carotenoid metabolism in plants is a complex process, and has been extensively characterized in a range of organisms providing an almost complete pathway for carotenogenesis and degradation (Cunningham and Gantt, 1998; Giuliano et al., 2008). The main steps of carotenoid metabolism in higher plants are briefly summarized below and presented in Figure 1.

Biosynthesis

Carotenoids are derived from the plastid-localized MEP pathway for which glyceraldehyde-3-phosphate and pyruvate act as initial substrates leading to the synthesis of GGPP, the common precursor for biosynthesis of carotenoids and several other terpenoid compounds (Farré et al., 2010; Rodriguez-Concepcion, 2010). The first committed step in the carotenoid biosynthesis pathway is condensation of two GGPP molecules by PSY to produce 15-*cis*-phytoene. Phytoene is converted into lycopene by two desaturation reactions catalyzed by PDS and ZDS. These enzymes give rise to poly-*cis* compounds which are converted to the all-*trans* form by ζ-carotene isomerase (ZISO) and CRTISO, as well as a light-mediated photo-isomerization. In bacteria, a single enzyme, CrtI, is believed to confer the same desaturation and isomerization reactions.

Lycopene constitutes a branching point in the pathway since it is the substrate of two competing cyclases, LCYB and LCYE. α-carotene is produced when LCYE and LCYB act together on the two ends of lycopene (β, ε-branch), whereas β-carotene is formed when LCYB acts alone (β, β-branch). Alpha-carotene and β-carotene are hydroxylated to produce lutein and zeaxanthin, respectively. These reactions are catalyzed by the β-ring carotene hydroxylase [HYDB, also known as non-heme di-iron β-carotene hydroxylase (BCH) or heme-containing cytochrome P450 β-ring hydroxylase (CYP97A and CYP97B)] and heme-containing cytochrome P450 carotene ε-ring carotene hydroxylase (CYP93C). Whereas lutein represents the natural end point of the β, ε-branch, zeaxanthin is further epoxidized by zeaxanthin epoxidase (ZEP) in a two-step reaction to produce violaxanthin via antheraxanthin. This reaction is reversed by violaxanthin deepoxidase (VDE) to give rise to the xanthophyll cycle for plants to adapt high light stress (Demmig-Adams and Adams, 2002). Violaxanthin is converted into neoxanthin by neoxanthin synthase (NXS), the final carotenoid of the β, β-branch of the classical biosynthetic pathway.

In some plants, the classical carotenoid biosynthesis pathway extends further to synthesize specialized ketocarotenoids. One such example is the red fruits of chili peppers, where the capsanthin and capsorubin are synthesized from antheraxanthin and violaxanthin by capsanthin-capsorubin synthase (CCS) enzyme (Gómez-García and Ochoa-Alejo, 2013). Another example is the ornamental plant *Adonis aestivalis* whose petals synthesize the red ketocarotenoid astaxanthin, which is usually found in microbes (Cunningham and Gantt, 2005). With progress in high-performance liquid chromatography-tandem mass spectrometric (HPLC-MS) and high-performance liquid chromatography-nuclear magnetic resonance (HPLC-NMR) technologies, more specialized ketocarotenoids will be detected, which will further enrich our knowledge of this pathway.

In grasses, PSY are encoded by three paralogous genes (*PSY1-3*; Dibari et al., 2012). *PSY1* is correlated with carotenoid accumulation in grain, *PSY2* is involved in protecting the photosynthetic apparatus from photo-oxidative degradation in green tissues, and *PSY3* is associated with root carotenogensis channeled into ABA formation, mainly responsing to abiotic stresses, such as drought and salt (Gallagher et al., 2004; Li et al., 2008; Welsch et al., 2008). *PSY* duplication has provided an opportunity for subfunctionalization whereby gene family members vary in tissue specificity of expression to control carotenogenesis

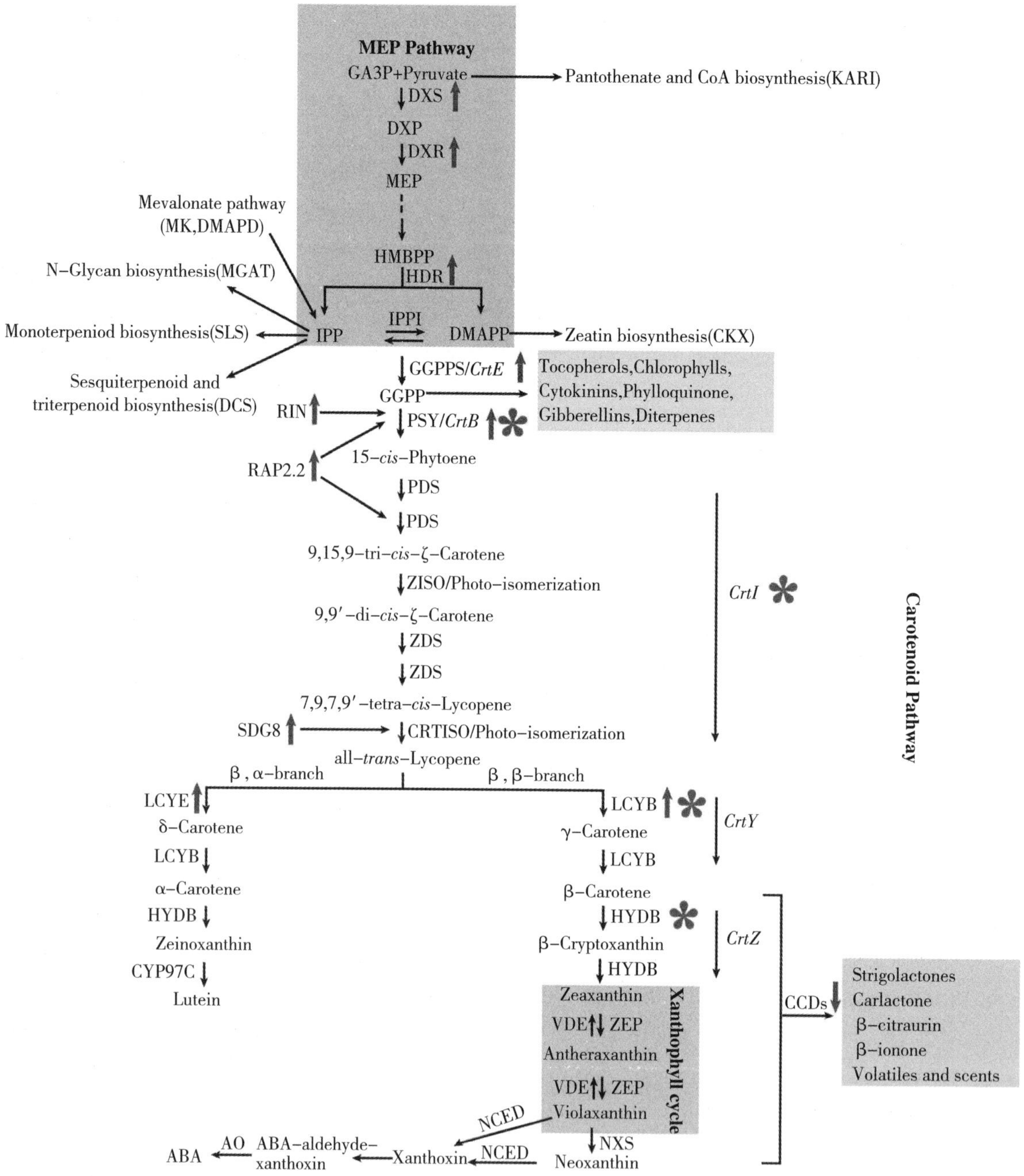

Figure 1 Carotenoid metabolism, regulation and genetic manipulation in higher plants. Names of bacterial enzymes are in italics. Candidate genes for carotenoid accumulation obtained by QTL analysis are displayed in parentheses and in red. Red upward pointing arrows, gene expression positively correlated with carotenoid biosynthesis; red downward pointing arrows, gene expression negatively correlated with carotenoid biosynthesis; green asterisk, main genetic manipulation nodes in staple cereals. Other MEP isoprenoid-derived metabolites and carotenoid cleavage products apocarotenoids are shown in the green box.

ABA, abscisic acid; AO, aldehyde oxidase; CCD, carotenoid cleavage dioxygenase; CKX, cytokinin oxidase/dehydrogenase; CrtB, bacterial phytoene synthase; CrtE, bacterial GGPP synthase; CrtI, bacterial phytoene desaturase/isomerase; CRTISO, carotene isomerase; CrtY, bacterial lycopene β-cyclase; CrtZ, bacterial β-carotene hydroxylase; CYP97C, heme-containing cytochrome P450 carotene ε-ring hydroxylase; DCS, delta-cadinene synthase; DMADP, dimethylallyl diphosphate; DXP, 1-deoxy-D-xylulose 5-phosphate; DXR, 1-deoxy-D-xylulose 5-phosphate reductoisomerase; DXS, 1-deoxyxylulose-5-phosphate synthase; GA3P, D-glyceraldehyde-3-phosphate; GGPP, geranylgeranyl diphosphate;

GGPPS, GGPP synthase; HDR, 1-hydroxy-2-methyl-2- (*E*) -butenyl 4-diphosphate reductase; HYDB, β-carotene hydroxylase [also known as non-heme di-iron β-carotene hydroxylase (BCH) and heme-containing cytochrome P450 β-ring hydroxylase (CYP97A and CYP97B)]; IPP, isopentenyl diphosphate; IPPI, IPP isomerase; KARI, ketol-acid reductoisomerase; LCYB, lycopene β-cyclase; LCYE, lycopene ε-cyclase; MEP, 2C-methyl-D-erythritol-4-phosphate; MGAT1, alpha-1, 3-mannosyl-glycoprotein 2-beta-*N*-acetylglucosaminyl-transferase; MK, mevalonate kinase; NCED, 9-*cis*-epoxycarotenoid dioxygenase; NXS, neoxanthin synthase; PDS, phytoene desaturase; PSY, phytoene synthase; RAP2.2, a member of the APETALA2 (AP2) / ethylene-responsive element-binding protein transcription factor family; RIN, MADS-box transcription factor RIPENING INHIBITOR; SDG8, SET2 histone methyltransferase; SLC, secologanin synthase; VDE, violaxanthin de-epoxidase; ZDS, ζ-carotene desaturase; ZEP, zeaxanthin epoxidase; ZISO, ζ-carotene isomerase.

independently of photosynthesis or in response to certain stresses (Li et al., 2008; Welsch et al., 2008; Arango et al., 2010).

Degradation

Carotenoid degradation can occur via non-specific mechanisms such as photo chemical oxidation or LOX (Siedow, 1991; Auldridge et al., 2006). However, specific tailoring of carotenoids is carried out by a family of CCDs, which appear to have different substrate preferences (Vallabhaneni and Wurtzel, 2009). The CCD gene family is divided into two types: nine-*cis*-epoxycarotenoid dioxygenases (NCEDs) catalyze both violaxanthin and neoxanthin to produce xanthoxin, the precursor of ABA (Seo and Koshiba, 2002; Walter et al., 2010), and CCDs that catalyze a vast array of different cleavage steps giving rise to apocarotenoids. For example, CCD1 is involved in β-ionone biosynthesis, whereas CCD7 and CCD8 are associated with strigolactone biosynthesis. These apocarotenoids are crucial for various biological processes in plants, such as regulation of growth and development and plant-insect interaction (Walter et al., 2010; Alder et al., 2012; Avendano-Vazquez et al., 2014).

Sequestration

Carotenoids are usually synthesized *de novo* in nearly all types of differentiated plastids of leaves, roots, flowers, fruits, and seeds, including chloroplasts, chromoplasts, amyloplasts, elaioplasts, leucoplasts, and etioplasts, but accumulate in large quantities in chloroplasts and chromoplasts (Howitt and Pogson, 2006; Cazzonelli and Pogson, 2010). Chloroplasts and chromoplasts differ considerably in the way they sequestrate end-product carotenoids. In chloroplasts, carotenoids are located in photosynthetic membranes and integrated with chlorophyll-binding proteins to form pigment-protein complexes (Vishnevetsky et al., 1999). Whereas, in chromoplasts, carotenoids are associated with polar lipids and carotenoid associated proteins to form carotenoid-lipoprotein sequestering contributes to a substructures (e.g., globules, crystals, membranes, fibrils, and tubules) to effectively sequester and retain a large quantity of carotenoids (Vishnevetsky et al., 1999; Egea et al., 2010; Li and Yuan, 2013).

To date, there is little understanding of carotenoid degradation. Much more effort to understand CCD gene family members, their substrates and products, is still needed. In addition, some acronyms of carotenogenes were confused in the previous literature, such as β-hydroxylases being replaced by BCH and HYD in rice (Du et al., 2010), crtRB1 and HYD in maize (Yan et al., 2010), BCH in *Arabidopsis* (Kim et al., 2009), and CHY in potato (Diretto et al., 2007), respectively. For a better understanding and communication, international efforts are needed to uniform the acronyms.

CAROTENOID REGULATORY MECHANISMS IN NON-GREEN PLANT TISSUES

Relatively little is known about the regulation of carotenogenesis in chloroplasts. Although expression of carotenoid genes does take place in etiolated plants, most carotenoid biosynthetic genes, including those in the MEP pathway, are activated during light-triggered de-etiolation (Giuliano et al., 2008; Cazzonelli and Pogson, 2010; Rodriguez-Concepcion, 2010). The phytochrome-interacting factor 1 (PIF1) is shown to bind to the *PSY* promoter and represses *PSY* expression under dark conditions. Toledo-Ortiz et al. (2010) indicated that light triggered the degradation of PIF1 by photoactivated phytochromes, which allowed *PSY* expression and subsequently rapid production of carotenoids. In addition, the relative concentration of zeaxanthin and violaxanthin in plant photosynthetic tissues is important in stimulating energy dissipation within light-harvesting antenna proteins through non-photochemical quenching to protect against photoinhibition. Under high light condition, violaxanthin is de-epoxidized into zeaxanthin by VDE to dissipate light energy, whereas the reverse reaction converts zeaxanthin to violaxanthin by ZEP under dark condition (Demmig-Adams and Adams, 2002). In conclusion, light played a significant role in regulation of carotenoid biosynthesis in green tissues, but how light ultimately regulates this process remains to be elucidated. Further researches are required to illustrate the carotenoid synthesis regulation in chloroplasts.

Regulatory mechanisms of carotenoid biosynthesis innon-green tissues are distinct from those in green tissues. Briefly, there are three major mechanisms affecting carotenoid accumulation in non-green plant tissues: (1) regulation of genes controlling carotenoid biosynthesis; (2) the regulation of genes for carotenoid degradation; and (3) the regulation of plastid development. Various lines of evidence show that the MEP pathway, GGPP pool, PSY and branch point enzymes might be key regulatory nodes for carotenoid content. They are discussed in detail below.

Regulation of Isoprenoid Precursor

Carotenoid biosynthesis requires an available source ofisoprenoid substrates derived from the MEP pathway, which is a key bottleneck influencing flux through the entire pathway (Farré et al., 2010; Rodriguez-Concepcion, 2010). In the MEP pathway, the transcript levels of DXS, 1-deoxy-D-xylulose 5-phosphate reductoisomerase (DXR) and 1-hydroxy-2-methyl-2-(*E*)-butenyl 4-diphosphate reductase (HDR) were positively correlated with carotenoid content in maize endosperm (Vallabhaneni and Wurtzel, 2009; Suwarno et al., 2015).

In addition to its role in carotenoid biosynthesis, GGPP is a precursor for synthesis of many other terpenoid compounds in plants. Therefore, the pool of GGPP represents the metabolic link between biosynthesis of carotenoids and other terpenoids, and is responsible for inter-pathway regulation via competition for GGPP. The expression level of GGPP synthase (*GGPPS*) was positively correlated with endosperm carotenoid content in maize (Vallabhaneni and Wurtzel, 2009; Suwarno et al., 2015). Another key regulatory issue is what mechanisms control the partitioning of precursors into various terpenoid pathways. There is clear evidence for multiple *GGPPS* genes in *Arabidopsis*, encoding dedicated enzymes for different branches of various terpenoid pathways (Okada et al., 2000).

Regulation of Carotenoid Biosynthesis

Phytoene synthase catalyzes the first committed step in carotenoid biosynthesis and is generally accepted as the most important regulatory node in the carotenoid biosynthesis pathway, whose transcripts were positively correlated with carotenoid accumulation (Cong et al., 2009; da Silva Messias et al., 2014). Moreover, *PSY* seems to be a key integrator for several signals regulating carotenoid biosynthesis. For example, blocking of the MEP pathway and loss-of-function of *PDS* result in down-regulation of *PSY*, whereas increased activity of *DXS* induces *PSY* expression in tomato (Rodriguez-Concepcion et al., 2001; Laule et al., 2003). Orange (OR) protein directly interacts with *PSY* to regulate carotenoid biosynthesis (Zhou et al., 2015). In addition, carotenoid metabolites also regulate *PSY* protein level and total carotenoid content (Kachanovsky et al., 2012; Arango et al., 2014). For example, expression of the *PSY* gene is positively up-regulated by ABA and has been associated with preharvest sprouting in cereals (Fang et al., 2008; Cazzonelli, 2011).

The cyclization of lycopene has a major role in modulating the β, β/β, ε branch ratio, suggesting that coordination between LCYE and LCYB activities may be necessary for regulation of metabolic flux through different branches of the carotenoid pathway (Cazzonelli et al., 2010; Farré et al., 2011). Over-expression of *LCYB* shifts the balance toward the β, β-branch, whereas over-expression of *LCYE* has the opposite effect (Rosati et al., 2000; D'Ambrosio et al., 2004). However, expression of *PSY1*, *CrtI*, and *LCYB* in transgenic maize endosperm increased β, β/β, ε ratio from 1.2 to 3.5 and also enhanced flux through the β, ε-branch of the pathway, producing almost 25 times more lutein than the normal level (Zhu et al., 2008). Naqvi et al. (2011) also found that when metabolic flux is shifted toward β-carotene there is still enough flux through the β, ε-branch to produce more lutein. These examples showed that regulation of the flux through different branches of the pathway was complex.

Some other carotenogenes also regulated carotenoid content. For example, viviparous mutants *vp*5, *vp*2, and *w*3 in maize have defective copies of the *PDS* gene and exhibit increased accumulation of phytoene (Matthews et al., 2003). High expression of the *ZDS* gene was consistent with accumulation of lycopene during carrot root development (Clotault et al., 2008). *ZISO* and *CRTISO* are essential for establishing an equilibrium between *cis*-and *trans*-carotenoid isomers (Chen et al., 2010; Yu et al., 2011). In addition, expression of *crtRB*1 was negatively correlated with β-carotene levels and positively correlated with zeaxanthin levels in maize (Yan et al., 2010; da Silva Messias et al., 2014).

Apart from the carotenogenes *per se*, transcriptional factors regulating carotenoid biosynthesis have been reported. Reduced transcript level of *RAP*2.2, a member of the APETALA2 (AP2) / ethylene-responsive element-binding protein transcription factor family, was accompanied by a significant decrease in transcript levels of both *PSY* and *PDS* with a concomitant 30% decrease in carotenoid content relative to wild-type (Welsch et al., 2007). The transcription factor RIN induces *PSY1* expression to regulate the flux of carotenoid biosynthesis in tomato (Martel et al., 2011). Moreover, epigenetic regulation was also considered important in carotenogenesis. A chromatin-modifying histone methyltransferase enzyme SDG8 (SET DOMAIN GROUP 8) maintains a transcriptionally permissive chromatin state surrounding the *CRTISO* and thus is able to regulate carotenoid content (Cazzonelli et al., 2009). Overexpression of *microRNA*156 in *Brassica napus* enhanced carotenoid content in seeds (Wei et al., 2010).

Regulation of Carotenoid Degradation

Recent studies have demonstrated that the carotenoid pool is determined in part by the rate of carotenoid degradation (Vallabhaneni and Wurtzel, 2009; Gayen et al., 2015). The expression of *CCD*1 or *CCD*4 was negatively correlated with carotenoid accumulation (Gonzalez-Jorge et al., 2013; da Silva Messias et al., 2014). It was shown that down-regulation of LOX enzyme activity reduces degradation of carotenoids in Golden Rice suggesting an effective tool to reduce large economic losses of biofortified rice seeds during storage (Gayen et al., 2015). Compared to carotenoid biosynthesis, little is known about the impact of carotenoid degradation on regulation of carotenoid accumulation, and much more work is needed to understand it.

Regulation of Carotenoid Sequestration

Various studies have shown that carotenoid accumulation is greatly modulated by size, number, and anatomical structure of the plastids in which carotenoid biosynthesis and storage occur. Organelle biogenesis is a major determinant of plastid size and storage compartment number, and affects carotenoid accumulation by providing a larger sink. CHCR (chromoplast-specific carotenoid-associated protein) enhances carotenoid content in high pigment tomato mutants (*hp*1, *hp*2, and *hp*3) due to increased chromoplast number and/or volume (Galpaz et al., 2008; Kilambi et al., 2013). A mutation in the *OR* gene led to differentiation of plastids to chromoplasts causing enhanced carotenoid accumulation in the curds of cauliflower (Lu et al., 2006). A change in chromoplast architecture is associated with carotenoid composition in *Capsicum* fruits (Kilcrease et al., 2013).

Esterification limits degradation of xanthophyllsand increases their sequestration within the chromoplast by increased lipophilic properties and integration into lipid-rich plastoglobules (Ariizumi et al., 2014; Mellado-Ortega and Hornero-Méndez, 2016). Moreover, it was suggested that carotenoid accumulation might be correlated with expression of genes influencing lipoprotein components of chromoplast structures, such as plastid-encoded acetyl coenzyme A, carboxylase D and Hsp21 (Neta-Sharir et al., 2005; Barsan et al., 2012; Carvalho et al., 2012).

Although significant progress has been made in understanding carotenoid regulatory mechanisms in plants, several key issues are yet to be addressed. Firstly, very little is known about the global regulatory mechanisms underlying carotenoid metabolism. Cross-talk between carotenoid biosynthesis and other pathways and how interaction responds to plant growth and development and environment remain unclear. Secondly, the molecular nature of regulation of metabolic feedback remains unknown. Finally, research on regulation of carotenoid biosynthesis has mostly focused on model species and such regulatory mechanisms in non-model species are not well documented, hence restricting a detailed understanding of regulation of carotenoid biosynthesis in specific crops.

GENETIC MANIPULATION OF CAROTENOID BIOSYNTHESIS IN STAPLE CEREALS

Maize, rice, and wheat comprise the main foods for human nutrition. However, carotenoid contents in the grains of these crops are usually low. Therefore, breeding staple cereals with high carotenoid content could have a huge impact on human health, without significantly altering current human diets. Such attempts to enhance carotenoid contents or improve carotenoid composition in staple cereals have been made, mainly based on metabolic engineering and marker-assisted breeding as described below.

Metabolic Engineering

Various metabolic engineering approaches have been made to increase the levels of nutritionally relevant carotenoids in staple cereals and to enable the use of plants as 'cell factories' for producing special carotenoids. Amplification of the rate-limiting enzyme with the highest flux control coefficient is the principal target for manipulation. Alternatively, it may be desirable to change the carotenoid composition or extend the classical carotenoid pathway in the tissue of interest.

A breakthrough in metabolic engineering of carotenoids for improved nutritional value of staple crops was achieved in rice, best-known as 'Golden Rice.' Here, daffodil *PSY* and *LCYB* genes together with the bacterial *CrtI* were transferred to a *japonica* rice cultivar in which the β-carotene content in the endosperm was 1.6 μg/g of seed dry weight, providing 10-20% of the recommended daily allowance (RDA) of β-carotene (Burkhardt et al., 1997; Ye et al., 2000). Further optimization of the pathway using the maize *PSY* gene driven by a rice glutelin promoter considerably increased carotenoid formation in transgenic rice endosperm, resulting in Golden Rice II lines with carotenoid levels up to 37 μg/g (Paine et al., 2005). Higher carotenoid accumulation was recently achieved through the combined expression of *ZmPSY*1, *PaCRTI* with *AtDXS* or *AtOR* in rice endosperm, suggesting that the supply of isoprenoid precursors and metabolic sink are important rate-limiting steps in carotenoid biosynthesis (Bai et al., 2016). Similarly, total carotenoid levels in wheat were enhanced by co-transformation with maize *PSY*1 and the bacterial *CrtI* gene, but the elevation of carotenoid content was only moderate compared with that in the donor wheat cultivar EM12 (Cong et al., 2009). In order to further enrich the provitamin A content in wheat grains, the bacterial *CrtB* and *CrtI* genes were co-transformed into cultivar Bobwhite (Wang et al., 2014), resulting in a total carotenoid content increase to 4.76 μg/g, a β-carotene increase to 3.21 μg/g, and a provitamin A content increase to 3.82 μg/g. Recently, higher levels of β-carotene accumulation up to 5.06 μg/g were obtained by simultaneously over-expressing *CrtB* and silencing carotenoid hydroxylase (Zeng et al., 2015b). Although the level was still insufficient to combat VAD, the progress was still important, as a small increase in carotenoid contents in wheat grains would have a large impact based on the huge daily consumption of wheat-based products throughout the world.

A wide variety of unusual keto-carotenoids and carotenoid intermediates, such as astaxanthin, adonixanthin, 3-hydroxye-chinenone, and echinenone have been engineered in transgenic maize plants with seed colors ranging from white and yellow to dark-red, despite the white-endosperm genetic background (Zhu et al., 2008). The carotenoid pathway in rice was recently further extended to form astaxanthin and 4-keto-α-carotene, with co-transformation of *ZmPSY*1, the bacterial *CrtI* and β-carotene ketolase genes (Breitenbach et al., 2014).

As already mentioned, most of the research on carotenoid manipulation in staple cereals has focused on a few main carotenogenes. In the future, manipulation of carotenoid biosynthesis could be extended to different regulatory nodes, such as the MEP pathway, carotenoid degradation, and sequestration. Moreover, the current status of metabolic engineering is somewhat restricted due to its reliance on gene-by-gene approaches. In other pathways, the focus has shifted from individual genes or collections thereof toward overarching regulatory mechanisms that may allow multiple genes in the pathway to be controlled simultaneously. Although enhancement of carotenoid biosynthesis by metabolic engineering proves to be a useful tool, the transgenic lines may induce hitherto undiscovered feedback mechanisms with unpredictable results. One of the major hurdles for commercialization of genetically engineered crops is the legal requirements and acceptance by consumers in various countries. Golden Rice has not yet been released in any country although daily consumption of 75 g of Golden Rice II grains can receive the RDA of β-carotene (Paine et al., 2005).

Marker-Assisted Breeding

Over the past decade, increasing carotenoid content in grainsof staple cereals such as rice, maize, and wheat, has been an important breeding objective. However, conventional breeding to select for QTL with positive effects on carotenoid levels is a slow and laborious process. The identification of rate-limiting steps, the elucidation of molecular basis of known QTL, or the characterization of new alleles for higher carotenoid content, will allow development of functional markers or gene-specific markers for a more efficient selection in breeding. Such functional markers allow breeders to select quantitative traits at the gene level rather than at the phenotypic level.

In maize, previous studies showed that two polymorphic sites within *PSY*1 each explained 7 and 8% of the total carotenoid variation (Harjes et al., 2008); four polymorphic sites in *LCYE* explained 58% of β, β/β, ε branch ratio variation and a threefold difference in provitamin A compounds (Yan et al., 2010); three polymorphisms in *crtRB*1 were significantly associated with variation in carotenoid content (Fu et al., 2013). Allele-specific markers of three key genes involved in maize endosperm carotenoid biosynthesis were developed to facilitate provitamin A biofortification in maize through marker-assisted selection (MAS). The effectiveness of these molecular markers was verified across diverse tropical yellow maize inbred lines (Azmach et al., 2013; Babu et al., 2013). A favorable *crtRB*1 allele was introgressed into seven elite inbred parents using a *crtRB*1-specific marker, and concentration of β-carotene among *crtRB*1-introgressed inbreds varied from 8.6 to 17.5 μg/g, with a maximum increase of up to 12.6-fold over recurrent parent (Muthusamy et al., 2014). Introgression of a favorable allele of the *crtRB1* gene using molecular markers also significantly increased provitamin A content in quality protein maize inbred lines (Liu et al., 2015). In rice, no carotenoids were detected in the endosperm due to lack of endosperm-specific *PSY* expression (Ye et al., 2000). Therefore, molecular marker-assisted breeding for rice carotenoid improvement is still not feasible. Although many molecular markers have been developed for genes involved in carotenoid biosynthesis in wheat as described below, there are no reports of higher carotenoid content wheat cultivars developed by marker-assisted breeding.

The objectives of Harvest Plus①, a worldwide collaboration that drives biofortification as a project within the Consultative Group of International Agricultural Research (CGIAR), are to breed more nutritious cultivars of staple food crops by conventional breeding technologies strengthened with molecular markers. Provitamin A-biofortified crops, including maize, cassava, and sweet potato, have been developed and released in Nigeria, Zambia, and Uganda. Eating orange sweet potato has been shown to improve vitamin A status of children.

The carotenoid biosynthesis is very complex, thereforemultiple genes must be taken into consideration during marker-assisted breeding in order to enhance the accuracy of prediction and selection. In addition, mutants with desirable carotenogenic properties generated by chemical treatment may provide new insights into carotenoid improvement in staple cereals that are not categorized as genetic manipulation and can be immediately introduced into breeding programs. Meanwhile, such mutants are not involved in the expensive and time-consuming gene transformation, and therefore, easy to be used in breeding programs.

CAROTENOIDS IN *Triticum* Spp.

Carotenoids, the main components of grain yellow pigment in wheat determine the flour color and affect both the nutritional value of the grain and its utility in different applications (Mares and Campbell, 2001). High yellow pigment is a very important quality parameter for pasta made from durum wheat and yellow alkaline noodles made from bread wheat, but low or medium levels of yellow pigment are preferred for Chinese white noodles and steamed bread produced by bread wheat. Thus, manipulations of yellow pigment in opposite directions are important breeding objectives in bread wheat and durum breeding programs. However, compared with maize and rice, carotenoid biosynthesis in wheat has received much less attention. Therefore, we provide a comprehensive overview of carotenoid biosynthesis in wheat in order to facilitate future studies of the carotenoid metabolism.

Carotenoid Profiles in Wheat

Lutein is the predominant carotenoid in wheat, and accountsfor 80-90% of total carotenoids along with small amounts of zeaxanthin, α-carotene, β-cryptoxanthin, and β-carotene (Abdel-Aal et al., 2007; Digesù et al., 2009). The pigments are variably distributed in the seed; the endosperm has the highest lutein content, whereas zeaxanthin and β-carotene are concentrated near the outer layers of the kernel (Hentschel et al., 2002; Borrelli et al., 2008). Although levels of carotenoids in wheat are low, there is significant genetic variation. Previous studies showed that primitive and wild relatives, landraces, and synthetic hexaploids usually accumulate higher levels of carotenoids. For example, einkorn ($2n=14$), and Khorasan and durum wheat ($2n=28$) contain higher levels of lutein (5.4-7.4 μg/g) compared to common wheat (1.9 μg/g; Hidalgo et al., 2006).

① http://www.harvestplus.org/

Carotenoid biosynthesis during grain development was examined using a doubled haploid (DH) bread wheat population (Howitt et al., 2009). During the early stages of grain development, carotenoids from the β, β-branch (zeaxanthin, antheraxanthin, and violaxanthin) were present at higher levels than those from the β, ε-branch (lutein). The highest amounts of lutein and zeaxanthin were detected at 10 days post anthesis (DPA). Although the level of lutein did not change significantly during endosperm development, carotenoids from the β, β-branch declined gradually and were undetectable in mature grains.

QTL Underpinning Carotenoids in Wheat

Although environmental factors play an important role in determining carotenoid contents in wheat, the genetic component is predominant and heritability is relatively high at 0.85-0.97 for YPC, a trait strictly related to carotenoids (Elouafi et al., 2001; Van Hung and Hatcher, 2011).

The genetic architecture of YPC was investigated through QTL analysis in both durum and bread wheat. QTL located in the telomeric regions of the long arms of the homeologous group 7 chromosomes, especially 7AL and 7BL, largely influenced YPC (Elouafi et al., 2001; Patil et al., 2008). Various minor QTLs were also detected on chromosomes of homeologous groups 2, 3 and 4, and chromosomes 1A, 1B, 5A, 5B, 6A, and 6B (Zhang et al., 2008; Blanco et al., 2011; Colasuonno et al., 2014). In addition, the 1BL.1RS wheat-rye translocation carried a major QTL for YPC and b^* explaining 25.4-32.2% of the phenotypic variance (Zhang et al., 2009; Zhai et al., 2016). Wheat cultivars with the 1BL.1RS translocation had higher total carotenoid contents (0.76 vs. 0.61 μg/g), lutein (0.46 vs. 0.40), zeaxanthin (0.08 vs. 0.07) and β-carotene (0.22 vs. 0.14) than those without the translocation (Li et al., 2016), an aspect that should be considered in breeding for higher provitamin A content in bread wheat.

Gene Cloning and Molecular Marker Development

Most of carotenogenes in wheat have been cloned and characterized. Briefly, the full-length genomic DNA sequence of *PSY1* was cloned, and two co-dominant markers (*YP7A* and *YP7B-1*) and two dominant markers (*YP7B-2* and *YP7B-3*) were developed for *PSY-A1* and *PSY-B1* (He et al., 2008, 2009). *YP7A* co-segregated with a QTL for YPC on chromosome 7AL and explained 20-28% of the phenotypic variance (He et al., 2008). Cultivars with *PSY-B1c* had the highest YPC (2.01 μg/g), followed by *PSY-B1a* (1.71 μg/g), whereas those with *PSY-B1b* had the lowest value (1.40 μg/g; He et al., 2009).

Dong (2011) cloned the full-length *PDS* gene and designed two complementary markers *YP4B-1* and *YP4B-2* corresponding to higher and lower YPC, respectively (no significant difference). The full-length genomic sequence of *ZDS* was cloned and co-dominant molecular markers *YP2A-1* and *YP2D-1* were developed for *ZDS-A1* and *ZDS-D1*, respectively (Zhang et al., 2011; Dong et al., 2012). *YP2A-1* and *YP2D-1* co-segregated with QTL for YPC on chromosome 2A and 2DL, respectively, explaining 11.3-18.4% of the phenotypic variance.

The entire sequence of the *LCYE* gene was isolated and located on homoeologous group 3 chromosomes, and it was identified as a candidate gene underlying QTL for lutein content on chromosome 3B (Howitt et al., 2009). Dong (2011) developed a co-dominant functional marker *YP3B-1* for *TaLCYE-B1*, but values of YPC from cultivars with *TaLCYE-B1a* were not significantly different from those with *TaLCYE-B1b*. Therefore, the effect of *TaLCYE-B1* on carotenoid contents in wheat grains need to be further investigated. *e-LCY3A-3*, a co-dominant functional marker, was developed based on *e-LYC3Aa* and *e-LYC3Ab* alleles (Crawford and Francki, 2013b). A highly significant ($P < 0.01$) association with QTL on chromosome 3A indicated that *e-LYC3A* is functionally associated with variation in b^*. The *TaLCYB* gene was cloned and shown to have a role in β-carotene biosynthesis using RNAi (Zeng et al., 2015a). In addition, *HYD*1, *HYD*2, and *HYE* were cloned and characterized (Kawaura et al., 2009; Qin et al., 2012). Information relating to these genes and molecular markers is provided in Table 1. The functional markers have been used in routine germplasm characterization and cultivar development.

Table 1 Summary of carotenogenic genes and molecular markers in bread wheat.

Enzyme	Gene	GenBank No.	Chromosomal location (IWGSC)	Marker	Allele	Fragment size (bp)	YPC
Phytoene synthase 1	*PSY1*	EF600063	7AL, 7BL, 7DL	*YP7A*	*PSY-A1a*/*PSY-A1c*	194	High
					PSY-A1b	213	Low
				YP7B-1	*PSY-B1a*	151	Medium
					PSY-B1b	156	Low
				YP7B-2	*PSY-B1c*	428	High
				YP7B-3	*PSY-B1d*	884	—
Phytoene desaturase	*PDS*	FJ517553	4AS, 4BL, 4DL	*YP4B-1*	*TaPDS-B1b*	562	High
				YP4B-2	*TaPDS-B1a*	382	Low
ζ-Carotene desaturase	*ZDS*	HQ703016	2AS, 2BS, 2DS	*YP2A-1*	*TaZDS-A1a*	183	Low
					TaZDS-A1b	179	High
				YP2D-1	*TaZDS-D1a*	No	High
					TaZDS-D1b	981	Low
Lycopene ε-cyclase	*LCYE*	EU649785	3AL, 3B, 3DL	*e-LCY3A-3*	*e-LCY3Aa*	537	—
					e-LCY3Ab	309 & 230	
				YP3B-1	*TaLCYE-B1a*	635	—
					TaLCYE-B1b	No	
Lycopene β-cyclase	*LCYB*	FJ814767	6AS, 6DS				
Carotenoid β-ring hydroxylase	*CHYB1*	JX171673	2AL, 2BL, 2DL				
	CHYB2	JX171670	6AL, 6BL, 6DL				
Carotenoid ε-ring hydroxylase	*CHYE*	AK334877	1AL, 1BL, 1DL				

IWGSC, International Wheat Genome Sequencing Consortium; YPC, yellow pigment content; —Unknown.

For carotenoid degradation, three copies of the *LOX*-1 gene (*LOX-B*1.1, *LOX-B*1.2, and *LOX-B*1.3) were cloned in durum wheat (Hessler et al., 2002; Verlotta et al., 2010). In bread wheat, the full-length genomic DNA sequence of *TaLOX-B*1 gene was cloned, and complementary markers *LOX*16 and *LOX*18 were developed (Geng et al., 2012). However, *CCD* sequences of wheat have not been reported to date.

The Molecular Basis of QTL for Carotenoid Content

With carotenogenes identified and functional markers developed, there is a growing interest in understanding the molecular basis of QTL underpin carotenoid content in wheat.

As expected, *PSY*1 gene was considered as a candidate gene responsible for YPC variation in wheat grains since *YP7A* and *YP7B* co-segregated with QTL for YPC on chromosomes 7AL and 7BL (He et al., 2008; Zhang and Dubcovsky, 2008; Singh et al., 2009). Other studies indicated that a second gene other than *PSY*1 in the distal regions of chromosomes 7A and 7B affects YPC (Singh et al., 2009; Crawford and Francki, 2013a). The geranylgeranyl transferase I α-subunit (*RGGT*) gene was mapped to distal regions on chromosomes 7BL and 7DL (Crawford et al., 2008). This gene encodes enzyme involved in the terpenoid backbone biosynthesis pathway, providing the precursor GGPP for carotenoid biosynthesis, and it could be a candidate for the additional gene. Moreover, a *Cat*3-*A*1 gene was co-located to the QTL for b* on 7AL, encoding a catalase enzyme which controls varying degrees of bleaching action on lutein by regulating hydrogen peroxide accumulation in developing wheat grain, and it could be another candidate for the additional gene (Crawford and Francki, 2013a; Li et al., 2015).

The *LCYE* gene was considered as a candidate gene for QTL affecting b* variation and lutein content on chromosomes 3A and 3B in bread wheat (Howitt et al., 2009; Crawford and Francki, 2013b). In addition, a QTL for pasta color on chromosome 4B was linked to a polymorphic deletion in *LOX-B*1, suggesting that it was associated with pigment degradation during pasta processing (Hessler et al., 2002).

With advances in genomics and bioinformatics, some other genes were found to be associated with carotenoid biosynthesis in wheat. A genome scan for QTL in durum and SNP homology prediction against annotated proteins in the wheat and *Brachypodium* genomes identified diphosphomevalonate decar-boxylase (*DMAPD*) and aldehyde oxidase (*AO*) co-located with the major QTL for YPC on chromosomes 5BL and 7AL, respectively (Colasuonno et al., 2014). Six candidate genes related to terpenoid backbone biosynthesis were within QTL intervals associated with four color-related traits in bread wheat (Zhai et al., 2016); these included genes for alpha-1, 3-mannosyl-glycoprotein 2-beta-*N*-acetylglucosaminyltransferase (*MGAT*1), mevalonate kinase (*MK*), delta-cadinene synthase(*DCS*), ketol-acid reductoisomerase (*KARI*), cytokinin oxidase/dehydrogenase (*CKX*), and secologanin synthase (*SLC*). All these genes further enrich carotenoid biosynthesis pathway (Figure 1).

Because quantification of carotenoids by HPLC is expensive and time-consuming, most studies of wheat carotenoid contents have depended on indirect parameters such as YPC and b*. In order to deepen understanding of the carotenoid metabolism in wheat, fast, cost-effective methods to detect individual carotenoids should be developed and improved, such as UPLC (ultra-high performance liquid chromatography), UPLC-MS and UPLC-NMR. Moreover, many QTLs affecting carotenoid content could not be explained by known genes. This provides opportunities to discover additional genes controlling carotenogenesis in wheat grain. With progress in next-generation DNA sequencing and SNP chips, it will be much easier to construct high-density genetic maps useful in detecting QTL for carotenoid content, identifying candidate genes, and map-based cloning of candidate genes.

FUTURE PROSPECTS

As discussed above, significant progress has been made in our understanding of carotenoid metabolism, genetic regulation, and genetic manipulation in higher plants. This has improved our capacity for breeding new cultivars with high carotenoid contents. Compared to other plants, there are still numerous unknown aspects on carotenoid biosynthesis in the staple cereals. Firstly, a more comprehensive and deeper understanding of carotenoid regulatory mechanisms will undoubtedly facilitate genetic manipulation to modify overall carotenoid contents and individual components with predictable outcomes. Secondly, genetic manipulations in crops were mainly focused on β-carotene enhancement to combat the VAD, but improvements in other carotenoids were rarely reported, even for lutein and zeaxanthin which play significant roles in promoting eye and skin health and in reducing the risk of several chronic diseases. Therefore, future studies should give more attention to improve other carotenoids or simultaneously engineer multiple carotenoid molecules. In addition, the carotenoid pathways in maize and rice have been extended to accumulate a wide variety of unusual keto-carotenoids, which could be exploited to other crop plants, including wheat.

New technologies provide novel opportunities for genetic manipulation of carotenoid biosynthesis in staple cereals. With progress in next-generation DNA sequencing and SNP chips, genomic selection is expected to play a key role in breeding programs (Varshney et al., 2014). KASP (Kompetitive Allele Specific PCR) technology with its much faster and higher detection accuracy offers cost-effective and scalable flexibility in application of gene-specific markers in breeding programs (Semagn et al., 2014). Development of practical breeding chips based on KASP markers and closely linked SNP markers from GWAS will be a big step forward in improving marker application in breeding high provitamin A-enriched cereals. New gene editing technologies, such as TALENs (transcription activator-like effector nucleases) and CRISPR (clustered regularly spaced palindromic repeat), are currently the most widely used methods for understanding gene function, and are emerging as transformative technologies for crop breeding due to ability to edit genomic sequences at defined sites rather than random introduction of foreign DNA (LaFountaine et al., 2015). We are strongly confident that provitamin A-enriched crops will be developed in the near future by application of improved genetic knowledge and new technologies.

ACKNOWLEDGMENTS

The authors are grateful to Prof. R. A. McIntosh, Plant Breeding Institute, University of Sydney, and Prof. J. B. Yan, National Key Laboratory of Crop Genetic Improvement, Huazhong Agricultural University, for review of this manuscript.

REFERENCES

Abdel-Aal, E. M. S., Young, J. C., Rabalski, I., Hucl, P., and Fregeau-Reid, J. (2007). Identification and quantification of seed carotenoids in selected wheat species. *J. Agric. Food Chem.* 55, 787-794. doi: 10.1021/jf062764p

Alder, A., Jamil, M., Marzorati, M., Bruno, M., Vermathen, M., Bigler, P., et al. (2012). The path from β-carotene to carlactone, a strigolactone-like plant hormone. *Science* 335, 1348-1351. doi: 10.1126/science.1218094

Arango, J., Jourdan, M., Geoffriau, E., Beyer, P., and Welsch, R. (2014). Carotene hydroxylase activity determines the levels of both α-carotene and total carotenoids in orange carrots. *Plant Cell* 26, 2223-2233. doi: 10.1105/tpc.113.122127

Arango, J., Wüst, F., Beyer, P., and Welsch, R. (2010). Characterization of phytoene synthases from cassava and their involvement in abiotic stress-mediated responses. *Planta* 232, 1251-1262. doi: 10.1007/s00425-010-1250-6

Ariizumi, T., Kishimoto, S., Kakami, R., Maoka, T., Hirakawa, H., Suzuki, Y., et al. (2014). Identification of the carotenoid modifying gene Pale Yellow Petal 1 as an essential factor in xanthophyll esterification and yellow flower pigmentation in tomato (*Solanum lycopersicum*). *Plant J.* 79, 453-465. doi: 10.1111/tpj.12570

Auldridge, M. E., McCarty, D. R., and Klee, H. J. (2006). Plant carotenoid cleavage oxygenases and their apocarotenoid products. *Curr. Opin. Plant Biol.* 9, 315-321. doi: 10.1016/j.pbi.2006.03.005

Avendano-Vazquez, A. O., Cordoba, E., Llamas, E., Roman, C. S., Nisar, N., De la Torre, S., et al. (2014). An uncharacterized apocarotenoid-derived signal generated in ζ-carotene desaturase mutants regulate leaf development and the expression of chloroplast and nuclear genes in *Arabidopsis*. *Plant Cell* 26, 2524-2537. doi: 10.1105/tpc.114.123349

Azmach, G., Gedil, M., Menkir, A., and Spillane, C. (2013). Marker-trait association analysis of functional gene markers for provitamin A levels across diverse tropical yellow maize inbred lines. *BMC Plant Biol.* 13: 227. doi: 10.1186/1471-2229-13-227

Babu, R., Rojas, N. P., Gao, S. B., Yan, J. B., and Pixley, K. (2013). Validation of the effects of molecular marker polymorphisms in LcyE and CrtRB1 on provitamin A concentrations for 26 tropical maize populations. *Theor. Appl. Genet.* 126, 389-399. doi: 10.1007/s00122-012-1987-3

Bai, C., Capell, T., Berman, J., Medina, V., Sandmann, G., Christou, P., et al. (2016). Bottlenecks in carotenoid biosynthesis and accumulation in rice endosperm are influenced by the precursor-product balance. *Plant Biotechnol. J.* 14, 195-205. doi: 10.1111/pbi.12373

Barsan, C., Zouine, M., Maza, E., Bian, W., Egea, I., Rossignol, M., et al. (2012). Proteomic analysis of chloroplast-to-chromoplast transition in tomato reveals metabolic shifts coupled with disrupted thylakoid biogenesis machinery and elevated energy-production components. *Plant Physiol*. 160, 708-725. doi: 10.1104/pp. 112.203679

Blanco, A., Colasuonno, P., Gadaleta, A., Mangini, G., Schiavulli, A., Simeone, R., et al. (2011). Quantitative trait loci for yellow pigment concentration and individual carotenoid compounds in durum wheat. *J. Cereal Sci*. 54, 255-264. doi: 10.1016/j.jcs. 2011.07.002

Borrelli, G. M., De Leonardis, A. M., Platani, C., and Troccoli, A. (2008). Distribution along durum wheat kernel of the components involved in semolina colour. *J. Cereal Sci*. 48, 494-502. doi: 10.1016/j.jcs. 2007.11.007

Breitenbach, J., Bai, C., Rivera, S. M., Canela, R., Capell, T., Christou, P., et al. (2014). A novel carotenoid, 4-keto-α-carotene, as an unexpected by-product during genetic engineering of carotenogenesis in rice callus. *Phytochemistry* 98, 85-91. doi: 10.1016/j.phytochem. 2013.12.008

Burkhardt, P. K., Beyer, P., Wunn, J., Kloti, A., Armstrong, G. A., Schledz, M., et al. (1997). Transgenic rice (*Oryza sativa*) endosperm expressing daffodil (*Narcissus pseudonarcissus*) phytoene synthase accumulates phytoene, a key intermediate of provitamin a biosynthesis. *Plant J*. 11, 1071-1078. doi: 10.1046/j.1365-313X. 1997.11051071.x

Carvalho, L. J. C. B., Lippolis, J., Chen, S., de Souza, C. R. B., Vieira, E. A., and Anderson, J. V. (2012). Characterization of carotenoid-protein complexes and gene expression analysis associated with carotenoid sequestration in pigmented cassava(*Manihot esculenta* Crantz)storage root. *Open Biochem. J*. 6,116-130. doi: 10.2174/1874091X01206010116

Cazzonelli, C. I. (2011). Carotenoids in nature: insights from plants and beyond. *Funct. Plant Biol*. 38, 833-847. doi: 10.1071/FP11192

Cazzonelli, C. I., Cuttriss, A. J., Cossetto, S. B., Pye, W., Crisp, P., Whelan, J., et al. (2009). Regulation of carotenoid composition and shoot branching in *Arabidopsis* by a chromatin modifying histone methyltransferase, SDG8. *Plant Cell* 21,39-53. doi:10.1105/tpc.108. 063131

Cazzonelli, C. I., and Pogson, B. J. (2010). Source to sink: Regulation of carotenoid biosynthesis in plants. *Trends Plant Sci*. 15, 266-274. doi: 10.1016/j.tplants. 2010.02.003

Cazzonelli, C. I., Roberts, A. C., Carmody, M. E., and Pogson, B. J. (2010). Transcriptional control of SET DOMAIN GROUP 8 and CAROTENOID ISOMERASE during *Arabidopsis* development. *Mol. Plant* 3, 174-191. doi: 10.1093/mp/ssp092

Chen, Y., Li, F., and Wurtzel, E. T. (2010). Isolation and characterization of the Z-ISO gene encoding a missing component of carotenoid biosynthesis in plants. *Plant Physiol*. 153, 66-79. doi: 10.1104/pp.110.153916

Clotault, J., Peltier, D., Berruyer, R., Thomas, M., Briard, M., and Geoffriau, E. (2008). Expression of carotenoid biosynthesis genes during carrot root development. *J. Exp. Bot*. 59,3563-3573. doi:10.1093/jxb/ern210

Colasuonno, P., Gadaleta, A., Giancaspro, A., Nigro, D., Giove, S., Incerti, O., et al. (2014). Development of a high-density SNP-based linkage map and detection of yellow pigment content QTLs in durum wheat. *Mol. Breed*. 34, 1563-1578. doi: 10.1007/s11032-014-0183-3

Cong, L., Wang, C., Chen, L., Liu, H. J., Yang, G. X., and He, G. Y. (2009). Expression of phytoene synthase1 and carotene desaturase crtI genes result in an increase in the total carotenoids content in transgenic elite wheat (*Triticum aestivum* L.). *J. Agric. Food Chem*. 57, 8652-8660. doi: 10.1021/jf9012218

Crawford, A. C., and Francki, M. G. (2013a). Chromosomal location of wheat genes of the carotenoid biosynthetic pathway and evidence for a catalase gene on chromosome 7A functionally associated with flour b* colour variation. *Mol. Genet. Genomics* 288, 483-493. doi: 10.1007/s00438-013-0767-3Crawford, A. C., and Francki, M. G. (2013b). Lycopene-ε-cyclase (e-LCY3A) is functionally associated with QTL for flour b* colour on chromosome 3A in wheat (*Triticum aestivum* L.). *Mol. Breed*. 31, 737-741. doi: 10.1007/s11032-012-9812-x

Crawford, A. C., Shaw, K., Stefanova, K., Lambe, W., Ryan, K., Wilson, R. H., et al. (2008). "A molecular toolbox for xanthophyll genes in wheat," in *Proceedings of the 11th International Wheat Genet Symposium*, Brisbane, QLD, 24-29

Cunningham, F. X., and Gantt, E. (1998). Genes and enzymes of carotenoid biosynthesis in plants. *Annu. Rev. Plant Physiol. Plant Mol. Biol*. 49, 557-583. doi: 10.1146/annurev. arplant. 49.1.557

Cunningham, F. X., and Gantt, E. (2005). A study in scarlet: enzymes of ketocarotenoid biosynthesis in the flowers of Adonis aestivalis. *Plant J*. 41, 478-492. doi: 10.1111/j.1365-313X.2004.02309.x da Silva Messias, R., Galli, V., Dos Anjos E Silva, S. D., and Rombaldi, C. V. (2014). Carotenoid biosynthetic and catabolic pathways: gene expression and carotenoid content in grains of maize landraces. *Nutrients* 6, 546-563. doi: 10.3390/nu6020546

D'Ambrosio, C., Giorio, G., Marino, I., Merendino, A., Petrozza, A., Salfi, L., et al. (2004). Virtually complete conversion of lycopene into β-carotene in fruits of tomato plants transformed with the tomato lycopene β-cyclase (tlcy-b) cDNA. *Plant Sci*. 166, 207-214. doi: 10.1016/j.plantsci.2003.09.015

Demmig-Adams, B., and Adams, W. W. III. (2002). Antioxidants in photosynthesis and human nutrition. Science 298, 2149-2153. doi: 10.1126/science.1078002

Dibari, B., Murat, F., Chosson, A., Gautier, V., Poncet, C., Lecomte, P., et al. (2012). Deciphering the genomic structure, function and evolution of carotenogenesis related phytoene synthase in grasses. BMC Genomics 13: 221. doi: 10.1186/1471-2164-13-221

Digesù, A. M., Platani, C., Cattivelli, L., Mangini, G., and Blanco, A. (2009). Genetic variability in yellow pigment components in cultivated and wild tetraploid wheats. J. Cereal Sci. 50, 210-218. doi: 10.1016/j.jcs.2009.05.002

Diretto, G., Welsch, R., Tavazza, R., Mourgues, F., Pizzichini, D., Beyer, P., et al. (2007). Silencing of beta-carotene hydroxylase increases total carotenoid and beta-carotene levels in potato tubers. *BMC Plant Biol*. 7: 11. doi: 10.1186/1471-2229-7-11

Dong, C. H. (2011). *Cloning of Genes Associated with Grain Yellow Pigment Content in Common Wheat and Development of Functional Markers*. master's thesis, Agricultural University of Hebei Province, Baoding

Dong, C. H., Ma, Z. Y., Xia, X. C., Zhang, L. P., and He, Z. H. (2012). Allelic variation at the TaZds-A1 locus on wheat chromosome 2A and development of a functional marker in common wheat. *J. Integr. Agric*. 11, 1067-1074. doi: 10.1016/S2095-3119 (12) 60099-9

Du, H., Wang, N., Cui, F., Li, X., Xiao, J., and Xiong, L. (2010). Characterization of the β-carotene hydroxylase gene DSM2 conferring drought and oxidative stress resistance by increasing xanthophylls and abscisic acid synthesis in rice. *Plant Physiol*. 154, 1304-1318. doi: 10.1104/pp.110.163741

Egea, I., Barsan, C., Bian, W., Purgatto, E., Latché, A., Chervin, C., et al. (2010). Chromoplast differentiation: current status and perspectives. *Plant Cell Physiol*. 51, 1601-1611. doi: 10.1093/pcp/pcq136

Elouafi, I., Nachit, M. M., and Martin, L. M. (2001). Identification of a microsatellite on chromosome 7B showing a strong linkage with yellow pigment in durum wheat (*Triticum turgidum* L. var. durum). *Hereditas* 135, 255-261. doi: 10.1111/j.1601-5223.2001.t01-1-00255.x

Fang, J., Chai, C. L., Qian, Q., Li, C. L., Tang, J. Y., Sun, L., et al. (2008). Mutations of genes in synthesis of the carotenoid precursors of ABA lead to pre-harvest sprouting and photo-oxidation in rice. *Plant J*. 54, 177-189. doi: 10.1111/j.1365-313X.2008.03411.x

Farré, G., Bai, C., Twyman, R. M., Capell, T., Christou, P., and Zhu, C. F. (2011). Nutritious crops producing multiple carotenoids-a metabolic balancing act. *Trends Plant Sci*. 16, 532-540. doi: 10.1016/j.tplants.2011.08.001

Farré, G., Sanahuja, G., Naqvi, S., Bai, C., Capell, T., Zhu, C. F., et al. (2010). Travel advice on the road to carotenoids in plants. *Plant Sci*. 179, 28-48. doi: 10.1016/j.plantsci.2010.03.009

Fraser, P. D., and Bramley, P. M. (2004). The biosynthesis and nutritional uses of carotenoids. *Prog. Lipid Res*. 43, 228-265. doi: 10.1016/j.plipres.2003.10.002

Fu, Z. Y., Chai, Y. C., Zhou, Y., Yang, X. H., Warburton, M. L., Xu, S. T., et al. (2013). Natural variation in the sequence of *PSY1* and frequency of favorable polymorphisms among tropical and temperate maize germplasm. *Theor. Appl. Genet*. 126, 923-935. doi: 10.1007/s00122-012-2026-0Gallagher, C. E., Matthews, P. D., Li, F., and Wurtzel, E. T. (2004). Gene duplication in the carotenoid biosynthetic pathway preceded evolution of the grasses. *Plant Physiol*. 135, 1776-1783. doi: 10.1104/pp.104.039818

Galpaz, N., Wang, Q., Menda, N., Zamir, D., and Hirschberg, J. (2008). Abscisic acid deficiency in the tomato mutant high-pigment 3 leading to increased plastid number and higher fruit lycopene content. *Plant J*. 53, 717-730. doi: 10.1111/j.1365-313X.2007.03362.x

Gayen, D., Ali, N., Sarkar, S. N., Datta, S. K., and Datta, K. (2015). Down-regulation of lipoxygenase gene reduces degradation of carotenoids of golden rice during storage. *Planta* 242, 353-363. doi: 10.1007/s00425-015-2314-4

Geng, H. W., He, Z. H., Zhang, L. P., Qu, Y. Y., and Xia, X. C. (2012). Development of functional markers for a lipoxygenase gene TaLox-B1 on chromosome 4BS in common wheat. *Crop Sci.* 52, 568-576. doi: 10.2135/cropsci2011.07.0365

Giuliano, G., Tavazza, R., Diretto, G., Beyer, P., and Taylor, M. A. (2008). Metabolic engineering of carotenoid biosynthesis in plants. *Trends Biotechnol.* 26, 139-145. doi: 10.1016/j.tibtech.2007.12.003

Gómez-García, M. R., and Ochoa-Alejo, N. (2013). Biochemistry and molecular biology of carotenoid biosynthesis in chili peppers (*Capsicum* spp.). *Int. J. Mol. Sci.* 14, 19025-19053. doi: 10.3390/ijms140919025

Gonzalez-Jorge, S., Ha, S. H., Magallanes-Lundback, M., Gilliland, L. U., Zhou, A., Lipka, A. E., et al. (2013). Carotenoid cleavage dioxygenase4 is a negative regulator of β-carotene content in *Arabidopsis* seeds. *Plant Cell* 25, 4812-4826. doi: 10.1105/tpc.113.119677

Harjes, C. E., Rocheford, T. R., Bai, L., Brutnell, T. P., Kandianis, C. B., Sowinski, S. G., et al. (2008). Natural genetic variation in lycopene epsilon cyclase tapped for maize biofortification. Science 319, 330-333. doi: 10.1126/science.1150255

Havaux, M. (2014). Carotenoid oxidation products as stress signals in plants. Plant J. 79, 597-606. doi: 10.1111/tpj.12386

He, X. Y., He, Z. H., Ma, W., Appels, R., and Xia, X. C. (2009). Allelic variants of phytoene synthase 1 (Psy1) genes in Chinese and CIMMYT wheat cultivars and development of functional markers for flour colour. *Mol. Breed.* 23, 553-563. doi: 10.1007/s11032-009-9255-1

He, X. Y., Zhang, Y. L., He, Z. H., Wu, Y. P., Xiao, Y. G., Ma, C. X., et al. (2008). Characterization of phytoene synthase 1 gene (Psy1) located on common wheat chromosome 7A and development of a functional marker. Theor. Appl. Genet. 116, 213-221. doi: 10.1007/s00122-007-0660-8

Hentschel, V., Kranl, K., Hollmann, J., Lindhauer, M. G., Bohm, V., and Bitsch, R. (2002). Spectrophotometric determination of yellow pigment content and evaluation of carotenoids by high-performance liquid chromatography in durum wheat grain. J. Agric. Food Chem. 50, 6663-6668. doi: 10.1021/jf025701p Hessler, T. G., Thomson, M. J., Benscher, D., Nachit, M. M., and Sorrells

M. E. (2002). Association of a lipoxygenase locus, Lpx-B1, with variation in lipoxygenase activity in durum wheat seeds. Crop Sci. 42, 1695-1700. doi: 10.2135/cropsci2002.1695

Hidalgo, A., Brandolini, A., Pompei, C., and Piscozzi, R. (2006). Carotenoids and tocols of einkorn wheat (*Triticum monococcum* ssp. monococcum L.). *J. Cereal Sci.* 44, 182-193. doi: 10.1016/j.jcs.2006.06.002

Howitt, C. A., Cavanagh, C. R., Bowerman, A. F., Cazzonelli, C., Rampling, L., Mimica, J. L., et al. (2009). Alternative splicing, activation of cryptic exons and amino acid substitutions in carotenoid biosynthetic genes are associated with lutein accumulation in wheat endosperm. *Funct. Integr. Genomics* 9, 363-376. doi: 10.1007/s10142-009-0121-3

Howitt, C. A., and Pogson, B. J. (2006). Carotenoid accumulation and function in seeds and non-green tissues. *Plant Cell Environ.* 29, 435-445. doi: 10.1111/j.1365-3040.2005.01492.x

Kachanovsky, D. E., Filler, S., Isaacson, T., and Hirschberg, J. (2012). Epistasis in tomato color mutations involves regulation of phytoene synthase 1 expression by cis-carotenoids. *Proc. Natl. Acad. Sci. U. S. A.* 109, 19021-19026. doi: 10.1073/pnas.1214808109

Kawaura, K., Mochida, K., Enju, A., Totoki, Y., Toyoda, A., Sakaki, Y., et al. (2009). Assessment of adaptive evolution between wheat and rice as deduced from full-length common wheat cDNA sequence data and expression patterns. *BMC Genomics* 10: 271. doi: 10.1186/1471-2164-10-271

Khoo, H. E., Prasad, K. N., Kong, K. W., Jiang, Y., and Ismail, A. (2011). Carotenoids and their isomers: color pigments in fruits and vegetables. *Molecules* 16, 1710-1738. doi: 10.3390/molecules16021710

Kilambi, H. V., Kumar, R., Sharma, R., and Sreelakshmi, Y. (2013). Chromoplast-specific carotenoid-associated protein appears to be important for enhanced accumulation of carotenoids in hp1 tomato fruits. Plant Physiol. 161, 2085-2101. doi: 10.1104/pp.112.212191

Kilcrease, J., Collins, A. M., Richins, R. D., Timlin, J. A., and O'Connell, M. A. (2013). Multiple microscopic approaches demonstrate linkage between chromoplast architecture and carotenoid composition in diverse Capsicum annuum fruit. Plant J. 76, 1074-1083. doi: 10.1111/tpj.12351

Kim, J., Smith, J., Tian, L., and DellaPenna, D. (2009). The evolution and function of carotenoid hydroxylases in Arabidopsis. Plant Cell Physiol. 50, 463-479. doi: 10.1093/pcp/pcp005

LaFountaine, J. S., Fathe, K., and Smyth, H. D. (2015). Delivery and therapeutic applications of gene editing technologies ZFNs, TALENs, and CRISPR/Cas9. *Int. J. Pharm.* 494, 180-194. doi: 10.1016/j. ijpharm. 2015. 08. 029

Laule, O., Fürholz, A., Chang, H. S., Zhu, T., Wang, X., Heifetz, P. B., et al. (2003). Crosstalk between cytosolic and plastidial pathways of isoprenoid biosynthesis in *Arabidopsis* thaliana. *Proc. Natl. Acad. Sci. U. S. A.* 100, 6866-6871. doi: 10.1073/pnas. 1031755100

Li, D. A., Walker, E., and Francki, M. G. (2015). Identification of a member of the catalase multigene family on wheat chromosome 7A associated with flour b* colour and biological significance of allelic variation. *Mol. Genet. Genomics* 290, 2313-2324. doi: 10.1007/s00438-015-1083-x

Li, F., Vallabhaneni, R., and Wurtzel, E. T. (2008). PSY3, a new member of the phytoene synthase gene family conserved in the Poaceae and regulator of abiotic stress-induced root carotenogenesis. *Plant Physiol.* 146, 1333-1345. doi: 10.1104/pp. 107. 111120

Li, L., and Yuan, H. (2013). Chromoplast biogenesis and carotenoid accumulation. *Arch. Biochem. Biophys.* 539, 102-109. doi: 10.1016/j. abb. 2013. 07. 002

Li, W. S., Zhai, S. N., Jin, H., Wen, W. E., Liu, J. D., Xia, X. C., et al. (2016). Genetic variation of carotenoids in Chinese bread wheat cultivars and the effect of the 1BL. 1RS translocation. *Front. Agric. Sci. Eng.* 3: 124-130. doi: 10.15302/J-FASE-2016094

Liu, L., Jeffers, D., Zhang, Y. D., Ding, M. L., Chen, W., Kang, M. S., et al. (2015). Introgression of the crtRB1 gene into quality protein maize inbred lines using molecular markers. *Mol. Breed.* 35: 154. doi: 10.1007/s11032-015-0349-7

Lu, S., Van Eck, J., Zhou, X., Lopez, A. B., O' Halloran, D. M., Cosman, K. M., et al. (2006). The cauliflower or gene encodes a DnaJ cysteine-rich domain-containing protein that mediates high levels of β-carotene accumulation. *Plant Cell* 18, 3594-3605. doi: 10.1105/tpc. 106. 046417

Mares, D., and Campbell, A. (2001). Mapping components of flour and noodle color in Australian wheat. *Aust. J. Agric. Res.* 52, 1297-1309. doi: 10.1071/AR01049

Martel, C., Vrebalov, J., Tafelmeyer, P., and Giovannoni, J. J. (2011). The tomato MADS-box transcription factor RIPENING INHIBITOR interacts with promoters involved in numerous ripening processes in a colorless nonripening-dependent manner. *Plant Physiol.* 157, 1568-1579. doi: 10.1104/pp. 111. 181107

Matthews, P. D., Luo, R., and Wurtzel, E. T. (2003). Maize phytoene desaturase and ζ-carotene desaturase catalyse a poly-Z desaturation pathway: implications for genetic engineering of carotenoid content among cereal crops. *J. Exp. Bot.* 54, 2215-2230. doi: 10.1093/jxb/ erg235

McGraw, K. J., and Toomey, M. B. (2010). Carotenoid accumulation in the tissues of zebra finches: predictors of integumentary pigmentation and implications for carotenoid allocation strategies. *Physiol. Biochem. Zool.* 83, 97-109. doi: 10.1086/648396

Mellado-Ortega, E., and Hornero-Méndez, D. (2016). Carotenoid evolution during short-storage period of durum wheat (Triticum turgidum conv. durum) and tritordeum (× *Tritordeum* Ascherson et Graebner) whole-grain flours. *Food Chem.* 192, 714-723. doi: 10.1016/j. foodchem. 2015. 07. 057

Moise, A. R., Al-Babili, S., and Wurtzel, E. T. (2013). Mechanistic aspects of carotenoid biosynthesis. Chem. Rev. 114, 164-193. doi: 10.1021/cr400106y Muthusamy, V., Hossain, F., Thirunavukkarasu, N., Choudhary, M., Saha, S., Bhat, J. S., et al. (2014). Development of β-Carotene rich maize hybrids through marker-assisted introgression of β-carotene hydroxylase allele. PLoS ONE 9: e113583. doi: 10.1371/journal. pone. 0113583

Naqvi, S., Zhu, C. F., Farre, G., Sandmann, G., Capell, T., and Christou, P. (2011). Synergistic metabolism in hybrid corn indicates bottlenecks in the carotenoid pathway and leads to the accumulation of extraordinary levels of the nutritionally important carotenoid zeaxanthin. *Plant Biotechnol. J.* 9, 384-393. doi: 10.1111/j. 1467-7652. 2010. 00554. x

Neta-Sharir, I., Isaacson, T., Lurie, S., and Weiss, D. (2005). Dual role for tomato heat shock protein 21: protecting photosystem II from oxidative stress and promoting color changes during fruit maturation. *Plant Cell* 17, 1829-1838. doi: 10.1105/tpc. 105. 031914

Niyogi, K. K. (2000). Safety valves for photosynthesis. *Curr. Opin. Plant Biol.* 3, 455-460. doi: 10.1016/S1369-5266 (00) 00113-8

Okada, K., Saito, T., Nakagawa, T., Kawamukai, M., and Kamiya, Y. (2000). Five geranylgeranyldiphosphate synthases expressed in different organs are localized into three subcellular compartments in *Arabidopsis*. *Plant Physiol.* 122, 1045-1056. doi: 10.1104/ pp. 122. 4. 1045

Paine, J. A., Shipton, C. A., Chaggar, S., Howells, R. M., Kennedy, M. J., Vernon, G., et al. (2005).

Improving the nutritional value of Golden Rice through increased pro-vitamin A content. *Nat. Biotechnol.* 23, 482-487. doi: 10.1038/nbt1082

Patil, R. M., Oak, M. D., Tamhankar, S. A., Sourdille, P., and Rao, V. S. (2008). Mapping and validation of a major QTL for yellow pigment content on 7AL in durum wheat (*Triticum turgidum* L. ssp. durum). *Mol. Breed.* 21, 485-496. doi: 10.1007/s11032-007-9147-1

Qin, X., Zhang, W., Dubcovsky, J., and Tian, L. (2012). Cloning and comparative analysis of carotenoid β-hydroxylase genes provides new insights into carotenoid metabolism in tetraploid (*Triticum turgidum* ssp. durum) and hexaploid (*Triticum aestivum*) wheat grains. *Plant Mol. Biol.* 80, 631-646. doi: 10.1007/s11103-012-9972-4

Rodriguez-Concepcion, M. (2010). Supply of precursors for carotenoidbio-synthesis in plants. *Arch. Biochem. Biophys.* 504, 118-122. doi: 10.1016/j.abb.2010.06.016

Rodriguez-Concepcion, M., Ahumada, I., Diez-Juez, E., Sauret-Gueto, S., Lois, L. M., Gallego, F., et al. (2001). 1-Deoxy-D-xylulose 5-phosphatereductoisomerase and plastid isoprenoid biosynthesis during tomato fruit ripening. *Plant J.* 27, 213-222. doi: 10.1046/j.1365-313x.2001.01089.x

Rosati, C., Aquilani, R., Dharmapuri, S., Pallara, P., Marusic, C., Tavazza, R., et al. (2000). Metabolic engineering of beta-carotene and lycopene content in tomato fruit. *Plant J.* 24, 413-420. doi: 10.1046/j.1365-313x.2000.00880.x

Sandmann, G. (2001). Genetic manipulation of carotenoid biosynthesis: strategies, problems and achievements. *Trends Plant Sci.* 16, 14-17. doi: 10.1016/S1360-1385(00)01817-3

Semagn, K., Babu, R., Hearne, S., and Olsen, M. (2014). Single nucleotidepolymorphism genotyping using Kompetitive Allele Specific PCR (KASP): overview of the technology and its application in crop improvement. *Mol. Breed.* 33, 1-14. doi: 10.1007/s11032-013-9917-x

Seo, M., and Koshiba, T. (2002). Complex regulation of ABA biosynthesis in plants. *Trends Plant Sci.* 7, 41-48. doi: 10.1016/S1360-1385(01)02187-2

Shewry, P. R. (2009). Wheat. *J. Exp. Bot.* 60, 1537-1553. doi: 10.1093/jxb/erp058 Siedow, J. N. (1991). Plant lipoxygenase: structure and function. *Annu. Rev. Plant Physiol. Plant Mol. Biol.* 42, 145-188. doi: 10.1146/annurev.pp.42.060191.001045

Singh, A., Reimer, S., Pozniak, C. J., Clarke, F. R., Clarke, J. M., Knox, R. E., et al. (2009). Allelic variation at Psy1-A1 and association with yellow pigment in durum wheat grain. *Theor. Appl. Genet.* 118, 1539-1548. doi: 10.1007/s00122-009-1001-x

Suwarno, W. B., Pixley, K. V., Palacios-Rojas, N., Kaeppler, S. M., andBabu, R. (2015). Genome-wide association analysis reveals new targets for carotenoid biofortification in maize. *Theor. Appl. Genet.* 128, 851-864. doi: 10.1007/s00122-015-2475-3

Toledo-Ortiz, G., Huq, E., and Rodriguez-Concepcion, M. (2010). Directregulation of phytoene synthase gene expression and carotenoid biosynthesis by phytochrome-interacting factors. *Proc. Natl. Acad. Sci. U. S. A.* 107, 11626-11631. doi: 10.1073/pnas.0914428107

Vallabhaneni, R., and Wurtzel, E. T. (2009). Timing and biosynthetic potential for carotenoid accumulationin genetically diverse germplasm of maize. *Plant Physiol.* 150, 562-572. doi: 10.1104/pp.109.137042

VanHung, P., and Hatcher, D. W. (2011). Ultra-performance liquid chromatography (UPLC) quantification of carotenoids in durum wheat: Influence of genotype and environment in relation to the colour of yellow alkaline noodles (YAN). *Food Chem.* 125, 1510-1516. doi: 10.1016/j.foodchem.2010.10.078

Varshney, R. K., Terauchi, R., and McCouch, S. R. (2014). Harvesting the promising fruits of genomics: applying genome sequencing technologies to crop breeding. PLoS Biol. 12: e1001883. doi: 10.1371/journal.pbio.1001883

Verlotta, A., De Simone, V., Mastrangelo, A. M., Cattivelli, L., Papa, R., and Trono, D. (2010). Insight into durum wheat Lpx-B1: a small gene family coding for the lipoxygenase responsible for carotenoid bleaching in mature grains. BMC Plant Biol. 10: 263. doi: 10.1186/1471-2229-10-263

Vishnevetsky, M., Ovadis, M., and Vainstein, A. (1999). Carotenoid sequestration in plants: the role of carotenoid-associated proteins. *Trends Plant Sci.* 4, 232-235. doi: 10.1016/S1360-1385(99)01414-4

Walter, M. H., Floss, D. S., and Strack, D. (2010). Apocarotenoids: hormones, mycorrhizal metabolites and aroma volatiles. *Planta* 232, 1-17. doi: 10.1007/s00425-010-1156-3

Wang, C., Zeng, J., Li, Y., Hu, W., Chen, L., Miao, Y. J., et al. (2014). Enrichment of provitamin A content in wheat (*Triticum aestivum* L.) by introduction of the bacterial carotenoid biosynthetic genes CrtB and CrtI. *J. Exp. Bot.* 65, 2545-2556. doi: 10.1093/jxb/eru138

Wei, S., Yu, B. Y., Gruber, M. Y., Khachatourians, G. G., Hegedus, D. D., and Hannoufa, A. (2010). Enhanced seed carotenoid levels and branching in transgenic *Brassica napus* expressing the *Arabidopsis* miR156b gene. *J. Agric. Food Chem.* 58, 9572-9578. doi: 10.1021/jf102635f

Welsch, R., Maass, D., Voegel, T., DellaPenna, D., and Beyer, P. (2007). Transcription factor RAP2.2 and its interacting partner SINAT2: stable elements in the carotenogenesis of *Arabidopsis* leaves. *Plant Physiol.* 145, 1073-1085. doi: 10.1104/pp.107.104828

Welsch, R., Wust, F., Bar, C., Al-Babili, S., and Beyer, P. (2008). A third phytoene synthase is devoted to abiotic stress-induced abscisic acid formation in rice and defines functional diversification of phytoene synthase genes. *Plant Physiol.* 147, 367-380. doi: 10.1104/pp.108.117028

Yan, J. B., Kandianis, C. B., Harjes, C. E., Bai, L., Kim, E. H., Yang, X., et al. (2010). Rare genetic variation at *Zea mays* crtRB1 increases β-carotene in maize grain. *Nat. Genet.* 42, 322-327. doi: 10.1038/ng.551

Ye, X., Al Babili, S., Kloti, A., Zhang, J., Lucca, P., Beyer, P., etal. (2000). Engineering the provitamin A (β-carotene) biosynthetic pathway into (carotenoid-free) rice endosperm. *Science* 287, 303-305. doi: 10.1126/science.287.5451.303

Yu, Q., Ghisla, S., Hirschberg, J., Mann, V., and Beyer, P. (2011). Plant carotene cis-trans isomerase CrtISO: a new member of the fad (red) -dependent flavoproteins catalyzing non-redox reactions. *J. Biol. Chem.* 286, 8666-8676. doi: 10.1074/jbc.M110.208017

Zeng, J., Wang, C., Chen, X., Zang, M. L., Yuan, C. H., Wang, X. T., et al. (2015a). The lycopene β-cyclase plays a significant role in provitamin A biosynthesis in wheat endosperm. *BMC Plant Biol.* 15: 112. doi: 10.1186/s12870-015-0514-5

Zeng, J., Wang, X. T., Miao, Y. J., Wang, C., Zang, M. L., Chen, X., et al. (2015b). Metabolic engineering of wheat provitamin A by simultaneously overexpressing CrtB and silencing carotenoid hydroxylase (TaHYD). *J. Agric. Food Chem.* 63, 9083-9092. doi: 10.1021/acs.jafc.5b04279

Zhai, S. N., He, Z. H., Wen, W. E., Jin, H., Liu, J. D., Zhang, Y., et al. (2016). Genome-wide linkage mapping of flour color-related traits and polyphenol oxidase activity in common wheat. *Theor. Appl. Genet.* 129, 377-394. doi: 10.1007/s00122-015-2634-6

Zhang, C. Y., Dong, C. H., He, X. Y., Zhang, L. P., Xia, X. C., and He, Z. H. (2011). Allelic variants at the TaZds-D1 locus on wheat chromosome 2DL and their association with yellow pigment content. *Crop Sci.* 51, 1580-1590. doi: 10.2135/cropsci2010.12.0689

Zhang, W., Chao, S., Manthey, F., Chicaiza, O., Brevis, J. C., Echenique, V., et al. (2008). QTL analysis of pasta quality using a composite microsatellite and SNP map of durum wheat. *Theor. Appl. Genet.* 117, 1361-1377. doi: 10.1007/s00122-008-0869-1

Zhang, W., and Dubcovsky, J. (2008). Association between allelic variation at the phytoene synthase 1 gene and yellow pigment content in the wheat grain. *Theor. Appl. Genet.* 116, 635-645. doi: 10.1007/s00122-007-0697-8

Zhang, Y. L., Wu, Y. P., Xiao, Y. G., He, Z. H., Zhang, Y., Yan, J., et al. (2009). QTL mapping for flour and noodle colour components and yellow pigment content in common wheat. *Euphytica* 165, 435-444. doi: 10.1007/s10681-008-9744-z

Zhou, X., Welsch, R., Yang, Y., Alvarez, D., Riediger, M., Yuan, H., et al. (2015). *Arabidopsis* OR proteins are the major posttranscriptional regulators of phytoene synthase in controlling carotenoid biosynthesis. *Proc. Natl. Acad. Sci. U. S. A.* 112, 3558-3563. doi: 10.1073/pnas.1420831112

Zhu, C. F., Naqvi, S., Breitenbach, J., Sandmann, G., Christou, P., and Capell, T. (2008). Combinatorial genetic transformation generates a library of metabolic phenotypes for the carotenoid pathway in maize. *Proc. Natl. Acad. Sci. U. S. A.* 105, 18232-18237. doi: 10.1073/pnas.0809737105

Effects of water deficit on breadmaking quality and storage protein compositions in bread wheat（*Triticum aestivum* L.）

Jiaxing Zhou,[a†] Dongmiao Liu,[a†] Xiong Deng,[a] Shoumin Zhen,[a]
Zhimin Wang[b] and Yueming Yan[a*]

* *Correspondence to*：*Y Yan*，*Laboratory of Molecular Genetics and Proteomics*，*College of Life Science*，*Capital Normal University*，*Beijing*，*China*.

E-mail：*yanym@cnu.edu.cn*

† *These authors contributed equally to this work*

[a] *College of Life Science*，*Capital Normal University*，*Beijing*，*China*

[b] *College of Agricultural and Biotechnology*，*China Agricultural University*，*Beijing*，*China*

Abstract：

BACKGROUND：Water deficiency affects grain proteome dynamics and storage protein compositions，resulting in changes in gluten viscoelasticity. In this study，the effects of field water deficit on wheat breadmaking quality and grain storage proteins were investigated.

RESULTS：Water deficiency produced a shorter grain-filling period，a decrease in grain number，grain weight and grain yield，a reduced starch granule size and increased protein content and glutenin macropolymer contents，resulting in superior dough properties and breadmaking quality. Reverse phase ultra-performance liquid chromatography analysis showed that the total gliadin and glutenin content and the accumulation of individual components were significantly increased by water deficiency. Two-dimensional gel electrophoresis detected 144 individual storage protein spots with significant accumulation changes in developing grains under water deficit. Comparative proteomic analysis revealed that water deficiency resulted in significant upregulation of 12 gliadins，12 high-molecular-weight glutenin subunits and 46 low-molecular-weight glutenin subunits. Quantitative real-time polymerase chain reaction analysis revealed that the expression of storage protein biosynthesis-related transcription factors *Dof* and *Spa* was upregulated by water deficiency.

CONCLUSION：The present results illustrated that water deficiency leads to increased accumulation of storage protein components and upregulated expression of *Dof* and *Spa*，resulting in an improvement in glutenin strength and breadmaking quality.

Keywords：bread wheat；water deficit；gliadins；glutenins，proteome

INTRODUCTION

The frequency and extent of extreme drought events are increasing worldwide along with the deteriorating climate. Plant productivity is severely limited by these environmental stresses. Wheat（*Triticum aestivum* L.）is a major grain crop widely planted around the world. Its seeds contain valuable nutrition for humans；therefore it serves as an important food source for the

Published in J. Sci Food Agric.，2018，98：4357-4368.

world's population. Water deficit has a significant impact on wheat production and results in the acceleration of grain filling and a decrease in yield.[1,2] Water deficit also affects the expression and accumulation of the grain proteome and causes significant changes in the com-position of storage proteins and gluten quality.[3,4]

Gliadins and glutenins are major storage proteins deposited in the wheat grain endosperm and are the main determinants of the viscoelastic properties of dough breadmaking quality.[5] The monomeric gliadins encoded by the *Gli-2* locus on the short arm of chromosomes 1 and 6, classified as /-, - and -gliadins based on their electrophoretic mobilities, are mainly linked to dough extensibility and ductility. The polymeric glutenins, which consist of high- and low-molecular-weight glutenin subunits (HMW-GS and LMW-GS respectively) encoded by the *Glu-1* and *Glu-3* loci on the long and short arms of chromosomes 1 and 6 respectively, provide dough viscoelasticity.[6-8] Although HMW-GS only account for 12% of total endosperm protein, they determine as much as two-thirds of breadmaking quality.[9-13] Particularly, HMW-GS interact with LMW-GS and gliadins via disulfide bonds and form glutenin macropolymers (GMPs), which are insoluble in sodium dodecyl sulfate (SDS) and closely linked to breadmaking quality.[14,15]

Wheat storage proteins are gradually accumulated during grain development, and a number of genes are involved in this process.[16,17] Adverse environmental factors have important effects on the composition of gliadins and glutenins and their relative proportions, which are closely related to flour quality.[18,19] Drought and heat stresses affect wheat quality by altering the concentrations of grain proteins and of the key gluten components HMW-GS and GMPs.[20] In the presence of fertilizer, high temperature or high temperature and fertilizer, the majority of HMW-GS, -gliadins and some-gliadins were increased in flour while the levels of several LMW-GS and a minor-gliadin were decreased.[21] The concentrations of protein and starch in seeds are important traits in the qualities of different wheat genotypes under drought and waterlogged conditions.[22] The changes in activity of the key regulatory enzymes important in starch and protein accumulation are different in response to different levels of water in the soil. Under drought stress, the activities of glutamine synthetase and glutamate pyruvic aminotransferase in wheat grains were slightly decreased whereas those of sucrose synthase, granule-bound starch synthase and soluble starch synthase were significantly decreased.[23]

In recent years, considerable work has been performed to investigate the drought response and defense mechanisms during wheat grain development at the proteome level.[19,20,24-29] These studies mainly focused on proteome changes in grain albumins and globulins and aimed to identify drought-tolerant proteins. More recently, our group showed that a few of the storage protein components in the mature grain endosperm were upregulated under water deficit, which could improve breadmaking quality.[4] However, comprehensive studies on the dynamic changes in individual gliadins and glutenins under drought stress and their effects on dough strength and breadmaking quality are still limited.

The compositions of wheat seed storage proteins are complicated and their synthesis and accumulation are controlled by a complex regulatory network.[30,31] Some important transcription factors and chaperonins are involved in the regulation of storage protein synthesis, including the transcription factors DNA binding with one finger (*Dof*) and storage protein activator (*Spa*). *Dof* is prominent in plants and belongs to a subfamily of zinc finger proteins. Generally, it consists of 200-400 amino acids with a con-served N-terminal domain and a variable C-terminal domain that can be regulated by different signals.[32] *Dof* plays a critical role in the expression of seed storage genes, ultimately regulating protein content in gramineous plants.[33-35] *Spa* belongs to the family of Opaque2 transcription factors and is involved in the expression of endosperm proteins. There is a conservative sequence called the endosperm box (EB) that is located in the transcription start site of endosperm proteins, which is related to the expression of the corresponding gene. The EB contains two *cis*-acting elements: the endosperm motif and the GCN4 motif.[36] *Spa* inter-acts with the GCN4 motif to activate gene expression. Thus *Spa* has important effects on the viscoelastic properties of dough and flour quality.[37]

In this study, we used a proteomic approach to reveal the effects of field water deficit on wheat breadmaking quality and the synthesis and accumulation of individual gliadins and glutenins. Changes in storage proteins in response to water deficit were identified and their contributions to gluten quality were estimated. The dynamic expression patterns of the *Dof* and *Spa* genes in developing grains under water deficit and their effects on storage protein synthesis were also investigated. Our results provide new insights into the molecular mechanisms of the effects of drought stress on storage protein synthesis and breadmaking quality.

MATERIALS AND METHODS

Plant materials, field trials and water deficit treatment

The Chinese elite bread wheat cultivar Zhongmai 175 used as the experimental material has high water and N fertilizer use efficiency, superior drought resistance and broad adaptation and high yield potential[38] and has been widely cultivated in the main winter wheat production areas of northern China in the past ten years. Field trials were performed at the experimental station of China Agricultural University Research Center, Wuqiao, Hebei Province, China (116°37′23″E, 37°6′ 02″N) during two consecutive growing seasons (2014—2015 and 2015—2016). The experimental design included control and water deficit treatment groups in three biological replicates, and each plot was 50 m^2. Cultivation management was the same as for local field production. The average annual sunshine was 2690 h, the average annual temperature was 12.6 ℃, the water table ranged from 6 to 9 m and the precipitation during wheat growing seasons 2014—2015 and 2015—2016 was 128 and 153.9 mm respectively at the experimental location. The experimental field met the minimum precipitation standard for normal wheat growth in northern China.

At the jointing and flowering stages, the control group was well watered as per local field cultivation with 750 $m^3 \cdot hm^{-2}$ of irrigation water twice, while the water deficit treatment group had no irrigation. Developing grains at 15, 20, 25 and 30 days post-anthesis (DPA) and mature grains (45 DPA) from the control and treatment groups were collected and stored at −80℃ prior to analysis.

Soil water content was measured in soil layers at depths of 0-200 cm (every 20 cm as one layer) in three replicates. Collected soil samples were immediately placed in aluminum boxes and dried to constant weight in an oven at 110 ℃. Soil water content (%) was calculated as $[(g_1-g_2)/(g_2-g_0)]\times 100$, where g_0 is the weight of the aluminum box, g_1 is the weight of the original soil sample plus the aluminum box and g_2 is the weight of the dried soil sample plus the aluminum box.

Agronomic traits, grain yield and quality testing

The main agronomic traits and quality properties from three bio-logical replicates were measured according to recent reports.[4,39] Agronomic traits included plant height, ear length, tiller number, unfruitful spikelets, thousand-grain weight and grain yield ($kg \cdot hm^{-2}$). Grain quality parameters included ash content, protein content, water absorption, rapid visco analyzer (RVA) parameters, loaf volume and appearance score and C-cell parameters related to bread inner structures. Student's t test was used for data evaluation.

GMP content determined by size exclusion high-performance liquid chromatography (SE-HPLC)

GMPs were separated and quantitated by SE-HPLC

based on Dong *et al.*[40] with minor modifications. The grains were ground with liquid nitrogen and oscillated with extraction buffer (0.05 mol · L^{-1} phosphate-buffered saline (PBS) with 0.07 mol · L^{-1} SDS) for 20 min. After centrifugation at 12 000×*g* for 15 min, the sediment was retained and mixed with extraction buffer (1 800 L) again. After ultrasonic disruption, the sample was kept on oscillating for 1 h and centrifuged at 12 000×*g* for 15 min. The supernatant was filtered into another centrifugal tube for HPLC analysis. An Agilent (California, USA) Bio SEC-5 column (5 m) was used with 0.05 mol · L^{-1} PBS and 3 mmol · L^{-1} SDS as mobile phase. The flow rate and injection volume were set as 0.5 mL · min^{-1} and 20 L respectively and the detection time was 15 min. GMP content was determined as the area under the peak.

Scanning electron microscopy (SEM) observation

The ultrastructure of mature grains from both treatment andcontrol groups was observed by SEM (S-4800 FE-SEM, Hitachi, Tokyo, Japan) according to Wang *et al.*[8]

Gliadin and glutenin accumulation determined by reverse phase ultra-performance liquid chromatography (RP-UPLC)

Gliadins and glutenins were extracted from three biological replicates and separated by RP-UPLC according to recent reports of Yu *et al.*,[31] Han *et al.*[41] and Yan *et al.*[42] with minor modifications. Wheat grains were crushed into powder and a 30 mg flour sample was used for glutenin extraction. First 0.5 mL of 0.04 mol · L^{-1} NaCl solution was added to remove albumin and globulin. After vortexing, the sample was centrifuged at 17 900×*g* for 10 min and the supernatant was discarded. Gliadins were extracted with 750 mL · L^{-1} ethanol by vortexing for 30 min. After centrifugation at 17 900 × *g* for 15 min, the supernatant was transferred to a new tube for RP-UPLC. Glutenins were extracted from the residue with 100 L of extraction buffer (550 mL · L^{-1} isopropanol, 0.08 mol · L^{-1} Tris-HCl, pH 8.0) with freshly added 0.065 mol · L^{-1} dithiothreitol (DTT) by stirring at 65 ℃ for 30 min and then alkylated by stirring with an equal volume of extraction buffer but replacing 0.065 mol · L^{-1} DTT with freshly added 14 mL · L^{-1} 4-vinylpyridine under the same incubation conditions. After centrifugation at 17 900 × *g* for 15 min, the supernatant was transferred to a new tube for sodium dodecyl sulfate polyacrylamide gel electrophoresis (SDS-PAGE) analysis. Glutenins were precipitated from glutenin extracts by adding acetone to a final concentration of 400 mL · L^{-1}. Precipitated HMW-GS were dissolved in 0.2 mL of 210 mL · L^{-1} acetonitrile (ACN) and 1 g · L^{-1} trifluoroacetic acid (TFA) and centrifuged at 17 900×*g* for 10 min, then used for RP-UPLC analysis within 24 h of extraction. RP-UPLC was performed using a Waters (Massachusetts, USA) Acquity UPLC system and a Waters 300SB C18 column (1.7 m). The column temperature was set at 45℃. The mobile phases were (A) doubly distilled water (ddH_2O) with 5 mmol · L^{-1} TFA and (B) ACN with 5 mmol · L^{-1} TFA. Gliadins (5 L) were injected and eluted at 0.35 mL · min^{-1} flow rate for a total elution and washing time of 25 min, while glutenins (10 L) were injected and eluted with a linear gradient of 21-47% B at 0.5 mL · min^{-1} flow rate for 20 min and then washed for 10 min. Protein peaks were detected by UV with absorbance areas at 210 nm.

Storage protein extraction, two-dimensional electrophoresis (2-DE) and image analysis

Grain gliadins and glutenins from three biological replicates were extracted according to Dupont *et al.*[43]. A 3 g seed sample was ground with liquid nitrogen, then 50 mL of 750 mL · L^{-1} ethanol was added to dissolve gliadins. After oscillation for 30 min, the sample was centrifuged at 17 900×*g* for 15 min. The supernatant was added with 200 mL of water or salt solution to achieve gliadin extraction. The sediment was used for glutenin extraction, and 50 mL of 550 mL · L^{-1} isopropyl alcohol was added to remove residual gliadin. After vortexing, the sample was heated at 65℃ for 30 min and centrifuged at 17 900×*g* for 10 min and the supernatant was discarded. This step was repeated twice. Subsequently, 10 mL of solution B (500 mL ·

L^{-1} isopropyl alcohol, 0.08 mol • L^{-1} Tris-HCl, pH 8.0) with 0.065 mol • L^{-1} DTT was added to the precipitate.

After vortexing, the sample was heated at 65 ℃ for 30 min. Then 10 mL of solution B with 14 mL • L^{-1} 4-vinylpyridine was added and the sample was heated at 65 ℃ for 30 min and centrifuged at 17 900 × *g* for 15 min. The supernatant was collected as the glutenin crude extract, and 25 mL of cold acetone was added to it.

After vortexing, the sample was stored at −20 ℃ overnight and then centrifuged at 17 900 × *g* for 5 min. The supernatant was discarded and the glutenin precipitate was dried at room temperature. The protein concentration was determined with a 2-D Quant Kit (GE Healthcare, Atlanta, Georgia, USA).

DE was performed according to Zhou *et al.*[44] with minor modifications. A total of 80 g of gliadins and 500 g of glutenins were used for 2-DE separation. pH 6-11 IPG strips (18 cm) and an IPGphor system (GE Healthcare) were used for isoelectric focusing. Protein samples were separated by 0.4 mol • L^{-1} SDS-PAGE, then visualized by colloidal Coomassie brilliant blue (CBB) staining (R-250/G-250 4:1) and destained with a solution of 3.4 mol • L^{-1} ethanol and 3.4 mol • L^{-1} acetic acid. The gel images were scanned by a GS-800™ Calibrated Imaging Densitometer (Bio-Rad, California, USA) and analyzed using ImageMaster™ 2D Platinum Version 7.0 (Amersham Biosciences, Atlanta, Georgia, USA). The abundance of protein spots with significant and reproducible changes matched in two consecutive growing seasons and $P<0.05$ by Student's t test were identified as differentially accumulated protein (DAP) spots and selected as 2-fold difference DAP for tandem mass spectrometry analysis.

Protein identification using matrix-assisted laser desorption/ionization time-of-flight tandem massspectrometry (MALDI-TOF/TOF-MS)

The recognized DAP spots were cut from the 2-DE gels and decolored by bleaching solution (0.0125 mol • L^{-1} NH_4HCO_3, 500 mL • L^{-1} ACN). After discarding the decoloring liquid, 100 L of ACN was added and then dried by vacuum dryer. Dried samples were digested with 7 L of diluted solvent (trypsin solution diluted with 25 mmol • L^{-1} NH4HCO3 to a final concentration of 15 ng • L^{-1}) and incubated at 37℃ for 16 h. The peptides were extracted with 0.44 mol • L^{-1} TFA, 500 mL • L^{-1} ACN and water at 37℃ for 1 h and dried by vacuum dryer. The dried peptide mixtures were completely dissolved in 2 L of a solution containing 0.008 mol • L^{-1} TFA mixed with 1 L of TFA, 500 L of ACN and 499 L of ddH_2O. Tryptic peptides were analyzed by MALDI-TOF/TOF-MS (4800 Proteomics Analyzer, Applied Biosystems, Framingham, MA, USA) and the MS/MS spectra were searched in the NCBI non-redundant green plant database. The peptide mass tolerance was set to 100 ppm and the fragment tolerance was set to ± 0.3 Da. Carbamidomethylation was set as fixed modification; oxidation of methionines was allowed as variable modification. All searches were evaluated based on the significant scores obtained from MASCOT. The Protein Score C. I. % and Total Ion Score C. I. % were both set above 95% and the significance threshold at $P < 0.05$ for the MS/MS based on Bian *et al.*[45]

mRNA extraction, cDNA synthesis and quantitative real-time polymerase chain reaction (qRT-PCR)

Total RNA was isolated from the seeds of Zhongmai 175 at different developmental stages using TRIzol Reagent (Invitrogen, California, USA). Reverse transcription reactions were carried out using a PrimeScript® RT Reagent Kit with gDNA Eraser (TaKaRa, Shiga, Japan) according to the manufacturer's instructions. Specific primers for *Dof* and *Dof* family genes were designed by the Primer3Plus online, and glyceraldehyde-3-phosphate dehydrogenase (GAPDH) was used as reference gene. The qRT-PCR system (20 μL volume) contained 10 L of 2×SYBR® Premix Ex Taq™ (TaKaRa), 2 μL of 50-fold diluted cDNA, 0.15 μL of each gene-specific primer and 7.7 μL of ddH_2O. The qRT-PCR procedure was as follows: 3

min at 94℃, 40 cycles of 15 s at 94℃, 15 s at different temperatures of annealing, 72 ℃ for 10 s, a melt curve from 65 to 94℃. The two-standard-curve relative quantitation method was established by five concentration gradients, and three biological replicates were used for each sample. PCRs were conducted in a CFX96 Real-time PCR Detection System (Bio-Rad). All data were analyzed with CFX Manager (Bio-Rad).

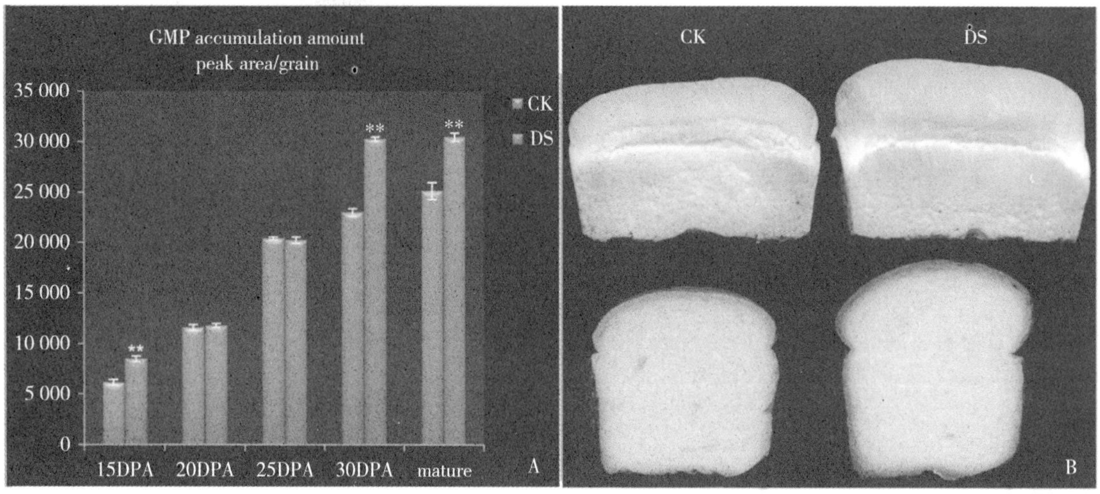

Fig. 1 (A) Dynamic changes in GMP content during grain development and (B) bread performance under water deficit in Chinese elite bread wheat cultivar Zhongmai 175. CK, control with normal irrigation; DS, water deficit treatment. The results presented are from the 2015-2016 growing season, which are well consistent with those from the 2014-2015 growing season.

RESULTS

Effects of water deficit on wheat agronomic traits and gluten quality

Changes in soil moisture content in the different soil layers during grain development were measured (Fig. S1). The results showed significant differences in moisture content between the control and treatment groups in soil layers at 20-200 cm in the two consecutive growing seasons (2014—2015 and 2015—2016). Severe drought occurred in soil layers at 0-60 cm and mild drought occurred in soil layers at 60-120 cm in the treatment group according to the grade of agricultural drought (GB/T 32136—2015).[46] In the treatment group, plant growth and earing were accelerated and grain filling and time to maturation were shortened (Fig. S2). Comparison of the main agronomic traits between the control and treatment groups is shown in Table S1. The results from both growing seasons showed that water deficit resulted in a significant increase in the number of unfruitful spikelets and a significant decrease in plant height, grain weight and grain yield. Particularly, the yield was decreased by 14.97 and 13.97% in the 2014—2015 and 2015—2016 growing seasons respectively.

Quality analysis showed that water deficit treatment caused significant changes in gluten quality parameters, including GMP content (Fig. 1A), basic quality parameters, RVA parameters, C-cell parameters, loaf volume and appearance score (Table 1, Fig. 1B). In the water deficit condition, GMP and protein contents, thickness, minimum viscosity, final viscosity, setback and peak viscosity were significantly increased. These led to significant improvements in loaf appearance and inner structure, including significantly increased slice area and circumference, slice brightness, loaf volume and score. The results from the two consecutive growing seasons were consistent with one another (Table 1).

Table 1 Quality parameters of dough and bread slices of Zhongmai 175 under drought stress (DS) and well-watered (CK) conditions in 2014—2015 and 2015—2016 growing seasons

Growing years	Treated group	Basic quality parameters						RVA parameters			
		Protein content	Water absorption (14%)	Ash content	Forming time	Stability time	Thickness	Minimum viscosity	Final viscosity	Setback	Peak viscosity
2014—2015	CK	7.93±0.01	52.35±0.15	0.50±0.02	1.95±0.05	2.55±0.05	488.50±6.50	1 861.50±32.5	3 247.50±0.50	1 386.00±32.00	2 674.50±3.50
	DS	8.34±0.09**	53.09±0.10**	0.42±0.01**	1.85±0.05	2.55±0.25	512.50±5.50**	2 153.50±4.50**	3 697.00±4.00**	1 543.00±0.50**	2 985.50±2.50**
2015—2016	CK	7.51±0.16	51.33±0.06	0.28±0.001	1.37±0.06	3.83±0.06	471±6.56	1 949.67±13.50	3 316.67±14.57	1 367±28.05	2 839±19.52
	DS	9.54±0.07**	53.27±0.06**	0.32±0.03	1.73±0.21 *	4.6±0.1**	500.33±2.08**	2 207±29.51**	3 728.67±18.61**	1 521.67±11.06**	3 071.67±11.50**

Growing years	Treated group	C-cell parameters							Loaf volume and appearance score		
		Slice area/px	Circumference/px	Slice brightness	Cell contrast	Cell extension	Cell diameter/px	Cell quantity	Wall thickness/px	Loaf volume (mL^3)	Score (100)
2014—2015	CK	240 941.50±688.50	1 824.00±12.00	126.60±0.60	0.65±0.003	1.51±0.01	20.06±0.03	2 099±32	3.65±0.04	653±4.58	42±3.00
	DS	243 243.50±268.50**	1 893.00±3.00 *	133.60±0.40**	0.68±0.000 5**	1.52±0.02	20.05±0.29	2111±5	3.62±0.01	714.33±5.86**	51.67±4.51 *
2015—2016	CK	237 895.67±444.00	1 802.33±1.53	140.33±0.76	0.73±0.001	1.74±0.006	16.60±0.14	2 495±31	3.35±0.03	707.33±7.51	56.67±2.08
	DS	280 754.33±2 557.50**	1 991.33±10.50**	147.33±0.78**	0.73±0.001	1.73±0.02	16.77±0.46	2 878±90**	3.29±0.04	822±12.53**	65.67±2.08**

Significance: ** $P<0.01$; * $P<0.05$. px, pixel.

Grain ultrastructural changes under water deficit

The dynamic changes in grain development and endosperm ultra-structure under well-watered and water deficit treatments were observed by SEM (Fig. 2). Water deficit accelerated grain filling and caused a decrease in grain size and weight (Fig. 2A). Both A-type starch granules with an oval shape and diameter greater than 10 m and B-type starch granules with a round shape and 5-10 m diameter could be clearly observed from 15 DPA to grain maturity by SEM (Fig. 2B). Starch granules of both types were significantly smaller under water deficit than under well-watered conditions during different grain developmental stages, which indicates that starch biosynthesis in the developing grains was influenced by water deficit treatment. The two consecutive growing seasons showed similar ultrastructural changes. Thus the decrease in starch biosynthesis is one of the factors responsible for the reduction in grain weight and yield under water deficit conditions.

Dynamic changes in gliadins and glutenins in developing grains under water deficit revealed by RP-UPLC

Our recent studies have demonstrated that using RP-UPLC as a new method for wheat storage protein separation and characterization has clear advantages in terms of separation time, resolution, reproducibility and reagent consumption compared with reverse phase high-performance liquid chromatography (RP-HPLC).[31,41,42] In this study, the dynamic accumulation patterns of gliadins, HMW-GS and LMW-GS from 15 DPA to grain maturity under well-watered and water deficit conditions were detected by RP-UPLC. The consecutive two-year field experiments showed similar results. In general, the storage proteins gradually accumulated from 15 DPA to grain maturity in both control and treatment groups, while a more significant increasing occurred under water deficit treatment during grain filling (Figs 3-5).

The accumulation of gliadins was detected by RP-UPLC and showed that /-, -and-gliadins were eluted at 5, 5-10 and 10-15 min respectively The /-, -and-gliadins gradually accumulated after 15 DPA and reached the highest level at grain maturity. The rapid accumulation stages occurred from 25 DPA to grain maturity (Fig. 3A). When subjected to water deficit treatment, the accumulation of total gliadins as well as different types of gliadins was significantly higher than that in the well-watered group, particularly at 25 and 30 DPA and grain maturity (Figs 3B-3E).

Total glutenins, HMW-GS and LMW-GS demonstrated a similar accumulation pattern as gliadins during grain development, with rapid accumulation from 20 DPA to grain maturity in both control and treatment groups (Fig. 4A). Under water deficit, the accumulation of total glutenins was significantly increased, particularly at 15 and 30 DPA and grain maturity (Fig. 4B). Meanwhile, the content of HMW-GS, LMW-GS and each type of HMW-GS showed a similar accumulation pattern (Fig. 5). In particular, all four HMW-GS (7+9 and 2+12) showed a significant increase at grain maturity under water deficit treatment, resulting in enhanced formation of GMPs and superior breadmaking quality.

Effects of water deficit on gliadin subproteome

2-DE was performed to reveal changes in the subproteome of seed gliadin and glutenin proteins from different grain developmental stages under water deficit (Fig. S3). The 2-DE results for the gliadin subproteome revealed 45 DAP spots during five developmental stages (Fig. 6A), and these proteins could be repeatedly detected in the two consecutive growing seasons. The heat maps of these proteins indicating their changes in accumulation in response to water deficit are shown in Fig. 6B. Detailed information regarding the changes in abundance of all protein spots is given in Table S2.

In total, eight accumulation patterns of storage proteins were detected and designated as patterns I-VIII. The accumulation pat-terns of almost 75% of the gliadin components were changed under water deficit. These protein spots were classified into seven expression patterns, excluding pattern II (Table 2). Pattern I included eight protein spots (17.8%), which exhibited gradually increased accumulation from 15 DPA to grain maturity. Pattern III included 14 (31.1%) protein spots and displayed an up-down accumulation trend with peak accumulation at 15 DPA and grain maturity. On the contrary, pattern IV with three (6%) protein spots showed a down-up trend, decreasing at the early stages and increasing slowly until grain maturity. Pattern V contained nine (20%) protein spots and showed an up-down-up accumulation, whereas pattern VI, the largest group with four (8%) protein spots, displayed a reverse down-up-down accumulation trend. Patterns VII and VIII had five (11%) and two (4%) protein spots respectively and exhibited an up-down-up-down and a down-up-down-up accumulated trend respectively. Among 45 DAPs, 40 (88.9%) showed an increase in accumulation at one or more developmental stages under water deficit. In particular, 12 DAPs (26.7%) were increased in mature grains, which could benefit increasing grain protein and GMP content and have important effects on dough properties.

All 45 DAPs were further subjected to MALDI-TOF/TOF-MS analysis, of which 19 DAPs, including three kinds of gliadin proteins, were successfully identified with a high degree of confidence (Table S3). These protein spots contained 13 /-gliadins and six-gliadins. Compared with the control group, the majority of these gliadin spots changed their accumulation patterns under water deficit. In particular, four /-gliadins (spots 4, 69, 85 and 100) and two-gliadins (spots 98 and 104) displayed upregulated accumulation, whereas nine /-gliadins (spots 5, 7, 10-12, 16, 46, 48 and 53) and four-gliadins (spots 9, 13, 49 and 50) showed downregulation in mature grains under water deficit (Table 2).

Effects of water deficit on glutenin subproteome

Figure 7 shows the 2-DE map of the glutenin subproteome and the changes in accumulation under water deficit. In total, 99 glutenin protein spots were repeatedly identified in the consecutive two-year field experiments, including 23 HMW-GS and 76 LMW-GS (Fig. 7A). HMW-GS in Zhongmai 175 included Null at *Glu-A1*, 1Bx7+1By9 at *Glu-B1* and 1Dx2+1Dy12 at *Glu-D1*, of which 1Bx7, 1By9, 1Dx2 and 1Dy12 consisted of six (64, 65, 78 and 88-90), seven (61-63, 76, 77, 94 and 96), one (66) and nine (54-60, 95 and 109) spots respectively. The subunits of 1Bx7 and 1By9 accounted for 76.92% of all HMW-GS components. Among the 76 LMW-GS spots, one m-type, two s-type and three i-type subunits were further identified by tandem MS (Table S3).

The accumulation abundance and patterns of 99 glutenin spots are shown in Tables 2 and S2 and Fig. 7B. Similarly to gliadins, glutenins exhibited eight accumulation patterns (I-VIII), and the majority were classified as patterns I, III, IV and V. Compared with the control group, 56.52% of the HMW-GS spots and 59.21% of the LMW-GS spots changed their accumulation patterns under water deficit. Among the 99 glutenin spots, 12 HMW-GS and 46 LMW-GS spots displayed upregulation under water deficit in mature grains (Table 2). In particular, water deficit induced the upregulation of some abundant LMW-B subunits such as spots 38, 44, 45, 47, 51, 52, 84, 85, 87 and 92. These glutenin components could contribute to the formation of stronger gluten and superior breadmaking quality (Table 1).

Transcriptional expression analysis of *Dof* and *Spa* genes in response to water deficit by qRT-PCR

The dynamic expression patterns of three *Dof* genes (*Dof-2*, *Dof-3* and *Dof-6*) and three *Spa* genes (*SpaA*, *SpaB* and *SpaD*) during seven stages of grain developmental (13, 15, 17, 20, 25 and 30 DPA and grain maturity) under water deficit were determined by qRT-PCR. The specific primers for the *Dof* and *Spa* genes are listed in Table S4. The optimal experiments showed high amplification specificity and efficiency (Fig. S4). The results showed that all six genes displayed a similar up-down expression trend during grain development and were generally upregulated under water deficit, particularly from 17 to 20 DPA. Again, the results from the two consecutive growing seasons were consistent (Fig. 8). The peak expression of the *Dof* and *Spa* genes at the middle stages of grain development under water deficit coincided with high levels of gliadins and glutenins (Figs 3 and 4), which could promote grain storage protein synthesis and improve gluten quality.

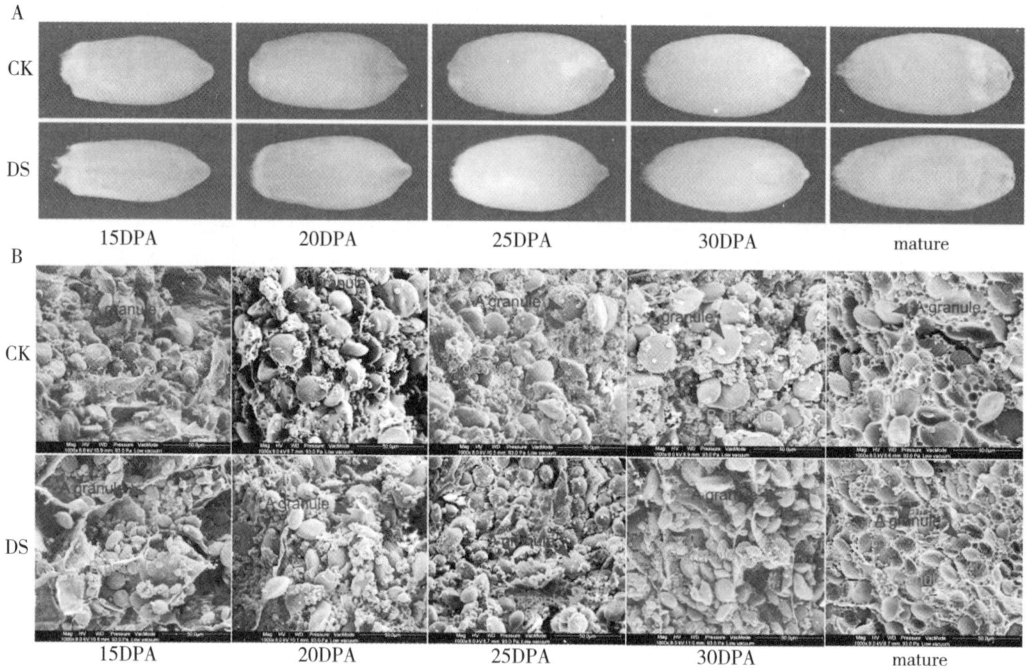

Fig. 2 (A) Grain phenotype changes at different developmental stages and (B) SEM images of developing grains from five periods of endosperm in control (CK) and drought stress (DS) groups. The scale bar is 50 m. The A and B starch granules are marked with blue and red respectively. The results presented are from the 2015—2016 growing season, which are well consistent with those from the 2014—2015 growing season.

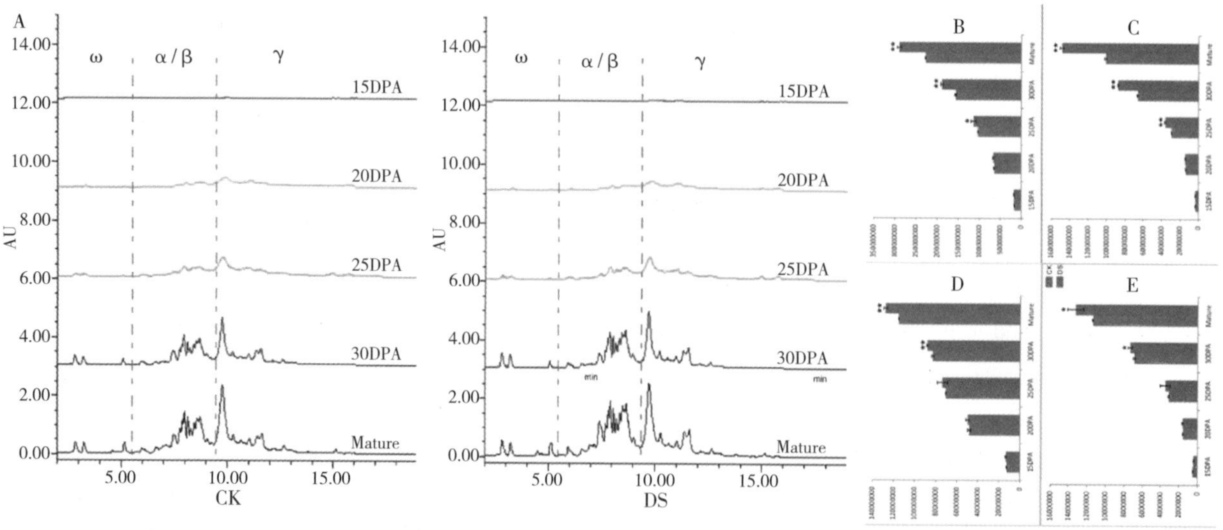

Fig. 3 Gliadin accumulation changes during grain development of Zhongmai 175 under water deficit revealed by RP-UPLC: A, RP-UPLC separation; B, total gliadins; C, /-gliadins; D, -gliadins; E, -gliadins. The RP-UPLC results are from the 2015—2016 growing season, which are well consistent with those from the 2014—2015 growing season. The results of B-E are represented by mean values from the two consecutive growing seasons.

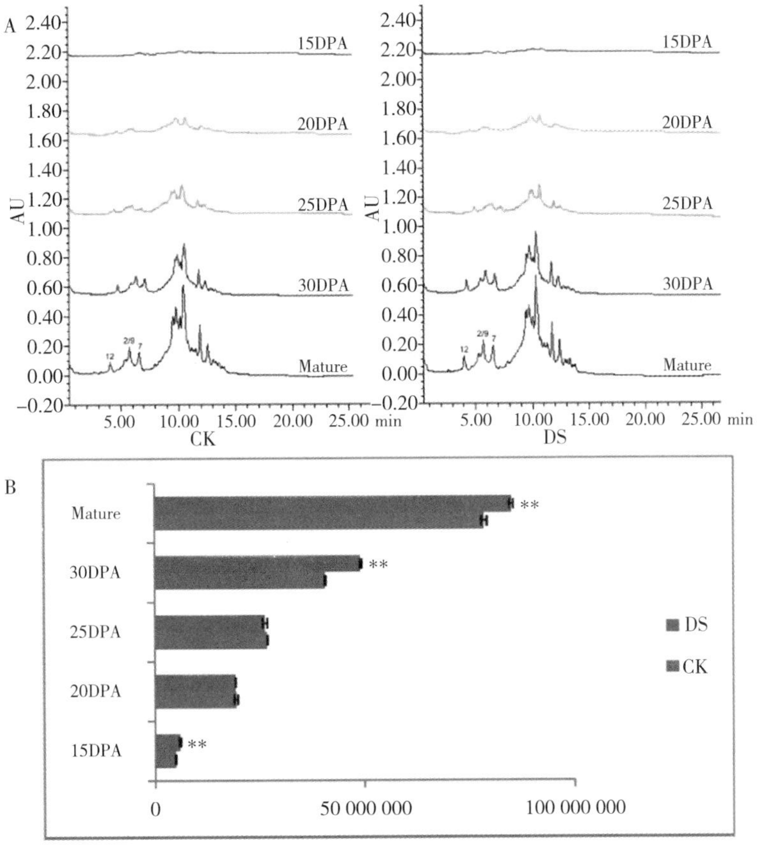

Fig. 4 Glutenin accumulation changes during grain development of Zhongmai 175 under water deficit as revealed by RP-UPLC: A, RP-UPLC separation; B, total glutenins. HMW-GS and LMW-GS are indicated. The RP-UPLC results are from the 2015—2016 growing season, which are well consistent with those from the 2014—2015 growing season. The results of B are represented by mean values from the two consecutive growing seasons.

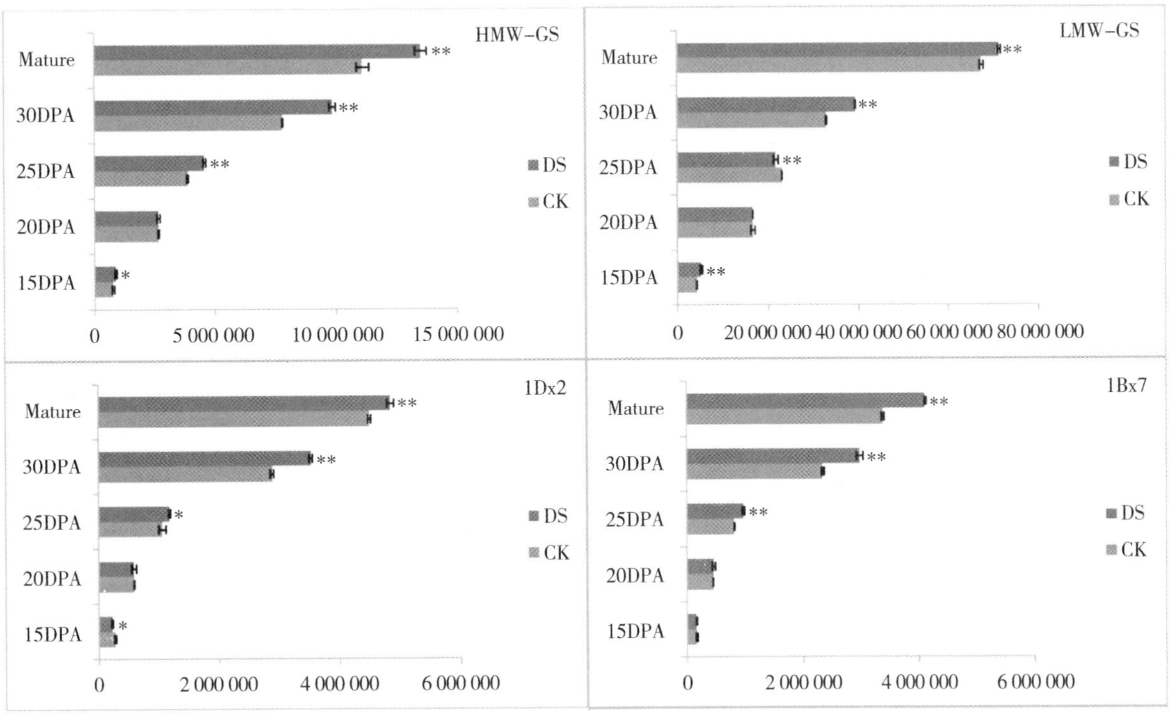

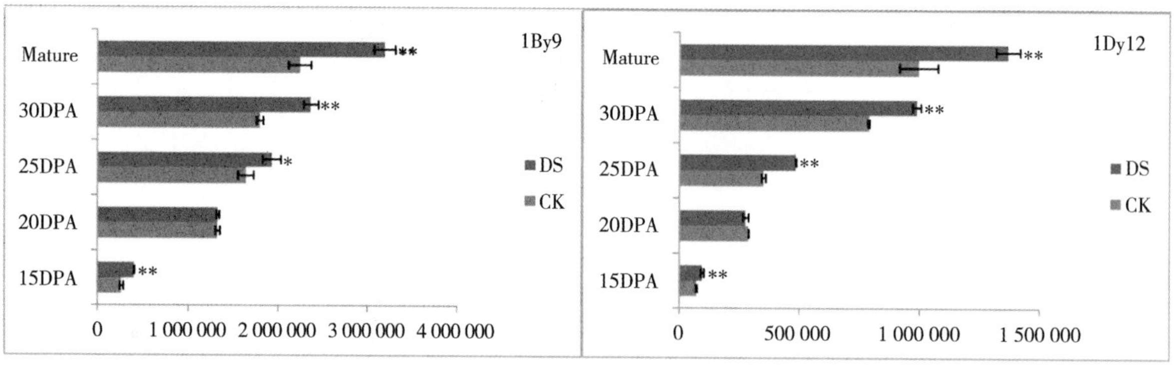

Fig. 5 Accumulation changes of total HMW-GS and LMW-GS and individual HMW-GS (1Dx2，1Bx7，1By9 and 1Dy12) during grain development of Zhongmai 175 under water deficit as revealed by RP-UPLC. The results are represented by mean values from the two consecutive growing seasons.

Table 2 Accumulation patterns of gliadin and glutenin protein spots during grain development in response to water deficit

Pattern	Gliadins		Glutenins			
			HMW-GS		LMW-GS	
	CK	DS	CK	DS	CK	DS
Ⅰ	i23，i49，i66，i72，i85，i94，i98	**i4**，i66，i72，**i84**，**i85**，i94，**i98**，i105	56	56，**77**	20，22，27，38，42，43，49，50，73，74，84，85，91	**16**，**20**，**27**，**38**，42，**43**，**49**，50，73，**84**，**85**，91，**97**，**102**
Ⅱ			89		24	15，24，75
Ⅲ	i16，i22，i35，i100，i101	i11，i22，**i23**，i26，i27，i28，i29，i35，i37，i41，i44，i46，i50，**i69**	54，55，64，88，95，109	54，**62**，64，65，**76**，**78**，**88**，**109**	3，10，13，15，17，18，31，35，41，45，47，75，79，83，87，93，97，98，99，102，103,105，106，107	**3**，**6**，9，18，19，**22**，32，34，**35**，**47**，48，**51**，74，**83**，**98**，**101**，**103**，104
Ⅳ	i4，i7，i12，i19，i27，i28，i41，i84，i104	i9，i48，i106	57，77	**89**，**96**	6，7，25，29，30，33，36，37，39，72，81	7，14，**25**，**29**，36，**37**，39，72，**92**，93，**105**
Ⅴ	i5，i6，i11，i14，i26，i29，i44，i53，i56，i102，i105	i5，**i6**，i10，i12，i14，i18，**i56**，**i100**，**i104**	65	**61**	8，9，16，21，28，32，44，48，92	5，**10**，**12**，**21**，41，**44**，**52**，**53**，79，**99**，106
Ⅵ	i18，i36，i37，i46，i50，i99	**i1**，i7，i16，i53	**59**，61，**66**，76，78，96	**59**，63，**66**，90	19，23，34，71，86，100，101，104，108	8，**13**，**17**，**40**，**71**，**81**，**86**，**107**
Ⅶ	i1，i10	i13,i36，i99，**i101**，i102	62，94	**55**，94，95	5，11，12，40，51，52	**11**，23，28，31，**100**，**108**
Ⅷ	i9，i13，i48，i69，i106	i19，i49	58，60，63，90	57，58，60	14，46，53	30，**33**，**45**，**46**，**87**

The upregulated accumulation protein spots under drought stress are marked in bold type. CK，control group；DS，drought stress group.

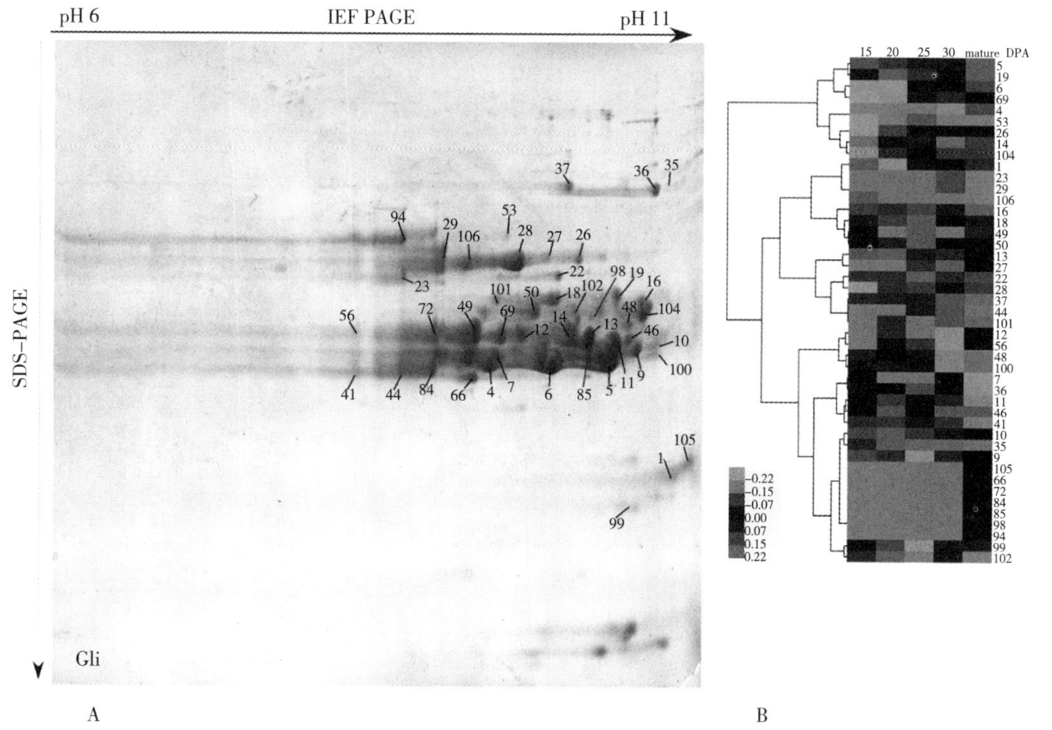

Fig. 6 2-DE map of gliadins and their accumulation patterns during grain development of Zhongmai 175 under water deficit. All gliadin DAPs were reproducibly detected by two consecutive growing seasons. The results of B are represented by mean values from the two consecutive growing seasons.

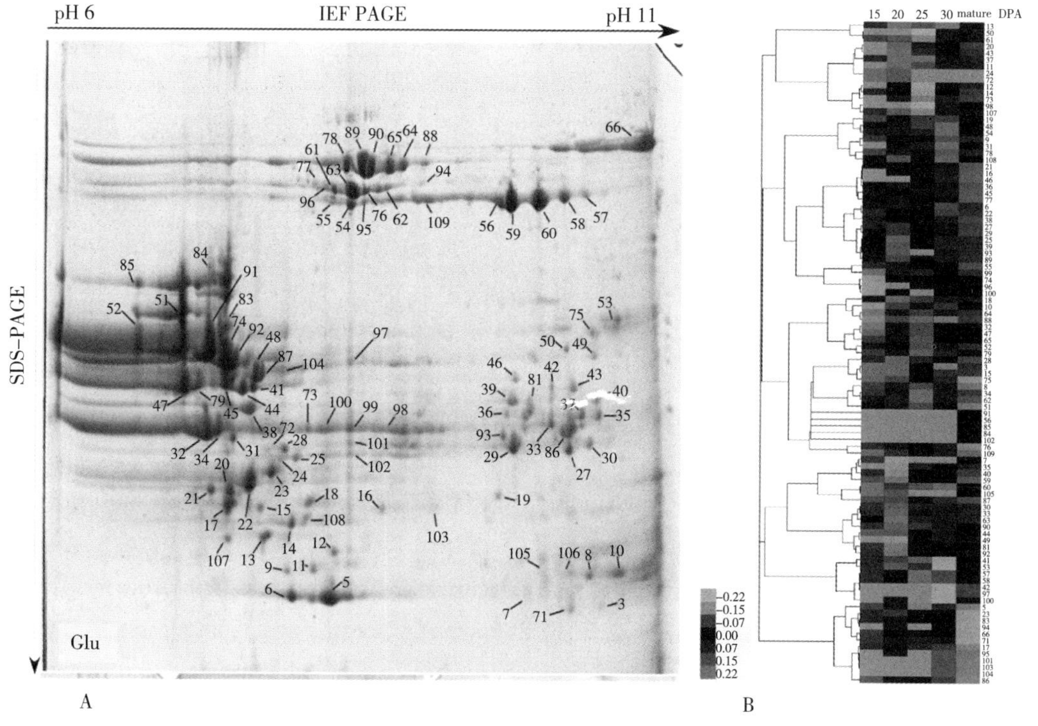

Fig. 7 2-DE map of glutenins and their accumulation patterns during grain development of Zhongmai 175 under water deficit. All glutenin DAPs were reproducibly detected by two consecutive growing seasons. The results of B are represented by mean values from the two consecutive growing seasons.

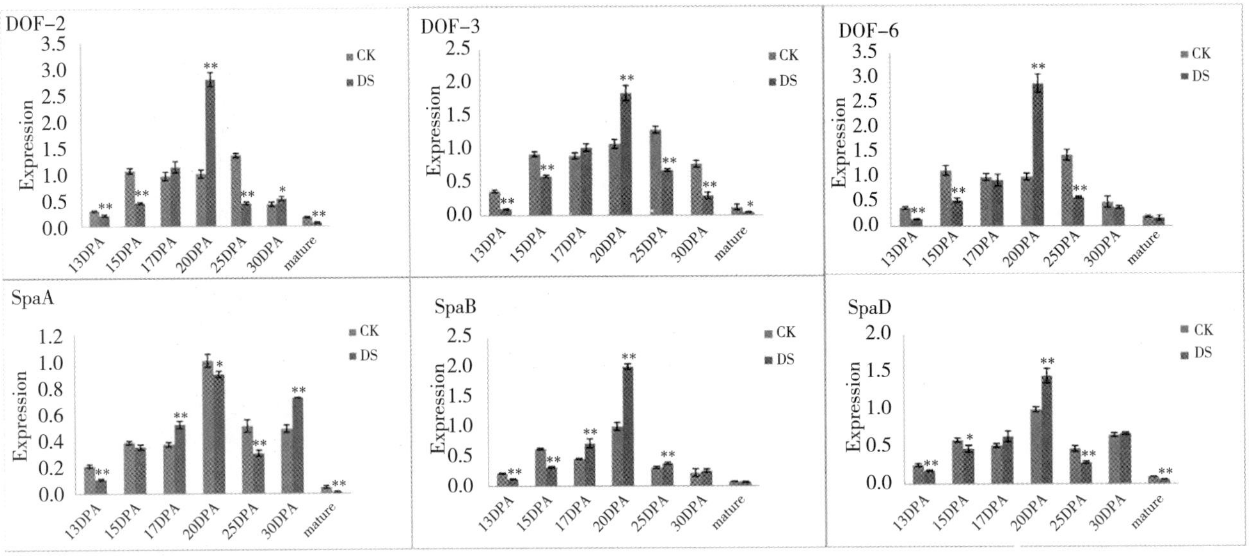

Fig. 8 Dynamic transcriptional expression patterns of *Dof* and *Spa* genes during grain development of Zhongmai 175 in response to water deficit by qRT-PCR. The results are represented by mean values from the two consecutive growing seasons.

DISCUSSION

Influence of water deficit on grain yield and breadmaking quality

It is widely accepted that drought stress causes grain filling shortening, sterile spikelets, increased grain weight and decreased yield[47] but promotes protein content and flour quality.[48] Our recent research from integrated proteome analysis of the wheat embryo and endosperm revealed that water deficit induced the up-regulation of some storage protein components and subsequently improved dough strength and breadmaking quality.[4] The results from ten different durum wheats showed that drought stress significantly reduced thousand-kernel weight and increased protein content.[49] Under drought and salt stresses, the protein content, gluten content and SDS sedimentation coefficient were increased in durum wheat.[50] Furthermore, both the genotype and environment have important influences on wheat quality. The susceptible and limited extent of drought-resistant varieties had a superior quality due to increased protein content, but the yield was sharply decreased. For highly drought-tolerant varieties of wheat, drought stress did not significantly change grain metabolism. The obvious increase in protein content was not observed and the quality did not dramatically change.[51] Our results from the two consecutive growing seasons showed that Zhongmai 175, which exhibits superior drought resistance, showed a significant decrease in growing period, spikelet grain number, thousand-kernel weight and grain yield but a significant increase in protein and GMP content, dough strength and bread-making quality under drought stress (Fig. 1, Tables 1 and S1). These results, generally consistent with previous reports, demonstrated that severe drought stress could improve protein content and breadmaking quality for superior drought-tolerant cultivars.

Wheat seeds are made of water, starch, protein, fat, cellulose and other chemicals, of which starch and protein are the main determinants for grain yield and quality formation. In the mature grains of cultivated cultivars, starch, which includes amylose and amylopectin, accounts for approximately 70% of the total content, while seed proteins consisting of albumins, globulins, gliadins and glutenins account for approximately 14%. Drought stress affects the synthesis and accumulation of grain reserve materials and usually causes seed shriveling and decreased moisture content.[44] The results from this study demonstrated that the sizes of both A-and B-granules were significantly decreased under water deficit (Fig. 2), which suggests that starch biosynthesis was clearly decreased. A previous study indicated that some key enzymes related to starch and protein synthesis had different sensitivities to drought stress. Particularly, the activities of sucrose synthase, granule-bound starch synthase and soluble starch synthase, which are closely associated with starch biosynthesis, were significantly decreased, while those of glutamine synthetase and glutamate pyruvic aminotransferase, which participate in protein synthesis, only showed a slight reduction under drought stress.[22,23] Thus, compared with protein synthesis, damage to starch biosynthesis had a greater effect when subjected to drought stress. This may explain why water deficit causes a decrease in starch biosynthesis and grain yield but an increase in protein content and breadmaking quality.

Influence of water deficit on subproteome of storage proteins

The *Gli-1* and *Glu-3* loci in common wheat are estimated to contain more than 50 gene copies.[52,53] Because the compositions of storage proteins are highly heterologous, their separation and characterization have generally been difficult. Therefore, to date, studies on the effects of drought stress on the subproteomes of gliadin and glutenin are limited. Along with the rapid progress of wheat genomics, the proteomic approach has become a powerful tool for studying the composition of the grain proteome as well as the effects of environmental stresses on their synthesis and flour quality.[4,18,19,28,29,43,54,55]

In general, different storage protein components showed distinct accumulation patterns under adverse environmental conditions. Dupont et al.[54,55] found that some high-molecular-weight glutenins and-gliadins showed upregulated patterns, while one LMW glutenin spot displayed downregulated expression under high temperature. Yang et al.[1] investigated the influences of temperature and drought stress on wheat grain proteins and found that five protein spots were storage proteins and one LMW glutenin was upregulated, whereas one-gliadin, two-gliadins and another LMW glutenin were downregulated. Our recent report found that some storage proteins such as HMW glutenins, globulins and avenin-like proteins showed upregulated expression under water deficit, which might benefit breadmaking quality.[4] In the current study, water deficit resulted in significant proteome changes in grain storage proteins and a significant increase in the content of both gliadin and glutenin (Figs 3 and 4). Particularly, 12 gliadins, 12 HMW-GS and 46 LMW-GS were significantly upregulated under water deficit (Figs 5 and 6). Most of the gliadin spots were identified as-gliadins (Table S3), which suggests that-gliadins were more sensitive to drought stress than other types of gliadins. Previous studies have also found that some-gliadins were significantly increased under high temperature stress.[54-56] Molecular characterization showed that gliadins contain cysteines that participate in GMP formation,[14] and their upregulated expression under water deficit could contribute to superior gluten quality. Meanwhile, some abundant B-subunits and i-and s-type subunits displayed significant upregulation under water-deficit. These protein subunits could be responsible for the significant improvement in grain protein and GMP content and overall breadmaking quality.

Expression of key transcription factors and storage protein accumulation under water deficit

Some transcription factors such as Dof and Spa are involved in the synthesis and regulation of storage proteins. Dof plays a key role in the synthesis and accumulation of wheat gluten[57] and can identify and combine with the AAAG sequence in the promoter of storage proteins in crops such as wheat, corn and rice. In addition, it can coordinate with other transcription factors to regulate the gene expression of storage proteins. Vicente-Carbajosa et al.[33] demonstrated that Dof participated in the regulation of protein content in maize endosperm by activating the expression of gliadins. In barley, the Dof transcription factor BPBF (barley prolamin-box binding factor) can activate transcription of the seed storage protein Hor2.[58] *Spa* recognizes and interacts with the GCN4 sequence in the EB of the storage protein promoter to activate transcription, which has important influences on the viscoelastic properties and quality of dough.[37] In this study, the transcription levels of six *Dof* and *Spa* transcription factors showed an up-down expression trend during grain development, particularly with peak expression at 17-25 DPA under water deficit (Fig. 8). This is consistent with the accumulation patterns of storage proteins, which suggests that the transcription factors *Dof* and *Spa* may play important roles in the transcriptional activation of storage protein genes. Their upregulated expression under water deficit could benefit the synthesis and accumulation of storage protein components and consequently promote protein content and breadmaking quality.

CONCLUSIONS

The results from the two consecutive growing seasons demonstrated that water deficit caused significant grain filling shortening, decreased grain number and weight, starch granule lessening and decreased yield, but an increase in protein and GMP content as well as superior gluten properties and breadmaking quality. RP-UPLC analysis revealed that total gliadin and glutenin contents as well as different types of gliadins, HMW-GS and LMW-GS were significantly increased under water deficit. Storage protein sub-proteome analysis demonstrated that 12 gliadins, 12 HMW-GS and 46 LMW-GS were significantly upregulated under water deficit. Particularly, considerable-gliadins and abundant LMW-B subunits showed significant upregulation, which could be responsible for the significant promotion of protein and GMP content and breadmaking quality. Six key *Dof* and *Spa* transcription factor genes also showed clear upregulation, consistent with the accumulation patterns of most storage protein components, which could play an important role in storage protein synthesis and breadmaking quality. Our results provide new insights into the molecular mechanisms of wheat storage protein synthesis and breadmaking quality under water deficit, as well as the theoretical basis for improving wheat quality in extreme climates.

ACKNOWLEDGEMENTS

This research was financially supported by grants from the National Natural Science Foundation of China (31471485) and the National Key R&D Program of China (2016YFD0100502). The English in this paper has been checked by at least two professional editors, both native speakers of English. For a certificate, please see http: //www. textcheck. com/certificate/so2hnM.

REFERENCES

[1] Yang J, Zhang J, Wang Z, Zhu Q and Liu L, Water deficit-induced senescence and its relationship to the remobilization of pre-stored carbon in wheat during grain filling. *Agron J* 93: 196-206 (2001).

[2] Fedoroff NV, Battisti DS, Beachy RN, Cooper PJM, Fischhoff DA, Hodges CN et al., Radically rethinking agriculture for the 21st century. *Science* 327: 833-834 (2010).

[3] Zhang X, Cai J, Wollenweber B, Liu F, Dai T, Cao W et al., Multiple heat and drought events affect grain yield and accumulations of high molecular weight glutenin subunits and glutenin macropolymers in wheat. *J Cereal Sci* 57: 134-140 (2013).

[4] Gu A, Hao P, Lv D, Zhen S, Bian Y, Ma C et al., Integrated proteome analysis of the wheat embryo and endosperm reveals central metabolic changes involved in the water deficit response during grain devel-opment. *J Agric Food Chem* 63: 8478-8487 (2015).

[5] Dong K, Ge P, Ma C, Wang K, Yan X, Gao L et al., Albumin and globulin dynamics during grain development of elite Chinese wheat cultivar Xiaoyan 6. *J Cereal Sci* 56: 615-622 (2012).

[6] Payne PI, Genetics of wheat storage proteins and the effect of allelic variation on bread-making quality. *Annu Rev Plant Physiol* 38: 141-153 (1987).

[7] Shewry PR and Halford NG, Cereal seed storage proteins: structures, properties and role in grain utilization. *J Exp Bot* 53: 947-958 (2002).

[8] Wang S, Yu Z, Cao M, Shen X, Li N, Li X et al., Molecular mechanisms of HMW glutenin subunits from

1Sl genome of Aegilops longissima positively affecting wheat breadmaking quality. *PLoS ONE* 8: e58947 (2013).

[9] Branlard G and Dardevet M, Diversity of grain protein and bread wheat quality: II. Correlation between high molecular weight subunits of glutenin and flour quality characteristics. *J Cereal Sci* 3: 345-354 (1985).

[10] Payne PI, Holt LM, Krattiger AF and Carrillo JM, Relationships between seed quality characteristics and HMW glutenin subunit composition determined using wheats grown in Spain. *J Cereal Sci* 7: 229-235 (1988).

[11] He ZH, Liu L, Xia XC, Liu JJ and Peña RJ, Composition of HMW and LMW glutenin subunits and their effects on dough properties, pan bread, and noodle quality of Chinese bread wheats. *Cereal Chem* 82: 345-350 (2005).

[12] Li Y, An X, Yang R, Guo X, Yue G, Fan R et al., Dissecting and enhancing the contributions of high-molecular-weight glutenin subunits to dough functionality and bread quality. *Mol Plant* 8: 332-334 (2015).

[13] Liu H, Wang K, Xiao L, Wang S, Du L, Cao X et al., Comprehensive identification and bread-making quality evaluation of common wheat somatic variation line AS208 on glutenin composition. *PLoS ONE* 11: e0146933 (2016).

[14] Wrigley CW, Giant proteins with flour power. *Nature* 381: 738-739 (1996).

[15] Don C, Mann G, Bekes F and Hamer RJ, HMW-GS affect the properties of glutenin particles in GMP and thus flour quality. *J Cereal Sci* 44: 127-136 (2006).

[16] Liu W, Zhang Y, Gao X, Wang K, Wang S, Zhang Y et al., Comparative proteome analysis of glutenin synthesis and accumulation in developing grains between superior and poor quality bread wheat cultivars. *J Sci Food Agric* 92: 106-115 (2012).

[17] Yu Y, Zhu D, Ma C, Cao H, Wang Y, Xu Y et al., Transcriptome analysis reveals key differentially expressed genes involved in wheat grain development. *Crop J* 4: 92-106 (2016).

[18] Dupont FM and Altenbach SB, Molecular and biochemical impacts of environmental factors on wheat grain development and protein synthesis. *J Cereal Sci* 38: 133-146 (2003).

[19] Altenbach SB, New insights into the effects of high temperature, drought and post-anthesis fertilizer on wheat grain development. *J Cereal Sci* 56: 39-50 (2012).

[20] Zhang Y, Huang X, Wang L, Wei L, Wu Z, You M et al., Proteomic analysis of wheat seed in response to drought stress. *J Integr Agric* 13: 919-925 (2014).

[21] Hurkman WJ, Tanaka CK, Vensel WH, Thilmony R and Altenbach SB, Comparative proteomic analysis of the effect of temperature and fertilizer on gliadin and glutenin accumulation in the developing endosperm and flour from *Triticum aestivum* L. cv. Butte 86. *Proteome Sci* 11: 8 (2013).

[22] Lan T, Jiang D, Xie Z, Dai T, Jing Q and Cao W, Effects of post-anthesis drought and waterlogging on grain quality traits in different specialty wheat varieties. *J Soil Water Conservat* 18: 193-196 (2004).

[23] Xie Z, Jiang D, Cao W, Dai T and Jing Q, Effects of post-anthesis soil water status on the activities of key regulatory enzymes of starch and protein accumulation in wheat grains. *J Plant Physiol Mol Biol* 29: 309-316 (2003).

[24] Hajheidari M, Eivazi A, Buchanan BB, Wong JH, Majidi J and Salekdeh GH, Proteomics uncovers a role for redox in drought tolerance in wheat. *J Proteome Res* 6: 1451-1460 (2007).

[25] Kamal AHM, Kim KH, Shin KH, Choi JS, Baik BK, Tsujimoto H et al., Abiotic stress responsive proteins of wheat grain determined using proteomics technique. *Aust J Crop Sci* 4: 196-208 (2010).

[26] Bazargani MM, Sarhadi E, Bushehri A-AS, Matros A, Mock H-P, Naghavi M-R et al., A proteomics view on the role of drought-induced senescence and oxidative stress defense in enhanced stem reserves remobilization in wheat. *J Proteomics* 74: 1959-1973 (2011).

[27] Ford KL, Cassin A and Bacic A, Quantitative proteomic analysis of wheat cultivars with differing drought stress tolerance. *Front Plant Sci* 2: 44 (2011).

[28] Ge P, Ma C, Wang S, Gao L, Li X, Guo G et al., Comparative proteomic analysis of grain development in two spring wheat varieties under drought stress. *Anal Bioanal Chem* 402: 1297-1313 (2012).

[29] Jiang S-S, Liang X-N, Li X, Wang S-L, Lv D-W, Ma C-Y et al., Wheat drought-responsive grain proteome analysis by linear and non-linear 2-DE and MALDI-TOF mass spectrometry. *Int J Mol Sci* 13:

16065-16083 (2012).

[30] Guo G, Lv D, Yan X, Subburaj S, Ge P, Li X et al., Proteome characterization of developing grains in bread wheat cultivars (*Triticum aestivum* L.). *BMC Plant Biol* 12: 147 (2012).

[31] Yu Z, Han C, Yan X, Li X, Jiang G and Yan Y, Rapid characterization of wheat low molecular weight glutenin subunits by ultraperformance liquid chromatography (UPLC). *J Agric Food Chem* 61: 4026-4034 (2013).

[32] Yanagisawa S, The Dof family of plant transcription factors. *Trends Plant Sci* 7: 555-560 (2002).

[33] Vicente-Carbajosa J, Moose SP, Parsons RL and Schmidt RJ, A maize zinc-finger protein binds the prolamin box in zein gene promotersand interacts with the basic leucine zipper transcriptional activator Opaque2. *Proc Natl Acad Sci USA* 94: 7685-7690 (1997).

[34] Dong G, Ni Z, Yao Y, Nie X and Sun Q, Wheat Dof transcription factor WPBF interacts with TaQM and activates transcription of an alpha-gliadin gene during wheat seed development. *Plant Mol Biol* 63: 73-84 (2007).

[35] Kawakatsu T, Yamamoto MP, Touno SM, Yasuda H and Takaiwa F, Compensation and interaction between RISBZ1 and RPBF during grain filling in rice. *Plant J* 59: 908-920 (2009).

[36] Kreis M, Shewry PR, Forde BG, Forde J and Miflin BJ, Structure and evolution of seed storage proteins and their genes with particular reference to those of wheat, barley and rye. *Oxford Surv Plant Mol Biol Cell Biol* 2: 253-317 (1985).

[37] Ravel C, Martre P, Romeuf I, Dardevet M, Ei-Malki R, Bordes J et al., Nucleotide polymorphism in the wheat transcriptional activator *Spa* influences its pattern of expression and has pleiotropic effects on grain protein composition, dough viscoelasticity, and grain hardness. *Plant Physiol* 151: 2133-2144 (2009).

[38] Chen XM, He ZH and Wang DS, The approval of new wheat variety Zhongmai 175. *China Seed Ind* 7: 69 (2009).

[39] Zhen S, Zhou J, Deng X, Zhu G, Cao H, Wang Z et al., Metabolite profiling of the response to high-nitrogen fertilizer during grain development of bread wheat (*Triticum aestivum* L.). *J Cereal Sci* 69: 85-94 (2016).

[40] Dong L, Li N, Lu X, Prodanovic S, Xu Y, Zhang W et al., Quality properties and expression profiling of protein disulfide isomerase genes during grain development of three spring wheat near isogenic lines. *Genetika* 48: 249-269 (2016).

[41] Han C, Lu X, Yu Z, Li X, Ma W and Yan Y, Rapid separation of seed gliadins by reversed-phase ultra performance liquid chromatography (RP-UPLC) and its application in wheat cultivar and germplasm identification. *Biosci Biotechnol Biochem* 79: 808-815 (2015).

[42] Yan X, Liu W, Yu Z, Han C, Zeller FJ, Hsam SLK et al., Rapid separation and identification of wheat HMW glutenin subunits by UPLC and comparative analysis with HPLC. *Aust J Crop Sci* 8: 140-147 (2014).

[43] Dupont FM, Vensel WH, Tanaka CK, Hurkman WJ and Altenbach SB, Deciphering the complexities of the wheat flour proteome using quantitative two dimensional electrophoresis, three proteases and tandem mass spectrometry. *Proteome Sci* 9: 10 (2011).

[44] Zhou J, Ma C, Zhen S, Cao M, Zeller FJ, Hsam SLK et al., Identification of drought stress related proteins from 1Sl (1B) chromosome substitution line of wheat variety Chinese Spring. *Bot Stud* 57: 20 (2016).

[45] Bian Y, Deng X, Yan X, Zhou J, Yuan L and Yan Y, Integrated proteomic analysis of Brachypodium distachyon roots and leaves reveals a synergistic network in the response to drought stress and recovery. *Sci Rep* 7: 46183 (2017).

[46] Ge Y, Chu L, Zhang G, Yu X and Zhang L, The standards for clarifying drought severity in northwestern Liaoning province. *J Irrigat Drain* 36: 115-120 (2017).

[47] Kobata T, Palta JA and Turner NC, Rate of development of post-anthesis water deficits and grain filling of spring wheat. *Crop Sci* 32: 1238-1242 (1992).

[48] Guttieri MJ, Stark JC, O'Brien K and Souza E, Relative sensitivity of spring wheat grain yield and quality parameters to moisture deficit. *Crop Sci* 41: 327-335 (2001).

[49] Rharrabti Y, Royo C, Villegas D, Aparicio N and García del Moral LF, Durum wheat quality in Mediterranean environments: I. Quality expression under different zones, latitudes and water regimes across

Spain. *Field Crops Res* 80: 123-131 (2003).

[50] Houshmand S, Arzani S and Mirmohammadi-Maibody SAM, Effects of salinity and drought stress on grain quality of durum wheat. *Commun Soil Sci Plant Anal* 45: 297-308 (2014).

[51] Mottaghi M, Najafian G and Bihamta MR, Effect of terminal drought stress on grain yield and baking quality of hexaploid wheat genotypes. *Iran J Crop Sci* 11: 290-306 (2009).

[52] Sabelli PA and Shewry PR, Characterization and organization of gene families at the *Gli-1* loci of bread and durum wheats by restriction fragment analysis. *Theor Appl Genet* 83: 209-216 (1991).

[53] Cassidy BG, Dvorak J and Anderson OD, The wheat low-molecular-weight glutenin genes: characterization of six new genes and progress in understanding gene family structure. *Theor Appl Genet* 96: 743-750 (1998).

[54] Dupont FM, Hurkman WJ, Vensel WH, Chan R, Lopez R, Tanaka CK et al., Differential accumulation of sulfur-rich and sulfur-poor wheat flour proteins is affected by temperature and mineral nutrition during grain development. *J Cereal Sci* 44: 101-112 (2006).

[55] Dupont FM, Hurkman WJ, Vensel WH, Tanaka C, Kothari KM, Chung OK et al., Protein accumulation and composition in wheat grains: effects of mineral nutrients and high temperature. *Eur J Agron* 25: 96-107 (2006).

[56] Majoul T, Bancel E, Triboï E, Ben Hamida J and Branlard G, Proteomic analysis of the effect of heat stress on hexaploid wheat grain: characterization of heat-responsive proteins from total endosperm. *Proteomics* 3: 175-183 (2003).

[57] She M, Ye X, Yan Y, Howit C, Belgard M and Ma W, Gene networks in the synthesis and deposition of protein polymers during grain development of wheat. *Funct Integr Genom* 11: 23-35 (2011).

[58] Mena M, Vicente-Carbajosa J, Schmidt RJ and Carbonero P, An endosperm-specific DOF protein from barley, highly conserved in wheat, binds to and activates transcription from the prolamin-box of a native B-hordein promoter in barley endosperm. *Plant J* 16: 53-62 (1998).

Dynamic metabolome profiling reveals significant metabolic changes during grain development of bread wheat (*Triticum aestivum* L.)

Shoumin Zhen,[a†] Kun Dong,[a†] Xiong Deng,[a] Jiaxing Zhou,[a] Xuexin Xu,[c] Caixia Han,[a] Wenying Zhang,[b] Yanhao Xu,[b] Zhimin Wang[c] and Yueming Yan[a,b*]

* *Correspondence to*: *Yueming Yan*, *Laboratory of Molecular Genetics and Proteomics*, *College of Life Science*, *Capital Normal University*, 100048 *Beijing*, *China*. *E-mail*: *yanym@cnu.edu.cn*

† *These authors contributed equally to this work*

[a] *College of Life Science*, *Capital Normal University*, 100048 *Beijing*, *China*

[b] *Hubei Collaborative Innovation Center for Grain Industry* (*HCICGI*), *Yangtze University*, 434025 *Jingzhou*, *China*

[c] *College of Agricultural and Biotechnology*, *China Agricultural University*, 100091 *Beijing*, *China*

Abstract:

BACKGROUND: Metabolites in wheat grains greatly influence nutritional values. Wheat provides proteins, minerals, B-group vitamins and dietary fiber to humans. These metabolites are important to human health. However, the metabolome of the grain during the development of bread wheat has not been studied so far. In this work the first dynamic metabolome of the developing grain of the elite Chinese bread wheat cultivar Zhongmai 175 was analyzed, using non-targeted gas chromatography/mass spectrometry (GC/MS) for metabolite profiling.

RESULTS: In total, 74 metabolites were identified over the grain developmental stages. Metabolite-metabolite correlation analysis revealed that the metabolism of amino acids, carbohydrates, organic acids, amines and lipids was interrelated. An integrated metabolic map revealed a distinct regulatory profile. The results provide information that can be used by metabolic engineers and molecular breeders to improve wheat grain quality.

CONCLUSION: The present metabolome approach identified dynamic changes in metabolite levels, and correlations among such levels, in developing seeds. The comprehensive metabolic map may be useful when breeding programs seek to improve grain quality. The work highlights the utility of GC/MS-based metabolomics, in conjunction with univariate and multivariate data analysis, when it is sought to understand metabolic changes in developing seeds.

Keywords: wheat; developing grain; metabolome; nutritional quality

Published in J. Sci Food Agric., 2016, 96: 3731-3740.

INTRODUCTION

Wheat (*Triticum aestivum* L.) is one of the most important crops worldwide, being a major food source because its grains are uniquely suitable for bread production. Wheat contributes essential amino acids, minerals, vitamins, beneficial phytochemicals and fiber to the human diet; these components are particularly enriched in whole-grain products.[1] Thus wheat production has expanded rapidly and many breeding programs seek to improve yield and nutritional value. An understanding of variation in and regulation of metabolism during grain development is essential if wheat quality is to be improved by either conventional breeding or genetic engineering.

Wheat grain development has three distinct phases, namely (1) cell division and differentiation, (2) grain filling and (3) desiccation/maturation.[2] Studies on wheat genomics[3] have made great progress in recent decades. Additionally, transcriptomics[4] and proteomics[5,6] have revealed how reserves are accumulated during grain development. However, unlike transcriptomics and proteomics which rely to a great extent on genome information, metabolomics can be applied to widely diverse species, with relatively little time required for re-optimizing protocols for a new species.[7] Additionally, transcriptomics and proteomics deal only with macromolecules; smaller materials playing important roles in seed development are not detected using these methods. Metabolite profiling is key for functional annotation of genes and allowing a comprehensive understanding of cellular responses to variations in biological conditions.[8] Many plant metabolites are essential human nutrients, but the study of biosynthesis of natural products is still in its infancy, because more than 90% of plant metabolites are unknown and only a small number of genes/enzymes involved in metabolism are identified in the plants *Arabidopsis* and rice.[9] Metabolomic approaches have been used to reveal the overall metabolic composition of and regulatory networks in plants, generating new information that may be valuable in breeding programs seeking to improve plant traits.[8] Thus, in recent years, metabolite profiling has been performed on many plant species, including *Arabidopsis*,[10] maize[11] and rice.[12] To date, several metabolic studies of wheat have appeared, but these were focused on the comparison of wheat grains from organic and conventional agriculture,[13] leaf metabolism under drought conditions[14] and metabolism of substantial equivalence of field-grown genetically modified wheat.[15] The study of molecular and biochemical impacts of environmental factors on wheat grain development and protein synthesis has also been carried out,[16] but the dynamic metabolome changes in developing wheat grains have not been described. Also, kernel metabolites can be used to generate bodily energy, including essential human nutrients, and serve as a resource for production of bioenergy.[17] Thus dynamic metabolome profiling during grain development would yield information facilitating an understanding of the biochemical mechanism of grain development, and data on nutritive quality.

Recently, metabolomic approaches have become important in crop breeding programs.[18] Identification and quantification of metabolites require sophisticated techniques such as mass spectrometry (MS) and/or nuclear magnetic resonance (NMR) spectroscopy, or may feature laser-induced fluorescence (LIF).[18] Each technique has specific advantages and disadvantages. NMR is highly selective and serves as the gold standard for structural elucidation, but it is rather insensitive. LIF is highly sensitive but lacks the chemical selectivity critical for structural identification.[19] MS offers a good combination of sensitivity and selectivity,[19] allowing detection and measurement of many primary and secondary metabolites at picomole-to-femtomole levels. MS is thus an important aid in metabolomics.[20,21] A gas chromatography/mass spectrometry (GC/MS) combination is ideal for analysis of both volatile and nonvolatile compounds after derivatization.[19]

Here we perform the first dynamic metabolome analysis of developing grains of the Chinese elite bread wheat cultivar Zhongmai 175, using a non-targeted GC/MS metabolome profiling approach. Our results reveal remarkable changes in and considerable variability of metabolite-metabolite correlations during wheat seed development, which will expand the under-standing of the internal developmental mechanism of developing grains.

MATERIALS AND METHODS

Plant materials

The Chinese elite winter wheat cultivarZhongmai 175 (*T. aestivum* L.) was grown in experimental fields of the China Agricultural University Research Center, Wuqiao, Hebei Province (116°37′23″E, 37°16′02″N) during the 2013-2014 growing season. Six biological replicate plots each 20 m^2 in area were planted. The average annual sunshine duration is 2 690 h, and the average annual temperature 12.6℃. The water table ranges from 6 to 9 m, and precipitation during the 2013-2014 season was 125 mm. Cultivation and management practices were identical to those applied in local fields. After flowering, the plants were marked with color lines, and the grains from middle ears at 10, 15, 20, 25 and 30 days post-anthesis (DPA) were collected at 09: 00-11: 00. The collected samples were quickly placed in liquid nitrogen and then stored −80℃ prior to analysis. The grains from 35, 40 and 45 DPA were also harvested for grain morphology observation, dry weight measurement or quality testing.

Grain morphology observation and dry weight measurement

The grain morphology from eight different developmental stages (10, 15, 20, 25, 30, 35, 40 and 45 DPA) was observed by stereo micro-scope, and dry weight was measured. The grains collected from different stages were put into a 65℃ oven and dried to constant weight. The dried grains were weighed and thousand-grain weight was calculated.

Gluten quality testing

Gluten quality parameters were tested according to Sun et al.[22] Flour moisture and ash contents (% dry basis) were deter-mined according to the American Association of Cereal Chemists Approved Methods (2000) 44-15A and 08-02 respectively. Protein content (%N×5.7, 14% moisture basis) was determined by nitrogen combustion analysis with an FP analyzer (LECO, St Joseph, MI, USA) calibrated against ethylenediaminetetraacetic acid (EDTA). Farinograph parameters were obtained using a 10 g Brabender Farinograh-E (Brabender® GmbH & Co., Duisburg, Germany) based on the American Association of Cereal Chemists Approved Method (2000) 54-21. Image analysis of crumb grain of bread was performed with C-Cell image analysis software and equipment (Calibre Control International Ltd, Warrington, UK). Slice brightness was used to describe the brightness of slices. Wall thickness, cell diameter, slice area and slice circumference were used to measure the cell properties.

Sample preparation and extraction prior to metabolome analysis

Grains (500 mg) of each sample with six biological replicates were put into six pre-cooled mortars, ground into fine powder with liquid nitrogen in 30 min (fast and evenly to avoid degradation) and lyophilized under vacuum. Then 100 mg amounts of powder were subjected to metabolite extraction with 1400 μL of cold pure methanol in test tubes, with vibration of the tubes for 30 s, and 60 μL of 0.2mg · μL^{-1} ribitol was added as interior label. Each mixture was shaken vigorously for 1 min at room temperature and then subjected to ultrasonic treatment for 15 min. Next, 750 μL of chloroform and 1 400 μL of water were added. After blending and centrifugation at 1 531 × *g* for 15 min, the supernatant (containing extracted metabolites) was transferred to another tube and dried with nitrogen. Following the addition of 60 μL of 15mg · mL^{-1} methoxypyridine solution to each sample, the tubes were vibrated for 30 s and left to stand for 16 h. Finally, 60 μL of bis (trimethylsilyl) triflouroacetamide (with 10mg • mL^{-1} trimethylchlorosilane) was added to each tube and left to react at room temperature for 60 min. After this reaction, the samples (six biological replicates) were prepared for subsequent GC/MS analysis.

Untargeted metabolomic analysis

Each seed extract was analyzed by GC/MS (Agilent 7890A/5975C, Agilent Technologies, Santa Clara, CA, USA). The samples were randomized to reduce system error. Data acquisition was done in one batch and finished in a single day. The GC column (HP-5MS capillary column, 30m × 250μm i. d., 0. 25μm; J&W Scientific, Folsom, CA, USA) contained 50mg • mL^{-1} phenylmethylsiloxane. Split sampling with an injection volume of 1 μL and a split ratio of 20: 1 was used. The temperatures of the injection port, ion source and connector were 280, 250 and 150℃ respectively. The column temperature was initially held at 40℃ for 5 min, then increased at 10℃ min^{-1} to 300℃ and held for 5 min. The carrier gas was helium at a flow rate of 1mL • min^{-1}. MS conditions were as follows: electrospray ionization (ESI); full scan mode; electron energy 70 eV; scanned area of quadrupole rod m/z 35-780.

Metabolites were identified by automated comparison of data with those in the NIST MS Search Program database (http: //www. nist. gov/srd/nist1a. htm) and the Wiley 9 metabolome database (http: //www. sisweb. com/software/ms/wiley. htm). We also used standard chemicals (Additional file2 Table S2 with red color) for qualitative confirmation of the identified chemicals based on their retention times and characteristic fragment ions. A combination of the chromatographic retention index and mass spectral signature was used to identify each metabolite. Along with these measures, we used quality control with a mixture of all 36 samples to test the stability of the system. Quality control of these chemicals would monitor deviations of the analytical results from the mixture and compare them with the errors caused by the analytical instrument itself. Ten injections with one quality control sample were done.

Construction of a metabolic network map and statistical analysis

Identified metabolites were mapped onto general biochemical pathways using the annotations of PMN (http: // www. plantcyc. org/) and KEGG.[23] A metabolic network map of wheat grain development incorporating identified and annotated metabolites was constructed with the aid of Cytoscape 3. 1. 0 (http: //www. cytoscape. org/). Missing values (if any) were assumed to be below the limits of detection and were imputed using functions implemented in XCMS (http: //xcmsonline. scripps. edu/) before the normalization step. R-language was used to perform one-way analysis of variance (ANOVA). In addition, to avoid false positive results in screening significant metabolites, the false discovery rate (FDR) significance criterion was performed with an FDR limit of 0. 05 to reduce the probability of false positives (Additional file4 Table S3). Cluster 3. 0 was next used to draw a heat map.[24] Principal component analysis (PCA) was performed with the aid of SIMCA-P11. 0 (http: //www. umetrics. com/SIMCA); the acquiescent set is Pareto-scaled, and key metabolites of developing grains were identified by partial least squares discriminant analysis (PLS-DA). Pearson's product moment correlations (r values) were calculated using Perl (http: //www. pm. org/). The corresponding P values and FDRs were calculated with the aid of the Cor. test function.[25] Cytoscape 3. 1. 0 (http: //www. cytoscape. org/) was used to draw the correlation figures.

RESULTS

Morphological characteristics and metabolome profilingof developing wheat grains

The changes in grain morphology and thousand-kernel weight (TKW) during eight different developmental periods of Zhongmai 175 are shown in Fig. 1. Main quality parameters were also tested (Additional file1 Table S1). These results indicated that the seeds grew normally.

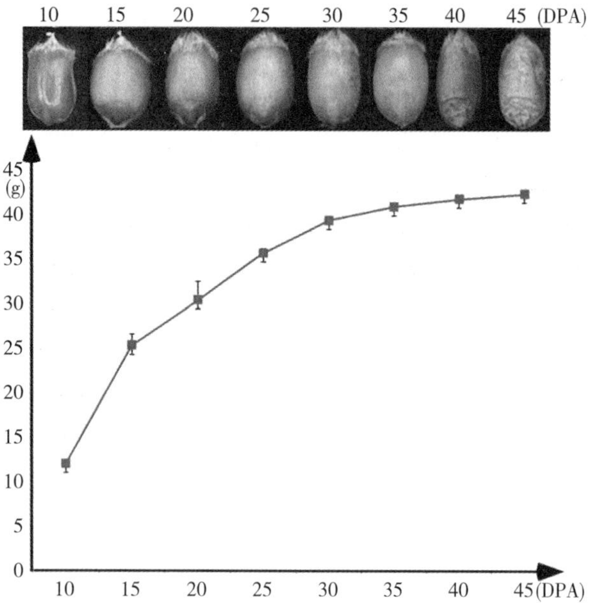

Fig. 1 Morphological characteristics and TKW changes of eight grain developmental stages in Zhongmai 175.

According to the results of library search of more than 80% similarity, if there was more than one chemical, then we further considered the retention index. A total of 74 metabolites were identified (Additional file2 Table S2). The total ion chromatogram is shown in Additional file3 Figure S1. ANOVA indicated that these metabolites changed significantly ($P < 0.05$, FDR < 0.05) during grain development (Additional file4 Table S3).

These 74 metabolites embraced most primary metabolic pathways and could be divided into seven major groups based on the characteristics of each chemical. The largest group contained 20 metabolites, thus 27.03% of all metabolites detected, and was amino acid metabolism. The second largest group contained 16 metabolites (21.62%) of carbohydrate metabolism. The third group had 14 metabolites (18.92%) of organic acid metabolism, followed by groups with ten metabolites (13.51%) of lipid or fatty acid metabolism, six metabolites (8.11%) of amine metabolism, five other chemicals (6.76%) and three metabolites (4.05%) of nucleotide metabolism (Fig. 2a).

PCA of metabolites

Our primary aim was to seek correlations between metabolite expression patterns in, and changes in the metabolism of developing grains. From PCA we can learn that PC1 accounted for 58.69% and PC2 for 23.12% of the variance, together amounting to 81.81%. Thus our PCA results indicated that the model is rather calculable (Fig. 2b). PCA showed that the expression patterns of the six developmental stages differed significantly. All 36 samples (6 periods × 6 biological replicates) fell into three classes based on caryopsis developmental times (Fig. 2b), corresponding to three distinct expression patterns: (1) 10 and 15 DPA, (2) 20, 25 and 30 DPA and (3) 45 DPA. The developmental stages of grain could also be separated into distinct clusters according to the results of PLS-DA (ten components; R2X=0.985, R2Y=0.979 and Q2 =0.908), which further revealed distinct metabolic alterations between the different stages. PLS-DA also revealed replicates, substrate expression patterns and the contributions of individual materials to seed development (Additional file5 Figure S2a). The validation plot strongly indicated that the PLS-DA model was valid (Additional file5 Figure S2b), since the Q2 regression line in blue had a negative intercept and all permuted R2 values in green on the left were lower than the original point of the R2 value on the right. This indicates that wheat grain development features three principal metabolic stages.

The contributions of individual metabolites to the PCA output of Fig. 1b were estimated from the PCA loading score plot (Fig. 2c and Additional file6 Table S4). Metabolites clustered around the origin of the plot contributed little to PCA separation. Outlying metabolites made greater impacts; these included acetamide, galactose, fructose, glucose, sucrose, pyroglutamic acid and serine.

Dynamic metabolic changes in developing grains

K-Median clustering analysis yielded an overview of dynamic metabolome changes during grain develop-

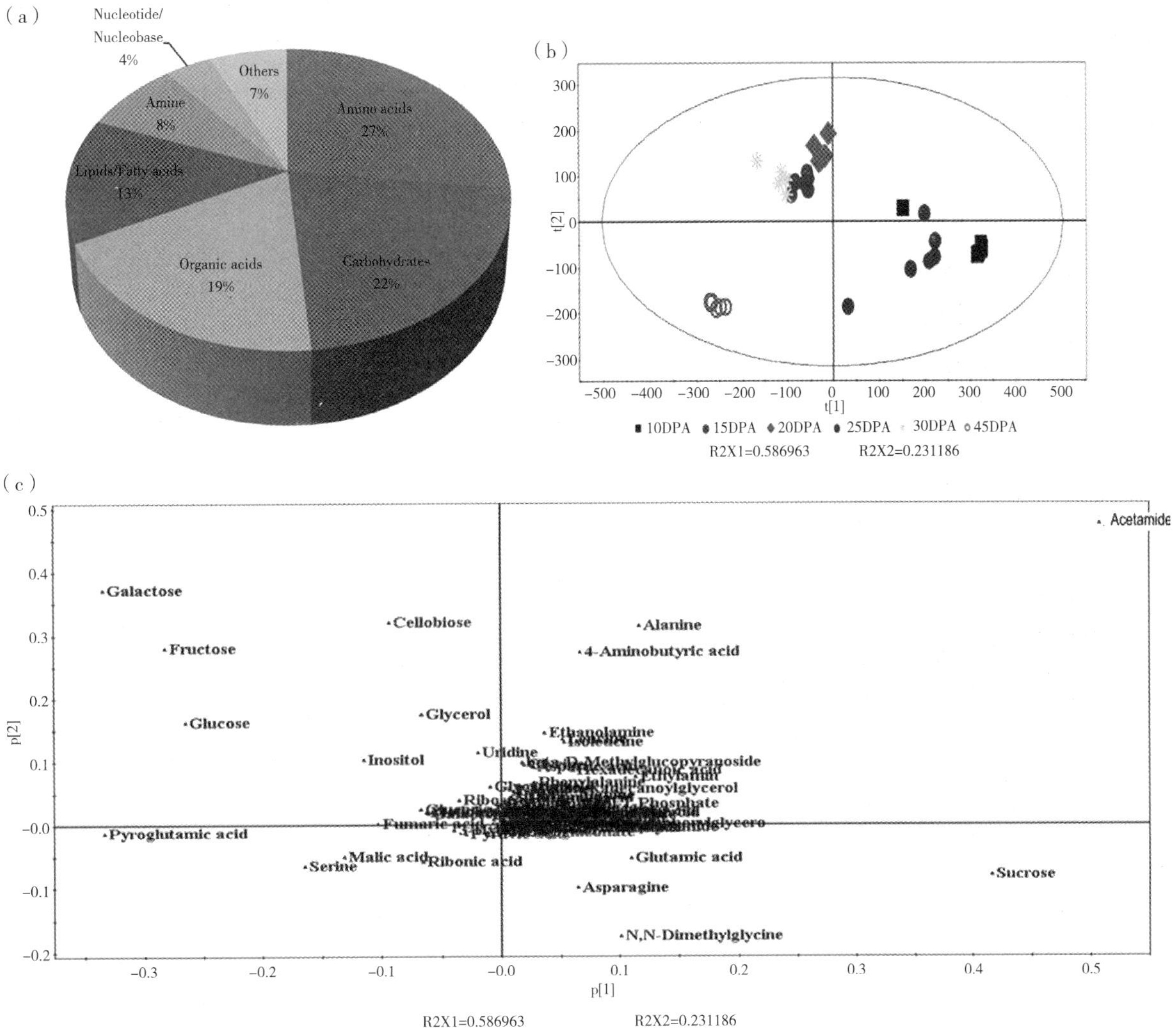

Fig. 2 Metabolites identified in developing seeds, with PCA featuring PCA loading. (a) Classification of 74 identified metabolites in developing seeds.
(b) PCA and (c) PCA loading of 74 metabolites of developing seeds. PC1, 58.69%; PC2, 23.12%.

ment. All 74 metabolites were evaluated and five expression patterns were discerned (Fig. 3 and Additional file7 Table S5). The first pattern (48.75% of all metabolites) was generally associated with up-regulation. At maturity (45 DPA), most metabolites were mildly down-regulated compared with the levels at 20-30 DPA. Most amino acids (17, 81%), all amines (six), five organic acids and five carbohydrates exhibited this pattern I. Notably, among the 36 metabolites of pattern I, 32 (all except nicotinic acid, heptanoic acid, uridine and glyceric acid) exhibited significant variations in level during grain development, indicating that these metabolites play very important roles in developing seeds. In particular, six of the eight essential amino acids exhibited the up-regulation pattern. Four lipids, one organic acid and one amino acid were included in pattern II, the up-down-up appearance of which was suggestive of an italicized letter 'N'. However, the levels of these six metabolites did not vary significantly during development, and all except proline were maximally expressed at 20 DPA. Pattern III, up-down-up-down in appearance, included 13 metabolites that were expressed at low levels at 20, 25 and 30 DPA. Pattern IV, the second largest group, included 15 metabolites: four carbohydrates, five organic acids, four others and two amino acids; down-regulation was evident. In other words, the 10 and 15 DPA levels were high but fell sharply thereafter. Pattern V contained four chemicals, namely mannose-6-phosphate, glycerol, threonic acid and ribose; an up-down expression pattern was evident during grain development.

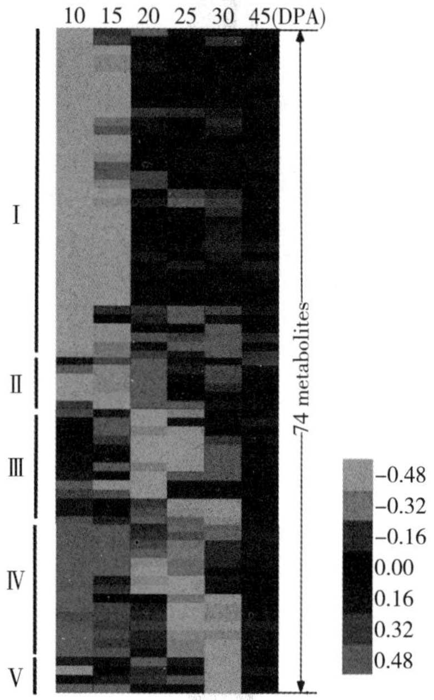

Fig. 3 Clustering analysis of metabolome data from various periods of grain development. Heat map representation of levels of 74 metabolites from developing seeds; five classes are evident upon clustering analysis. Each line in the heat map represents a metabolite. The deeper the red color, the higher the level of that metabolite in developing seeds. Similarly, the deeper the green color, the lower the level of that metabolite in developing seeds.

Metabolite-metabolite correlation analysis

Metabolite-metabolite correlation analysis has been usedto reveal potentially important regulatory features of plant seeds. [17,26] Pearson's pairwise correlation coefficients were calculated for each pair of identified metabolites and visualized on a heat map (Fig. 4). A total of 5746 correlations were found, ranging in value from 1 for galactose and fructose to -0.91 for putrescine and fumaric acid (Additional file8 Table S6). Apart from the metabolite-metabolite correlations based on the six periods, we also did the correlations based on the three groups, namely 10 and 15 DPA, 20, 25 and 30 DPA and 45 DPA (Additional file 9 Figure S3 and Additional files10-12 Tables S7-S9), to detect the changes during grain development.

Significant correlations ($r^2 > 0.64$ ($r > 0.8$ or $r < -0.8$) and FDR < 0.05) were further detected in the study (Fig. 5). Of these, 117 were positive and only 17 negative (Additional file13 Table S10). Notably, amino acids dominated the significant metabolite-metabolite correlations (61 in number), of which 42 were between two amino acids and 19 between an amino acid and another metabolite. Notably also, 12 of the 20 basic amino acids were involved in 134 significant correlations, whereas another six were not involved in any significant correlation. Tyrosine, isoleucine and lysine were significantly intercorrelated. We found four significant correlations between amino acids and amines, eight between amino acids and organic acids, three between amino acids and lipids and two between amino acids and carbohydrates.

The second metabolic group exhibiting notable metabolite correlations was the carbohydrates. Over 25 significant correlations were noted, including six between two carbohydrates and 19 between a carbohydrate and a non-carbohydrate. In addition, the six amines were involved in 17 significant correlations, three nucleotides in 14 and six lipids in eight (Fig. 5). Of the 16 negative correlations, asparagine had three with three carbohydrates, while norvaline had six (with fumaric acid, glucose, malic acid, sedoheptulose, serine and monomethyl phosphate).

Comprehensive study of metabolic pathways involved in wheat grain development

To fully understand the regulation of metabolic pathways involved in wheat grain development, we constructed a metabolome pro-file with the aid of PMN and KEGG (Fig. 6). A total of 74 metabolites of known structure were identified, including amino acids and their derivatives, carbohydrates, lipids and nucleotides, and these metabolites embraced most key metabolic pathways and reflected the physiological state and nutritional value of the developing wheat grain.

Most identified substances (with the exception of several alkanes) were mapped to the network. We found that most transformations of amino acids of the network changed significantly during grain development. Additionally, 16 carbohydrates were distributed to mapped pathways, including glycolysis and the tricarboxylic acid (TCA) cycle; all play vital roles in kernel development. Particularly, glucose and fructose assumed central positions in carbohydrate metabolism (Fig. 6), suggesting that these sugars are important in terms of energy production. Of the six amines identified, three (ethanolamine, ethylamine and acetamide) are involved in lipid metabolism.

As we know, the synthesis of uracil nucleotides needs aspartic acid as the raw material, and the degradation of uracil will produce some alanine. In this study we found the following correlations between them: for uracil and aspartic acid, $r = 0.8$ and FDR $= 7.93 \times 10^{-8}$; for uracil and alanine, $r = 0.82$ and FDR $= 2.13 \times 10^{-8}$ (Additional file13 Table S10). We can learn that the two pairs of chemicals are closely correlated with each other in the network. We provide indirect evidence for these phenomena.

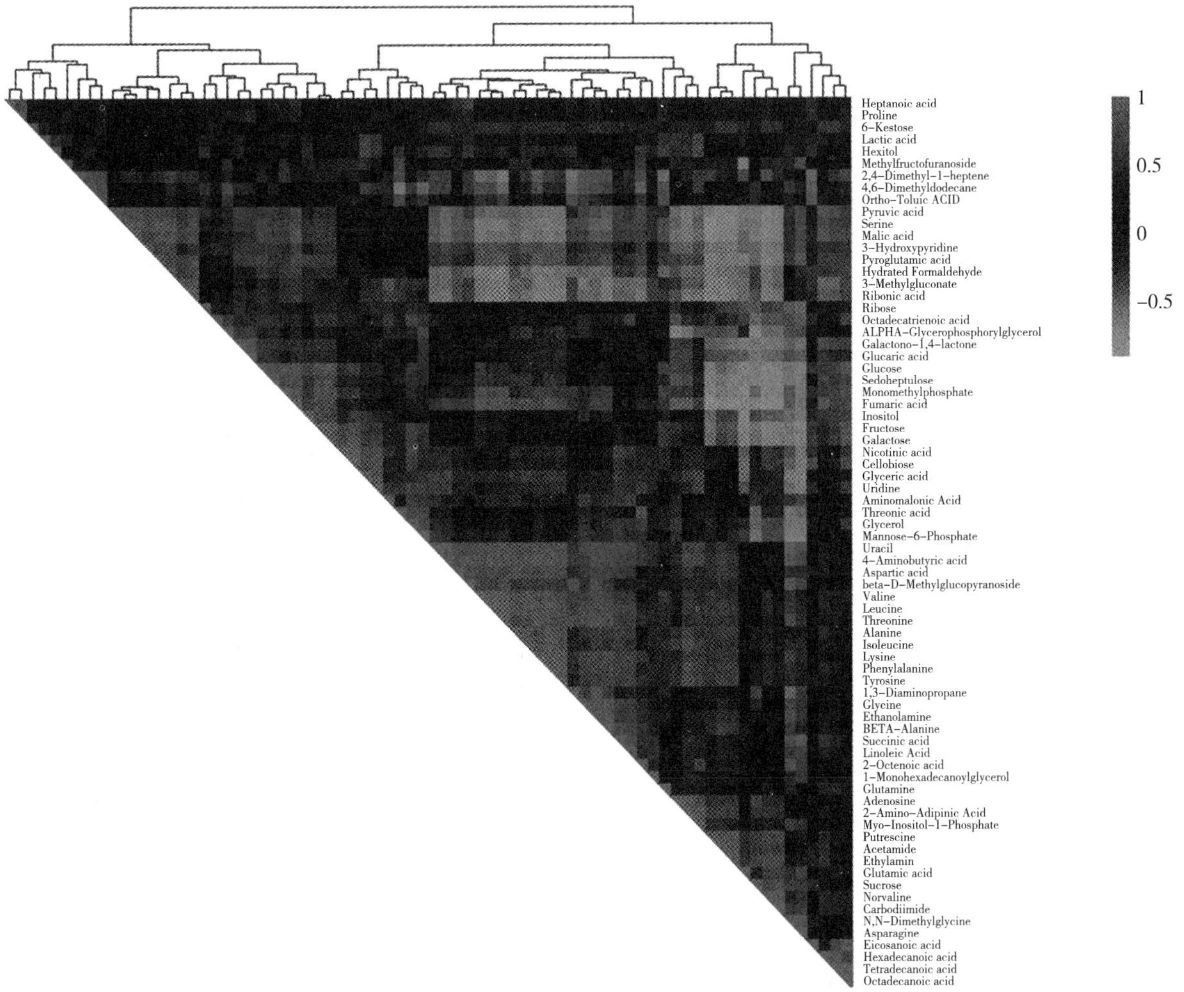

Fig. 4 Metabolite-metabolite correlations in developing seeds of Zhongmai 175. The vertical axis and horizontal axis show different metabolites grouped by pathway information. The r and P values of all correlations are shown in distinct colors.

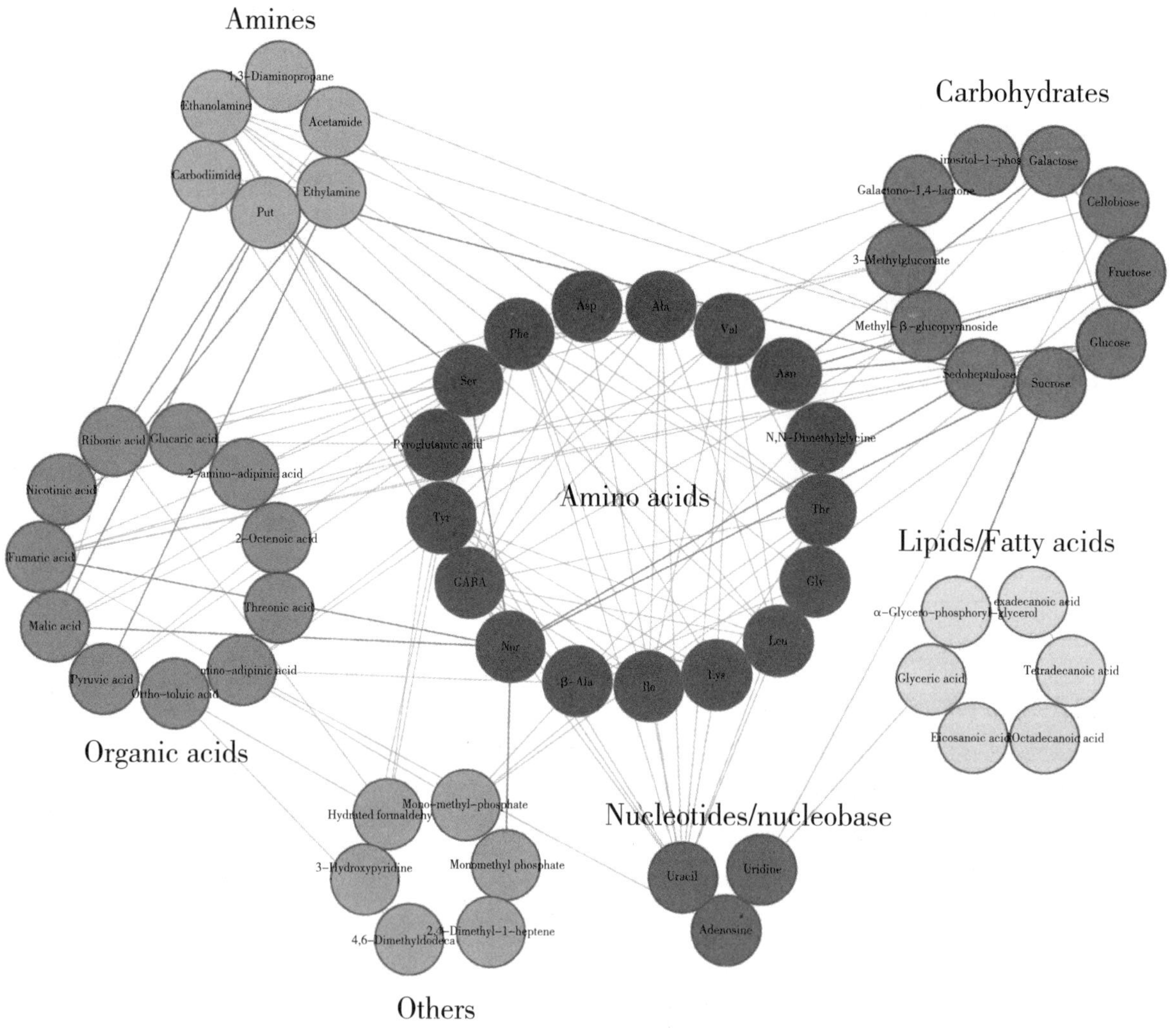

Fig. 5 Map of significant seed metabolite-metabolite correlations. Significant correlations had r^2 values > 0.64 and FDRs < 0.05. Metabolites are represented by circles, and metabolites in the same metabolic function groups are identically colored. Connecting lines indicate correlations. Positive correlations are shown in gray and negative correlations in red.

DISCUSSION

We obtained a dynamic metabolome profile of wheat grain development using a standard non-targeted GC/MS approach. Here we discuss the principal metabolome changes found.

Amino acid metabolism

Wheat is considered as a good source of proteins, minerals, B-group vitamins and dietary fiber according to previousstudies.[27] Various amino acids accumulate in seeds during grain development.[28] About 70% of all amino acid metabolites (including 14 of the 20 basic amino acids) detected in our present study varied significantly in their accumulation levels during kernel development (Additional file2 Table S2 and Additional file14 Table S11). This is similar to previous findings in *Arabidopsis* seeds.[10] Amino acid expression exhibited a particular pattern (Fig. 3) featuring a significant change between two adjacent periods of seed development (Additional file14 Table S11). Amino acids not only play important roles in maintaining osmotic potential during seed desiccation but also serve as precursors of protein synthesis during germination.[12] In particular, alanine constitutes a significant proportion of the free amino acid pool under both anaerobic and other stress conditions.[29] Asparagine serves

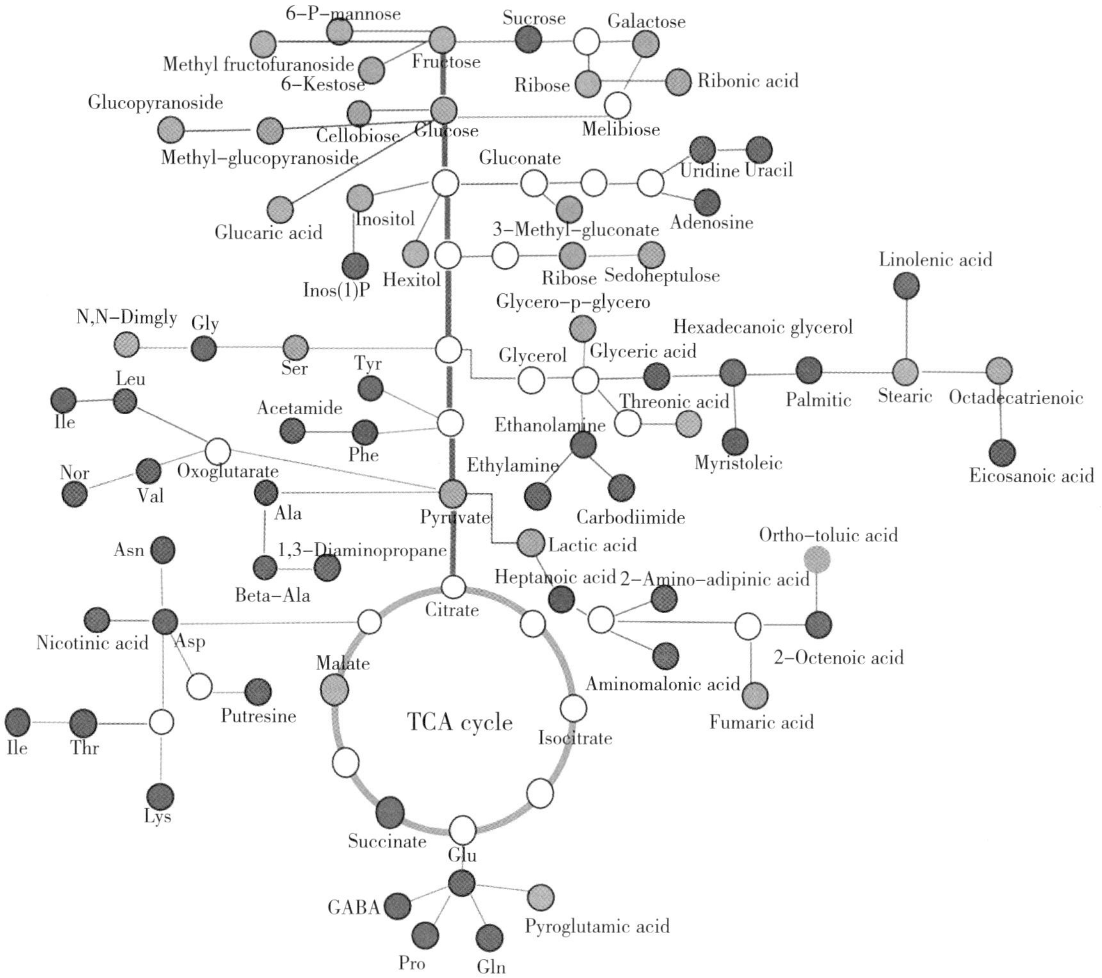

Fig. 6 Metabolic network of wheat seeds. The red circles denote chemicals that underwent up-regulation totally during grain development, and the green circles denote metabolites that underwent down-regulation during grain development by and large in our study. The white circles denote chemicals that had previously been placed in the network.

to both store and transport nitrogen[30] from roots to leaves and from leaves to developing seeds, being thereafter rapidly metabolized to glutamate and other amino acids during germination.[31] We found (Additional file14 Table S11) that the levels of alanine and asparagine changed significantly between 15-20 and 25-30 DPA. In general, the levels of these two amino acids rose steadily to 30 DPA, suggesting that they were extensively transformed into other amino acids during early periods and later accumulated in mature seeds to aid in the synthesis of storage proteins.

Most (~85%) of the identified amino acids exhibited two distinct metabolic phases during kernel development, as did the two amino acids discussed above. One phase was evident between 15 and 20 DPA and featured rapid up-regulation; the other was evident between 30 DPA and maturation (desiccation) and featured an element of down-regulation. This indicates that nutrients are accumulated rapidly during grain filling, at which time photosynthesis is vigorous and the TCA cycle active. However, during maturation, the non-protein amino acid-aminobutyric acid (GABA) also declined mildly. This decline may fuel the TCA cycle via the GABA shunt (Fig. 3).[32,33] Previously it was found that several amino acids were synthesized during seed desiccation and fell in level during early seed germination.[10] The translational processes of early germination explain this phenomenon.[34] Seed maturation was associated with significant reductions in the levels of most sugars, organic acids and amino acids, suggesting the efficient incorporation thereof into storage reserves.[10] In our present study the accumulation levels of most amino acids (alanine, aspartic acid, glutamine and isoleucine) declined during desiccation because they were used to synthesize storage proteins. According to Ma et al.,[35] high-molecular-weight glutenin subunits (HMW-GS) mainly affect dough elasticity, while low-molecular-weight glutenin subunits (LMW-GS) influence dough viscosity. In the maturity period the expression quantity of glutamine, a major component of glutenin proteins, was 26.45, rather lower than 30 DPA (Additional file7 TableS5), implying its incorporation into glutenins in the mature grains. Thus these metabolites set up an important base for wheat gluten quality (Additional file1 Table S1). The farinograph analysis showed that these amino acids maybe contribute to the parameters (wheat gluten, gluten index, protein content, etc.).

The eight essential amino acids for humans are Lys, Try, Phe, Met, Thr, Ile, Leu and Val. These amino acids cannot be synthesized by animals. We identified six thereof (Phe, Ile, Leu, Lys, Thr and Val) in developing grains, wherein their levels increased gradually (Fig. 3 and Additional file7 Table S5). Extensive accumulation during early development allowed rapid sequestering in storage protein at later stages, contributing to the nutritive value of grain. Apart from that, the extensive involvement of amino acid metabolites in significant metabolite-metabolite correlations (Fig. 6) also indicates that amino acids play a dominant role in the regulation of wheat seed metabolism.

Carbohydrate metabolism

Seed development is tightly linked to carbohydrate metabolism.[17] Maize kernels accumulate sucrose, stachyose and raffinose during development.[17] Sucrose plays important roles in regulation of the C/N balance of plant cells[36] and can also be used to synthesize starch.[37] Glucose and fructose are both products of photosynthesis. They are present during the early stages of seed development but disappear as the seed matures.[38]

We detected two distinct patterns of sugar expression during seed development. Raffinose, sucrose and glucopyranoside rose gradually in level from 10 to 30 DPA and then declined slightly to the time of maturation (Fig. 3). This pattern is reminiscent of that of Arabidopsis seeds, which accumulate reserves from 14 to 17 DPA.[10] Weschke et al.[37] showed that starch biosynthesis required sucrose to be imported into the endosperm, both because sucrose was the direct precursor of starch and to promote storage-associated processes. Chen et al.[39] found that the high-level expression of biosynthetic genes during intermediate developmental stages of Chinese spring wheat played a major role in the production of new B-granules. High-level expression of raffinose and sucrose during grain filling may assist starch biosynthesis. However, fructose, glucose and sedoheptulose fell steadily in level over the entire course of grain development, possibly because most of these simple sugars were used to synthesize starch,[40] the synthesis of which rose gradually during the grain-filling period of rice seeds, as revealed by proteomics.[41] Thus sugar levels decline steadily and starch synthesis rises slowly during the entire course of seed development. Starch is the most significant form of carbon reserve in plants in terms of the amount made and the universality of its distribution among different plant species.[42] Thus, when the grains have matured, the decrease in sugars may be a result of their assembling into starch, which would improve the conformation of starch quality of wheat flour.

Energy metabolism

Glycolysis, the TCA cycle and the mitochondrial electron transport chain are essential to provide energy for many cellular functions.[43] The TCA cycle plays a central role in catabolism of organic fuel molecules, including glucose and other sugars, fatty acids and certain amino acids.

Weidentified several organic acids, including fumaric

acid, malic acid, glucaric acid and pyruvic acid, all of which play important roles in the TCA cycle. These organic acids are taken up by mitochondria, interconverted and used to produce energy and reducing power. Ultimately, each pyruvate molecule yields 15 ATP equivalents.[44] Synthesis of these organic acids was down-regulated during the course of seed development (Fig. 3). At the beginning of seed filling, the levels were rather high, but they decreased later. At the proteome level, all TCA cycle-related proteins were expressed at high levels from 6 to 15 DPA but then decreased sharply.[5] In rice, Xu et al.[41] found that the levels of proteins involved in the TCA cycle tended to peak at early stages; the levels of TCA cycle-related organic acids and proteins were correlated at both the metabolome and proteome levels. This implies that the TCA cycle is very active during early developmental stages, to provide the required energy.

Lipid metabolism

Plant seeds store significant amounts of neutral lipids, including triacylglycerols (TAGs), in cytosolic lipid droplets. Most lipids are mobilized immediately after germination to fuel seedling growth and development prior to establishment of photosynthesis.[44] For these reasons, lipids play important roles during seed development.

We detected several lipidic materials. The fatty acids linoleic acid, tetradecanoic acid, eicosanoic acid, octadecanoic acid, octadeca-trienoic acid and hexadecanoic acid play important roles in human health. Of these, linoleic acid changed significantly in level during seed development; rapid up-regulation of synthesis was evident to 20 DPA, followed by a sharp downturn maintained to the time of maturation (Fig. 3). We speculate that the fatty acid may become integrated into the oil body (a nutritive store). Most lipids were deposited before the mid-stage of grain development, as is also true of oats.[45] According to Baud et al.,[46] we can also learn that, in the second stage, storage oils and proteins were accumulated in large quantities, accounting for approximately 40% of dry matter each at the end of this stage. The TAGs of developing seeds are assembled in the endoplasmic reticulum from acyl-CoAs and glycerol, via the conserved 'Kennedy' pathway.[44] Thus the decline in seed glycerol levels from 20 DPA to the time of maturation may be explained by the fact that glycerol was either used to produce energy or stored in the oil body.

CONCLUSIONS

In summary, our metabolome approach identified dynamic changes in metabolite levels and correlations among such levels in developing seeds. Our comprehensive metabolic map may be useful when breeding programs seek to improve grain quality. This work highlights the utility of GC/MS-based metabolomics, in conjunction with univariate and multivariate data analysis, when it is sought to understand metabolic changes in developing seeds. However, the exact molecular mechanisms underlying the observed changes remain to be elucidated. In future, the metabolic map will be improved by addition of more x-omic data (from the metabolome, transcriptome, genome and proteome), to further explore interrelationships among different substances.

ACKNOWLEDGEMENTS

This research was financially supported by grants from theNational Natural Science Foundation of China (31471485), Natural Science Foundation of Beijing City and the Key Developmental Project of Science and Technology from Beijing Municipal Commission of Education (KZ201410028031). Thanks for the help of data analysis from Bionovogene (Suzhou).

REFERENCES

[1] Shewry PR, Wheat. J Exp Bot 60: 1537-1553 (2009).

[2] Shewry PR, Mitchell RAC, Tosi P, Wan YF, Underwood C, Lovegrove A, et al., An integrated study of grain development of wheat (cv. Hereward). J Cereal Sci 56: 21-30 (2012).

[3] Pfeifer M, Kugler KG, Sandve SR, Zhan BJ, Rudi HD, Hvidsten TR et al., Genome interplay in the grain transcriptome of hexaploid bread wheat. Science 345: 1250091 (2014).

[4]Wan Y, Poole RL, Huttly AK, Toscano-Underwood C, Feeney K, Welham S et al., Transcriptome analysis of grain development in hexaploid wheat. *BMC Genom* 9: 121 (2008) .

[5] Guo GF, Lv DW, Yan X, Subburaj S, Ge P, Li XH et al., Proteome char-acterization of developing grains in bread wheat cultivars (Triticum aestivum L.). BMC Plant Biol 12: 147 (2012).

[6] Nadaud I, Girousse C, Debiton C, Chambon C, Bouzidi MF, Martre P, et al., Proteomic and morphological analysis of early stages of wheat grain development. Proteomics 10: 2901-2910 (2010).

[7] Stitt M and Fernie AR, From measurements of metabolites to metabolomics: an 'on the fly' perspective illustrated by recent studies of carbon-nitrogen interactions. Curr Opin Biotechnol 14: 136-144 (2003).

[8] Schauer N and Fernie AR, Plant metabolomics: towards biolog-ical function and mechanism. Trends Plant Sci 11: 508-516 (2006).

[9] Qi XQ and Zhang DB, Plant metabolomics and metabolic biology. JIntegr Plant Biol 56: 814-815 (2014).

[10] Fait A, Angelovici R, Less H, Ohad I, Wochniak EU, Fernie AR, et al., Arabidopsis seed development and germination is associated with temporally distinct metabolic switches. Plant Physiol 142: 839-854 (2006).

[11] Riedelsheimer C, Lisec J, Eysenberg AC, Sulpice R, Flis A, Grieder C, et al., Genome-wide association mapping of leaf metabolic pro-files for dissecting complex traits in maize. Proc Natl Acad Sci 109: 8872-8877 (2012).

[12] Hu CY, Shi JX, Quan S, Cui B, Kleessen S, Nikoloski Z, et al., Metabolic variation between japonica and indica rice cultivars as revealed by non-targeted metabolomics. Sci Rep 4: 5067 (2014).

[13] Bonte A, Neuweger H, Goesmann A, Thonar C, Mäder P, Langenkämper G, et al., Metabolite profiling on wheat grain to enable a distinction of samples from organic and conventional farming systems. J Sci Food Agric 94: 2605-2612 (2014).

[14] Bowne JB, Erwin TA, Juttner J, Schnurbusch T, Langridge P, Bacic A, et al., Drought responses of leaf tissues from wheat cultivars of differing drought tolerance at the metabolite level. Mol Plant 5: 418-429 (2012).

[15] Baker JM, Hawkins ND, Ward JL, Lovegrove A, Napier JA, Shewry PR, et al., A metabolomic study of substantial equivalence of field-grown genetically modified wheat. Plant Biotechnol J 4: 381-392 (2006).

[16] Dupont FM and Altenbach SB, Molecular and biochemical impacts of environmental factors on wheat grain development and protein synthesis. J Cereal Sci 38: 133-146 (2003).

[17] Rao J, Cheng F, Hu CY, Quan S, Lin H, Wang J, et al., Metabolic map of mature maize kernels. Metabolomics 10: 775-787 (2014).

[18] Kusano M, Fukushima A, Redestig H and Saito K, Metabolomic approaches toward understanding nitrogen metabolism in plants. J Exp Bot 62: 1439-1453 (2011).

[19] Lei ZT, Huhman DV and Sumner LW, Mass spectrometry strategies in metabolomics. J Biol Chem 286: 25435-25442 (2011).

[20] Sumner LW, Mendes P and Dixon RA, Plant metabolomics: large-scale phytochemistry in the functional genomics era. Phytochemistry 62: 817-836 (2003).

[21] Bedair M and Sumner LW, Current and emerging mass-spectrometry technologies for metabolomics. Trends Anal Chem 27: 238-250 (2008).

[22] Sun H, Yan S, Jiang W, Li G and MacRitchie F, Contribution of lipid to physicochemical properties and-Mantou-making quality of wheat flour. Food Chem 121: 332-337 (2010).

[23] Kanehisa M and S Goto, KEGG: Kyoto Encyclopedia of Genes and Genomes. Nucleic Acids Res 28: 27-30 (2000).

[24] De Hoon MJL, Imoto S, Nolan J and Miyano S, Open source clustering software. Bioinformatics 20: 1453-1454 (2004).

[25] Stamova BS, Roessner U, Suren S, Chingcuanco DL, Bacic A and Beckles DM, Metabolic profiling of transgenic wheat over-expressing the high-molecular-weight Dx5 glutenin subunit. Metabolomics 5: 239-252 (2009).

[26] Lin H, Rao J, Shi JX, Hu CY, Cheng F, Wilson ZA, et al., Seed metabolomic study reveals significant metabolite variations and correlations among different soybean cultivars. JIntegr Plant Biol 56: 826-

836 (2014).

[27] Shewry PR, Improving the protein content and composition of cereal grain. J Cereal Sci 46: 239-250 (2007).

[28] Obendorf RL, Oligosaccharides and galactosyl cyclitols in seed desiccation tolerance. Seed Sci Res 7: 63-74 (1997).

[29] Sousa CAF and Sodek L, Alanine metabolism and alanine amino transferase activity in soybean (Glycine max) during hypoxia of the root system and subsequent return to normoxia. Environ Exp Bot 50: 1-8 (2003).

[30] Sebastia CH, Marsolais F, Saravitz C, Israel D, Dewey RE and Huber SC, Free amino acid profiles suggest a possible role for asparagine in the control of storage-product accumulation in developing seeds of low-and high-protein soybean lines. J Exp Bot 56: 1951-1963 (2005).

[31] Gaufichon L, Cren MR, Rothstein SJ, Chardon F and Suzuki A, Biological functions of asparagine synthetase in plants. Plant Sci 179: 141-153 (2010).

[32] Angelovici R, Galili G, Fernie AR and Fait A, Seed desiccation: a bridge between maturation and germination. Trends Plant Sci 15: 211-218 (2010).

[33] FaitA, Fromm H, Walter D, Galili G and Fernie AR, Highway or byway: the metabolic role of the GABA shunt in plants. Trends Plant Sci 13: 14-19 (2007).

[34] Rajjou L, Gallardo K, Debeaujon I, Vandekerckhove J, Job C and Job D, The effect of-amanitin on the Arabidopsis seed proteome highlights the distinct roles of stored and neosyn-thesized mRNAs during germination. Plant Physiol 134: 1598-1613 (2004).

[35] Ma WJ, Appels R, Bekes F, Larroque O, Morell MK and Gale KR, Genetic characterisation of dough rheological properties in a wheat doubled haploid population: additive genetic effects and epistatic interactions. Theor Appl Genet 111: 410-422 (2005).

[36] Zheng ZL, Carbon and nitrogen nutrient balance signaling in plants. Plant Signal Behav 4: 584-591 (2009).

[37] Weschke W, Panitz R, Sauer N, Wang Q, Neubohn B, Weber H, et al., Sucrose transport into barley seeds: molecular characterization of two transporters and implications for seed development and starch accumulation. Plant J 21: 455-457 (2000).

[38] Handley LW, Pharr DM and McFeeters RF, Carbohydrate changes during maturation of cucumber fruit: implications for sugar metabolism and transport. Plant Physiol 72: 498-502 (1983).

[39] Chen GX, Zhu JT, Zhou JW, Subburaj S, Zhang M, Han CX, et al., Dynamic development of starch granules and the regulation of starch biosynthesis in Brachypodium distachyon: comparison with common wheat and Aegilops peregrina. BMC Plant Biol 14: 198 (2014).

[40] Tiessen A, Hendriks JHM, Stitt M, Branscheid A, Gibon Y, Farré EM, et al., Starch synthesis in potato tubers is regulated by post-translational redox modification of ADP-glucose pyrophosphorylase: a novel regulatory mechanism linking starch synthesis to the sucrose supply. Plant Cell 14: 2191-2213 (2002).

[41] Xu BS, Li T, Deng ZY, Chong K, Xue YB and Wang T, Dynamic proteomic analysis reveals a switch between central carbon metabolism and alcoholic fermentation in rice filling grains. PlantPhysiol 148: 908-925 (2008).

[42] Martin C and Smith AM, Starch biosynthesis. Plant Cell 7: 971-985 (1995).

[43] Fernie A, Carrari F and Sweetlove LJ, Respiratory metabolism: glycolysis, the TCA cycle and mitochondrial electron transport. Curr Opin Plant Biol 7: 254-261 (2004).

[44] Chapman KD, Dyer JM and Mullen RT, Biogenesis and functions of lipid droplets in plants. J Lipid Res 53: 215-226 (2012).

[45] Banas A, Debski H, Banas W, Heneen WK, Dahlqvist A, Bafor M, et al., Lipids in grain tissues of oat (Avena sativa): differences in content, time of deposition, and fatty acid composition. J Exp Bot 58: 2463-2470 (2007).

[46] Baud S, Boutin JP, Miquel M, Lepiniec L and Rochat C, An integrated overview of seed development in Arabidopsis thaliana ecotype WS. Plant Physiol Biochem 40: 151-160 (2002).

Effects of water-deficit and high-nitrogen treatments on wheat resistant starch crystalline structure and physicochemical properties

Jian Xia[a,1], Dong Zhu[a,1], Hongmiao Chang[a], Xing Yan[b,*], Yueming Yan[a,c,*]

[a] *College of Life Science, Capital Normal University*, 100048 *Beijing, China*

[b] *College of Global Change and Earth System Science, Beijing Normal University, Beijing* 100875, *China*

[c] *Hubei Collaborative Innovation Center for Grain Industry (HCICGI), Yangtze University*, 434025 *Jingzhou, China*

* Corresponding authors

E-mail addresses: yanxing@bnu.edu.cn (X. Yan), yanym@cnu.edu.cn (Y. Yan).

[1] Contributed equally to this work.

Abstract: This work investigated the effects of water-deficit and high-nitrogen (N) treatments on wheat resistant starch (RS) formation, molecular structure, and physicochemical properties. The results of consecutive 2-year field experiments revealed that water deficit significantly reduced starch granule number and diameter, amylose, RS content, RS particle size distribution, and physicochemical properties, including peak and trough viscosities, oil absorption capacity, and freeze-thaw stability. Water deficit also altered the long-and short-range structures of RS. In contrast, high-N fertilizer application significantly improved the RS content, long-and short-range structures, and physicochemical properties. Pearson correlation analysis revealed that RS content was positively correlated with total starch, amylose, rapidly digesting starch, 90th percentile of RS particle size, relative crystallinity, infrared 1047/1022 cm^{-1} ratio, peak and breakdown viscosities, oil absorption capacity, and freeze-thaw stability, and was negatively correlated with slowly digestible starch content, 1022/995 cm^{-1} ratio, and final viscosity.

Keywords: High-N fertilizer, Molecular structures, Physicochemical properties, Water deficit, Wheat resistant starch

1 Introduction

Wheat (*Triticum aestivum* L.) is the staple food and principal protein source for approximately 35% of the global population. Starch, the most abundant carbohydrate and energy source for humans, is the main constituent of wheat endosperm, accounting for 65—75% of the total dry grain weight. Triticeae contain A-type (diameter > 10μm) and B-type (diameter 5—10μm) starch granules, which are initiated at 4-14 and 10-16 days post-anthesis (DPA), respectively (Cao et al.,

Published in Carbohydrate Polymers, 2020, 234: 115905.

2015). Amylose and amylopectin, two starch polysaccharides, are both composed of hundreds of D-glucose units linked to each other with α- (1, 4) glycoside bonds and branches with α- (1,6) glycoside bonds. These vary in structural order and physicochemical properties across botanical species, plant varieties, and genotypes (Xia, Zhu, Wang, Cui, & Yan, 2018) and due to various environmental factors (biotic and abiotic stresses) (Dai, Yin, & Wang, 2009; Yu et al., 2016). Therefore, the composition and content of seed starch are closely related to wheat grain yield, nutritional properties, and processing quality.

In terms of digestibility behaviour, starch can be classified into rapidly digestible starch (RDS), slowly digestible starch (SDS), and resistant starch (RS) (Englyst, Kingman, & Cummings, 1992). RS, the sum of undegraded and unabsorbed starch after ingestion, is fermented into short-chain fatty acids (SCFAs) in the colon (Englyst et al., 1992). Depending upon the botanical source and nature, RS can be further divided into five types: RS_1, physically inaccessible starch; RS_2, uncooked starch granules; RS_3, retrograded starch; RS_4, chemically modified starch; and RS_5, amylaselipid complexes (Morell, Konik-Rose, Ahmed, Li, & Rahman, 2004; Xia et al., 2018). SCFAs are the functional form of RS; they directly provide 5—10% of energy to the body and help prevent non-communicable diseases such as type 2 diabetes, colon and rectal cancer, and chronic kidney and liver disease (Lockyer & Nugent, 2017). Therefore, enriching the RS component in cereal crops or starchy foodstuffs is a feasible way to improve human health.

With ongoing climate change, drought is a common abiotic stress and has a significant negative impact on wheat growth and grain yield (Deng et al., 2018). Under water-deficit conditions, starch biosynthesis is impeded, leading to significant reductions in total starch, the amylopectin content, and the amylose/amylopectin ratio (Chen et al., 2017; Lu et al., 2014). Water-deficit conditions significantly decrease the starch granule size distribution, number of B-type starch granules, and ratio of A-/B-type starch granules in wheat endosperm (Fábián, Jäger, Rakszegi, & Barnabás, 2011; Lu et al., 2014; Zhou et al., 2018). Water-deficit conditions also significantly reduce the RDS and SDS contents of retrograded starches, degree of starch hydrolysis, and expression of the main starch biosynthesis enzymes (Chen et al., 2017; Yu et al., 2016). Similar results have been obtained for other crops, including maize (Mohammadkhani & Heidari, 2008), barley (Worch et al., 2011), sorghum (Yi et al., 2014), and soybean (Liu, Jensen, & Andersen, 2004).

Nitrogen (N) is a critical inorganic nutrient in the synthesis of amino acids, chlorophyll, nucleotides, and numerous other metabolites and cellular components. N fertilizer can improve the tillering rate and amount and size of spikes, and promote grain weight and quality (Zhen et al., 2016). High N uptake increases the protein components of albumins, globulins, glutenins, and gliadins, and the activity of key starch biosynthesis enzymes in seeds (Zhen et al., 2017; Zhen, Deng, Li, Zhu, & Yan, 2018). High-N fertilizer produces more and larger A- and B-type starch granules, which improve the total starch content, grain yield, and bread-making quality (Zhen et al., 2017). Under high-N conditions, total starch and amylopectin contents increase significantly, whereas the ratio of amylose to amylopectin decreases significantly (Fu et al., 2008). Furthermore, treatment with high-N fertilizer alters the amylopectin branch ratio and starch properties including digestibility, pasting properties, and thermal properties (Gous, Warren, Mo, Gilbert, & Fox, 2015).

However, changes in the content, molecular structure, and physicochemical properties of wheat RS under water-deficit and high-N fertilizer conditions are still not clear. Therefore, this work is the first comprehensive investigation of how water-deficit and high-N fertilizer application affect wheat RS formation, molecular structure, and physicochemical properties. The results provide new insights into the mechanisms of crop RS formation under abiotic stresses.

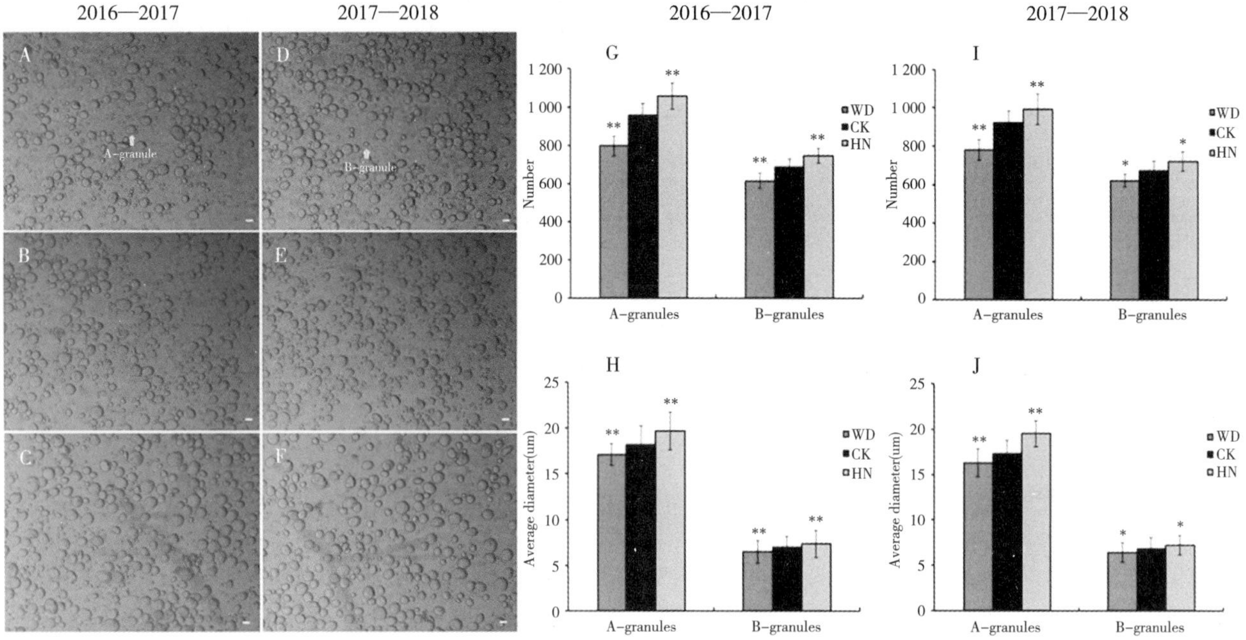

Fig. 1 Changes in the number and diameter of purified starch granules from the elite Chinese bread wheat cultivar Zhongmai 175 under water-deficit and high-N fertilizer treatments in the 2016—2017 and 2017—2018 growing seasons. A-C are the purified starch granules from the control (CK), water-deficit (WD) and high-N fertilizer (HN) treatments in 2016—2017, respectively. D-F are the respective values for 2017—2018. The scale bar represents 20μm and A-and B-granules are indicated. G-H are the number and diameter of A-and B-granules in 2016—2017, respectively, and I-J are the values for 2017—2018.

Table 1 Chemical composition of wheat starches with water-deficit and high-N fertilizer treatments

Treatments	TS (%)	AM (%)	AP (%)	RDS (%)	SDS (%)	NRS (%)	RS (%)
2016—2017							
CK	63.71± 0.59	21.92± 0.34	41.80± 1.06	31.79± 0.34	24.70± 1.13	1.19± 0.092	8.28± 0.11
WD	60.23± 0.20*	18.46± 0.12**	41.77± 0.14	29.70± 1.05*	24.22± 1.28	1.10± 0.147*	7.35± 0.04**
HN	70.47± 1.04**	24.29± 0.39**	46.17± 0.31**	38.44± 0.70**	21.62± 0.68**	1.28± 0.055*	9.42± 0.04**
2017—2018							
CK	63.04± 0.53	21.30± 0.23	41.75± 0.76	31.30± 0.91	24.86± 0.75	1.12± 0.023	7.89± 0.12*
WD	59.87± 0.90*	18.38± 0.64**	41.48± 0.91	28.63± 0.84**	24.66± 0.97	1.06± 0.031*	7.27± 0.35
HN	71.69± 0.75**	23.71± 0.59**	47.99± 1.26**	37.56± 0.29**	22.42± 0.56**	1.25± 0.037**	9.03± 0.19**

CK, control group; WD, water-deficit group; HN, high-nitrogen group; TS, total starch; AM, amylose; AP, amylopectin; AM/AP, amylose/amylopectin ratio; RDS, rapidly digesting starch; SDS, slowly digesting starch; NRS, native resistant starch; RS, resistant starch.

* $p < 0.05$.

** $p < 0.01$.

Table 2 Changes in RS particle size distribution and crystalline structures under water-deficit and high-N fertilizer conditions

Treatments	d (0.1) (μm)	d (0.5) (μm)	d (0.9) (μm)	D (3, 2) (μm)	D (4, 3) (μm)	RC (%)	IR_1	IR_2
2016—2017								
CK	95.95± 0.21	262.0± 1.41	560.0± 14.1	180.0± 0.00	299.5± 0.71	6.80± 0.23	1.064± 0.023 6	0.946± 0.047 8

(continued)

Treatments	d (0.1) (μm)	d (0.5) (μm)	d (0.9) (μm)	D (3, 2) (μm)	D (4, 3) (μm)	RC (%)	IR_1	IR_2
WD	73.75± 0.21**	246.5± 3.54*	108.5± 0.71	108.5± 0.71**	306.0± 14.14	3.28± 0.26**	1.022± 0.003 3*	1.026± 0.012 5*
HN	138.0± 4.24**	423.5± 2.12**	824.5± 4.94**	178.5± 5.66	458.0± 1.41**	7.83± 0.01*	1.119± 0.014 3*	0.840± 0.031 7*
2017—2018								
CK	96.9± 0.14	267.0± 1.41	575.5± 17.68	181.0± 0.00	307.5± 5.66	6.52± 0.13	1.076± 0.001 2	0.933± 0.042 1
WD	73.5± 0.71**	245.5± 3.54*	108.5± 0.71	108.5± 0.925**	293.5± 13.44	3.36± 0.32**	1.042± 0.019 4*	1.006± 0.008 8*
HN	139.0± 2.83**	430.0± 1.41**	841.5± 5.01**	179.5± 2.12	466.5± 2.12**	7.39± 0.06*	1.084± 0.001 5**	0.733 ± 0.011 8**

CK, control group; WD, water-deficit group; HN, high-nitrogen group.

d (0.1), d (0.5), and d (0.9) are the granule sizes at which 10%, 50%, and 90% of all the granules by volume are smaller, respectively; D (3, 2) is the surface area weighted mean diameter; D (4, 3) is the volume weight mean diameter, RC, relative crystallinity; IR, infrared ratio, IR_1, the proportion of absorbance 1047/1022 cm^{-1}; IR_2, the proportion of absorbance 1022/995 cm^{-1}.

* $p < 0.05$.

** $p < 0.01$.

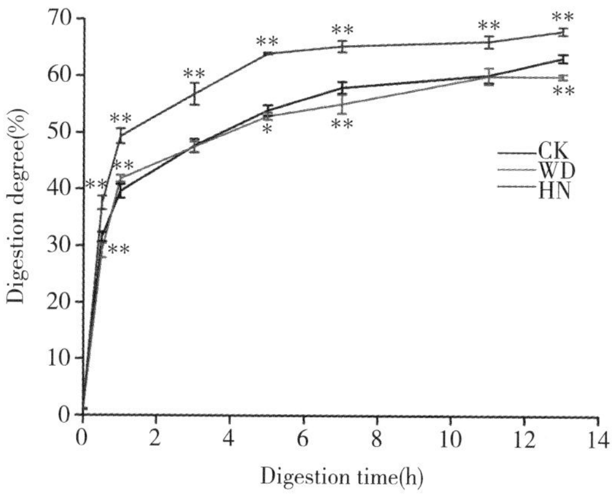

Fig. 2 Hydrolysis curves of wheat starch under water-deficit and high-N fertilizer conditions. The data shown in the figure are the averages from two consecutive wheat growing seasons.

2 Materials and methods

2.1 Plant materials, field trial and treatments

The elite Chinese bread wheat cultivar Zhongmai 175 (*Triticum aestivum* L.) was used as material, which has high water and fertilizer use efficiency, high yielding potential, and broad adaptation, and is widely cultivated in the winter wheat region of northern China plain (He et al., 2015). Field experiments were performed in the Wuqiao experimental station (China Agricultural University) during two consecutive wheat growing seasons (2016—2017 and 2017—2018).

A randomized block design with three biological replicates (each plot with 30 m^2) was performed, and the base nitrogen fertilizer application was 120kg · hm^{-2} urea $(NH_2)_2CO$. Three different treatments were conducted: the normal control group (CK) with watering 75 mm water at heading and flowering stages and 120kg · hm^{-2} nitrogen fertilizer as local field cultivation; water-deficit group (WD) without watering at heading and flowering stages (rainfed), and with normal nitrogen fertilizer; high-nitrogen group (HN) with normal watering at heading and flowering stages and 240kg · hm^{-2} nitrogen fertilizer.

2.2 Starch granule purification, microscopic observation and total starch extraction

Wheat starch granules were purified from mature seeds using the procedure described by Cao et al. (2015). The morphology of starch granules under three field treatments was observed by using an inverted microscopy (DMi8, Leica, Germany) with bright field light. Total starch was extracted by using 30 mL wash buffer [55 mM Tris-HCl, pH 6.8; 2.3% (w/v) sodium dodecyl sulfate; 1% (w/v) dithiothreitol; and 10% (v/ v) glycerol]. The starch was released from amyloplasts by using ultrasonic cell pulverizer (SCIENTZ-IID, SCIENTZ, China) at a power of 20 W for 20 s, the mixed solution was centrifuged at 3500 × g for 5 min. This starch released procedure was repeated twice, then wash buffer was removed. The total starch powder was obtained and air-dried at room temperature, and then stored at −20℃ prior to use.

2.3 Measurement of total starch, amylose and amylopectin content

The content of total starch and amylose content was measured *in vitro* by Resistant Starch Assay Kit and Amylose/Amylopectin Assay Kit (Megazyme Int. Ireland Ltd. Co., Wicklow, Ireland), respectively. The content of amylopectin was designated as total starch content minus amylose content.

2.4 RS preparation

RS was prepared based on the method of Ashwar et al. (2016) with minor modifications. The prepared RS samples were dried in an 45℃ oven to constant weight, put them in a ground mill to pass through a 1.0 mm sieve, and then used for molecular structure and physicochemical properties characterization.

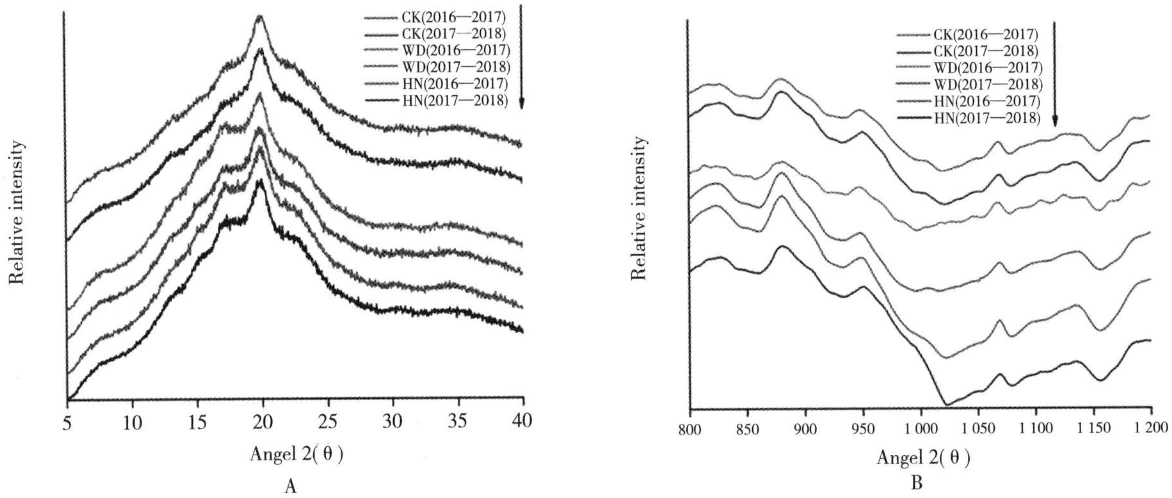

Fig. 3 Changes in the long-and short-range crystalline structures of resistant starch from water-deficit and high-N treatments. (A) X-ray powder diffraction (XRD) patterns. (B) Fourier-transform infrared spectroscopy (FTIR) spectra.

Table 3 Changes in RS physicochemical properties under water-deficit and high-N fertilizer conditions

Treatments	PV (cP)	TV (cP)	BV (cP)	FV (cP)	SV (cP)	PT (min)	OAC ($g \cdot g^{-1}$)	FTS (%)
2016—2017								
CK	25.67± 0.58	23.12± 1.00	2.67± 0.58	26.00± 1.73	3.25± 1.00	2.00± 0.52	1.93± 0.038	6.33± 0.34
WD	16.67± 1.53**	14.67± 1.53**	2.00± 0.00**	54.67± 0.58**	40.25± 1.73**	6.60± 0.00**	1.19± 0.140**	5.40± 0.46**
HN	45.67± 0.58**	5.13± 0.00**	40.67± 0.58**	26.67± 0.58	21.67± 0.58**	3.14± 0.07**	2.02± 0.016*	7.04± 0.18**
2017—2018								
CK	25.67± 0.58	22.67± 1.53	3.00± 1.00	25.67± 0.58	3.14± 1.00	1.90± 0.03	1.97± 0.080	6.37± 0.36
WD	14.67± 1.53**	12.67± 1.53**	2.00± 0.00**	54.00± 3.00**	41.33± 1.53**	6.62± 0.10**	1.15± 0.245**	5.59± 0.25**
HN	45.12± 2.00**	6.14± 0.00**	44.00± 2.00**	25.00± 1.00	22.15± 1.00**	2.91± 0.03**	2.16± 0.079**	8.12± 0.81**

CK, control group; WD, water-deficit group; HN, high-nitrogen group; PV, peak viscosity; TV, trough viscosity; BV, breakdown viscosity; FV, final iscosity; SV, setback viscosity; PT, pasting temperature; OAC, oil absorption capacity; FTS, freeze-thaw stability.

* $p < 0.05$.

** $p < 0.01$.

2.5 Determination of RDS, SDS, RS and in vitro starch digestibility

Starch nutrition fractions were evaluated *in vitro* by using Resistant Starch Assay Kit (Megazyme Int. Ireland Ltd. Co., Wicklow, Ireland) based on McCleary, McNally, and Rossiter (2002). The percentage of starch solubilized and hydrolyzed to D-glucose at 20 min (RDS), 2 h (SDS) and 16 h (RS) was assessed. The starch (100 mg, dry base, db) was hydrolyzed by pancreatic α-amylase (30U · mL^{-1}) and amyloglucosidase (AMG, 300U · mL^{-1}) with constant shaking (200 rpm) at 37℃ for different hydrolysis times. Then, digestible starch portions were converted gradually to D-glucose, and ethanol (99% v/v) was added to inactivate enzymes. After centrifuging at 1 500×g for 10 min, the supernatant solution was collected, and the rest of pellets was re-suspended by industrial methylated spirits (~95% v/v ethanol plus 5% v/v methanol) for twice to make D-glucose completely dissolved. The amount of D-glucose dissolved in the above three steps (*in vitro* starch digestibility) was measured with glucose oxidase/peroxidase reagent after centrifuging. In addition, pellets through 16 h partial solubilization were dissolved completely in KOH solution (2 M), and then transformed into D-glucose by AMG (3300U · mL) to determine RS content. The amount of D-glucose was converted by the absorbance value of glucose oxidase/peroxidase reagent at 510 nm.

In vitro starch digestibility was studied using HCl to substitute pancreatic α-amylase and AMG, the other procedures of digestibility were followed above. The digestion time points at 0.5, 1, 3, 5, 7, 11 and 13 h were chosen, and the starch digestion rate was computed by measuring the content of D-glucose in the supernatant fraction of the digested starch. The kit mentioned above was used to determine D-glucose content.

2.6 RS molecular structural analysis

2.6.1 Determination of RS particle size distribution

The RS particle size distribution of all measurements was calculated by using malvern laser diffraction particle size analyzer (Mastersizer 2000, Malvern, England). The particle size was defined based on the 10th percentile [d (0.1)], median [d (0.5)], 90th percentile [d (0.9)], surface-weighted mean [D (3,2)], and volume-weighted mean [D (4,3)] according to Lin et al. (2016).

2.6.2 X-ray powder diffraction (XRD)

Long-range crystalline structure of RS samples was analyzed with an XRD (D8 ADVANCE, Bruker, Germany). The following working conditions were used: targeting the diffraction angle (2θ) from 5° to 45° with a constant scanning rate of 0.5° · min^{-1}, 75% relative humidity and voltage of 40 kV and 30 mA. Spectral analysis and crystallinity determination were carried out by using MDI Jade 5 software.

2.6.3 Fourier-transform infrared spectroscopy (FTIR)

Short-range crystalline structure of RS samples was analyzedby FTIR system (Nicolet 6700, ThermoFisher, USA). The 2 mg dried test specimen was ground and mixed well with 150 mg KBr. Tests were completed at room temperature, the scanning range of FTIR spectrum is 1200-800 cm^{-1} at a resolution of 4 cm^{-1}. The amplitudes of absorbance for each RS samples at 1047, 1022 and 995 cm^{-1} were extracted, and starch surface short-range ordered structures of the test specimens were estimated refer to Hu et al. (2018).

2.7 Determination of RS physicochemical properties

2.7.1 Rapidvisco analyzer analysis

Pasting properties of RS samples were detected by using arapid visco analyzer (RVA, MCR302, Anton Paar, Austria). The heating and cooling procedure described earlier (Kong, Zhu, Sui, & Bao, 2015) was as follows: 3 g (db) RS samples were dispersed into 25 mL ddH_2O, hold at 50℃ for 1 min, heated to 95℃ within 7.5 min, equilibrated for 5 min, cooled to 50℃ in 7.5 min and equilibrated for 3 min.

2.7.2 Oil absorption capacity

The method described by Ashwar et al. (2016) was used to analyze the oil absorption capacity of specimens. Starch paste (starch: mustard oil, 1 : 8 w/v) vortexed for 30 min, and then centrifuged at 3000×g for 10 min and poured out the supernatant. The oil absorption capacity is equivalent to the percentage increase in sediment quality.

2.7.3 Freeze-thaw stability

Freeze-thaw stability was analyzed based on Ashwar et

al. (2016) with minor modifications. Mixed starch (db) with water in a ratio of 1 g/9 mL was put in a 90℃ water shaking bath for 30 min, cooled to 4℃ in refrigerator and frozen at −20℃ for 24 h, and then thawed at room temperature for 2 h. This step was repeated five times. After centrifuging at 1000×g for 20 min, the percentage dehydration of starch was used as the index to estimate the freeze-thaw stability that resisted the unpopular physical variation when freezing-to-thawing.

2.8 Statistical analysis

The statistical analyses were performed in three biological replicates by using SPSS 19 (IBM, Chicago, USA). Results were reported as mean ± standard deviations for each sample. A comparison of the means was ascertained by Turkey's test at 5% and 1% levels of significance using one-way analysis of variance (ANOVA). A two-tailed Pearson correlation test was performed to reveal the correlations among RS-related parameters including structural and physicochemical properties.

3 Results and discussion

3.1 Effects of water deficit and treatment with high-N fertilizer on starch granule formation, starch chemical composition, and hydrolysis

Observation of purified starch granules after different treatments under ordinary light revealed clear separation of spherical or ellipsoid starch granules of different sizes (A-granules>10μm, B-granules 5～10μm; Fig. 1A-F). Under water deficit, the number and diameter of both A-and B-granules were significantly reduced, whereas the granule number and diameter increased markedly under high-N conditions (Fig. 1G-J). The results from the two growing seasons were very consistent.

Table 1 lists changes in total starch, amylose, amylopectin, and RS content under water-deficit and high-N treatments in the two consecutive growing seasons; these were similar in both growing seasons. Water deficit caused significant decreases in total starch (average-5.25%) and amylose (average-14.75%) contents, while amylopectin content did not change significantly. In contrast, high-N fertilizer significantly increased total starch, amylose, and amylopectin contents by more than 10%. These results demonstrate that both water-deficit and high-N treatments have significant effects on total starch and amylose contents, while amylopectin content is not susceptible to drought stress.

Unlike RS, both RDS and SDS are digested or assimilated completely (within 20 and 120 min, respectively) in the small intestine. Analysis revealed that water deficit reduced and high-N fertilizer promoted RDS, native NRS, and RS contents. The SDS content changed only slightly under water-deficit, but decreased significantly under high-N conditions. High-N conditions dramatically increased RDS and RS contents by 20.46% and 14.11%, respectively (Table 2).

Fig. 2 presents the hydrolysis properties of starches with different treatments. When digested by HCl, the hydrolysis of starch specimens from the three groups was rapid before 7 h and slower from 7 to 13 h. Compared with the control, the degree of starch hydrolysis was considerably enhanced in the high-N group, but was generally decreased significantly under water deficit, except for a significant increase at 1-3 h and no significant difference at 5 and 11 h. The amylose/amylopectin ratio, granule surface area and size, crystallinity, porosity, and integrity affect the starch digestibility rate in vitro (Blazek & Gilbert, 2010; da Rosa Zavareze & Dias, 2011). Our results demonstrated that both water-deficit and high-N treatment significantly affected starch granule number and size, and chemical composition, which is reasonable given the differences in starch hydrolysis under different conditions.

3.2 Characterisation of RS particle size distribution and long-and shortrange crystalline structures under different treatments

High-N treatment promoted and water-deficit treatment

reduced RS particle size. In particular, water deficit significantly decreased the diameters at d (0.1), d (0.5), and D (3, 2), while the measured values related to RS particle size all increased markedly with high-N fertilizer application (Table 2).

The long-range crystalline structure of RS samples, including X-ray diffractograms (XRD) (Fig. 3A) and relative crystallinity (Table 2), can be used to classify the crystalline structure into A-, B-, C-, and V-types (Fábián et al., 2011). Water-deficit treatment reduced and high-N treatment increased the relative crystallinity, but did not alter the XRD pattern of RS. With both treatments, RS had a B + V-type polymorph diffraction feature with a strong diffraction peak at 20.2° (2θ) and weak peaks at 6.3°, 17.2°, and 36-38° (2θ). The intensity at 20.2° (2θ) probably represented the single helices of linear starch chains crystallites, rather than amylose-lipid complexes (Varatharajan, Hoover, Liu, & Seetharaman, 2010). The characteristic diffraction peaks at 17.2° (2θ) gradually became sharppointed with water-deficit treatment and ill-defined with high-N treatment. Consistent with the changes in the diffraction peaks, the relative crystallinity decreased from 6.66% (control) to 3.32% (water-deficit treatment) and increased from 6.66% to 7.61% (high-N treatment) (Table 2). These altered peak intensities and relative crystallinities suggest that the long-range crystalline structure of RS changed with different treatments, consistent with the changes in the RS particle size distribution. Compared with the high-N condition, the lower relative crystallinity under water deficit was mainly attributed to changes in the amylose content, the amylose/ amylopectin ratio, and the larger proportion of short/long amylopectin branches (Hyunseok & Kerryc, 2010).

Fig. 3B presents the infrared spectra under water-deficit and high-N treatments; these reflect the short-range crystalline structure of RS at a particular absorption range (1200-800 cm^{-1}). The series of bands in this area is chiefly caused by CeO and CCe stretching vibration and is susceptible to variation in the carbohydrate physical state, which reflects the ordered information on the starch molecular chain structure (Lin et al., 2016). The RS specimens from different treatments shared a similar deconvolution spectrum and had identical primary characteristic peaks, with no fresh absorption peaks. The crystalline and amorphous regions in RS were seen in the infrared spectrum at 1047 and 1022 cm^{-1}, respectively. The structural order degree of RS and amount of ordered crystalline to amorphous domains can be quantified using the infrared ratio at 1047/1022 cm^{-1} (Sharma, Yadav, Singh, & Tomar, 2015). The intensity ratios 1047/1022 cm^{-1} (IR_1) and 1022/995 cm^{-1} (IR_2) are useful indices of FTIR data compared with other measures of starch conformation. The 1047/1022 cm^{-1} ratio of absorbance can be used to quantify the degree of order, while that of 1022/995 cm^{-1} serves as a measure of the proportion of amorphous to ordered carbohydrate structure (Qin et al., 2012). Our results indicate that high-N fertilizer significantly promoted the 1047/1022 cm^{-1} ratio and reduced the 1022/995 cm^{-1} ratio, while opposite changes were seen under water deficiency (Table 2). These results are consistent with those of Qin et al. (2012), who reported that high amylose led to high and low infrared 1045/1022 and 1022/995 cm^{-1} ratios, respectively. Thus, the RS from the high-N group had a more ordered short-range structure compared with the water-deficit treatment, in agreement with the changes in relative crystallinity. The disruption of RS short-range crystalline structures by water deficit might result from excessive chain mobility (Hu et al., 2018).

3.3 Changes of RS physicochemical properties in response to different treatments

Table 3 lists changes in the physicochemical properties of wheat RS under different treatments in two consecutive growing seasons. Pasting property parameters include peak (PV), trough (TV), breakdown (BV), final (FV), and setback (SV) viscosities and pasting temperature (PT). The pasting properties of starch are related to granule size, amylase content and molecular size, amylopectin chain-length distribution,

Table 4 Pearson correlation coefficients between RS-related parameters

	RS	TS	AM	AP	RDS	SDS	NRS	d (0.9)	RC	IR_1	IR_2	PV	BV	FV	PT	OAC
TS	0.965**															
AM	0.978**	0.942**														
AP	0.860*	0.955**	0.799*													
RDS	0.980**	0.987**	0.939**	0.934**												
SDS	−0.861*	−0.882*	−0.748*	−0.914**	−0.925**											
NRS	0.691	0.596	0.614	0.522	0.622	−0.587										
d (0.9)	0.941**	0.926**	0.989**	0.780*	0.908**	−0.684	0.516									
RC	0.907**	0.849*	0.973**	0.658	0.838**	−0.581	0.529	0.982**								
IR_1	0.914**	0.847*	0.927**	0.693	0.871*	−0.730	0.427	0.911**	0.906**							
IR_2	−0.877*	−0.965**	−0.883*	−0.945**	−0.914**	0.773*	−0.460	−0.899**	−0.807*	−0.761*						
PV	0.979**	0.994**	0.959**	0.928**	0.995**	−0.886**	0.582	0.941**	0.875*	0.889**	−0.940**					
BV	0.895**	0.960**	0.822*	0.988**	0.959**	−0.958**	0.539	0.790*	0.678	0.762*	−0.917**	0.946**				
FV	−0.792*	−0.752*	−0.902**	−0.548	−0.715	0.399	−0.407	−0.941**	−0.972**	−0.808*	−0.758*	−0.773*	−0.543			
PT	−0.636	−0.564	−0.780*	−0.320	−0.529	0.176	−0.303	−0.830*	−0.900**	−0.698	0.573	−0.595	−0.315	0.967**		
OAC	0.842*	0.820*	0.935**	0.640	0.786*	−0.493	0.434	0.970**	0.980**	0.831*	−0.823*	0.837*	0.632	−0.992**	−0.932**	
FTS	0.867*	0.951**	0.890**	0.914**	0.892**	−0.720	0.463	0.913**	0.832*	0.753*	−0.996**	0.925**	0.879*	−0.800*	−0.631	0.857*

* $p < 0.05$, ** $p < 0.01$

RS, resistant starch; TS, total starch; AM, amylose; AP, amylopectin; RDS, rapidly digesting starch; SDS, slowly digesting starch; NRS, native resistant starch; RC, relative crystallinity; d (0.9): the granule sizes at which 90% of all the granules by volume are smaller; IR, infrared ratio, IR_1, the proportion of absorbance 1047/1022 cm^{-1}; IR_2: the proportion of absorbance 1022/995 cm^{-1}; PV, peak viscosity; BV, breakdown viscosity; FV, final viscosity; PT, pasting temperature; OAC, oil absorption capacity; and FTS, freeze-thaw stability.

and amylase-lipid complex (Gous et al., 2015; Zi et al., 2019). Peak viscosity represents the swelling power of starch granules and generally determines the foodstuff quality and texture (Thitisaksakul, Jiménez, Arias, & Beckles, 2012), which are positively correlated with large starch granules and negatively correlated with small starch granules (Singh, Singh, Singh, & Singh, 2008). The results indicate that high-N treatment significantly increased PV, BV, SV, and PT, and reduced TV. However, water-deficit treatment significantly promoted FV, SV, and PT, and decreased PV, TV, and BV.

Oil absorption capacity (OAC) analysis of RS revealed that water-deficit and high-N treatment had negative and positive effects on OAC, respectively (Table 3). In the control group, OAC ranged from 1.93 to 1.97 $g \cdot g^{-1}$; it decreased significantly by about 40% under water deficit and increased by 7.15% under high-N treatment. Generally, oil absorption occurs because oil is adsorbed between the molecular structures of starch or amylose-lipid complexes (Ashwar et al., 2016).

Freeze-thaw stability (FTS) is a manifestation of starch retrogradation, which is also a critical consideration in the processing and transportation of starchy foodstuffs (Ashwar et al., 2016). The temperature fluctuation of starchy foods after freezing-thawing will result in water loss, skin cracking, hardening, and brittleness during storage, transportation, and sale (Wang et al., 2013). These phenomena are due to the enhanced molecular associations between starch chains and pushing water outward from the gel (Ashwar et al., 2016; Morris, 1990). As shown in Table 3, water-deficit treatment reduced, and high-N treatment promoted, the freeze-thaw stability of RS. A previous study demonstrated that a higher amylose content led to a higher dehydration percentage of RS; the increase or decrease of starch syneresis might be related to interactions between amylose and certain gums (Charoenrein & Preechathammawong, 2012).

3.4 Correlation analysis among RS-related parameters

The Pearson correlation coefficients among 17 RS-related parameters (Table 4) revealed that the RS content is significantly positively correlated with total starch, amylose, RDS, 90th percentile of RS particle size, relative crystallinity, 1047/1022 cm^{-1} ratio, peak and breakdown viscosities, OAC, and FTS; RS content is significantly negatively correlated with SDS, the 1022/995 cm^{-1} ratio, and final viscosity, and not correlated with native RS or pasting temperature. In particular, the RS content was highly positively correlated with amylose content ($R^2 = 0.978$, $p < 0.01$), consistent with previous reports focusing on cereal starch (Bao, Zhou, Xu, He, & Park, 2017; Huang et al., 2015; Morales-Medina, del Mar Muñío, Guadix, & Guadix, 2014). Amylose and RS contents had strong and consistent correlations with other parameters. In rice, almost all physicochemical properties of starch are affected by amylose content (Qin et al., 2012).

4 Conclusions

Water-deficit and high-N fertilizer treatment had negative and positive effects, respectively, on RS formation, molecular structures, and physicochemical properties. Under water-deficit conditions, RS-related parameters decreased significantly, including amylose and RS contents, RS particle size distribution, relative crystallinity, 1047/1022 cm^{-1} ratio, peak and trough viscosities, oil absorption capacity, and freeze-thaw stability. Conversely, high-N fertilizer significantly promoted these parameters. Although XRD and FTIR of RS revealed similar patterns and deconvolution spectra, long-and short-range RS structures were significantly disordered by water deficit and improved by the high-N condition. Pearson correlation analysis revealed that RS content was positively correlated with total starch, amylose, RDS, 90th per-centile of RS particle size, relative crystallinity, the 1047/1022 cm^{-1} ratio, peak and breakdown viscosities, oil absorption capacity, and freeze-thaw stability, and was negatively correlated with SDS, the 1022/995 cm^{-1} ratio, and final viscosity.

Acknowledgements

This research was financially supported by grants from the National Key R & D Program of China (2016YFD0100502) and the National Natural Science Foundation of China (31971931). The English in this document has been checked by at least two professional editors, both native speakers of English. For a certificate, please see: http://www.textcheck.com/certificate/r2QAYH.

References

Ashwar, B. A., Gani, A., Wani, I. A., Shah, A., Masoodi, F. A., & Saxena, D. S. (2016). Production of resistant starch from rice by dual autoclaving-retrogradation treat-ment: *In vitro* digestibility, thermal and structural characterization. *Food Hydrocolloids*, 56, 108-117. https://doi.org/10.1016/j.foodhyd.2015.12.004.

Bao, J. S., Zhou, X., Xu, F. F., He, Q., & Park, Y. J. (2017). Genome-wide association study of the resistant starch content in rice grains. *Starch/Stärke*, 69 (7-8), 1600343. https://doi.org/10.1002/star.201600343.

Blazek, J., & Gilbert, E. P. (2010). Effect of enzymatic hydrolysis on native starch granule structure. *Biomacromolecules*, 11 (12), 3275-3289. https://doi.org/10.1021/bm101124t.

Cao, H., Yan, X., Chen, G. X., Zhou, J. W., Li, X. H., Ma, W. J., et al. (2015). Comparative proteome analysis of A-and B-type starch granule-associated proteins in bread wheat (*Triticum aestivum* L.) and *Aegilops crassa*. *Journal of Proteomics*, 112, 95-112. https://doi.org/10.1016/j.jprot.2014.08.002.

Charoenrein, S., & Preechathammawong, N. (2012). Effect of waxy rice flour and cassava starch on freeze-thaw stability of rice starch gels. *Carbohydrate Polymers*, 90 (2), 1032-1037. https://doi.org/10.1016/j.carbpol.2012.06.038.

Chen, G. X., Zhen, S. M., Liu, Y. L., Yan, X., Zhang, M., & Yan, Y. M. (2017). *In vivo* phosphoproteome characterization reveals key starch granule-binding phosphoproteins involved in wheat water-deficit response. *BMC Plant Biology*, 17 (1), 168. https://doi.org/10.1186/s12870-017-1118-z.

da Rosa Zavareze, E., & Dias, A. R. G. (2011). Impact of heat-moisture treatment and annealing in starches: A review. *Carbohydrate Polymers*, 83 (2), 317-328. https://doi.org/10.1016/j.carbpol.2010.08.064.

Dai, Z., Yin, Y., & Wang, Z. (2009). Starch granule size distribution from seven wheat cultivars under different water regimes. *Cereal Chemistry*, 86 (1), 82-87. https://doi.org/10.1094/CCHEM-86-1-0082.

Deng, X., Liu, Y., Xu, X. X., Liu, D. M., Zhu, G. R., Yan, X., et al. (2018). Comparative proteome analysis of wheat flag leaves and developing grains under water deficit. *Frontiers in Plant Science*, 9, 425. https://doi.org/10.3389/fpls.2018.00425.

Englyst, H. N., Kingman, S. M., & Cummings, J. H. (1992). Classification and measure-ment of nutritionally important starch fractions. *European Journal of Clinical Nutrition*, 46, S33-S50. https://doi.org/10.1128/IAI.01649-06.

Fábián, A., Jäger, K., Rakszegi, M., & Barnabás, B. (2011). Embryo and endosperm development in wheat (*Triticum aestivum* L.) kernels subjected to drought stress. *Plant Cell Reports*, 30 (4), 551-563. https://doi.org/10.1007/s00299-010-0966-x.

Fu, X. L., Wang, C. Y., Guo, T. C., Zhu, Y. J., Ma, D. Y., & Wang, Y. H. (2008). Effects of water-nitrogen interaction on the contents and components of protein and starch in wheat grains. *The Journal of Applied Ecology*, 19 (2), 317-322. https://doi.org/10.3724/SP.J.1005.2008.01083.

Gous, P. W., Warren, F., Mo, O. W., Gilbert, R. G., & Fox, G. P. (2015). The effects of variable nitrogen application on barley starch structure under drought stress. *Journal of the Institute of Brewing*, 121 (4), 502-509. https://doi.org/10.1002/jib.260.

He, Z. H., Chen, X. M., Wang, D. S., Zhang, Y., Xiao, Y. G., Li, F. J., et al. (2015). Characterization of wheat cultivar Zhongmai 175 with high yielding potential, high water and fertilizer use efficiency, and broad adaptability. *Scientia Agricultura Sinica*, 48(17), 3394-3403. https://doi.org/10.3864/j.issn.0578-1752.2015.17.007.

Hu, X. T., Guo, B. Z., Liu, C. M., Yan, X. Y., Chen, J., Luo, S. J., et al. (2018). Modification of potato starch by using superheated steam. *Carbohydrate Polymers*, 198, 375-384. https://doi.org/10.1016/

j. carbpol. 2018. 06. 110.

Huang, J., Shang, Z. Q., Man, J. M., Liu, Q. Q., Zhu, C. J., & Wei, C. X. (2015). Comparison of molecular structures and functional properties of high-amylose starches from rice transgenic line and commercial maize. *Food Hydrocolloids*, 46, 172-179. https://doi. org/10. 1016/j. foodhyd. 2014. 12. 019.

Hyunseok, K., & Kerryc, H. (2010). Physicochemical properties and amylopectin fine structures of A-and B-type granules of waxy and normal soft wheat starch. *Journal of Cereal Science*, 51 (3), 256-264. https://doi. org/10. 1016/j. jcs. 2009. 11. 015.

Kong, X. L., Zhu, P., Sui, Z. Q., & Bao, J. S. (2015). Physicochemical properties of starches from diverse rice cultivars varying in apparent amylose content and gelatinisation temperature combinations. *Food Chemistry*, 172, 433-440. https://doi. org/10. 1016/j. foodchem. 2014. 09. 085.

Lin, L. S., Guo, D. W., Zhao, L. X., Zhang, X. D., Wang, J., Zhang, F. M., et al. (2016). Comparative structure of starches from high-amylose maize inbred lines and their hybrids. *Food Hydrocolloids*, 52, 19-28. https://doi. org/10. 1016/j. foodhyd. 2015. 06. 008.

Liu, F., Jensen, C. R., & Andersen, M. N. (2004). Drought stress effect on carbohydrate concentration in soybean leaves and pods during early reproductive development: Its implication in altering pod set. *Field Crops Research*, 86 (1), 1-13. https://doi. org/10. 1016/S0378-4290 (03) 00165-5.

Lockyer, S., & Nugent, A. P. (2017). Health effects of resistant starch. *Nutrition Bulletin*, 42 (1), 10-41. https://doi. org/10. 1111/nbu. 12244.

Lu, H. F., Wang, C. Y., Guo, T. C., Xie, Y. X., Feng, W., & Li, S. Y. (2014). Starch com-position and its granules distribution in wheat grains in relation to post-anthesis high temperature and drought stress treatments. *Starch/Stärke*, 66 (5-6), 419-428. https://doi. org/10. 1002/star. 201300070.

McCleary, B. V., McNally, M., & Rossiter, P. (2002). Measurement of resistant starch by enzymatic digestion in starch and selected plant materials: Collaborative study. *Journal of AOAC International*, 85 (5), 1103-1111. https://doi. org/10. 1080/0963748021000044787.

Mohammadkhani, N., & Heidari, R. (2008). Drought-induced accumulation of soluble sugars and proline in two maize varieties. *World Applied Sciences Journal*, 3 (3), 448-453.

Morales-Medina, R., del Mar Muñío, M., Guadix, E. M., & Guadix, A. (2014). Production of resistant starch by enzymatic debranching in legume flours. *Carbohydrate Polymers*, 101, 1176-1183. https://doi. org/10. 1016/j. carbpol. 2013. 10. 027.

Morell, M. K., Konik-Rose, C., Ahmed, R., Li, Z. Y., & Rahman, S. (2004). Synthesis of resistant starches in plants. *Journal of AOAC International*, 87 (3), 740-748. https://doi. org/10. 1080/09637480410001734003.

Morris, V. J. (1990). Starch gelation and retrogradation. *Trends in Food Science & Technology*, 1, 2-6. https://doi. org/10. 1016/0924-2244 (90) 90002-G.

Qin, F. L., Man, J. M., Cai, C. H., Xu, B., Gu, M. H., Zhu, L. J., et al. (2012). Physicochemical properties of high-amylose rice starches during kernel development. *Carbohydrate Polymers*, 88 (2), 690-698. https://doi. org/10. 1016/j. carbpol. 2012. 01. 013.

Sharma, M., Yadav, D. N., Singh, A. K., & Tomar, S. K. (2015). Effect of heat-moisture treatment on resistant starch content as well as heat and shear stability of pearl millet starch. *Agricultural Research*, 4 (4), 411-419. https://doi. org/10. 1007/s40003-015-0177-3.

Singh, S., Singh, G., Singh, P., & Singh, N. (2008). Effect of water stress at different stages of grain development on the characteristics of starch and protein of different wheat varieties. *Food Chemistry*, 108 (1), 130-139. https://doi. org/10. 1016/j. foodchem. 2007. 10. 054.

Thitisaksakul, M., Jiménez, R. C., Arias, M. C., & Beckles, D. M. (2012). Effects of environmental factors on cereal starch biosynthesis and composition. *Journal of Cereal Science*, 56 (1), 67-80. https://doi. org/10. 1016/j. jcs. 2012. 04. 002.

Varatharajan, V., Hoover, R., Liu, Q., & Seetharaman, K. (2010). The impact of heat-moisture treatment on the molecular structure and physicochemical properties of normal and waxy potato starches. *Carbohydrate Polymers*, 81, 466-475. https://doi. org/10. 1016/j. carbpol. 2010. 03. 002.

Wang, L., Xie, B., Xiong, G., Wu, W., Wang, J., Qiao, Y., et al. (2013). The effect of freeze-thaw cycles on microstructure and physicochemical properties of four starch gels. *Food Hydrocolloids*, 31 (1), 61-67. https://doi. org/10. 1016/j. foodhyd. 2012. 10. 004.

Worch, S., Rajesh, K., Harshavardhan, V. T., Pietsch, C., Korzun, V., Kuntze, L., et al. (2011). Haplotyping, linkage mapping and expression analysis of barley genes regulated by terminal drought stress influencing seed quality. *BMC Plant Biology*, 11 (1), 1-14. https://doi.org/10.1186/1471-2229-11-1.

Xia, J., Zhu, D., Wang, R. M., Cui, Y., & Yan, Y. M. (2018). Crop resistant starch and genetic improvement: A review of recent advances. *Theoretical and Applied Genetics*, 131 (12), 2495-2511. https://doi.org/10.1007/s00122-018-3221-4.

Yi, B., Zhou, Y. F., Gao, M. Y., Zhang, Z., Han, Y., Yang, G. D., et al. (2014). Effect of drought stress during flowering stage on starch accumulation and starch synthesis enzymes in sorghum grains. *Journal of Integrative Agriculture*, 13 (11), 2399-2406. https://doi.org/10.1016/S2095-3119 (13) 60694-2.

Yu, X. R., Li, B., Wang, L. L., Chen, X. Y., Wang, W. J., Gu, Y. J., et al. (2016). Effect of drought stress on the development of endosperm starch granules and the composition and physicochemical properties of starches from soft and hard wheat. Journal of the Science of Food and Agriculture, 96 (8), 2746-2754. https://doi.org/10.1002/jsfa.7439.

Zhen, S. M., Deng, X., Li, M. F., Zhu, D., & Yan, Y. M. (2018). 2D-DIGE comparative proteomic analysis of developing wheat grains under high-nitrogen fertilization revealed key differentially accumulated proteins that promote storage protein and starch biosynthesis. *Analytical and Bioanalytical Chemistry*, 410 (24), 6219-6235. https://doi.org/10.1007/s00216-018-1230-4.

Zhen, S. M., Deng, X., Zhang, M., Zhu, G. R., Lv, D. M., Wang, Y. P., et al. (2017). Comparativephosphoproteomic analysis under high-nitrogen fertilizer reveals central phosphoproteins promoting wheat grain starch and protein synthesis. *Frontiers in Plant Science*, 8, 67. https://doi.org/10.3389/fpls.2017.00067.

Zhen, S. M., Zhou, J. X., Deng, X., Zhu, G. R., Cao, H., Wang, Z. M., et al. (2016). Metabolite profiling of the response to high-nitrogen fertilizer during grain development of bread wheat (Triticum aestivum L.). Journal of Cereal Science, 69, 85-94. https://doi.org/10.1016/j.jcs.2016.02.014.

Zhou, J. X., Liu, D. M., Deng, X., Zhen, S. M., Wang, Z. M., & Yan, Y. M. (2018). Effects of water deficit on breadmaking quality and storage protein compositions in bread wheat (*Triticum aestivum* L.). *Journal of the Science of Food and Agriculture*, 98 (11), 4357-4368. https://doi.org/10.1002/jsfa.8968.

Zi, Y., Shen, H., Dai, S., Ma, X., Ju, W., Wang, C. G., et al. (2019). Comparison of starch physicochemical properties of wheat cultivars differing in bread-and noodle-making quality. *Food Hydrocolloids*, 93, 78-86. https://doi.org/10.1016/j. foodhyd. 2019. 02. 014.

图书在版编目（CIP）数据

中麦175选育与主要特性解析 / 何中虎主编. —北京：中国农业出版社，2023.8
ISBN 978-7-109-31022-3

Ⅰ.①中… Ⅱ.①何… Ⅲ.①小麦—选择育种—研究
Ⅳ.①S512.103

中国国家版本馆CIP数据核字（2023）第155008号

中麦175选育与主要特性解析
ZHONGMAI175 XUANYU YU ZHUYAO TEXING JIEXI

中国农业出版社出版
地址：北京市朝阳区麦子店街18号楼
邮编：100125
责任编辑：李　梅
版式设计：王　晨　　责任校对：吴丽婷
印刷：北京通州皇家印刷厂
版次：2023年8月第1版
印次：2023年8月北京第1次印刷
发行：新华书店北京发行所
开本：889mm×1168mm　1/16
印张：22.75　　插页：1
字数：1000千字
定价：298.00元